AUG 2005

Electrical Studies for Trades

Third Edition

Stephen L. Herman

D0080708

WITHDRAWN
From Toronto Public Library

THOMSON
DELMAR LEARNING

Australia Canada Mexico Singapore Spain United Kingdom United States

Electrical Studies for Trades, 3e
Stephen L. Herman

Vice President, Technology and Trades SBU:
Alar Elken

Editorial Director:
Sandy Clark

Acquisitions Editor:
Steve Helba

Development Editor:
Dawn Daugherty

Marketing Director:
David Garza

Channel Manager:
Dennis Williams

Marketing Coordinator:
Stacey Wiktorek

Production Director:
Mary Ellen Black

Production Manager:
Andrew Crouth

Production Editor:
Dawn Jacobson

Technology Project Manager:
Kevin Smith

Editorial Assistant:
Dawn Daugherty

COPYRIGHT © 2006 Thomson Delmar Learning. Thomson, the Star Logo, and Delmar Learning are trademarks used herein under license.

Printed in the United States of America
1 2 3 4 5 XX 07 06 05

For more information contact Thomson Delmar Learning
Executive Woods
5 Maxwell Drive, PO Box 8007,
Clifton Park, NY 12065-8007
Or find us on the World Wide Web at

www.delmarlearning.com

ALL RIGHTS RESERVED. No part of this work covered by the copyright hereon may be reproduced in any form or by any means—graphic, electronic, or mechanical, including photocopying, recording, taping, Web distribution, or information storage and retrieval systems—without the written permission of the publisher.

For permission to use material from the text or product, contact us by
Tel. (800) 730-2214
Fax (800) 730-2215

www.thomsonrights.com

Library of Congress Cataloging-in-Publication Data

Herman, Stephen L.
Electrical studies for trades / Stephen L. Herman.-- 3rd ed.
p. cm.
Includes index.
ISBN 1-4018-9797-5
1. Electric apparatus and appliances.
2. Electric engineering. I. Title.
TK452.H47 2006
621.3--dc22
2005043952

NOTICE TO THE READER

Publisher does not warrant or guarantee any of the products described herein or perform any independent analysis in connection with any of the product information contained herein. Publisher does not assume, and expressly disclaims, any obligation to obtain and include information other than that provided to it by the manufacturer.

The reader is expressly warned to consider and adopt all safety precautions that might be indicated by the activities herein and to avoid all potential hazards. By following the instructions contained herein, the reader willingly assumes all risks in connection with such instructions.

The publisher makes no representation or warranties of any kind, including but not limited to, the warranties of fitness for particular purpose or merchantability, nor are any such representations implied with respect to the material set forth herein, and the publisher takes no responsibility with respect to such material. The publisher shall not be liable for any special, consequential, or exemplary damages resulting, in whole or part, from the readers' use of, or reliance upon, this material.

Contents

Preface

Many technical fields require a working knowledge of electricity, such as air conditioning and refrigeration, automotive repair, electrical apprenticeship, carpentry, building maintenance, construction, and appliance repair. ***Electrical Studies for Trades,*** Third Edition, is written for technicians who are not electricians but who must have a practical working knowledge of electricity in their chosen field.

This text assumes that the reader has no knowledge of electricity. ***Electrical Studies for Trades,*** Third Edition, begins with atomic structure and basic electricity. Because they are used in so many applications in the electrical field, extended coverage of resistors has been included in the third edition. The text progresses through Ohm's Law calculations and series, parallel, and combination circuits. These concepts are presented in an easy-to-follow, step-by-step manner. The math level is kept to basic algebra and trigonometry. It is not the intent of this text to present electricity from a purely mathematical standpoint, but rather to explain it in an easy-to-read, straightforward manner using examples and illustration.

Electrical Studies for Trades, Third Edition, includes concepts of inductance and capacitance in alternating-current circuits. Both single-phase and three-phase power systems are covered. The text discusses transformers, three-phase motors, and single-phase motors. Common measuring instruments such as the voltmeter, ammeter, and ohmmeter are covered. The third edition also includes information on oscilloscopes because many circuits require the use of this instrument in troubleshooting.

Electrical Studies for Trades, Third Edition, provides information on basic wiring practices such as connection of electrical outlets and switch connections. Detailed explanations for the connection of single-pole, three-way, and four-way switches are presented in an easy-to-follow, step-by-step manner. The third edition also includes information on ground fault interrupters, arc-fault interrupters, and light dimmers. The final unit includes information on motor control schematics and wiring diagrams.

The author and Thomson Delmar Learning would like to acknowledge and thank the reviewers for the many suggestions and comments given during the development of this third edition. Thanks go to:

Marvin Moak
Hinds Community College
Raymond, MS

Larry Snyder
Red Rocks Community College
Lakewood, CO

Randy Ludington
Guilford Technical Community College
Jamestown, NC

Wes Evans
Truckee Meadows Community College
Reno, NV

Unit 1

Atomic Structure

objectives

fter studying this unit, you should be able to:

- List the three major parts of an atom.
- State the law of charges.
- Discuss the law of centrifugal force.
- Discuss the differences between conductors, insulators, and semiconductors.

Electricity is the driving force that provides most of the power for the industrialized world. It is used to light homes, cook meals, heat and cool buildings, drive motors, and supply the ignition for most automobiles. The technician who understands electricity can seek employment in almost any part of the world.

direct current

alternating current

uni-directional

bi-directional

Electrical sources are divided into two basic types, **direct current** (DC) and **alternating current** (AC). Direct current is **unidirectional,** which means that it flows in only one direction. The first part of this text will be devoted mainly to the study of direct current. Alternating current is **bi-directional,** which means that it reverses its direction of flow at regular intervals. The latter part of this text is devoted mainly to the study of alternating current.

EARLY ELECTRICAL HISTORY

Although the practical use of electricity has become common within the last hundred years, it has been known as a force for much longer. The Greeks discovered electricity about 2,500 years ago. They noticed that when amber was rubbed with other materials, it became charged with an unknown force. This force had the power to attract other objects, such as dried leaves, feathers, bits of cloth, or other lightweight materials. The Greeks called amber *elektron.* The word *electric* was derived from this word because like amber, it had the ability to attract other objects. This mysterious force remained a curious phenomenon until other people began to conduct experiments about 2,000 years later. In the early 1600s, William Gilbert discovered that materials other than amber could be charged to attract other objects. He called materials that could be charged *electriks* and materials that could not be charged *nonelektriks.*

repulsion

attraction

About 300 years ago, a few men began to study the behavior of various charged objects. In 1733, a Frenchman named Charles DuFay found that a piece of charged glass would repel some charged objects and attract others. These men soon learned that the force of **repulsion** was just as important as the force of **attraction.** From these experiments, two lists were developed, *Figure 1-1.* Any material in list A would attract any of the materials in list B. All materials in list A would repel each other, and all the ma-

LIST A	LIST B
Glass (rubbed on silk)	Hard rubber (rubbed on wool)
Glass (rubbed on wool or cotton)	Block of sulfur (rubbed on wool or fur)
Mica (rubbed on cloth)	Most kinds of rubber (rubbed on cloth)
Asbestos (rubbed on cloth or paper)	Sealing wax (rubbed on silk, wool, or fur)
Stick of sealing wax (rubbed on wool)	Glass or mica (rubbed on dry wool)
	Amber (rubbed on cloth)

Figure 1-1 List of charged materials.

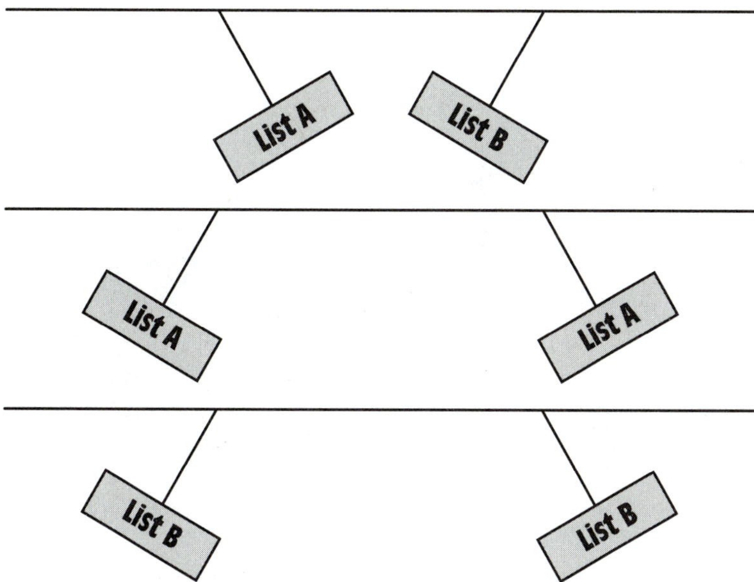

Figure 1-2 Unlike charges attract and like charges repel.

terials in list B would repel each other, *Figure 1-2.* Various names were suggested for the materials in lists A and B. Any opposite-sounding names such as east and west, north and south, male and female could have been chosen. Benjamin Franklin named the materials in list A **positive** and the materials in list B **negative.** These names are still used today. The first item in each list was used as a standard for determining if a charged object was positive or negative. Any object repelled by a piece of glass rubbed on silk had a positive charge, and any item repelled by a hard rubber rod rubbed on wool had a negative charge.

positive

negative

ATOMS

To understand electricity, it is necessary to start with the study of atoms. The **atom** is the basic building block of the universe. All **matter** is made from a combination of atoms. Matter is any substance that has mass and occupies space. Matter can exist in any of three states: solid, liquid, or gas. Water, for example, can exist as a solid in the form of ice, as a liquid, or as a gas in the form of steam, *Figure 1-3.* An atom is the smallest part of an **element.** A chart listing both natural and artificial elements is shown in *Figure 1-4.* The three principal parts of an atom are the **electron, neutron,** and **proton.** *Figure 1-5* illustrates these parts of the atom. It is theorized that electrons, protons, and neutrons are actually made of smaller particles called *quarks.*

atom

matter

element

electron

neutron

proton

Figure 1-3 Water can exist in three states depending on temperature and pressure.

nucleus

atomic number

Notice that the proton has a positive charge, the electron has a negative charge, and the neutron has no charge. The neutron and proton combine to form the **nucleus** of the atom. Since the neutron has no charge, the nucleus will have a net positive charge. The number of protons in the nucleus determines the element of an atom. Oxygen, for example, contains eight protons in its nucleus, and gold contains seventy-nine. The **atomic number** of an element is the same as the number of protons in the nucleus. The lines of force produced by the positive charge of the proton extend outward in all directions, *Figure 1-6.* The nucleus may or may not contain as many neutrons as protons. For example, an atom of helium contains two protons and two neutrons in its nucleus. An atom of copper contains twenty-nine protons and thirty-five neutrons, *Figure 1-7.*

The electron orbits around the outside of the nucleus. Notice that the electron is shown to be larger than the proton in Figure 1-5. Actually, the electron is about three times larger than a proton. The estimated size of a proton is 0.07 trillionth of an inch in diameter, and the estimated size of an electron is 0.22 trillionth of an inch in diameter. Although the electron is larger in size, the proton weighs about 1,840 times more than an electron. It is like comparing a soap bubble to a piece of buckshot. This means that the proton is a very massive particle as compared to the electron. Since the electron exhibits a negative charge, the lines of force come from all directions, *Figure 1-8.*

ATOMIC NUMBER	NAME	VALENCE ELECTRONS	SYMBOL	ATOMIC NUMBER	NAME	VALENCE ELECTRONS	SYMBOL	ATOMIC NUMBER	NAME	VALENCE ELECTRONS	SYMBOL
1	Hydrogen	1	H	37	Rubidium	1	Rb	73	Tantalum	2	Ta
2	Helium	2	He	38	Strontium	2	Sr	74	Tungsten	2	W
3	Lithium	1	Li	39	Yttrium	2	Y	75	Rhenium	2	Re
4	Beryllum	2	Be	40	Zirconium	2	Zr	76	Osmium	2	Os
5	Boron	3	B	41	Niobium	1	Nb	77	Iridium	2	Ir
6	Carbon	4	C	42	Molybdenum	1	Mo	78	Platinum	1	Pt
7	Nitrogen	5	N	43	Technetium	2	Tc	79	Gold	1	Au
8	Oxygen	6	O	44	Ruthenium	1	Ru	80	Mercury	2	Hg
9	Fluorine	7	F	45	Rhodium	1	Rh	81	Thallium	3	Tl
10	Neon	8	Ne	46	Palladium	–	Pd	82	Lead	4	Pb
11	Sodium	1	Na	47	Silver	1	Ag	83	Bismuth	5	Bl
12	Magnesium	2	Ma	48	Cadmium	2	Cd	84	Polonium	6	Po
13	Aluminum	3	Al	49	Indium	3	In	85	Astatine	7	At
14	Silicon	4	Si	50	Tin	4	Sn	86	Radon	8	Rd
15	Phosphorus	5	P	51	Antimony	5	Sb	87	Francium	1	Fr
16	Sulfur	6	S	52	Tellurium	6	Te	88	Radium	2	Ra
17	Chlorine	7	Cl	53	Iodine	7	I	89	Actinium	2	Ac
18	Argon	8	A	54	Xenon	8	Xe	90	Thorium	2	Th
19	Potassium	1	K	55	Cesium	1	Cs	91	Protactinium	2	Pa
20	Calcium	2	Ca	56	Barium	2	Ba	92	Uranium	2	U
21	Scandium	2	Sc	57	Lanthanum	2	La				
22	Titanium	2	Ti	58	Cerium	2	Ce		Artifical Elements		
23	Vanadium	2	V	59	Praseodymium	2	Pr				
24	Chromium	1	Cr	60	Neodymium	2	Nd	93	Neptunium	2	Np
25	Manganese	2	Mn	61	Promethium	2	Pm	94	Plutonium	2	Pu
26	Iron	2	Fe	62	Samarium	2	Sm	95	Americium	2	Am
27	Cobalt	2	Co	63	Europium	2	Eu	96	Curium	2	Cm
28	Nickel	2	Ni	64	Gadolinium	2	Gd	97	Berkelium	2	Bk
29	Copper	1	Cu	65	Terbium	2	Tb	98	Californium	2	Cf
30	Zinc	2	Zn	66	Dysprosium	2	Dy	99	Einsteinium	2	E
31	Gallium	3	Ga	67	Holmium	2	Ho	100	Fermium	2	Fm
32	Germanium	4	Ge	68	Erbium	2	Er	101	Mendelevium	2	Mv
33	Arsenic	5	As	69	Thulium	2	Tm	102	Nobelium	2	No
34	Selenium	6	Se	70	Ytterbium	2	Yb	103	Lawrencium	2	Lw
35	Bromine	7	Br	71	Lutetium	2	Lu				
36	Krypton	8	Kr	72	Hafnium	2	Hf				

Figure 1-4 Table of elements.

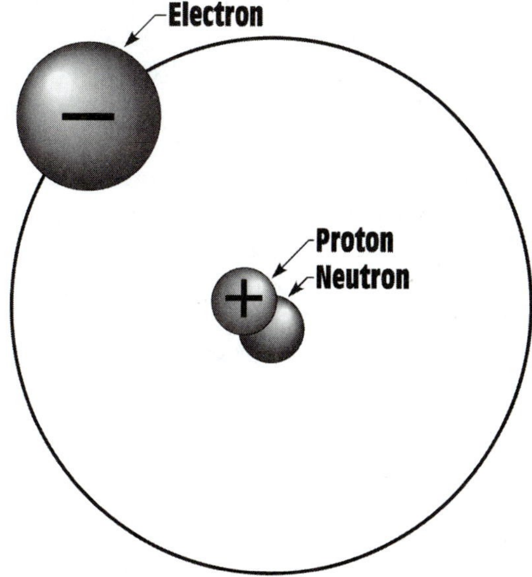

Figure 1-5 The three principal parts of an atom.

Figure 1-6 The lines of force extend outward.

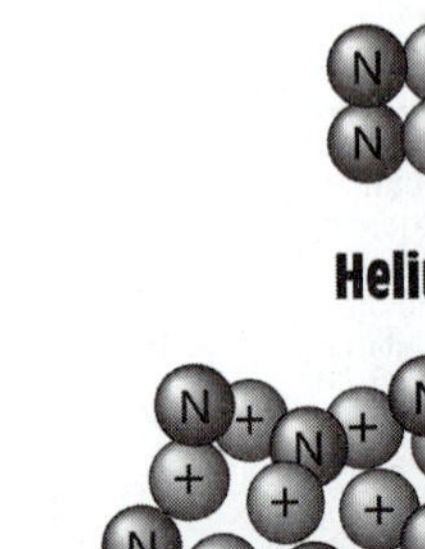

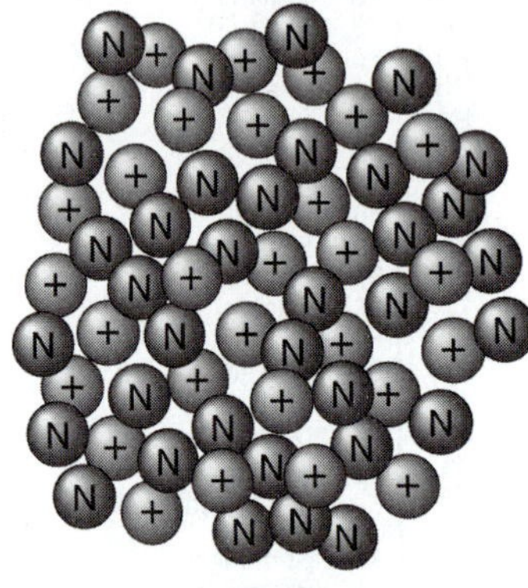

Figure 1-7 The nucleus may or may not contain the same number of protons and neutrons.

Electron

Figure 1-8 The lines of force come inward.

Opposite charges attract and like charges repel.

THE LAW OF CHARGES

To understand atoms, it is necessary first to understand two basic laws of physics. One of these is the law of charges, which states that **opposite charges attract and like charges repel.** *Figure 1-9* illustrates this principle. In Figure 1-9, charged balls are suspended from strings. Notice that the two balls that contain opposite charges are attracted to each other. The two positively charged balls and the two negatively charged balls are repelled from each other. The reason for this is that a basic law of physics states that lines of force can never cross each other. The outward-going lines of force of a positively charged object combine with the inward-going lines of force of a negatively charged object, *Figure 1-10.* This produces an attraction between the two objects. If two objects with like charges come in proximity with each other, the lines of force repel, *Figure 1-11.* Since the nucleus has a net positive charge and the electron has a negative charge, the electron is attracted to the nucleus.

Because the nucleus of an atom is formed from the combination of protons and neutrons, one might ask why the protons of the nucleus do not repel each other since they all have the same charge. Two theories attempt to explain this. The first theory, which is no longer supported, asserted that the force of gravity held the nucleus together. Neutrons, like protons, are

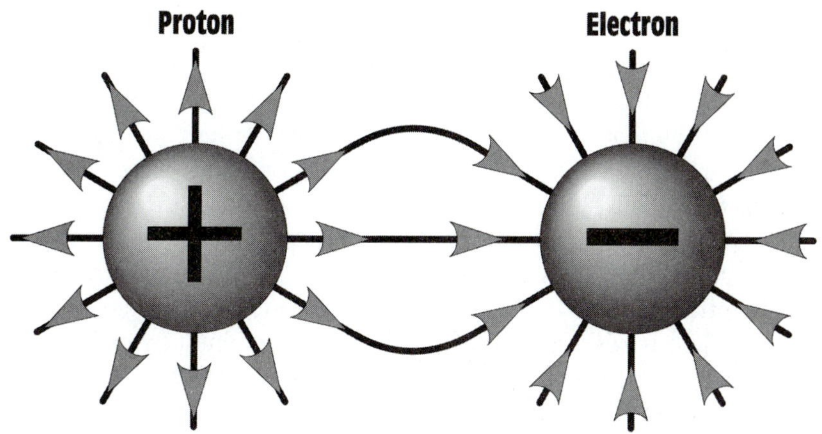

Figure 1-10 Unlike charges attract each other.

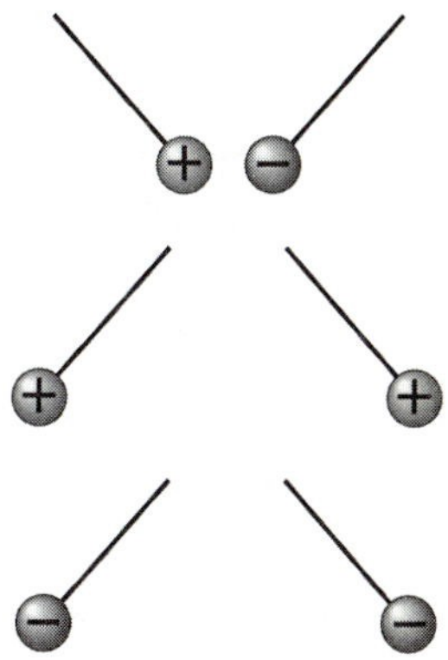

Figure 1-9 Unlike charges attract and like charges repel.

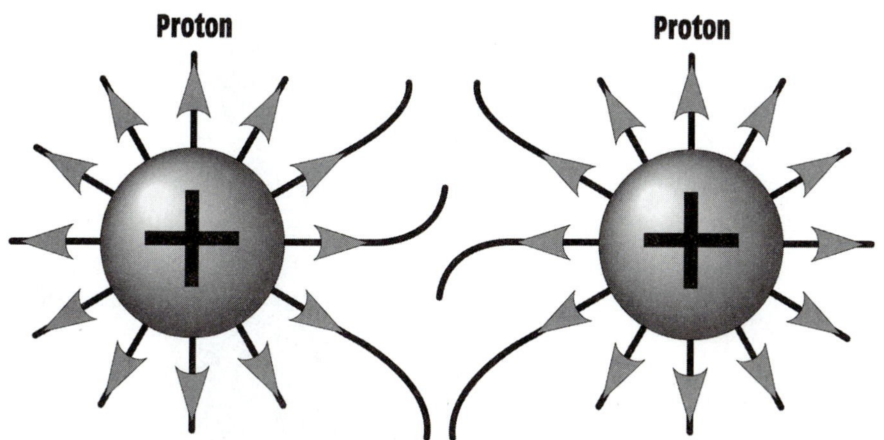

Figure 1-11 Like charges repel each other.

extremely massive particles. It was first theorized that the gravitational attraction caused by their mass overcame the repelling force of the positive charges. By the mid 1930s, however, it was known that the force of gravity could not hold the nucleus together. According to Coulomb's Law, the electromagnetic force in helium is about 1.1×10^{36} times greater than Newton's Law of gravitational force. In 1947 the Japanese physicist, Hideki Yukawa, introduced the second theory by identifying a subatomic particle that acts as a mediator to hold the nucleus together. The particle is a quark known as a *gluon*. The force of the gluon is about 10^2 times stronger than the electromagnetic force.

CENTRIFUGAL FORCE

centrifugal force

The law of centrifugal force states that a spinning object will pull away from its center point and that the faster it spins, the greater the centrifugal force.

Another concept important to the understanding of atoms is **centrifugal force.** According to Newton's first law of motion, a body continues in a state of rest or uniform motion in a straight line unless acted upon by an external force. A free electron (one that has not been captured by an atom) will tend to travel in a straight line. If the negatively charged electron comes close enough to an atom, the attraction of the positively charged nucleus will pull it into orbit. It is this attraction that holds the electron in orbit around the nucleus. This is very similar to someone flying a model airplane, *Figure 1-12.* The wires attached to the airplane prevent it from flying away. If the force of attraction were to be suddenly interrupted, the electron would move off in a straight line.

Figure 1-12 Centrifugal force causes a spinning object to pull away from its axis point.

ELECTRON ORBITS

electron orbit

Atoms have a set number of electrons that can be contained in one orbit, or shell, called an **electron orbit,** *Figure 1-13.* The number of electrons that can be contained in any one shell is found by the formula ($2N^2$). The

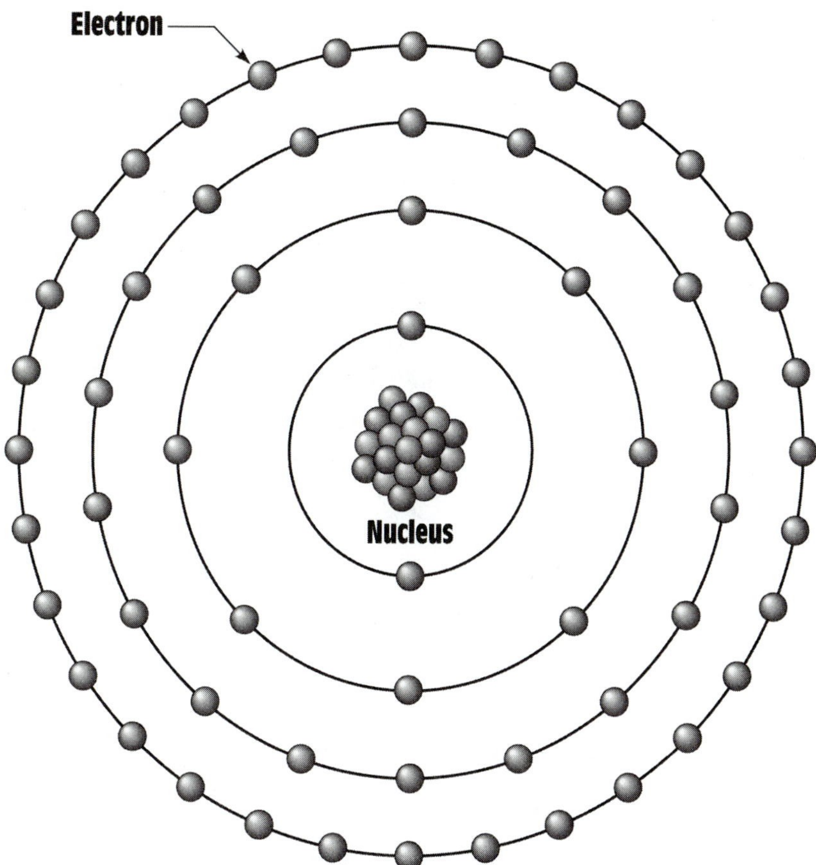

Figure 1-13 Electron orbits.

letter *N* represents the number of the orbit, or shell. For example, the first orbit can hold no more than two electrons.

$$2 \times (1)^2 \text{ or}$$

$$2 \times 1 = 2$$

The second orbit can hold no more than eight electrons.

$$2 \times (2)^2 \text{ or}$$

$$2 \times 4 = 8$$

The third orbit can contain no more than eighteen electrons.

$$2 \times (3)^2 \text{ or}$$

$$2 \times 9 = 18$$

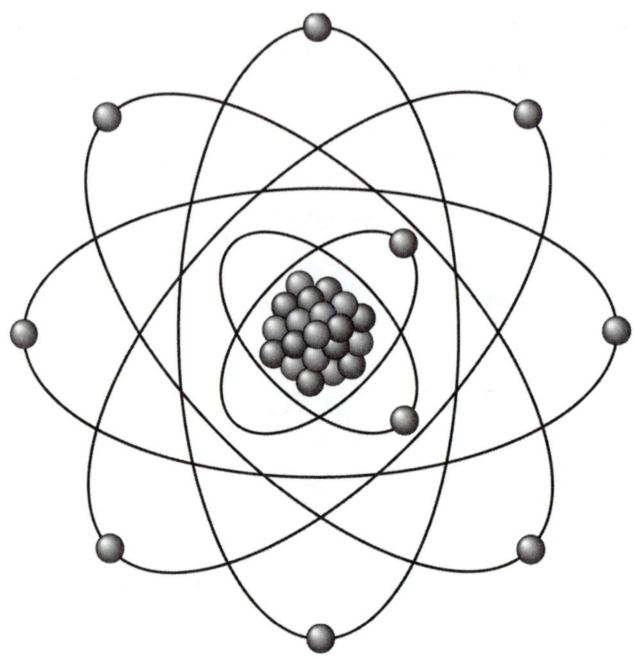

Figure 1-14 Electrons orbit the nucleus in a circular fashion.

The fourth and fifth orbits can hold no more than thirty-two electrons. Thirty-two is the maximum number of electrons that can be contained in any orbit.

$$2 \times (4)^2 \text{ or}$$

$$2 \times 16 = 32$$

Although atoms are often drawn flat, as illustrated in Figure 1-13, the electrons orbit around the nucleus in a circular fashion, as shown in *Figure 1-14.* The electrons travel at such a high rate of speed that they form a shell around the nucleus. This is similar to a golf ball that is surrounded by a tennis ball that is surrounded by a basketball. For this reason, electron orbits are often referred to as *shells.*

VALENCE ELECTRONS

valence electrons

conductor

The outer shell of an atom is known as the *valence* shell. Electrons located in the outer shell of an atom are known as **valence electrons,** *Figure 1-15.* The valence shell of an atom cannot hold more than eight electrons. It is the valence electrons that are of primary concern in the study of electricity, because it is these electrons that explain much of electrical theory. A **conductor,** for instance, is made from a material that contains one

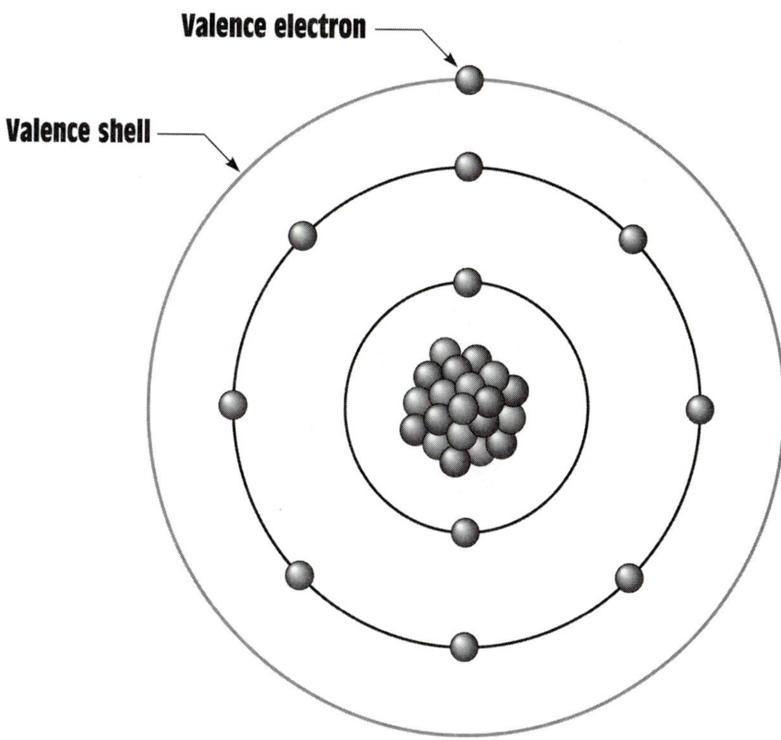

Figure 1-15 The electrons located in the outer orbit of an atom are valence electrons.

or two valence electrons. Conductors are materials that permit electrons to flow through them easily. When an atom has only one or two valence electrons, the electrons are loosely held by the atom and are easily given up for current flow. Silver, copper, gold, and platinum all contain one valence electron and are excellent conductors of electricity. Silver is the best natural conductor of electricity, followed by copper, gold, and aluminum. Aluminum, which contains three valence electrons, is a better conductor of electricity than platinum, which contains only one valence electron. An atom of copper is shown in *Figure 1-16*. Although it is known that atoms that contain few valence electrons are the best conductors, it is not known why some of these materials are better conductors than others.

ELECTRON FLOW

Electricity is the flow of electrons. It is produced by the knocking of the electrons of an atom out of orbit by another electron. *Figure 1-17* illustrates this action. When an atom contains only one valence electron, it is easily given up when struck by another electron. The striking electron

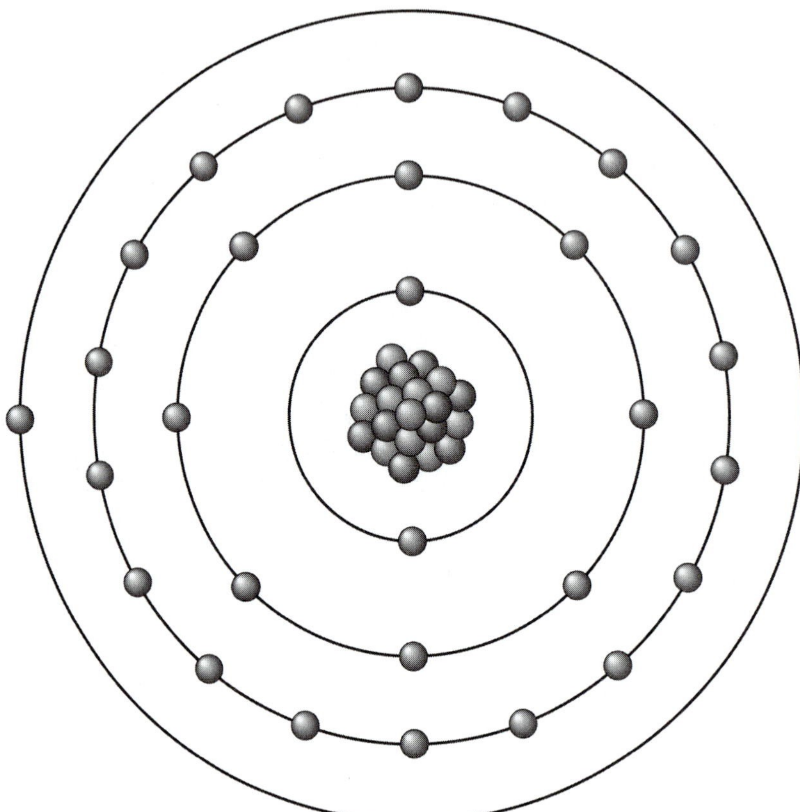

Figure 1-16 A copper atom contains twenty-nine electrons and has one valence electron.

gives its energy to the electron being struck. The striking electron settles into orbit around the atom, and the electron that was struck moves off to strike another electron. This same action can often be seen in the game of pool. If the moving cue ball strikes a stationary ball exactly right, the energy of the cue ball is given to the stationary ball. The stationary ball then moves off with most of the energy of the cue ball, and the cue ball stops moving, *Figure 1-18*. The stationary ball did not move off with *all* of the energy of the cue ball; it moved off with most of the energy of the cue ball. Some of the energy of the cue ball was lost to heat when it struck the stationary ball. This is true when one electron strikes another. This is the reason why a wire heats when current flows through it. If too much current flows through a wire, it will overheat and become damaged and possibly become a fire hazard.

If an atom that contains two valence electrons is struck by a moving electron, the energy of the striking electron is divided between the two valence electrons, *Figure 1-19*. If the valence electrons are knocked out of or-

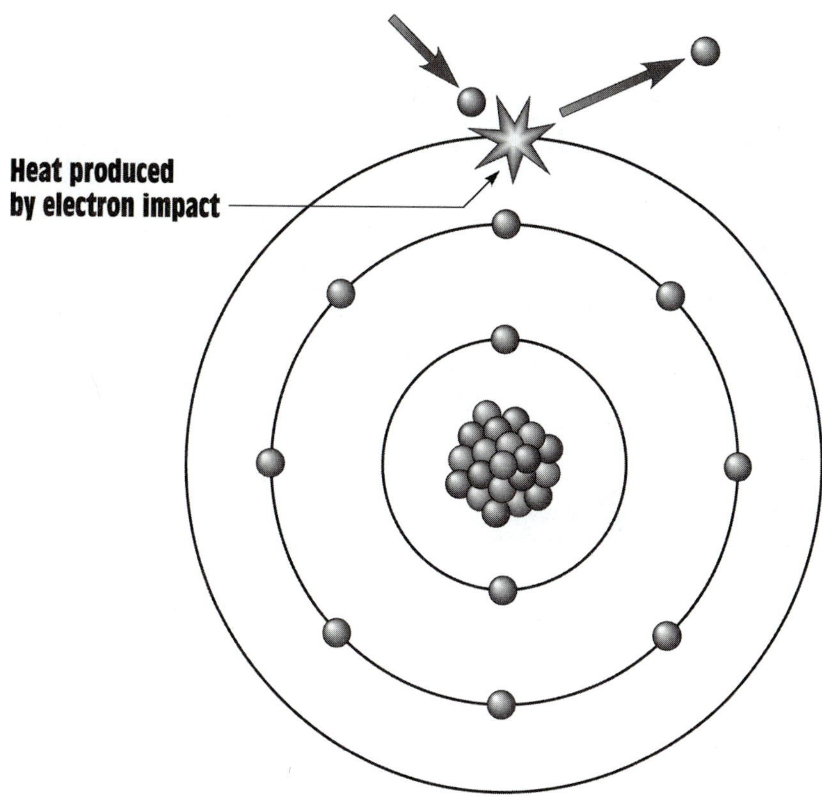

Figure 1-17 An electron of one atom knocks an electron of another atom out of orbit.

bit, they will contain only half the energy of the striking electron. This action can also be seen in the game of pool, *Figure 1-20.* If a moving cue ball strikes two stationary balls at the same time, the energy of the cue ball is divided between the two stationary balls. Both stationary balls will move, but with only half the energy of the cue ball.

INSULATORS

insulators

Materials that are made from atoms that contain seven or eight valence electrons are known as **insulators.** Insulators are materials that resist the flow of electricity. When the valence shell of an atom is full, or almost full, of electrons, the electrons are tightly held and not easily given up. Some good examples of insulator materials are rubber, plastic, glass, and wood. *Figure 1-21* illustrates what happens when a moving electron strikes an atom that contains eight valence electrons. The energy of the moving electron is divided so many times that it has little effect on the atom. An atom that has seven or eight valence electrons is extremely stable and does not easily give up an electron.

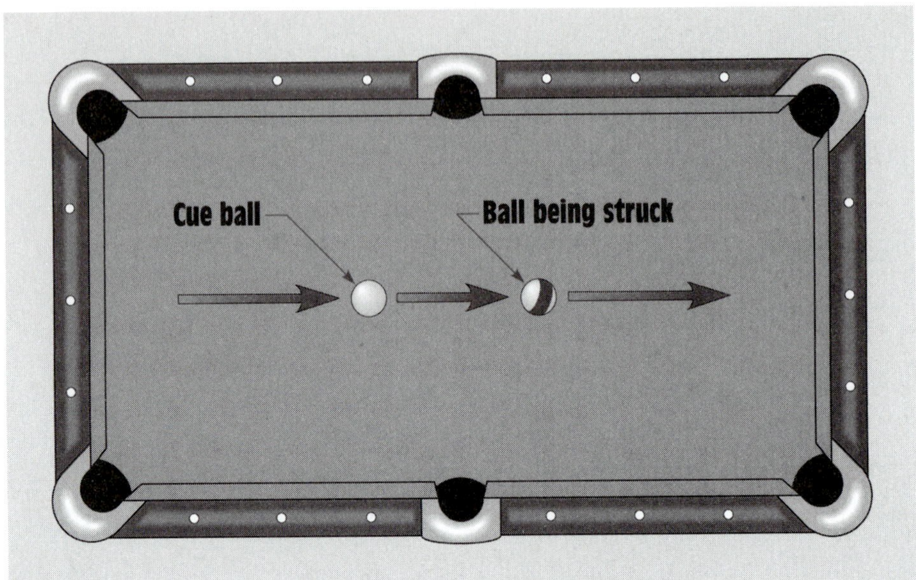

Figure 1-18 The energy of the cue ball is given to the ball being struck.

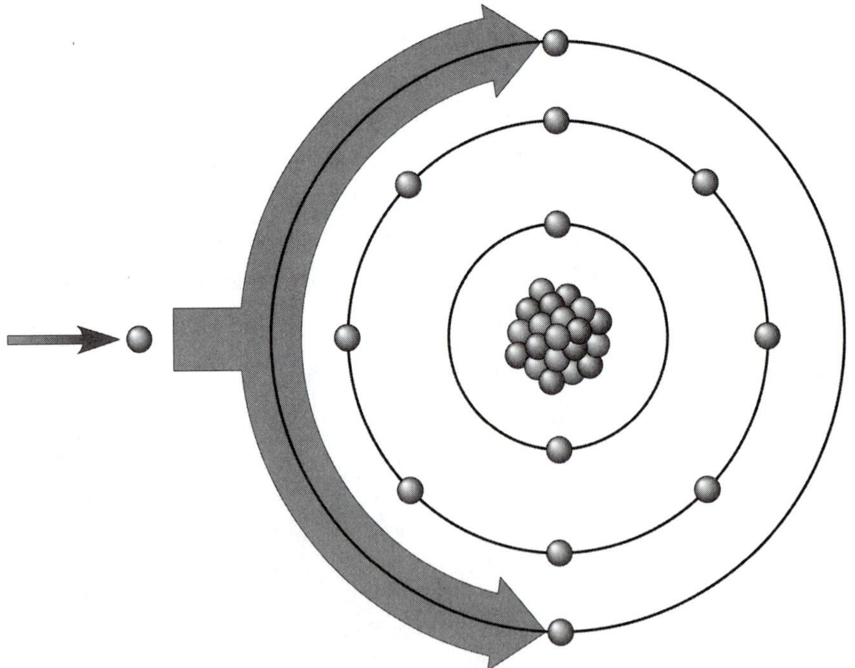

Figure 1-19 The energy of the striking electron is divided.

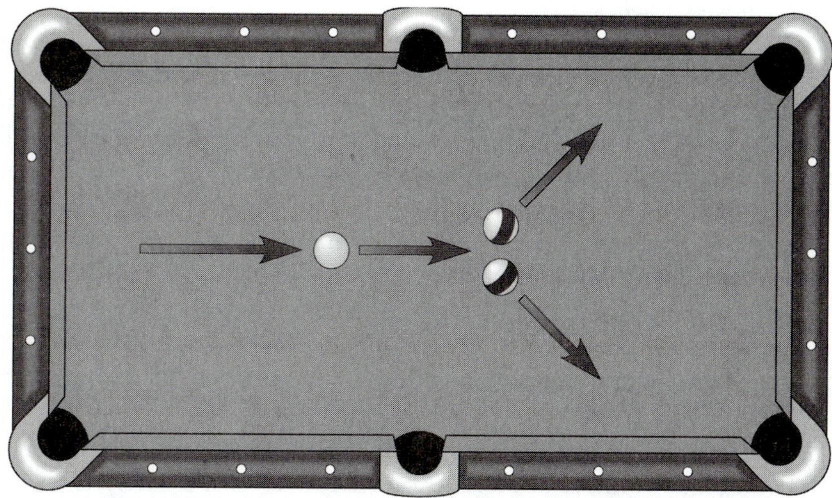

Figure 1-20 The energy of the cue ball is divided between the two other balls.

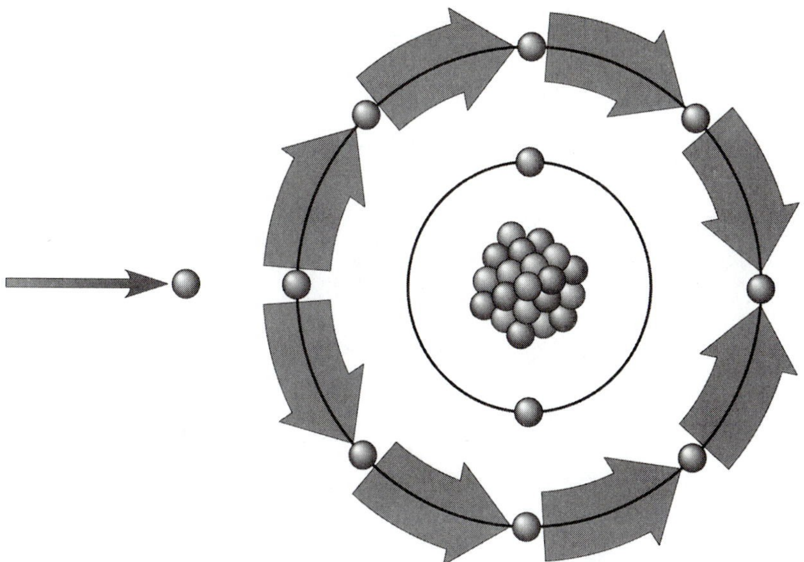

Figure 1-21 The energy of the striking electron is divided among the eight electrons.

SEMICONDUCTORS

semi-conductors

Semiconductors are materials that are neither good conductors nor good insulators. They contain four valence electrons, *Figure 1-22*. Semiconductors are also characterized by the fact that as they are heated, their resistance decreases. This is the opposite of a conductor, which *increases* its resistance with an increase of temperature. Semiconductors have become extremely important in the electrical industry since the invention of the transistor in 1947. All solid-state devices, such as diodes, transistors, and integrated circuits, are made from combinations of semiconductor materials. The two most common materials used in the production of electronic components are silicon and germanium. Of the two, silicon is used more often because of its ability to withstand heat.

Metal oxides are also popular for the production of semiconductor devices. Metal oxide semiconductor (MOS) devices are used where low power drain is essential. These devices are used in electronic watches, computer memories, calculators, and hundreds of other low-current devices.

MOLECULES

molecules

Although all matter is made from atoms, atoms should not be confused with **molecules,** which are the smallest part of a compound. Water, for ex-

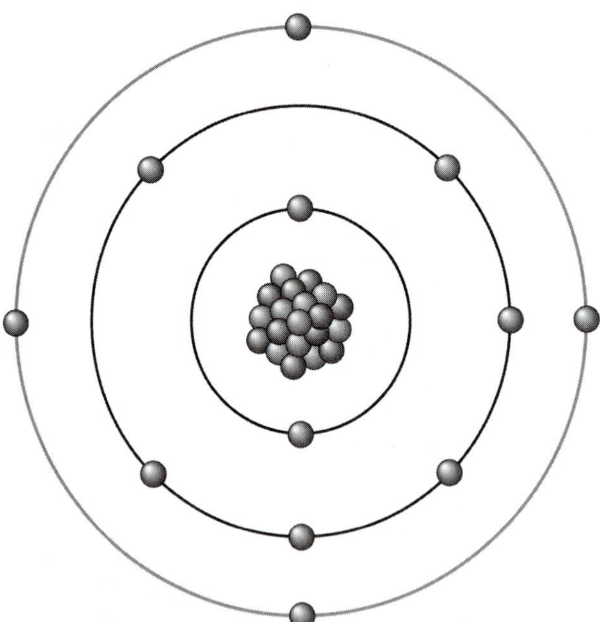

Figure 1-22 Semiconductors contain four valence electrons.

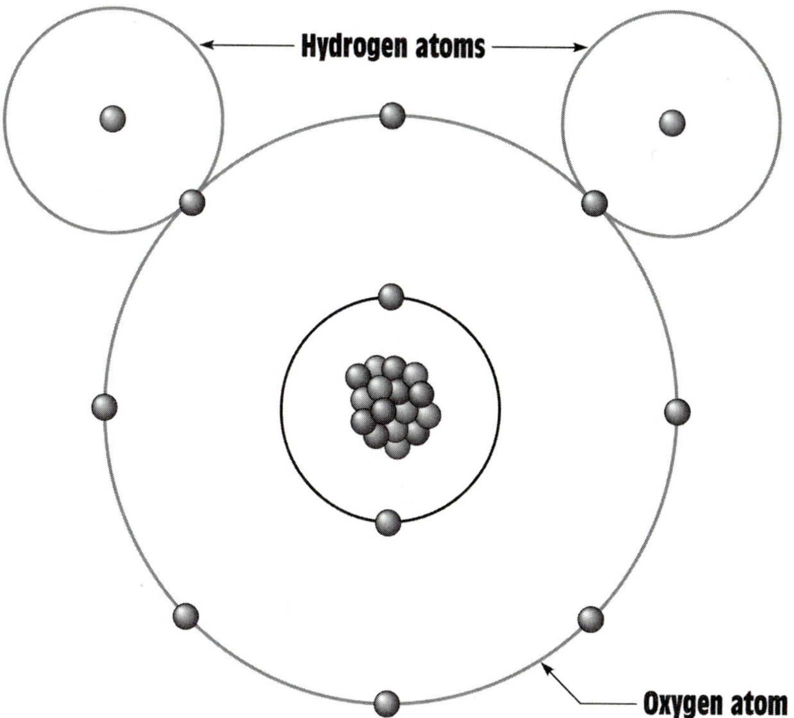

Figure 1-23 Water molecule.

ample, is a compound, not an element. The smallest particle of water that can exist is a molecule, which is made of two atoms of hydrogen and one atom of oxygen, H_2O, *Figure 1-23*. If the molecule of water is broken apart, it becomes two hydrogen atoms and one oxygen atom, but it is no longer water.

SUMMARY

1. The atom is the smallest part of an element.
2. The three basic parts of an atom are the proton, electron, and neutron.
3. Protons have a positive charge, electrons have a negative charge, and neutrons have no charge.
4. Valence electrons are the electrons located in the outer orbit of an atom.
5. Conductors are materials that provide an easy path for electron flow.

6. Conductors are made from materials that contain one, two, or three valence electrons.
7. Insulators are materials that do not provide an easy path for the flow of electrons.
8. Insulators are generally made from materials that contain seven or eight valence electrons.
9. Semiconductors contain four valence electrons.
10. Semiconductors are used in the construction of all solid-state devices, such as diodes, transistors, and integrated circuits.
11. Molecules are the smallest part of a compound.

REVIEW QUESTIONS

1. What are the three subatomic parts of an atom and what charge does each carry?
2. How many times larger is an electron than a proton?
3. How many times greater is the weight of a proton than an electron?
4. State the law of charges.
5. What force keeps the electron from falling into the nucleus of the atom?
6. How many valence electrons are generally contained in materials that are used for conductors?
7. How many valence electrons are generally contained in materials that are used for insulators?
8. What is electricity?
9. What is a gluon?
10. It is theorized that electrons, protons, and neutrons are actually formed from a combination of smaller particles. What are these particles called?

Unit 2

Electrical Quantities, Ohm's Law, and Resistors

objectives

*fter studying this unit, you should be able to:*

- Define a coulomb.
- State the definition of *amp*.
- State the definition of *volt*.
- State the definition of *ohm*.
- State the definition of *watt*.
- Compute different electrical values using Ohm's Law formulas.
- Discuss different types of electric circuits.
- Select the proper Ohm's Law formula from a chart.
- List the major types of fixed resistors.
- Determine the resistance of a resistor using the color code.
- Determine if a resistor is operating within its power rating.
- Connect a variable resistor used as a potentiometer.

Electricity has a standard set of values. Before technicians can work with electricity, they must have a knowledge of these values and how to use them. Since the values of electrical measurement have been standardized, they are understood by everyone who uses them. For instance, carpenters use a standard system for measuring length, such as the inch, foot, meter, or centimeter. Imagine what a house would look like that was constructed by two carpenters that used different lengths of measure for an inch or foot. The same holds true for people who work with electricity. The standards of measure must be the same for everyone. Meters should be calibrated to indicate the same quantity of current flow, or voltage, or resistance. A volt, amp, or ohm is the same for everyone who uses it, regardless of where in the world they live and work.

coulomb

Coulomb's Law of electrostatic charges states that the force of electrostatic attraction or repulsion is directly proportional to the product of the two charges and inversely proportional to the square of the distance between them.

THE COULOMB

The first electrical quantity to be discussed is the **coulomb.** A coulomb is a quantity measurement of electrons. One coulomb contains 6.25×10^{18} or 6,250,000,000,000,000,000 electrons. To better understand the number of electrons contained in a coulomb, think of comparing one second to two hundred billion years. Since the coulomb is a quantity measurement, it is similar to a quart, gallon, or liter. It takes a certain amount of liquid to equal a liter, just as it takes a certain amount of electrons to equal a coulomb.

The coulomb is named for a scientist named Coulomb who lived in the 1700s. Coulomb experimented with electrostatic charges and developed a law dealing with the attraction and repulsion of these forces. This law, known as **Coulomb's Law of electrostatic charges, states that the force of electrostatic attraction or repulsion is directly proportional to the product of the two charges and inversely proportional to the square of the distance between them.** The number of electrons contained in the coulomb was actually determined by the average charge of an electron.

THE AMP

amp

The next electrical measurement to be discussed is the **amp** or *ampere.* The ampere is named for a scientist named Andre Ampere who lived in the late 1700s and early 1800s. Ampere is most famous for his work dealing with electromagnetism, which will be discussed in a later chapter in this text. *The amp is defined as one coulomb per second.* Notice that the definition of an amp involves a quantity measurement, the coulomb, combined with a time

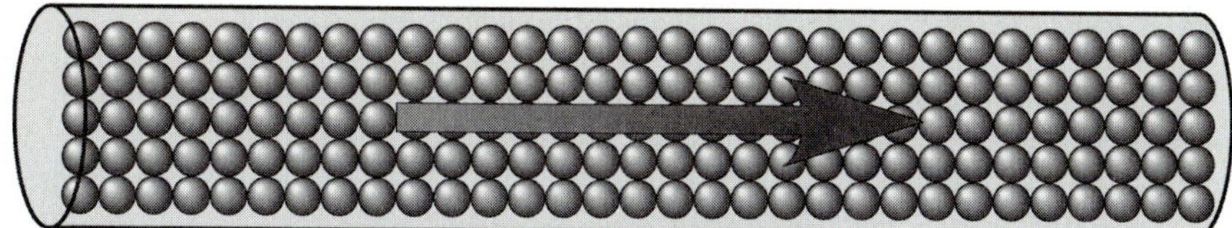

Figure 2-1 One ampere equals one coulomb per second.

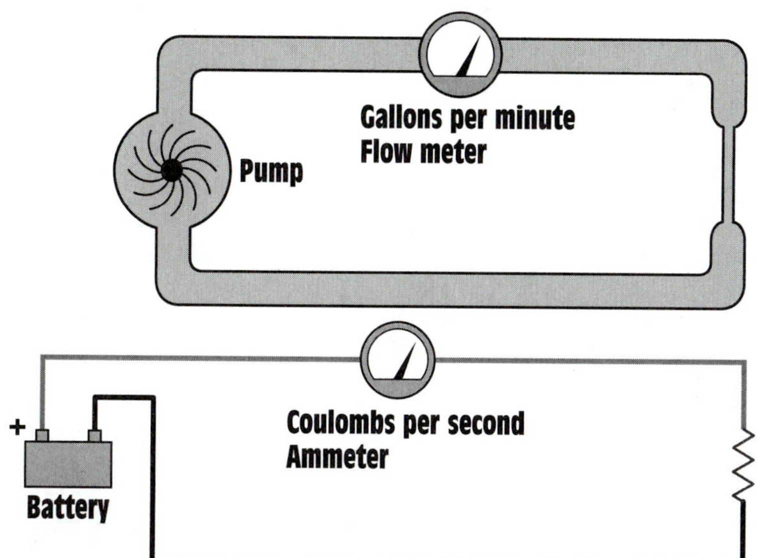

Figure 2-2 Current in an electrical circuit can be compared to flow rate in a water system.

measurement, the second. One amp of current flows through a wire when one coulomb flows past a point in one second, *Figure 2-1*. The ampere is a measurement of the actual amount of electricity that is flowing through a circuit. In a water system, it would be comparable to gallons per minute or gallons per second, *Figure 2-2*. The letter *I*, which stands for *intensity* of current, or the letter *A*, which stands for *amp*, is predominately used to represent current flow in algebraic formulas. This text will use the letter *I* to represent current.

THE ELECTRON THEORY

There are actually two theories concerning current flow. One theory is known as the **electron theory,** and states that since electrons are negative particles, current flows from the most negative point in the circuit to the most positive. The electron theory is the most widely accepted as being correct and is used throughout this text.

electgron theory

conven-
tional
current
flow theory

THE CONVENTIONAL CURRENT THEORY

The second theory is known as the **conventional current flow theory.** This theory is older than the electron theory, and it states that current flows from the most positive point to the most negative. Although it has been established that the electron theory is probably correct, the conventional current theory is still used to a large extent. There are several reasons for this. Most electronic circuits use the negative terminal as ground or common. When this is done, the positive terminal is considered to be above ground, or hot. It is easier for most people to think of something flowing down rather than up, or from a point above ground to ground. An automobile electrical system is a good example of this type of circuit. Most people consider the positive battery terminal to be the hot terminal.

Many of the people who work in the electronics field prefer the conventional current flow theory because all the arrows on the semiconductor symbols point in the direction of conventional current flow. If the electron flow theory is used, it must be assumed that current flows against the arrow, *Figure 2-3*. Another reason why many people prefer using the con-

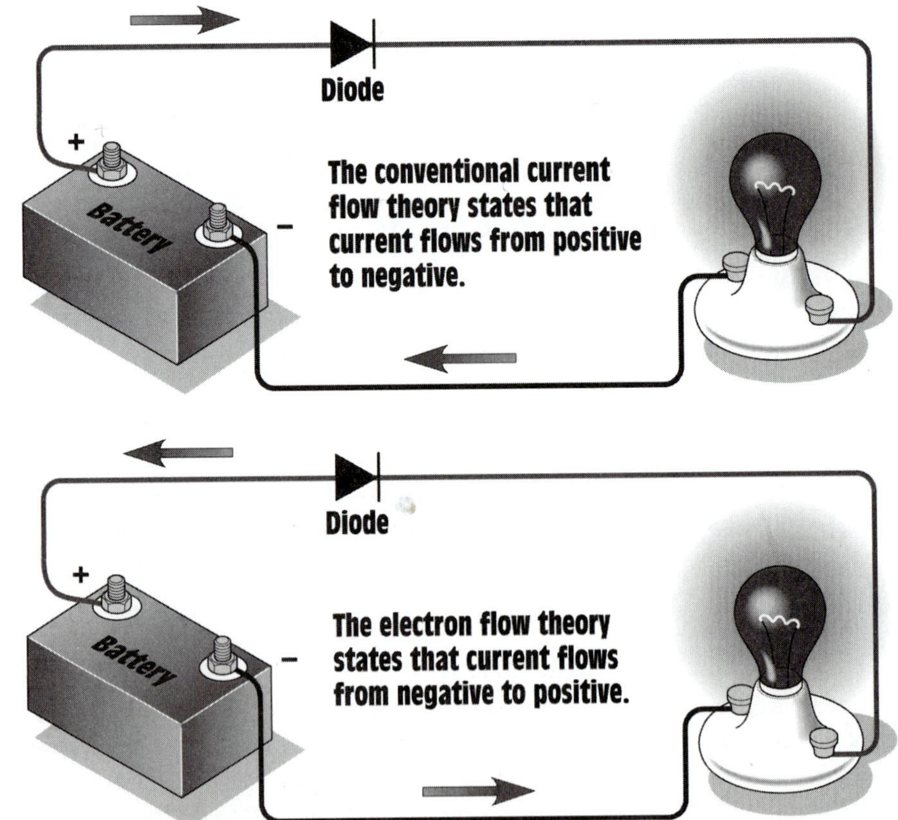

Figure 2-3 Conventional current flow theory and electron flow theory.

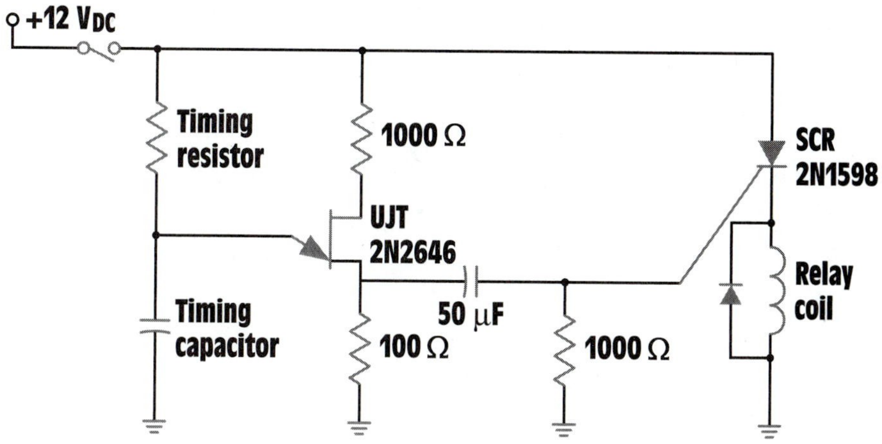

Figure 2-4 On-delay timer.

ventional current flow theory is that most electronic schematics are drawn in a manner assuming current to flow from the more positive to the more negative source, *Figure 2-4*. In this schematic, the positive voltage point is shown at the top of the schematic, and negative (ground) is shown at the bottom. When tracing the flow of current through a circuit, most people find it easier to understand something flowing from top to bottom rather than bottom to top.

SPEED OF CURRENT

Before it is possible to determine the speed of current flow through a wire, we must first establish exactly what is being determined. As stated previously, current is a flow of electrons through a conductive substance. Assume for a moment that it is possible to remove a single electron from a wire and identify it by painting it red. If it were possible to observe the progress of the identified electron as it moved from atom to atom, you would see that a single electron moves rather slowly, *Figure 2-5*. It is estimated that a single electron moves at a rate of about three inches per hour at one ampere of current flow.

The *impulse* of electricity, however, is extremely fast. Assume for a moment that a pipe has been filled with ping-pong balls, *Figure 2-6*. If another ball is forced into one end of the pipe, the ball at the other end of the pipe is forced out. Each time a ball enters one end of the pipe, another ball is forced out the other end. This same principle is true for electrons in a wire. There are billions of electrons that exist in a wire. If an electron enters one end of a wire, another electron is forced out the other end of the wire.

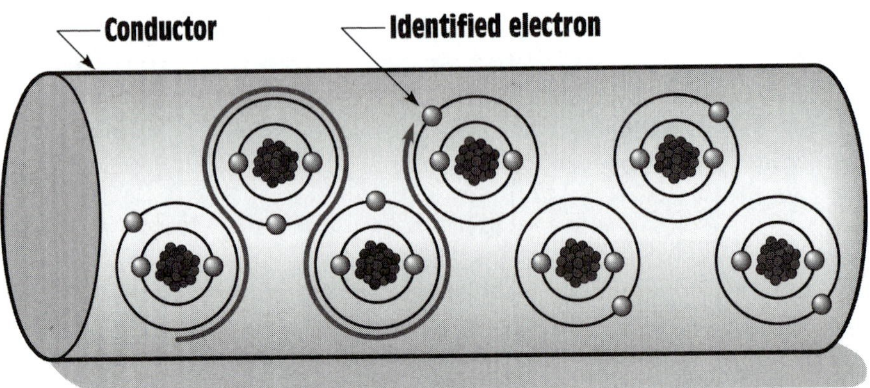

Figure 2-5 Electrons moving from atom to atom.

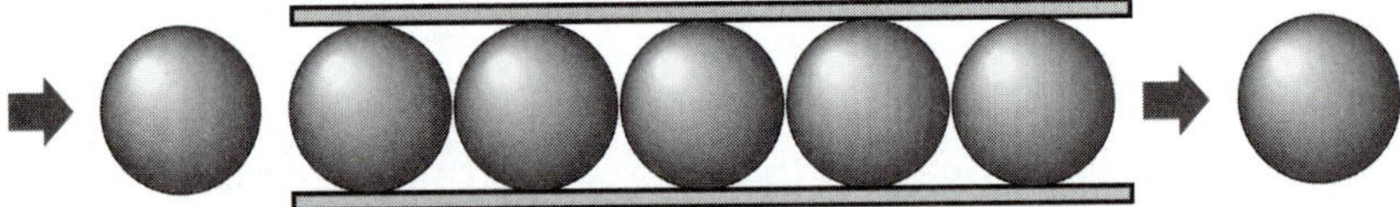

Figure 2-6 When a ball is pushed into one end, another ball is forced out the other end.

For many years it was assumed that the speed of the electrical impulse had a theoretical limit of 186,000 miles per second, or 300,000,000 meters per second, which is the speed of light. In recent years, however, it has been shown that the impulse of electricity can actually travel faster than light. Assume that a wire is long enough to be wound around the earth ten times. If a power source and switch were to be connected at one end of the wire, and a light at the other end, *Figure 2-7,* you would see that the light would turn on immediately when the switch was closed. It would take light approximately 1.3 seconds to travel around the earth ten times.

BASIC ELECTRIC CIRCUITS

complete path

A **complete path** must exist before current can flow through a circuit, *Figure 2-8*. A complete circuit is often referred to as a *closed* circuit, because the power source, conductors, and load form a closed loop. In Figure 2-8 a lamp is used as the load. Although the load offers resistance to the circuit and limits the amount of current that can flow, do not confuse *load* with *resistance*. A load can be anything that permits current to flow. The greater the amount of current flow, the greater the load. Since current flow is inversely proportional to resistance, more current will flow when circuit resistance is decreased, not increased. If the switch is opened, there is no longer a closed loop and no current can flow. This is often referred to as

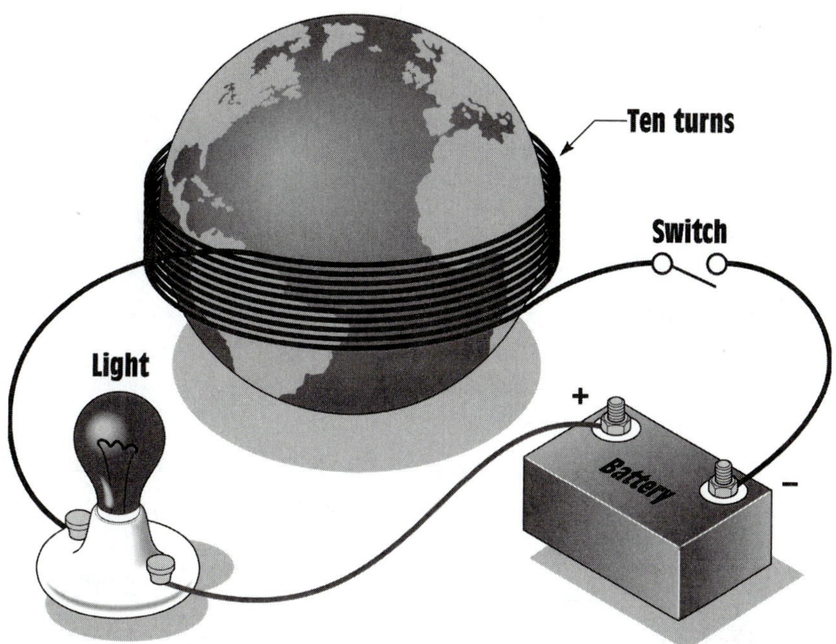

Figure 2-7 The impulse of electricity can travel faster than light.

an *incomplete* or *open* circuit. When the switch is opened, an infinite amount of resistance is added to the circuit and no current can flow.

Another type of circuit is the *short* circuit. The definition of a short circuit is a circuit that has very little or no resistance to limit the flow of current. A short circuit generally occurs when the conductors leading from and back to the power source become connected, *Figure 2-9*. In this example, a separate current path has been established to bypass the load. Since the load is the device that limits the flow of current, when it is bypassed, an excessive amount of current can flow. Short circuits generally cause a fuse to blow or a circuit breaker to open. If the circuit has not been protected by a fuse or circuit breaker, a short circuit can damage equipment, melt wires, and start fires.

Another type of circuit that is often confused with a short circuit is a *grounded* circuit. Grounded circuits can cause an excessive amount of current flow just as a short circuit can. Grounded circuits occur when a path, other than that intended, is established to ground. Many circuits contain an extra conductor called the **grounding conductor.** A typical 120 volt appliance circuit is shown in *Figure 2-10*. In this circuit, the *ungrounded* or *hot* conductor is connected to the fuse or circuit breaker. The hot conductor supplies power to the load. The grounded conductor, or **neutral conductor,** provides the return path and completes the circuit back to the

grounding conductor

neutral conductor

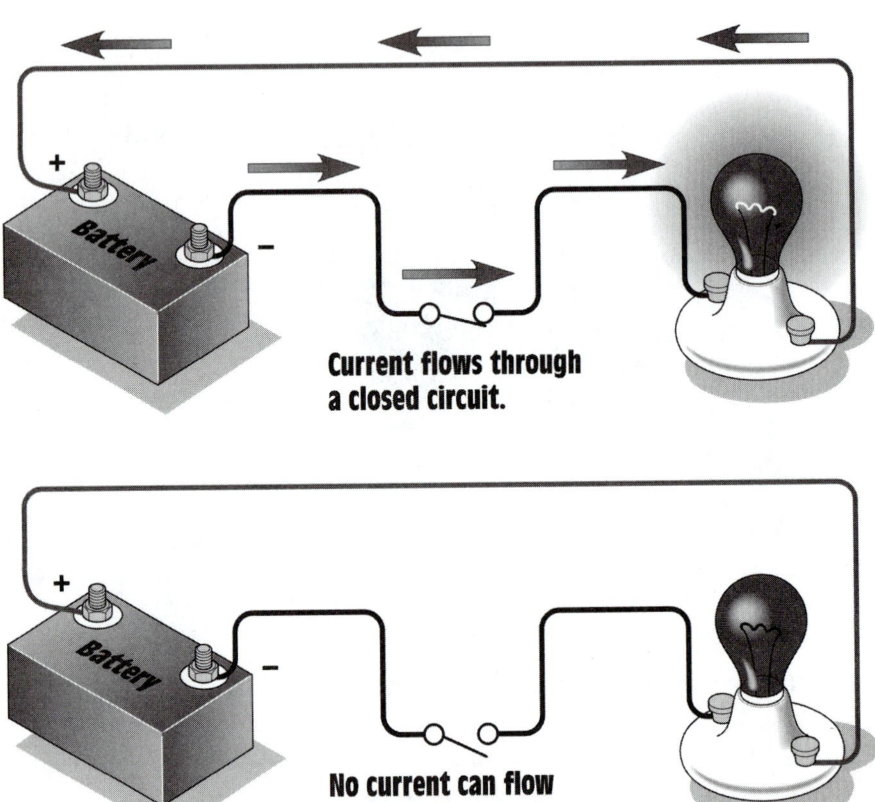

Figure 2-8 Current flows only through a closed circuit.

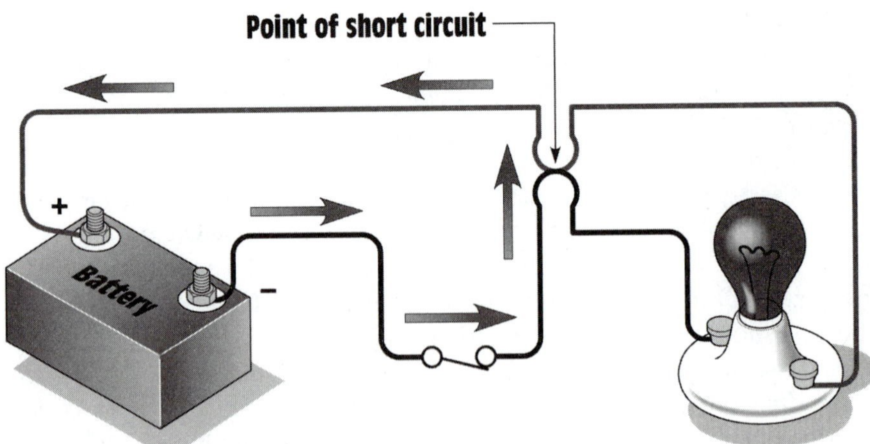

Figure 2-9 A short circuit bypasses the load and permits too much current to flow.

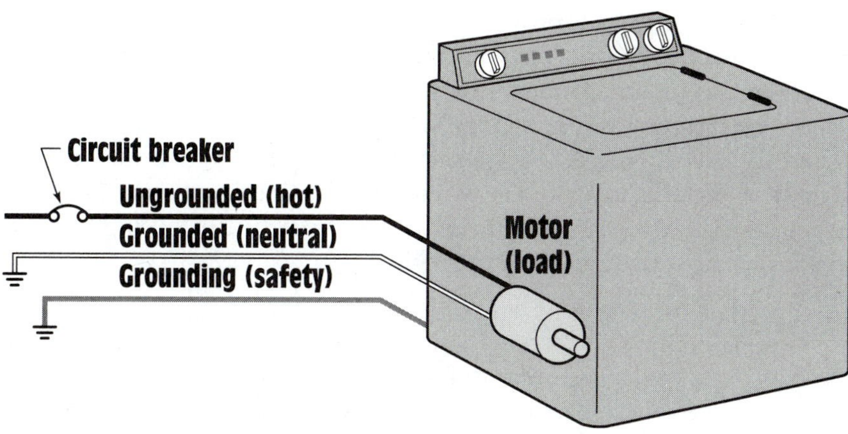

Figure 2-10 120 V appliance circuit.

power source. The grounding conductor is generally connected to the case of the appliance to provide a low resistance path to ground. Although both the neutral and grounding conductors are grounded at the power source, the grounding conductor is not considered to be a circuit conductor. The reason for this is that the only time current will flow through the grounding conductor is when a circuit fault develops. In normal operation, current flows through the hot and neutral conductors only.

The grounding conductor is used to help prevent a shock hazard in the event that the ungrounded, or hot, conductor comes in contact with the case or frame of the appliance, *Figure 2-11*. This condition could occur in several ways. In this example it will be assumed that the motor winding becomes damaged and makes connection to the frame of the motor. Since the frame of the motor is connected to the frame of the appliance, the

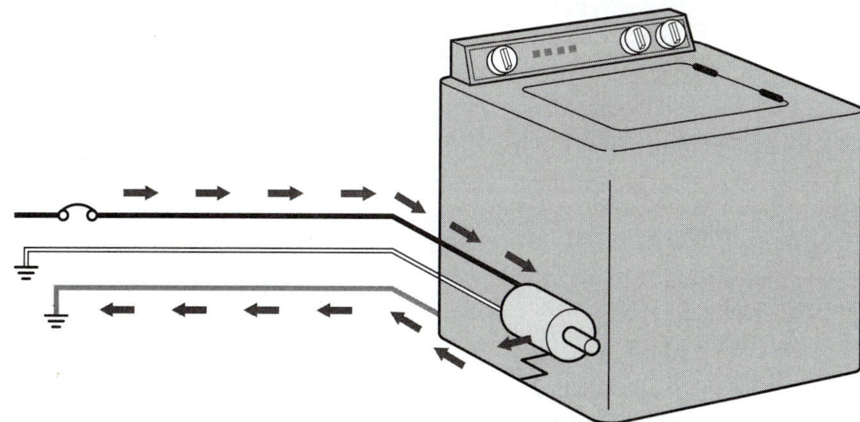

Figure 2-11 The grounding conductor provides a low-resistance path to ground.

The grounding prong of a plug should never be cut off or bypassed.

grounding conductor will provide a circuit path to ground. If enough current flows, the circuit breaker will open. Imagine this condition occurring without the grounding conductor being connected to the frame of the appliance. The frame of the appliance would become *hot* and anyone touching the case and a grounded point, such as a water line, would complete the circuit to ground. The person would receive an electrical shock and perhaps an injury as well. For this reason, **the grounding prong of a plug should never be cut off or bypassed.**

THE VOLT

electromotive force

volt

Voltage is actually defined as **electromotive force** or EMF. It is the force that pushes the electrons through a wire and is often referred to as *electrical pressure*. A **volt** is the amount of potential necessary to cause one coulomb to produce one joule of work. One thing to remember is that voltage cannot flow. To say that voltage flows through a circuit is like saying that pressure flows through a pipe. Pressure can push water through a pipe, and it is correct to say that water flows through a pipe, but it is not correct to say that pressure flows through a pipe. The same is true for voltage. Voltage pushes current through a wire, but voltage cannot flow through a wire. In a water system, the voltage could be compared to the pressure of the system, *Figure 2-12*.

Voltage is often thought of as the potential to do something. For this reason it is frequently referred to as *potential,* especially in older publications

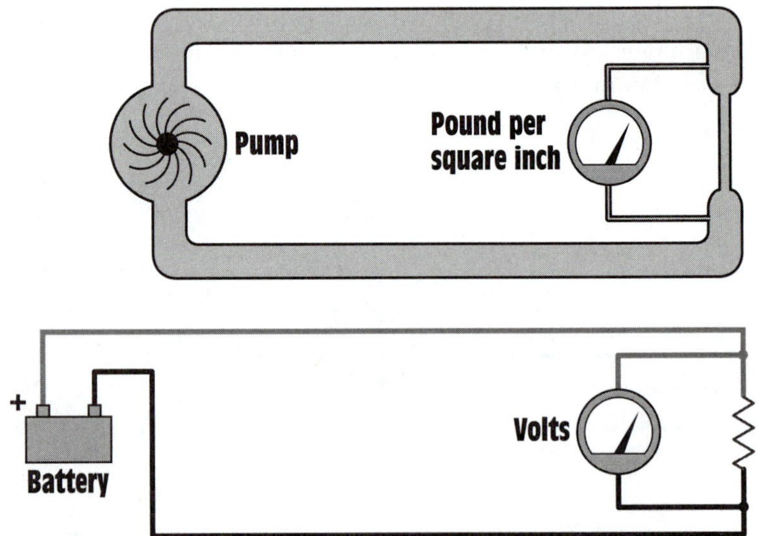

Figure 2-12 Voltage in an electrical circuit can be compared to pressure in a water system.

and service manuals. Voltage must be present before current can flow, just as pressure must be present before water can flow. A voltage, or potential, of 120 volts is present at a common wall outlet, but there is no flow until some device is connected and a complete circuit exists. The same is true in a water system. Pressure is present, but water cannot flow until the valve is opened and a path is provided to a region of lower pressure. The letter *E,* which stands for EMF, or the letter *V,* which stands for volt, is generally used to represent voltage in an algebraic formulas. This text uses the letter *E* to represent voltage in algebraic formulas.

THE OHM

An **ohm** is the measurement of **resistance** to the flow of current. The ohm is named for a German scientist named Georg S. Ohm. The symbol used to represent an ohm, or resistance, is the Greek letter omega (Ω). The letter *R,* which stands for resistance, is used to represent ohms in algebraic formulas. The voltage of the circuit must overcome the resistance before it can cause electrons to flow through it. Without resistance, every electrical circuit would be a short circuit. All electrical loads, such as heating elements, lamps, motors, transformers, and so on, are measured in ohms. In a water system, a reducer can be used to control the flow of water. In an electrical circuit, a resistor can be used to control the flow of electrons. *Figure 2-13* illustrates this concept.

ohm

resistance

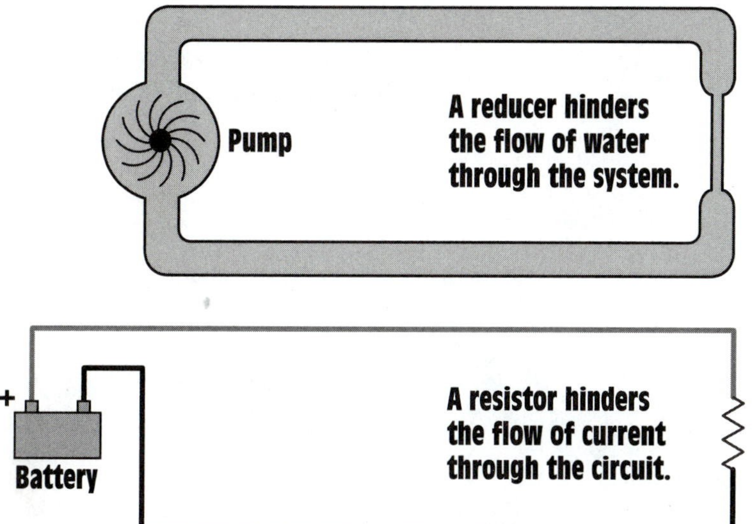

Figure 2-13 A resistor in an electrical circuit can be compared to a reducer in a water system.

To understand the effect of resistance on an electric circuit, consider a person running along a beach. As long as the runner stays on the hard, compact sand, he can run easily along the beach. This is like current flowing through a good conductive material, such as a copper wire. Now imagine the runner wades out into the water until it is knee deep. He will no longer be able to run along the beach as easily because of the resistance of the water. Now assume that the runner wades out into the water until it is waist deep. His ability to run along the beach will be hindered to a greater extent because of the increased resistance of the water against his body. The same is true for resistance in an electric circuit. The higher the resistance, the greater the hindrance to current flow.

Whenever current flows through a resistance, heat is produced.

Another fact that an electrician should be aware of is that **whenever current flows through a resistance, heat is produced,** *Figure 2-14.* This is the reason why wire becomes warm when current flows through it. The filament of an incandescent lamp becomes extremely hot, and the elements of an electric range become hot, because of the resistance of the element.

impedance

A term that has a similar meaning as resistance is **impedance.** Impedance is most often used in calculations of alternating current rather than direct current. Impedance will be discussed to a greater extent later in this text.

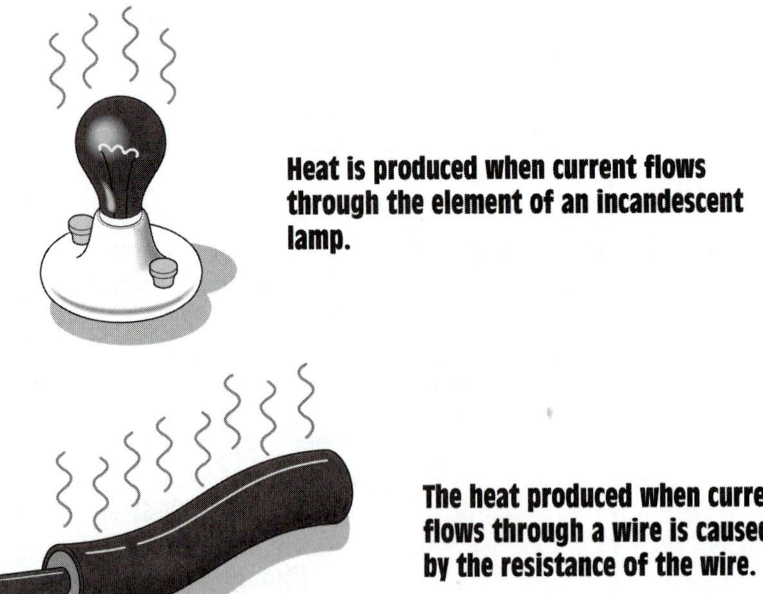

Figure 2-14 Heat is produced when current flows through resistance.

THE WATT

watt

power

Wattage is a measure of the amount of power that is being used in the circuit. The **watt** is named in honor of an English scientist named James Watt. In algebraic formulas, it is generally represented by the letter *P,* for power, or *W,* for watts.

A very important concept that should be understood concerning **power** in an electrical circuit is that before true power, or watts, can exist, there must be some type of energy change or conversion. Electricity is a form of pure energy, and in accord with physical laws, energy cannot be created or destroyed, but its form can be changed. A watt is a measure of this change. When current flows through a resistive element, electrical energy is converted into heat energy. The electric power supplied to a motor is converted into mechanical energy, *Figure 2-15.*

Watts in an electic circuit can be computed using several formulas. Probably the most common formula is: watts equals volts times amps, *Figure 2-16.*

$$P = E \times I$$

Assume that a circuit has a voltage of 120 volts and a current draw of 3 amperes. If the current is flowing through a resistive load, 360 watts of electrical energy is being converted into heat (120 V × 3 A = 360 W). Power can also be computed by dividing the square of the voltage by the resistance.

$$P = \frac{E^2}{R}$$

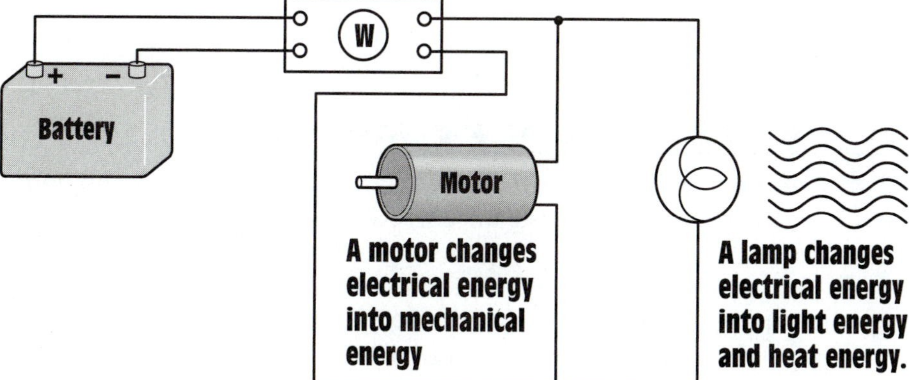

Figure 2-15 Watts are produced when electrical energy is converted into some other form.

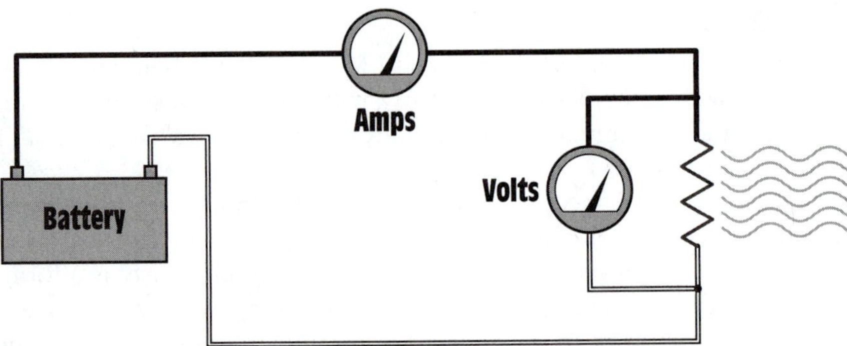

Figure 2-16 Amps times volts equals watts.

Example: A 24-Ω resistor is connected to a 6-volt battery. How much power is converted into heat?

$$P = \frac{6 \times 6}{24}$$

$$p = 1.5\ W$$

A third formula used to compute the power dissipation of an electric circuit is the current squared multiplied by the resistance.

$$P = I^2 \times R$$

Example: A 15-Ω resistor has a current of 2.5 amperes flowing through it. How much power is being dissipated into heat by the resistor?

$$P = 2.5 \times 2.5 \times 15$$

$$P = 93.75\ W$$

OTHER MEASURES OF POWER

horsepower

The watt is not the only unit of power measure. Many years ago, James Watt decided that to sell his steam engines, he would have to rate their power in terms that the average person could understand. He decided to compare his steam engines with the horses he hoped to replace with his engines. After experimentation, Watt found that the average horse, working at a steady rate, could do 550-foot pounds of work per second. A foot-pound is the amount of force required to raise a one pound weight one foot. This rate of doing work is the definition of a **horsepower** (hp).

$$1\ hp = 550\ ft\text{-}lb$$

Horsepower can also be expressed as 33,000 foot-pounds per minute (550 ft-lb × 60 = 33,000).

$$1 \text{ hp} = 33{,}000 \text{ ft-lb/min}$$

It was later computed that the amount of electrical energy needed to produce one horsepower was 746 watts.

$$1 \text{ hp} = 746 \text{ W}$$

BTU (British thermal unit)

Another measure of energy frequently used in the English system of measure is the **BTU (British thermal unit).** A BTU is defined as the amount of heat required to raise the temperature of one pound of water one degree Fahrenheit. In the metric system, the *calorie* is used to measure heat instead of the BTU. A calorie is the amount of heat needed to raise the temperature of one gram of water one degree Celsius.

joule

The **joule** is the metric equivalent of the watt. A joule is defined as one newton per meter. A newton is a force of 100,000 dynes, or about $3^1/2$ ounces, and a meter is about 39 inches. The joule can also be expressed as the amount of work done by one coulomb flowing through a potential of one volt. The joule can also be expressed as the amount of work done by one watt for one second.

$$1 \text{ joule} = 1 \text{ watt/sec.}$$

The chart in *Figure 2-17* gives some common conversions for different quantities. These quantities can be used to calculate different values.

1 Horsepower =	746 Watts
1 Horsepower =	550 Ft-lb./s
1 Watt =	0.00134 Horsepower
1 Watt =	3.412 BTU/hr
1 Watt/s =	1 Joule
1 BTU/s =	1.055 Watts
1 Cal/s =	4.19 Watts
1 Ft-lb./s =	1.36 Watts
1 BTU =	1050 Joules
1 Joule =	0.2389 Cal
1 Cal =	4.186 Joules

Figure 2-17 Common power units.

Example 1

An elevator must lift a load of 4,000 pounds to a height of 50 feet in 20 seconds. How much horsepower is required to operate the elevator?

Solution:

Find the amount of work that must be performed, and then convert that to horsepower.

$$4000 \text{ lb} \times 50 \text{ ft} = 200{,}000 \text{ ft-lb}$$

$$\frac{200{,}000 \text{ ft-lb}}{20 \text{ s}} = 10{,}000 \text{ ft-lb/s}$$

$$\frac{10{,}000 \text{ ft-lb/s}}{550 \text{ ft-lb/s}} = 18.18 \text{ Hp}$$

Example 2

A water heater contains 40 gallons of water. Water weighs 8.34 pounds per gallon. The present temperature of the water is 68°F. The water must be raised to a temperature of 160°F in one hour. How much power will be required to raise the water to the desired temperature?

Solution:

First determine the weight of the water in the tank, because a BTU is the amount of heat required to raise the temperature of one pound of water one degree Fahrenheit.

$$40 \text{ gal} \times 8.34 \text{ lb/gal} = 333.6 \text{ lb}$$

The second step is to determine how many degrees of temperature the water must be raised. This will be the difference between the present temperature and the desired temperature.

$$160°\text{F} - 68°\text{F} = 92°\text{F}$$

The amount of heat required in BTUs will be the product of the pounds of water and the desired increase in temperature.

$$333.6 \text{ lb} \times 92° = 30{,}691.2 \text{ BTU}$$

$$1 \text{ W} = 3.412 \text{ BTU/hr}$$

Therefore:

$$\frac{30{,}691 \text{ BTU}}{3.412 \text{ BTU/hr}} = 8995.1 \text{ W/hr}$$

OHM'S LAW

Ohm's Law

Ohm's Law is named for Dr. Ohm, who discovered that all electrical quantities are proportional to each other and can, therefore, be expressed as mathematical formulas. In its simplest form, Ohm's Law states that **it takes one volt to push one amp through one ohm.** Ohm's Law is actually a statement of proportion. Ohm discovered that if the resistance of a circuit remained constant and the voltage increased, there was a corresponding proportional increase of current. If the resistance remained constant and the voltage decreased, there would be a proportional decrease of current. He also found that if the voltage remained constant and the resistance increased, there would be a decrease of current. If the voltage remained constant and the resistance decreased, there would be an increase of current. This led Ohm to the conclusion that **in a DC circuit, the current is directly proportional to the voltage and inversely proportional to the resistance.**

It takes one volt to push one amp through one ohm.

In a DC circuit, the current is directly proportional to the voltage and inversely proportional to the resistance.

Since Ohm's Law is a statement of proportion, it can be expressed as an algebraic formula when standard values such as the volt, amp, and ohm are used. The three basic Ohm's Law formulas are shown.

$$E = I \times R$$

$$I = \frac{E}{R}$$

$$R = \frac{E}{I}$$

E = EMF, or voltage
I = intensity of current, or amperage
R = resistance

The first formula states that the voltage can be found if the current and resistance are known. Voltage is equal to amps multiplied by ohms. For example, assume a circuit has a resistance of 50 Ω and a current flow through it of 2 A. The voltage connected to this circuit is 100 V.

$$E = I \times R$$

$$E = 2 \times 50$$

$$E = 100 \text{ V}$$

The second formula states that the current can be found if the voltage and resistance are known. In the example shown, 120 V is connected to a resistance of 30 Ω. The amount of current flow will be 4 A.

$$I = \frac{E}{R}$$

$$I = \frac{120}{30}$$

$$I = 4$$

The third formula states that if the voltage and current are known, the resistance can be found. Assume a circuit has a voltage of 240 V and a current flow of 10 A. The resistance in the circuit is 24 Ω.

$$R = \frac{E}{I}$$

$$R = \frac{240}{10}$$

$$R = 24\ \Omega$$

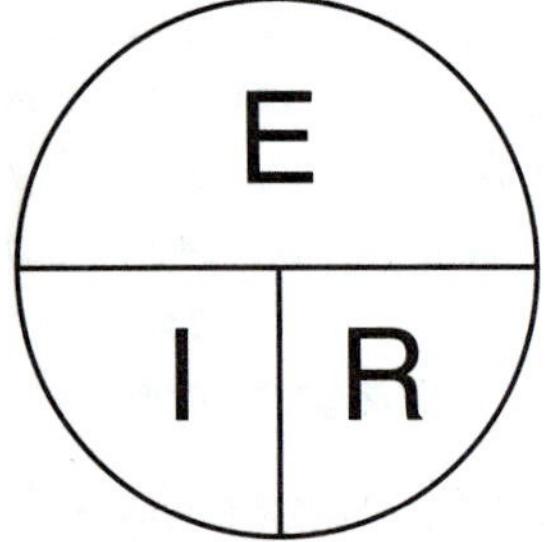

Figure 2-18 Chart for finding values of voltage, current, and resistance.

Figure 2-18 shows a simple chart that can be a great help when trying to remember an Ohm's Law formula. To use the chart, cover the quantity that is to be found. For example, if the voltage, E, is to be found, cover the E on the chart. The chart now shows the remaining letters I and R, *Figure 2-19,* thus E = I × R. If the current is to be found, cover the I on the chart. The chart now shows I = E × R. If the resistance is to be found, cover the R on the chart. The chart now shows R = E × I . A similar chart for determining formulas for power (watts), voltage, and current is shown in *Figure 2-20.* This chart is used in the same manner as the chart shown in Figure 2-18.

A larger chart, which shows the formulas needed to find watts, voltage, amperage, and resistance, is shown in *Figure 2-21*. The letter *P* (power) is used to represent the value of watts. Notice that this chart is divided into four sections, and each section contains three different formulas. To use this chart, find the section that contains the quantity to be found, and then choose the proper formula from the given quantities.

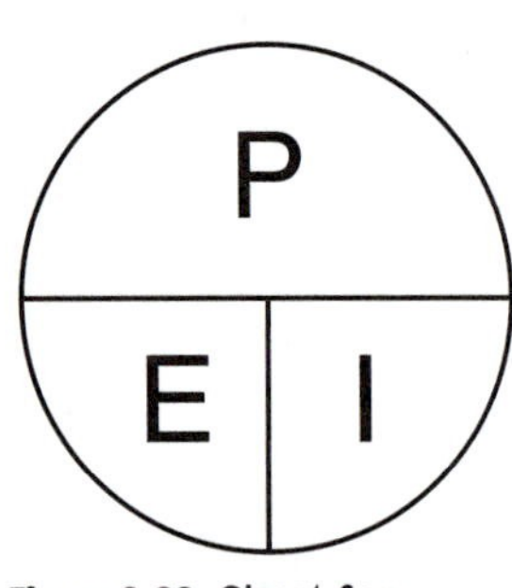

Figure 2-20 Chart for finding values of power, voltage, and current.

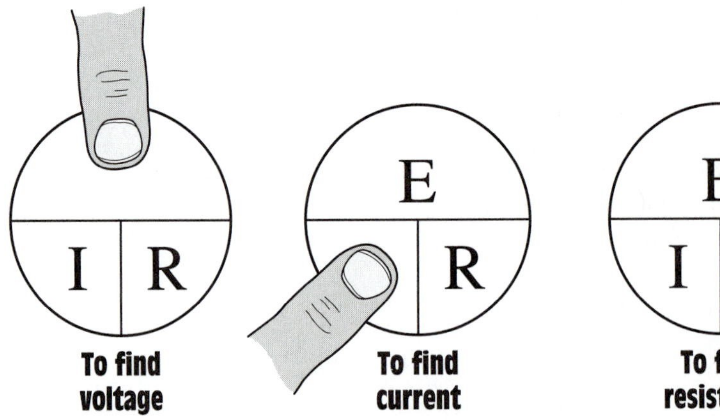

Figure 2-19 Using the Ohm's Law chart.

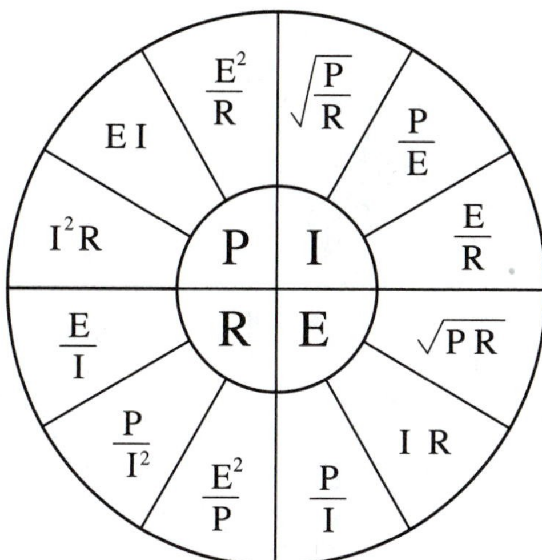

Figure 2-21 Formula chart for finding values of voltage, current, resistance, and power.

Example 3

An electric iron is connected to 120 V and has a current draw of 8 A. How much power is used by the iron?

Solution:

The quantity to be found is watts, or power. The known quantities are voltage and amperage. The proper formula to use is shown in *Figure 2-22*.

$$P = EI$$

$$P = 120 \times 8$$

$$P = 960\ W$$

Example 4

An electric hair dryer has a power rating of 1,000 W. How much current will it draw when connected to 120 V?

Solution:

The quantity to be found is amperage or current. The known quantities are power and voltage. To solve this problem, choose the formula shown in *Figure 2-23*.

$$I = \frac{P}{E}$$

$$I = \frac{1000}{120}$$

$$I = 8.33\ A$$

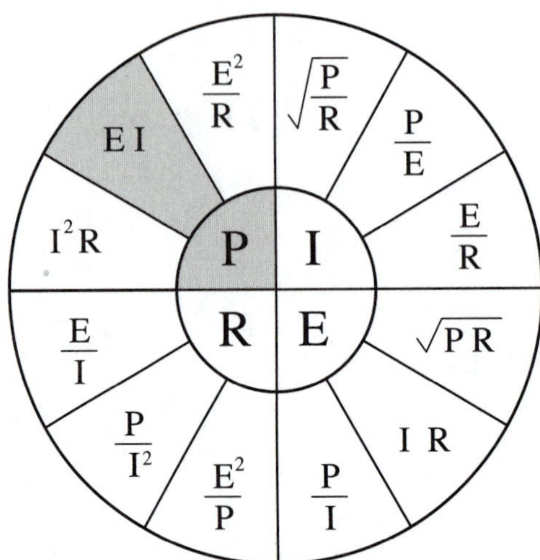

Figure 2-22 Finding power when voltage and current are known.

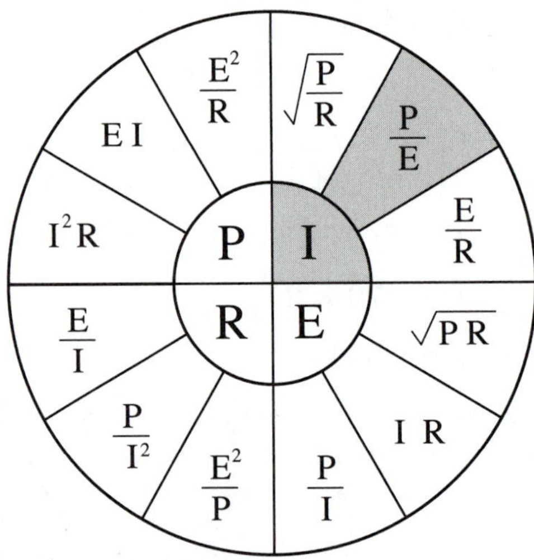

Figure 2-23 Finding current when power and voltage are known.

Example 5

An electric hotplate has a power rating of 1,440 W and a current draw of 12 A. What is the resistance of the hotplate?

Solution:

The quantity to be found is resistance, and the known quantities are power and current. Use the formula shown in *Figure 2-24*.

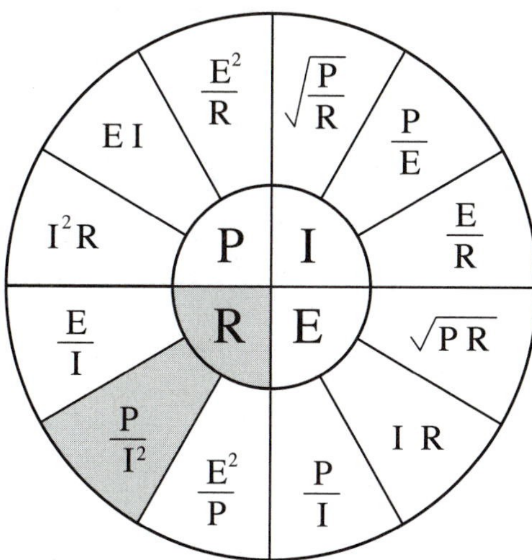

Figure 2-24 Finding resistance when power and current are known.

$$R = \frac{P}{I^2}$$

$$R = \frac{1440}{12 \times 12}$$

$$R = \frac{1440}{144}$$

$$R = 10\ \Omega$$

METRIC UNITS

Metric units of measure are used in the electrical field just as they are in most scientific fields. A special type of metric notation known as *engineering notation* is used in electrical measurements. Engineering notation is the same as any other metric measure, except that engineering notation is in steps of one thousand instead of ten. The chart in *Figure 2-25* shows standard metric units. The first step above the base unit is deka, which means 10; the second unit is hecto, which means 100; and the third unit is kilo, which means 1,000. The first unit below the base unit is deci, which means 1/10; the second unit is centi, which means 1/100; and the third unit is milli, which means 1/1000.

Kilo	1000
Hecto	100
Deka	10
Base unit	1
Deci	1/10 or 0.1
Centi	1/100 or 0.01
Milli	1/1000 or 0.001

Figure 2-25 Standard units of metric measure.

The chart in *Figure 2-26* shows engineering units. The first unit above the base unit is kilo, or 1,000; the second unit is mega, or 1,000,000; and

ENGINEERING UNIT	SYMBOL	MULTIPLY BY	
Tera	T	1,000,000,000,000	$\times 10^{12}$
Giga	G	1,000,000,000	$\times 10^{9}$
Mega	M	1,000,000	$\times 10^{6}$
Kilo	k	1,000	$\times 10^{3}$
Base unit		1	
Milli	m	0.001	$\times 10^{-3}$
Micro	μ	0.000,001	$\times 10^{-6}$
Nano	n	0.000,000,001	$\times 10^{-9}$
Pico	p	0.000,000,000,001	$\times 10^{-12}$

Figure 2-26 Standard units of engineering notation.

the third unit is giga, or 1,000,000,000. Notice that each unit is 1,000 times greater than the previous unit. The chart also shows that the first unit below the base unit is milli, or 1/1000; the second unit is micro, represented by the Greek letter mu (μ), or 1/1,000,000; and the third unit is nano, or 1/1,000,000,000.

Metric units are used in almost all scientific measurements for ease of notation. It is much easier to write a value such as 10MΩ than it is to write 10,000,000 ohms, or to write 0.5 ns than to write .000,000,000,5 second. Once the metric system has been learned, measurements such as 47 kilohms (kW) or 50 milliamps (MA) become commonplace to the technician.

RESISTORS

Resistors are one of the most common components found in electrical circuits. The unit of measure for resistance (R) is the *ohm,* which was named for a German scientist named Georg S. Ohm. The symbol used to represent resistance is the Greek letter omega (Ω). Resistors come in various sizes, types, and ratings to accommodate the needs of almost any circuit applications.

USES OF RESISTORS

Resistors are commonly used to perform two functions in a circuit. One is to limit the flow of current through the circuit. In *Figure 2-27* a 30-Ω re-

Figure 2-27 Resistor used to limit the flow of current.

sistor is connected to a 15-V battery. The current in this circuit is limited to a value of 0.5 A.

$$I = \frac{E}{R}$$

$$I = \frac{15}{30}$$

$$I = 0.5\ A$$

If this resistor were not present, the circuit current would be limited only by the resistance of the conductor, which would be very low, and a large amount of current would flow. Assume for example that the wire has a resistance of 0.0001 Ω. When the wire is connected across the 15-V power source, a current of 150,000 A would try to flow through the circuit (15/0.0001 = 150,000). This is commonly known as a **short circuit.**

short circuit

The second principal function of resistors is to produce a **voltage divider.** The three resistors shown in *Figure 2-28* are connected in series with a 17.5-V battery. If the leads of a volunteer were connected between different points in the circuit, it would indicate the following voltages:

voltage divider

A to B, 1.5 V

A to C, 7.5 V

A to D, 17.5 V

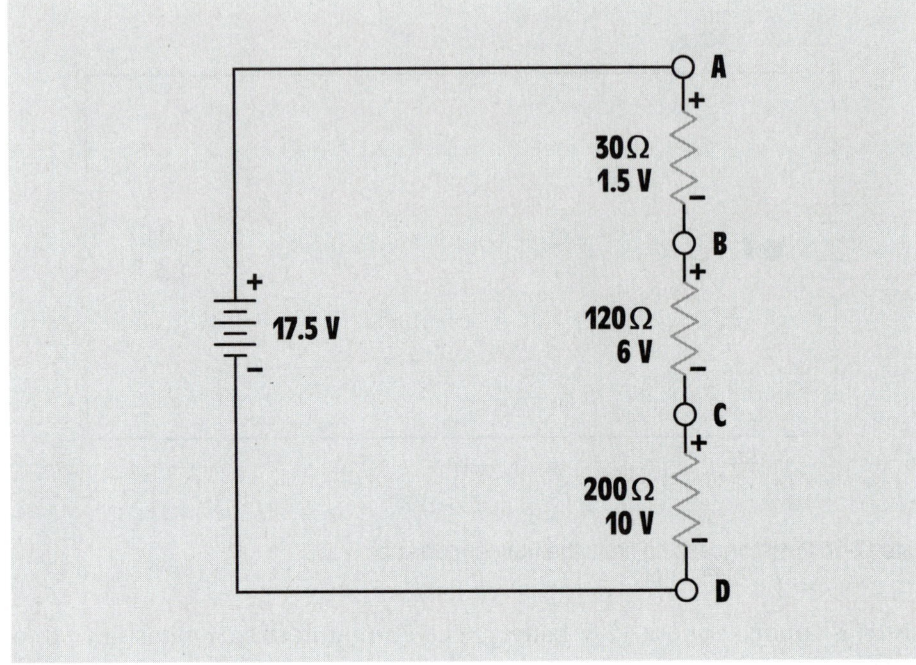

Figure 2-28 Resistors used as a voltage divider.

B to C, 6 V

B to D, 16 V

C to D, 10 V

By connecting resistors of the proper value, almost any voltage desired can be obtained. Voltage dividers were used to a large extent in vacuum tube circuits many years ago. Voltage divider circuits are still used today in applications involving field effect transistors (FETs) and in multirange voltmeter circuits.

fixed resistors

composition carbon resistor

FIXED RESISTORS

Fixed resistors have only one ohmic value, which cannot be changed or adjusted. There are several different types of fixed resistors One of the most common types of fixed resistors is the **composition carbon resistor.** Carbon resistors are made from a compound of carbon graphite and a resin bonding material. The proportions of carbon and resin material determine the value of resistance. This compound is enclosed in a case of nonductive material with connecting leads, *Figure 2-29.*

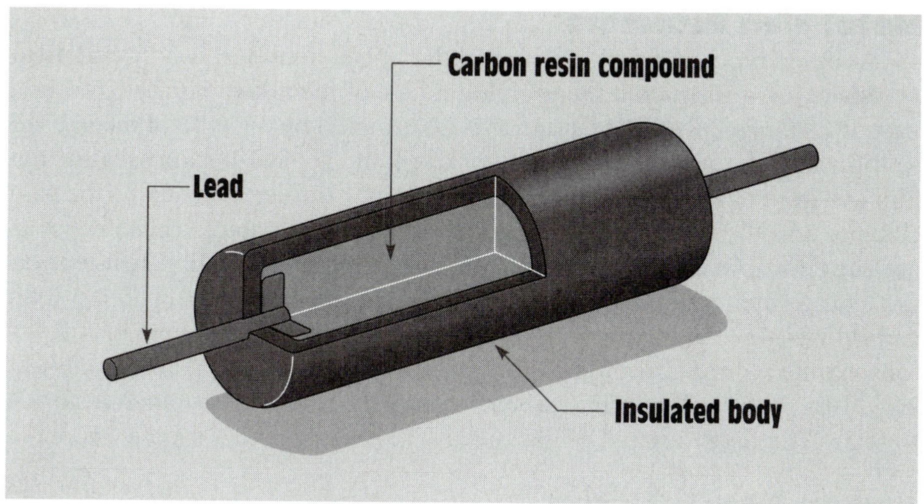

Figure 2-29 Composition carbon resistor.

Carbon resistors are very popular for most applications because they are inexpensive and readily available. They are made in standard values that range from about 1 Ω to about 22 MΩ (M represents meg), and they can be obtained in power ratings of 1/8, 1/4, 1/2, 1, and 2 W. The power rating of the resistor is indicated by its size. A 1/2-W resistor is approximately 3/8 in. in length and 1/8 in. in diameter. A 2-W resistor has a length of approximately 11/16 in. and a diameter of approximately 5/16 in, Figure 2-30. The 2-W resistor is larger than the 1/2-W or 1-W because it must have a larger surface area to be able to dissipate more heat. Although carbon resistors have a lot of desirable characteristics, they have one characteristic that is not desirable. Carbon resistors will change their value with age or if they are overheated. Carbon resistors genially increase instead of decrease in value.

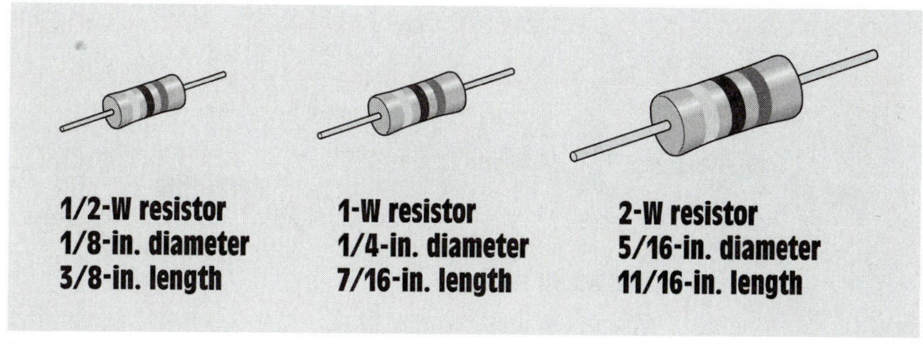

Figure 2-30 Power rating is indicated by size.

metal film resistors

tolerance

carbon film resistor

metal glaze resistor

Metal Film Resistors

Another type of fixed resistor is the metal film resistor. **Metal film resistors** are constructed by applying a film of metal to a ceramic rod in a vacuum, *Figure 2-31*. The resistance is determined by the type of metal used to form the film and the thickness of the film. Typical thicknesses for the film are from 0.00001 to 0.00000001 in. Leads are then attached to the film coating, and the entire assembly is covered with a coating. These resistors are superior to carbon resistors in several respects. Metal film resistors do not change their value with age, and their tolerance is generally better than carbon resistors. **Tolerance** indicates the plus and minus limits of a resistor's ohmic value. Carbon resistors commonly have a tolerance range of 20%, 10%, or 5%. Metal film resistors generally range in tolerance from 2% to 0.1%. The disadvantage of the metal film resistor is that it costs more.

Carbon Film Resistors

Another type of fixed resistor that is constructed in a similar manner is the **carbon film resistor.** This resistor is made by coating a ceramic rod with a film of carbon instead of metal. Carbon film resistors are less expensive to manufacture than metal film resistors and can have a higher tolerance rating than composition carbon resistors.

Metal Glaze Resistors

The **metal glaze resistor** is also a fixed resistor, similar to the metal film resistor. This resistor is made by combining metal with glass. The compound is then applied to a ceramic base as a thick film. The resistance is

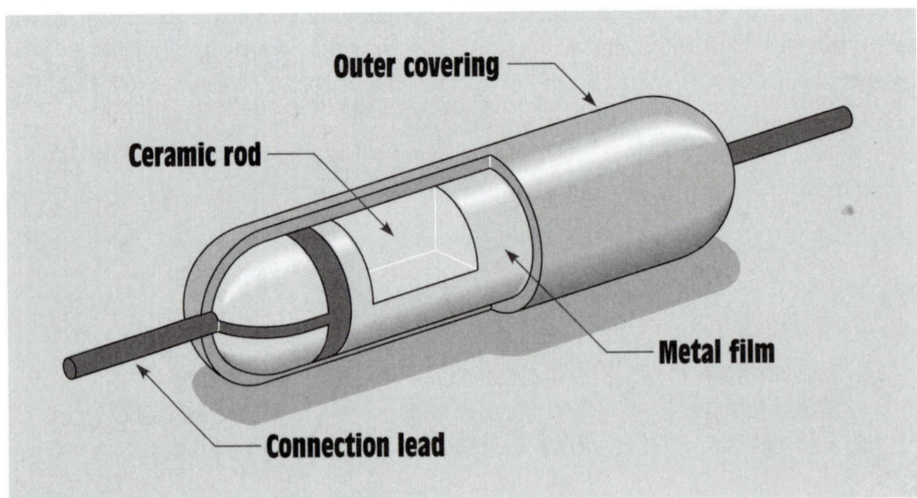

Figure 2-31 Metal film resistor.

determined by the amount of metal used in the compound. Tolerance ratings of 2% and 1% are common.

Wire Wound Resistors

wire wound resistors

Wire wound resistors are fixed resistors that are made by winding a piece of resistive wire around a ceramic core, *Figure 2-32*. The resistance of a wire wound resistor is determined by three factors:

1. the type of material used to make the resistive wire,
2. the diameter of the wire, and
3. the length of the wire.

Wire wound resistors can be found in various case styles and sizes. These resistors are generally used when a high power rating is needed. Wire wound resistors can operate at higher temperatures than any other type of resistor. A wire wound resistor that has a hollow center is shown in *Figure 2-33*. This type of resistor should be mounted vertically and not horizontally. The center of the resistor is hollow for a very good reason. When the resistor is mounted vertically, the heat from the resistor produces a chimney effect and causes air to circulate through the center, *Figure 2-34*. This increase of air flow dissipates heat at a faster rate to help keep the resistor from overheating. The disadvantages of wire wound resistors is that they are expensive and generally require a large amount of space for mounting. They can also exhibit an amount of inductance in circuits that operate at high frequencies. This added inductance can cause problems to the rest of the circuit. Inductance will be covered in later units.

THE COLOR CODE

The values of a resistor can often be determined by the color code. Many resistors have bands of color that are used to determine the resistance

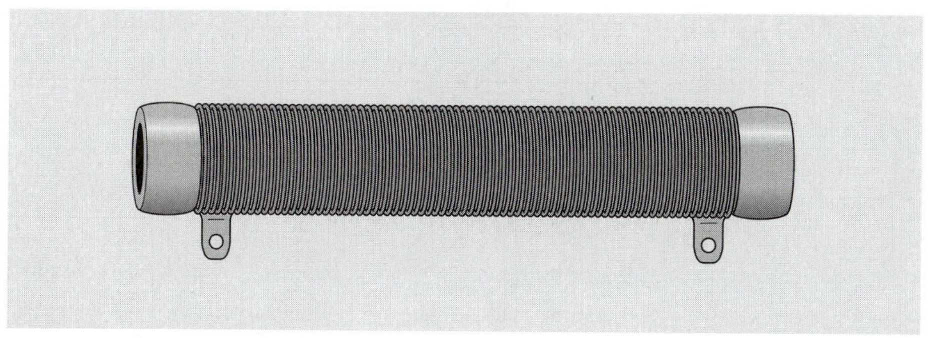

Figure 2-32 Wire wound resistor.

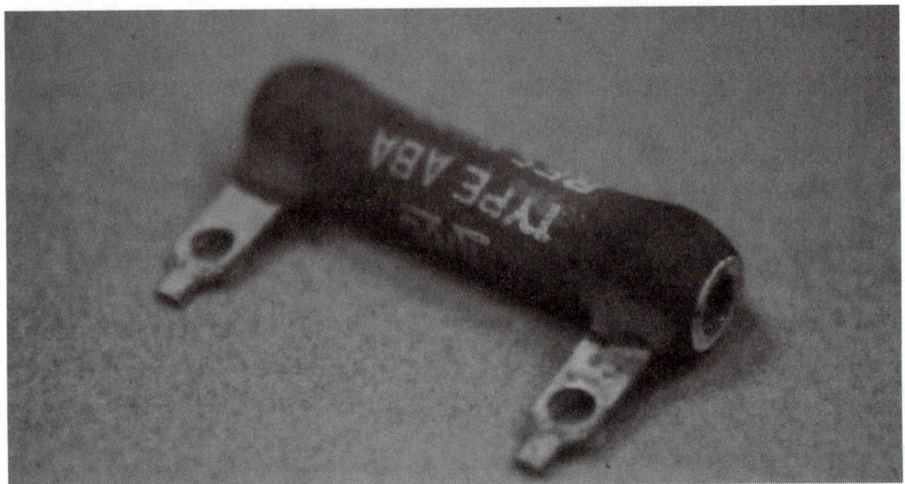

Figure 2-33 Wire wound resistor with hollow core.

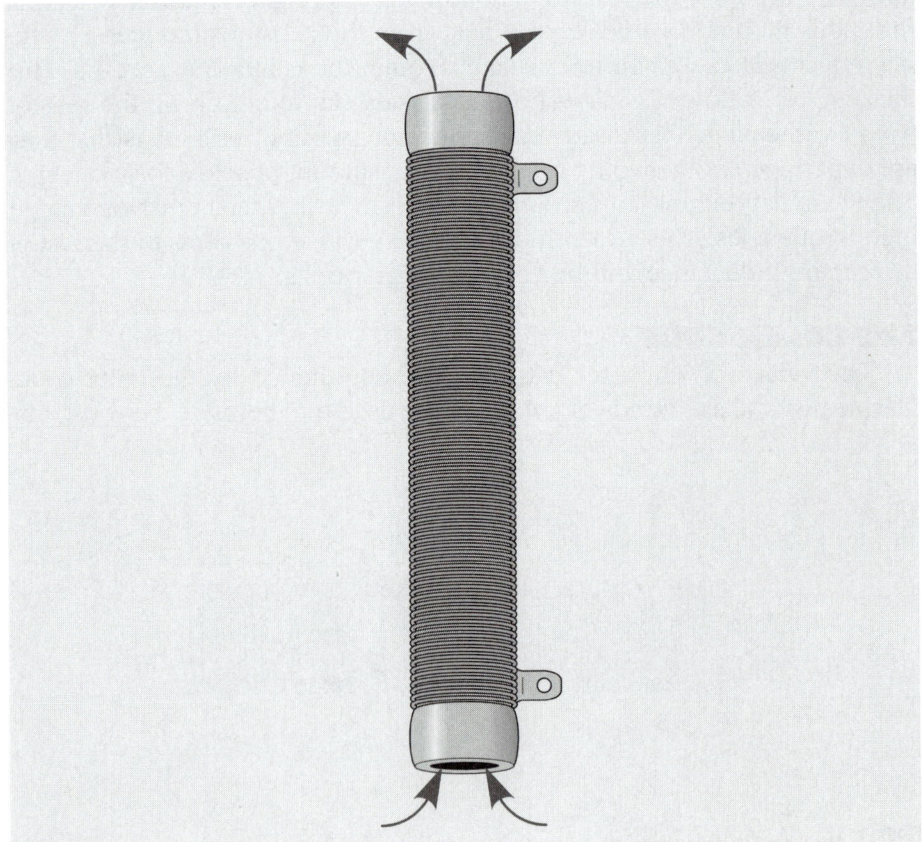

Figure 2-34 Air flow helps cool the resistor.

value, tolerance, and, in some cases, reliability. Resistors typically contain from three to five bands of color. Resistors with three bands of color have a tolerance of plus or minus 20%. Resistors that have tolerance rating from 10% to 2% have four bands of color. Resistors with a tolerance rating of 1% and some military resistors use five bands of color.

The color bands represent numbers. Each color represents a different numerical value, *Figure 2-35*. The chart shown in Figure 2-35 lists the color and number value assigned to each color. The resistor shown in Figure *2-36* illustrates how to determine the resistance value and the tolerance of a four-band resistor. The first two bands represent number values, the second band is the multiplier, and the fourth band is the tolerance.

For example, assume a resistor has color bands of brown, green, red, and silver (Figure 2-36). The first two bands represent the numbers 1 and 5 (brown is 1 and green is 5). The third band is red, which has a number value of 2. The number 15 should be multiplied by 10^2, or 100. The value of the resistor is 1,500 Ω. Another method, which is simpler to understand, is to add the number of zeroes indicated by the multiplier band to the combined first two numbers. The multiplier band in this example is red, which has a numeric value of 2. Add two zeroes to the first two numbers. The number 15 becomes 1,500.

The fourth band is the tolerance band. The tolerance band in this example is silver, which means ±10%. This resistor should be 1,500 Ω plus or minus 10%. To determine the value limits of this resistor, find 10% of 1,500.

$$1500 \times 0.10 = 150$$

Color	Value	Tolerance
Black	0	No Color Band +/- 20%
Brown	1	Silver +/- 10%
Red	2	Gold +/- 5%
Orange	3	Red +/- 2%
Yellow	4	Brown +/- 1%
Green	5	
Blue	6	
Violet	7	
Gray	8	
White	9	

Figure 2-35 Color code chart.

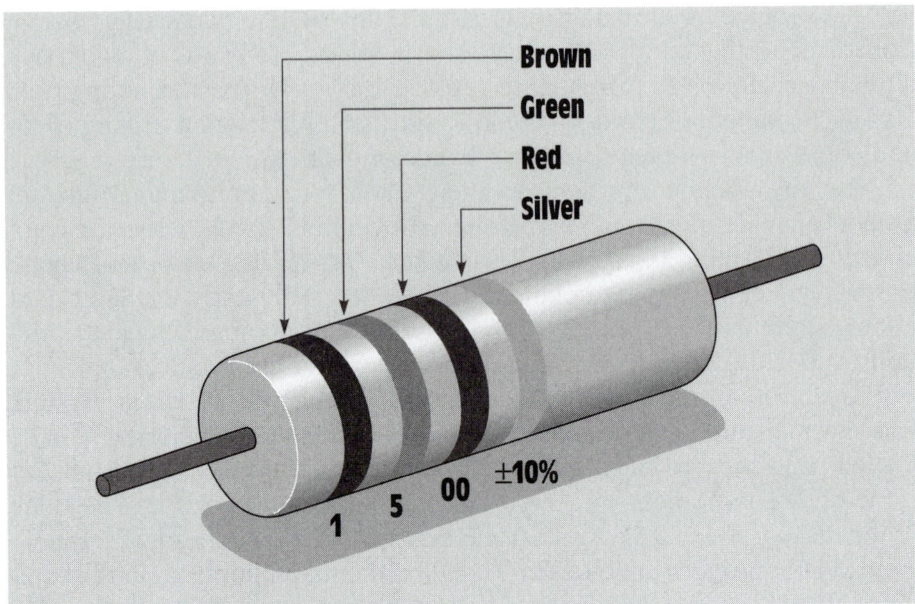

Figure 2-36 Determining resistor values using the color code.

The value can range from 1500 + 10%, or 1500 + 150 = 1650 Ω to 1500 − 10% or 1500 − 150 = 1350 Ω.

Resistors that have a tolerance of ±1%, as well as some military resistors, contain five bands of color.

Example 6

The resistor shown in *Figure 2-37* contains the following bands of color:

first band = brown

second band = black

third band = black

fourth band = brown

fifth band = brown

Solution:

The brown fifth band indicates that this resistor has a tolerance of ±1%. To determine the value of a 1% resistor, the first three bands are numbers, and the fourth band is the multiplier. In this example, the first band is brown, which has a number value of 1. The next two bands are black, which represent a number value of 0. The fourth band is brown, which means add one 0 to the first three numbers. The value of this resistor is 1000 Ω ±1%.

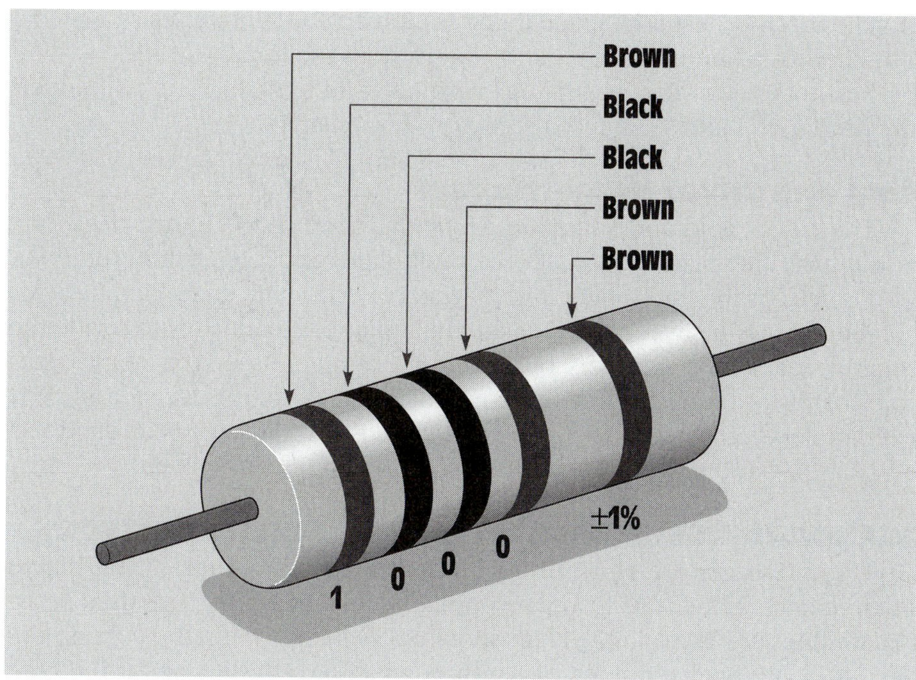

Figure 2-37 Determining the value of a ±1% resistor.

Example 7

A five-band resistor has the following color bands:

first band = red

second band = orange

third band = violet

fourth band = red

fifth band = brown

Solution:

The first three bands represent number values. Red is 2, orange is 3, and violet is 7. The fourth band is the multiplier; in this case, red represents 2. Add two zeroes to the number 237. The value of the resistor is 23,700 Ω. The fifth band is brown, which indicates a tolerance of ±1%.

Military resistors, also often have five bands of color. These resistors are read in the same manner as a resistor with four bands of color. The fifth band can represent different things. A fifth band of orange or yellow is used to indicate reliability. Resistors with a fifth band of orange have a reliability good enough to be used in missile systems, and a resistor with a fifth band of yellow can be used in space-flight equipment. A military resistor with a fifth band of white indicates that the resistor has solderable leads.

Resistors with tolerance ratings ranging from 0.5% to 0.1% will generally have their values printed directly on the resistor.

Gold and Silver as Multipliers

The colors gold and silver are generally found in the fourth band of a resistor, but they can be used in the multiplier band also. When the color gold is used as the multiplier band it means to divide the combined first two numbers by 10. If silver is used as the multiplier band it means to divide the combined first two numbers by 100. For example, assume a resistor has color bands of orange, white, gold, gold. The value of this resistor is 3.9 Ω with a tolerance of ±5% (orange = 3; white = 9; gold means to divide 39 by 10 = 3.9; and gold in the fourth band means ±5% tolerance).

STANDARD RESISTANCE VALUES OF FIXED RESISTORS

Fixed resistors are generally produced in standard values. The higher the tolerance value, the fewer resistance values available. Standard resistance values are listed in the chart shown in *Figure 2-38.* In the column under 10% only twelve values of resistors are listed. these standard values, however, can be multiplied by factors of ten. Notice that one of the standard values listed is 33 Ω. There are also standard values in 10% resistors of 0.33, 3.3, 330, 3300, 33,000, 330,000, and 3,300,000 gV. The 2% and 5% column shows twenty-four resistor values, and the 1% column lists ninety-six values. All of the values listed in the chart can be multiplied by factors of ten to obtain other resistance values.

POWER RATINGS

Resistors also have a power rating in watts that should not be exceeded or the resistor will be damaged. The amount of heat that must be dissipated by (given off to the surrounding air) the resistor can be determined by the use of the following formulas.

$$P = I^2R$$

$$P = \frac{E^2}{R}$$

$$P = EI$$

Example 8

The resistor shown in *Figure 2-39* has a value of 100 Ω and a power rating of 1/2 W. If the resistor is connected to a 10-V power supply, will it be damaged?

STANDARD RESISTANCE VALUES (Ω)									
0.1% 0.25% 0.5%	1%	0.1% 0.25% 0.5%	1%	0.1% 0.25% 0.5%	1%	0.1% 0.25% 0.5%	1%	0.1% 0.25% 0.5%	1%
10.0	10.0	17.2	–	29.4	29.4	50.5	–	86.6	86.6
10.1	–	17.4	17.4	29.8	–	51.1	51.1	87.6	–
10.2	10.2	17.6	–	30.1	30.1	51.7	–	88.7	88.7
10.4	–	17.8	17.8	30.5	–	52.3	52.3	89.8	–
10.5	10.5	18.0	–	30.9	30.9	53.0	–	90.9	90.9
10.6	–	18.2	18.2	31.2	–	53.6	53.6	92.0	–
10.7	10.7	18.4	–	31.6	31.6	54.2	–	93.1	93.1
10.9	–	18.7	18.7	32.0	–	54.9	54.9	94.2	–
11.0	11.0	18.9	–	32.4	32.4	55.6	–	95.3	95.3
11.1	–	19.1	19.1	32.8	–	56.2	56.2	96.5	–
11.3	11.3	19.3	–	33.2	33.2	56.9	–	97.6	97.6
11.4	–	19.6	19.6	33.6	–	57.6	57.6	98.8	–
11.5	11.5	19.8	–	34.0	34.0	58.3	–		
11.7	–	20.0	20.0	34.4	–	59.0	59.0		
11.8	11.8	20.3	–	34.8	34.8	59.7	–		
12.0	–	20.5	20.5	35.2	–	60.4	60.4		
12.1	12.1	20.8	–	35.7	35.7	61.2	–		
12.3	–	21.0	21.0	36.1	–	61.9	61.9		
12.4	12.4	21.3	–	36.5	36.5	62.6	–		
12.6	–	21.5	21.5	37.0	–	63.4	63.4		
12.7	12.7	21.8	–	37.4	37.4	64.2	–	2%,5%	10%
12.9	–	22.1	22.1	37.9	–	64.9	64.9	10	10
13.0	13.0	22.3	–	38.3	38.3	65.7	–	11	–
13.2	–	22.6	22.6	38.8	–	66.5	66.5	12	12
13.3	13.3	22.9	–	39.2	39.2	67.3	–	13	–
13.5	–	23.2	23.2	39.7	–	68.1	68.1	15	15
13.7	13.7	23.4	–	40.2	40.2	69.0	–	16	–
13.8	–	23.7	23.7	40.7	–	69.8	69.8	18	18
14.0	14.0	24.0	–	41.2	41.2	70.6	–	20	–
14.2	–	24.3	24.3	41.7	–	71.5	71.5	22	22
14.3	14.3	24.6	–	42.2	42.2	72.3	–	24	–
14.5	–	24.9	24.9	42.7	–	73.2	73.2	27	27
14.7	14.7	25.2	–	43.2	43.2	74.1	–	30	–
14.9	–	25.5	25.5	43.7	–	75.0	75.0	33	33
15.0	15.0	25.8	–	44.2	44.2	75.9	–	36	–
15.2	–	26.1	26.1	44.8	–	76.8	76.8	39	39
15.4	15.4	26.4	–	45.3	45.3	77.7	–	43	–
15.6	–	26.7	26.7	45.9	–	78.7	78.7	47	47
15.8	15.8	27.1	–	46.4	46.4	79.6	–	51	–
16.0	–	27.4	27.4	47.0	–	80.6	80.6	56	56
16.2	16.2	27.7	–	47.5	47.5	81.6	–	62	–
16.4	–	28.0	28.0	48.1	–	82.5	82.5	68	68
16.5	16.5	28.4	–	48.7	48.7	83.5	–	75	–
16.7	–	28.7	28.7	49.3	–	84.5	84.5	82	82
16.9	16.9	29.1	–	49.9	49.9	85.6	–	91	–

Figure 2-38 Standard resistance values.

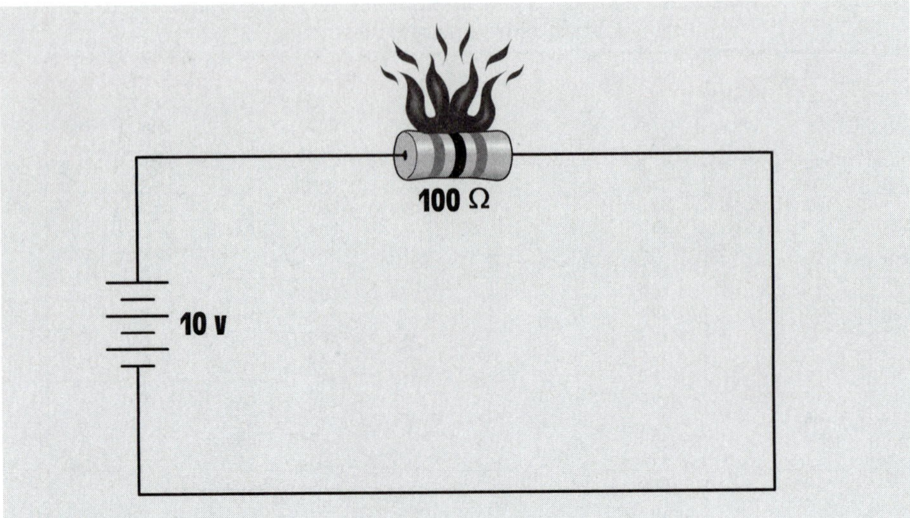

Figure 2-39 Exceeding the power rating causes damage to the resistor.

Solution:

Using the formula $P = \frac{E^2}{R}$ determine the amount of heat that will be dissipated by the resistor.

$$P = \frac{E^2}{R}$$

$$P = \frac{10 \times 10}{100}$$

$$P = \frac{100}{100}$$

$$P = 1\ \text{W}$$

Since the resistor ha a power rating of 1/2 W, and the amount of heat that will be dissipated is 1 W, the resistor will be damaged.

variable resistor

VARIABLE RESISTORS

A **variable resistor** is a resistor whose values can be changed or varied over a range. Variable resistors can be obtained in different case styles and power ratings. *Figure 2-40* illustrates how a variable resistor is constructed. In this example, a resistive wire is wound in a circular pattern, and a sliding tap makes contact with the wire. The value of a resistance can be adjusted between one end of the resistive wire and the sliding tap. If the resistive wire has a total value of 100 Ω, the resistor can be set between the values of 0 and 100 Ω.

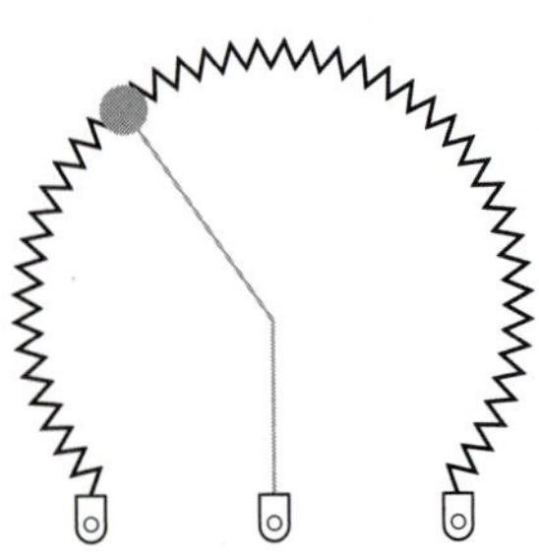

Figure 2-40 Variable resistor.

A variable resistor with three terminals is shown in *Figure 2-41*. This type of resistor has a wiper arm inside the case that makes contact with the resistive element. The full resistance value is between the two outside terminals, and the wiper arm is connected to the center terminal. The resistance between the center terminal and either of the two outside terminals can be adjusted by turning the shaft and changing the position of the wiper arm. Wire wound variable resistors of this type can be obtained also, *Figure 2-42*. The advantage of the wire wound type is a higher power rating.

The resistor shown in Figure 2-41 can be adjusted from its minimum to maximum value by turning the control approximately three-quarters of a turn. In some types of electrical equipment this range of adjustment may be too coarse to allow for sensitive adjustments. When this becomes a problem, a multiturn resistor (*Figure 2-43*) can be used. **Multiturn variable resistors** operate by moving the wiper arm with a screw of some number of turns. They generally range from three turns to ten turns. If a ten-turn variable resistor is used, it will require ten turns of the control knob to move the wiper from one end of the resistor to the other end instead of three-quarters of a turn.

multiturn variable resistors

Figure 2-41 Variable resistors with three terminals. *(Courtesy of Allen Bradley Co., Inc., a Rockwell International Company)*

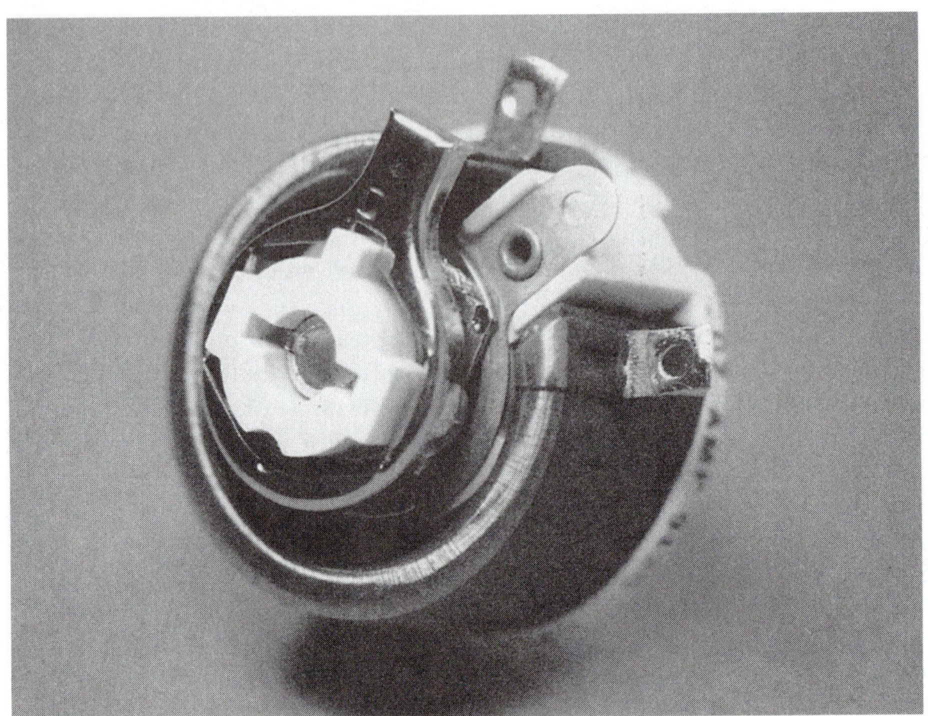

Figure 2-42 Wire wound variable resistor.

Figure 2-43 Multiturn variable resistor.

Variable Resistor Terminology

pot

rheostat

potentio-meter

Variable resistors are known by several common names. The most popular name is **pot,** which is shortened from the word *potentiometer.* Another common name is **rheostat.** A rheostat is actually a variable resistor that has two terminals. They are used to adjust the current in a circuit to a certain value. A potentiometer is a variable resistor that has three terminals. Potentiometers can be used as rheostats by only using two of their three terminals. A **potentiometer** describes how a variable resistor is used rather than some specific type of resistor. The word *potentiometer* comes from the word *potential,* or voltage. A potentiometer is a variable resistor used to provide a variable voltage, as shown in *Figure 2-44.* In this example, one end of a variable resistor is connected to +12 V, and the other end is connected to ground. The middle terminal, or wiper, is connected to the positive terminal of a volunteer and the negative led is connected to ground. If the wiper is moved to the upper end of the resistor, the volunteer will indicate a potential of 12 V. If the wiper is moved to the bottom, the volunteer will indicate a value of 0 V. The wiper can be adjusted to provide any value of voltage between 12 and 0 V.

SCHEMATIC SYMBOLS

Electrical schematics use symbols to represent the use of a resistor. Unfortunately, the symbol used to represent a resistor is not standard. *Figure 2-45* illustrates several schematic symbols used to represent both fixed and variable resistors.

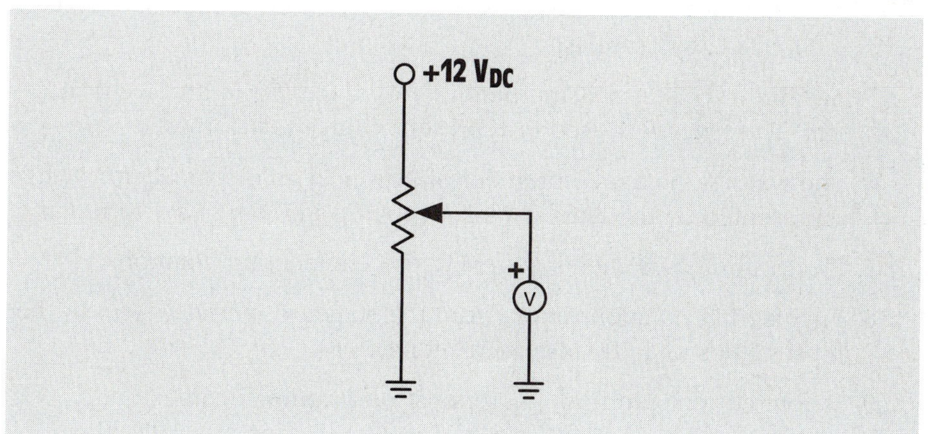

Figure 2-44 Variable resistor used as a potentiometer.

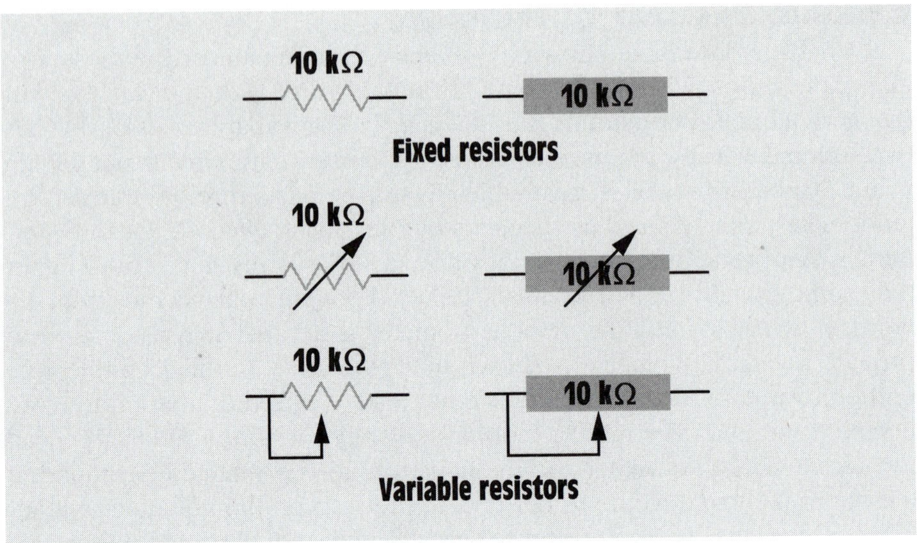

Figure 2-45 Schematic symbols used to represent resistors.

SUMMARY

1. A coulomb is a quantity measurement of electrons.
2. The definition of an amp (A) is one coulomb per second.
3. The letters *I,* stands for intensity of current flow, or *A,* which stands for amps, are often used in Ohm's Law formulas.
4. Voltage is referred to as electrical pressure, potential difference, or electromotive force. The letters *E* and *V* are used to represent voltage in Ohm's Law formulas.
5. An ohm (Ω) is a measurement of resistance (R) in an electrical circuit. The letter *R* is used to represent Ohm's Law formulas.
6. The watt (W) is a measurement of power in an electrical circuit. It is represented by the letters *W* or *P* (power) in Ohm's Law formulas.
7. Electrical measurements generally use engineering notation.
8. Engineering notation differs from the standard metric system in that it uses steps of 1,000 instead of steps of 10.
9. Before current can flow, there must be a compete circuit.
10. A short circuit has little or no resistance.

11. Resistors are used in two main applications: as voltage dividers and to limit the flow of current in a circuit.
12. The value of a fixed resistor cannot be changed.
13. There are several types of fixed resistors, such as composition carbon, metal film, and wire wound.
14. Carbon resistors change their value with age or if overheated.
15. Metal film resistors never change their value, but they are more costly than carbon resistors.
16. The advantage of wire wound resistors is their high power ratings.
17. Resistors often have bands of color to indicate their resistance value and tolerance.
18. Resistors are produced in standard values. The number of values between 0 Ω and 100 Ω is determined by the tolerance.
19. Variable resistors can change their value within the limit of their full value.
20. A potentiometer is a variable resistor used as a voltage divider.

REVIEW QUESTIONS

1. What is a coulomb?
2. What is an amp?
3. Define voltage.
4. Define ohm.
5. Define watt.
6. An electric heating element has a resistance of 16 Ω and is connected to a voltage of 120 V. How much current will flow in this circuit?
7. How many watts of heat are being produced by the heating element in question 6?
8. A 240-V circuit has a current flow of 20 A. How much resistance is connected to the circuit?
9. An electric motor has an apparent resistance of 15 Ω. If 8 A of current are flowing through the motor, what is the connected voltage?

10. A 240-V air conditioning compressor has an apparent resistance of 8 Ω. How much current will flow in the circuit?
11. How much power is being used by the compressor in question 8?
12. A 5-kW electric heating unit is connected to a 240-V. line. What is the current flow in the circuit?
13. If the voltage in question 12 is reduced to 120 V, how much current would be needed to reduce the same amount of power?
14. Is it less expensive to operate the electric heating unit in question 12 on 240 V or 120 V?
15. Name three types of fixed resistors.
16. What is the advantage of a metal film resistor over a carbon resistor?
17. What is the advantage of a wire wound resistor?
18. How should tubular wire wound resistors be mounted and why?
19. A half-watt 2,000-Ω resistor has a current flow of .01 amps through it. Is this resistor operating within its power rating?
20. A one-watt 350-ohm resistor is connected to 24 volts. Is this resistor operating within its power rating?
21. A resistor has color bands of orange, blue, yellow, and gold. What is the resistance and tolerance of this resistor?
22. A 10,000-ohm resistor has a tolerance of 5%. What are the minimum and maximum ratings of this resistor?
23. Is 51,000 ohms a standard value for a 5% resistor?
24. What is a potentiometer?

Practice Problems

OHM'S LAW

1. A 1,500-W heating element has a resistance of 38.4 Ω. How much current is flowing through the heating element?
2. How much voltage is applied to the heating element in question 1?
3. A 100-W lamp is connected to a 120-V circuit. How much current will flow through the lamp?
4. A 15-W soldering iron is connected to a 120-V circuit. What is the resistance of the iron?

5. Four 300-W heating elements are connected to a 277-V line. How much current will flow in the circuit?
6. A 1.5K-Ω resistor has a current of 8 ma flowing through it. How much voltage is applied to the resistor?
7. Change 0.00045 A into milliamps (ma).
8. Change 0.00045 A into microamps (μa).
9. An alternator produces 850 megawatts (Mw) at a voltage of 13,800 V. What is the current output of the alternator?
10. Change 2542 V into kilovolts (kV).

RESISTORS

Practice Problems

1st Band	2nd Band	3rd Band	4th Band	Value	%Tol
Red	Yellow	Brown	Silver		
				6800 Ω	5
Orange	Orange	Orange	Gold		
				12 Ω	2
Brown	Green	Silver	Silver		
				1.8 MΩ	10
Brown	Black	Yellow	None		
				10kΩ	5
Violet	Green	Black	Red		
				4.7 kΩ	20
Gray	Red	Green	Red		
				5.6 kΩ	2

3 **Unit**

Static Electricity

objectives

After studying this unit, you should be able to:

- Discuss the nature of static electricity.
- Use an electroscope to determine unknown charges.
- Discuss lightning protection.
- List nuisance charges of static electricity.
- List useful charges of static electricity.

Static electrical charges are a common occurrence in everyday life. Almost everyone has received a shock after walking across a carpet and then touching a metal object, or sliding across seat covers in a car and touching the door handle, *Figure 3-1*. Almost everyone has combed their hair with a hard rubber or plastic comb and then used the comb to attract small pieces of paper or other lightweight objects. Static electric charges are the reason why clothes stick together when they are taken out of a clothes dryer. Lightning is without doubt the greatest display of a static electric discharge.

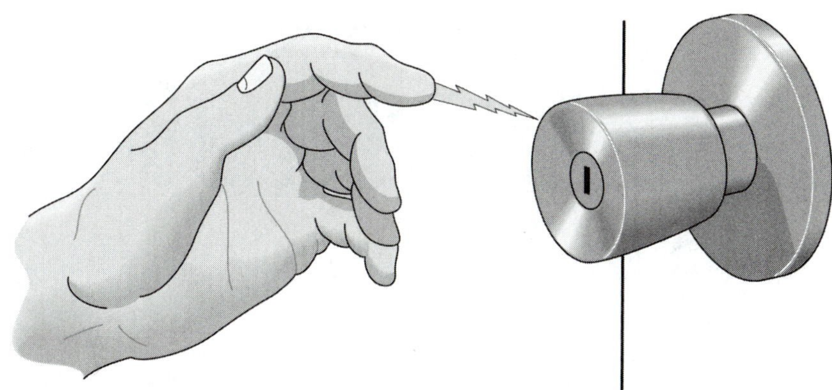

Figure 3-1 Static electric charges can cause a painful shock.

STATIC ELECTRICITY

Although static charges can be a nuisance, they can also be beneficial. Copy machines, for example, operate on the principle of static electricity. Grains of sand receive a static charge to make them stand apart and expose a sharper edge when sandpaper is made, *Figure 3-2*. Electronic air filters

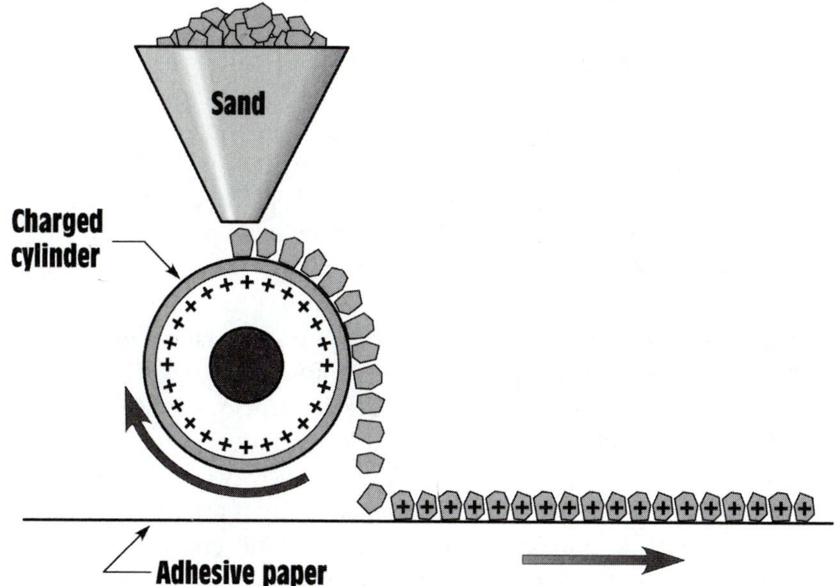

Figure 3-2 Grains of sand receive a charge to help them stand apart.

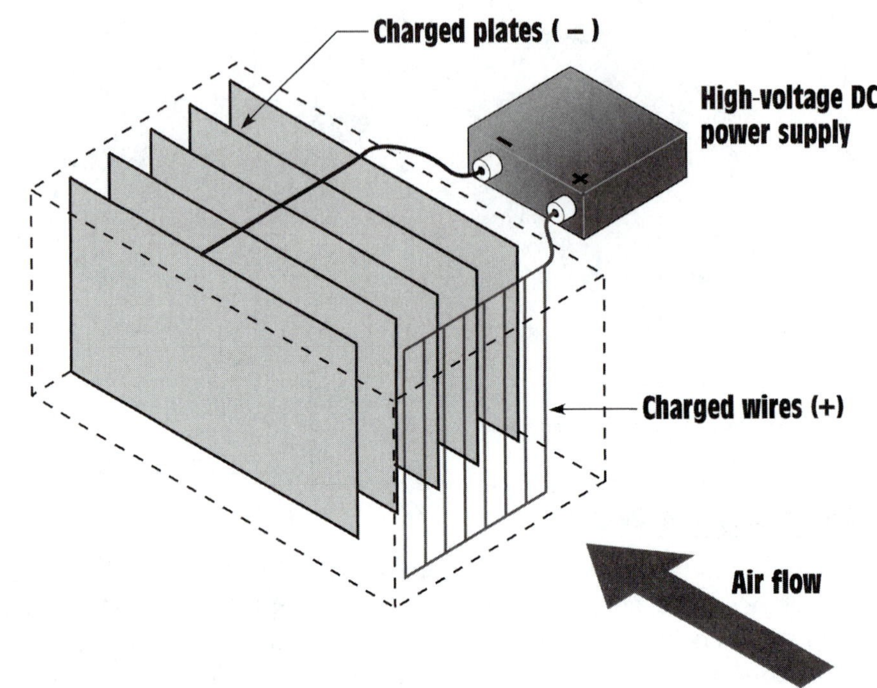

Figure 3-3 Electronic air cleaner.

precipitators

(precipitators) use static charges to attract small particles of smoke, dust, and pollen, *Figure 3-3*. The precipitator uses a high-voltage DC power supply to provide a set of wires with a positive charge, and a set of plates with a negative charge. As a blower circulates air through the unit, small particles receive a positive charge as they move across the charged wires. The charged particles are then attracted to the negative plates. The negative plates hold the particles until the unit is turned off and the plates are cleaned.

static

electrostatic charges

The word **static** means not moving or sitting still. Static electricity refers to electrons that are sitting still and not moving. Static electricity is, therefore, a charge and not a current. **Electrostatic charges** are built up on insulator materials because insulators are the only materials that can hold the electrons stationary; they do not permit them to flow to a different location. A static charge can be built up on a conductor only if the conductor is electrically insulated from surrounding objects. A static charge can be either positive or negative. If an object has a lack of electrons, it will have a positive charge, and if it has an excess of electrons, it will have a negative charge.

CHARGING AN OBJECT

The charge that accumulates on an object is determined by the materials used to produce the charge. If a hard rubber rod is rubbed on a piece of wool, the wool will deposit excess electrons on the rod and give it a negative charge. If a glass rod is rubbed on a piece of wool, electrons will be removed from the rod, thus producing a positive charge, *Figure 3-4*.

THE ELECTROSCOPE

electro-
scope

An early electrical instrument that can be used to determine the electrostatic charge of an object is the **electroscope,** *Figure 3-5*. An electroscope is a metal ball attached to the end of a metal rod. The other end of the rod is attached to two thin metal leaves. The metal leaves are inside a transparent container that permits the action of the leaves to be seen by an

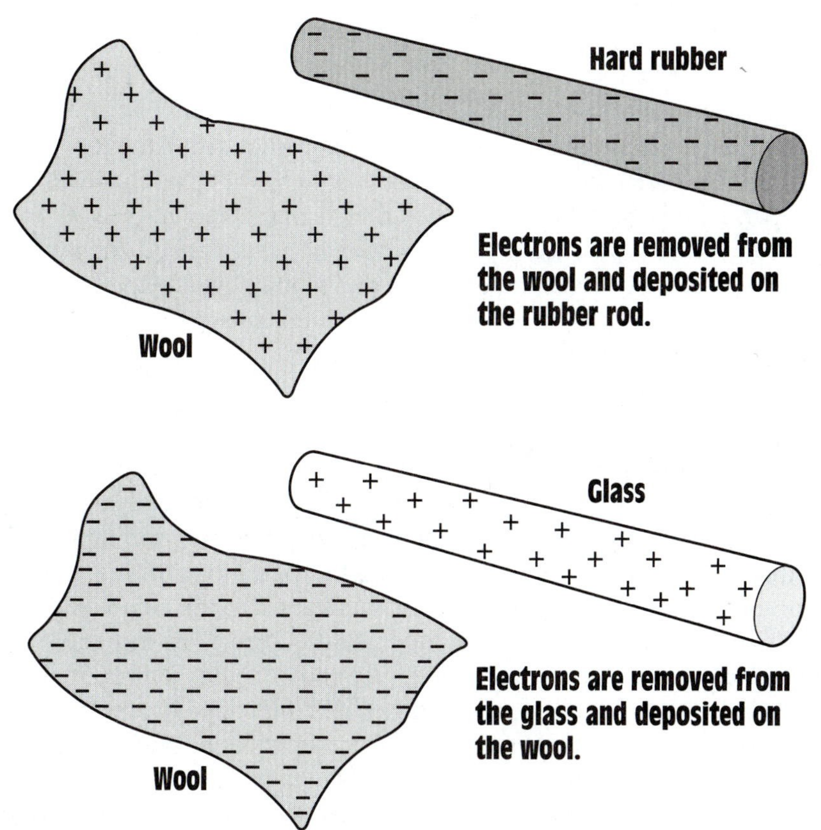

Figure 3-4 Producing a static charge.

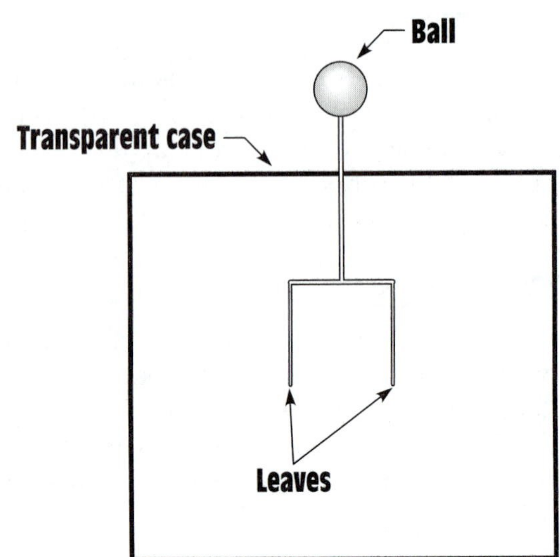

Figure 3-5 An electroscope.

observer. The metal rod is insulated from the box. The metal leaves are placed inside a container so that air currents cannot affect their movement.

Before the electroscope can be used, it must first be charged. This is done by touching the ball with an object that has a known charge. For this example, assume that a hard rubber rod has been rubbed on a piece of wool to give it a negative charge. When the rubber rod is wiped against the metal ball, excess electrons are deposited on the metal surface of the electroscope. Since both of the metal leaves now have an excess of electrons, they repel each other, as shown in *Figure 3-6*.

TESTING AN OBJECT

A charged object can now be tested to determine if it has a positive or negative polarity. Assume a ballpoint pen is charged by rubbing the plastic body through a person's hair. Now bring the pen close to, but not touching, the ball and observe the action of the leaves. If the pen has taken on a negative charge, the leaves will move farther apart, as shown in *Figure 3-7*. The field caused by the negative electrons on the pen repel electrons from the ball. These electrons move down the rod to the leaves. This causes the leaves to become more negative and to repel each other more, forcing them to move farther apart.

If the pen has a positive charge, the leaves will move closer together when the pen is moved near the ball, *Figure 3-8*. This action is caused by

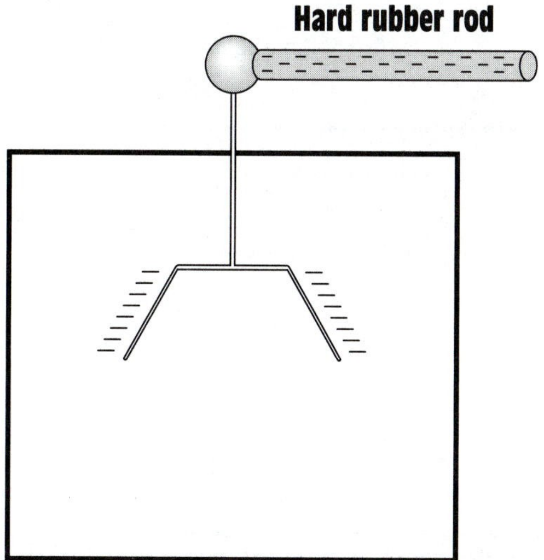

Figure 3-6 The electroscope is charged with a known static charge.

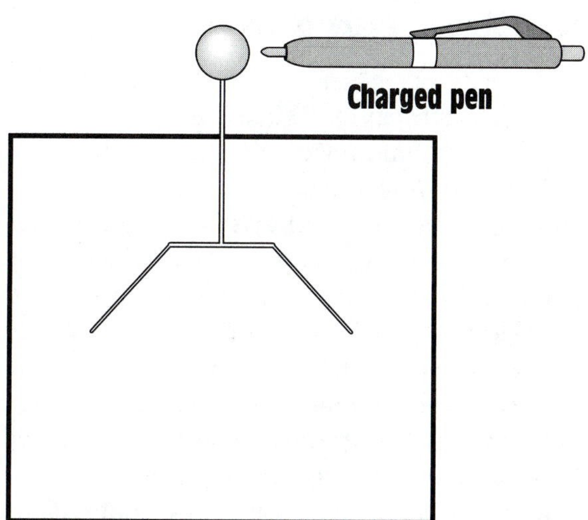

Figure 3-7 The leaves are deflected farther apart, indicating that the object has a negative charge.

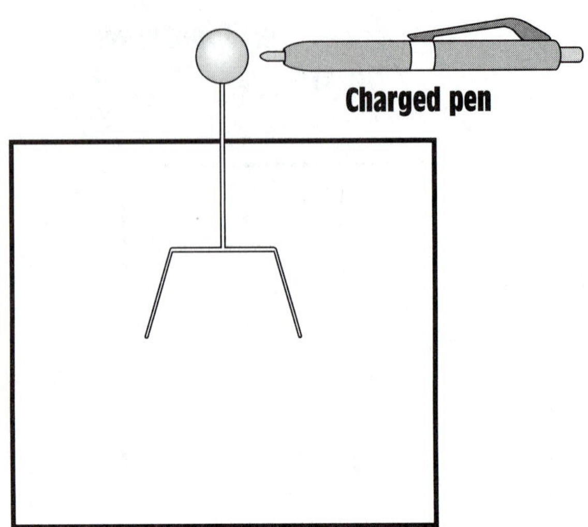

Figure 3-8 The leaves move closer together, indicating that the object has a positive charge.

the positive field of the pen attracting electrons. When electrons are attracted away from the leaves, they become less negative and move closer together. If the electroscope is charged with a positive charge in the beginning, a negatively charged object will cause the leaves to move closer together and a positively charged object will cause the leaves to move farther apart.

STATIC ELECTRICITY IN NATURE

lightning

thunder-cloud

lightning bolt

When static electricity occurs in nature, it can be harmful. The best example of natural static electricity is **lightning.** A static charge builds up in clouds that contain a large amount of moisture as they move through the air. It is theorized that the movement causes a static charge to build up on the surface of drops of water. Large drops become positively charged and small drops become negatively charged. *Figure 3-9* illustrates a typical **thundercloud.** Notice that both positive and negative charges can be contained in the same cloud. Most lightning discharges take place within the cloud. Lightning discharges can also take place between different clouds, between a cloud and the ground, and between the ground and the cloud, *Figure 3-10.* Whether a **lightning bolt** travels from the cloud to the ground or from the ground to the cloud is determined by which contains the negative charge and which contains the positive. Current will always flow from negative to positive. If a cloud is negative and an object on the ground is positive, the lightning discharge will travel from the cloud to the ground. If

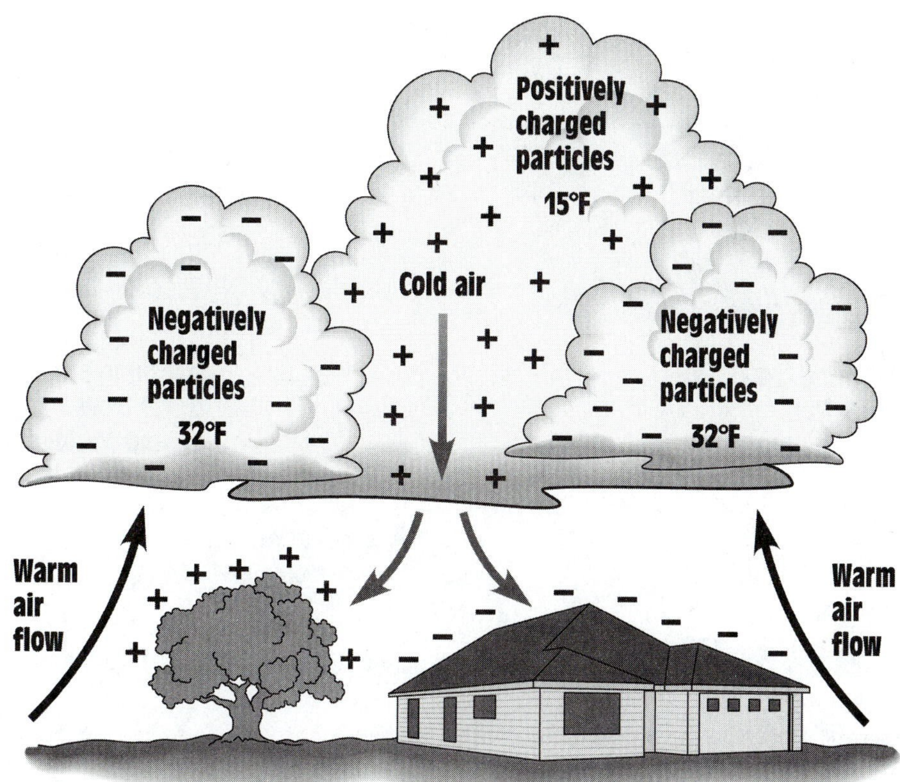

Figure 3-9 The typical thundercloud contains both negatively and positively charged particles.

the cloud has a positive charge and the object on the ground has a negative charge, the discharge will be from the ground to the cloud. A lightning bolt has an average voltage of about 15,000,000 V.

LIGHTNING PROTECTION

lightning rods

Lightning rods are sometimes used to help protect objects from lightning. Lightning rods work by providing an easy path to ground for current flow. If the protected object is struck by a lightning bolt, the lightning rod bleeds the lightning discharge to ground before harm can occur to the protected object, *Figure 3-11*. Lightning rods were first invented by Benjamin Franklin.

lightning arrestor

Another device used for lightning protection is the **lightning arrestor.** The lightning arrestor is grounded at one end, and the other end is brought close to, but does not touch, the object to be protected. If the protected

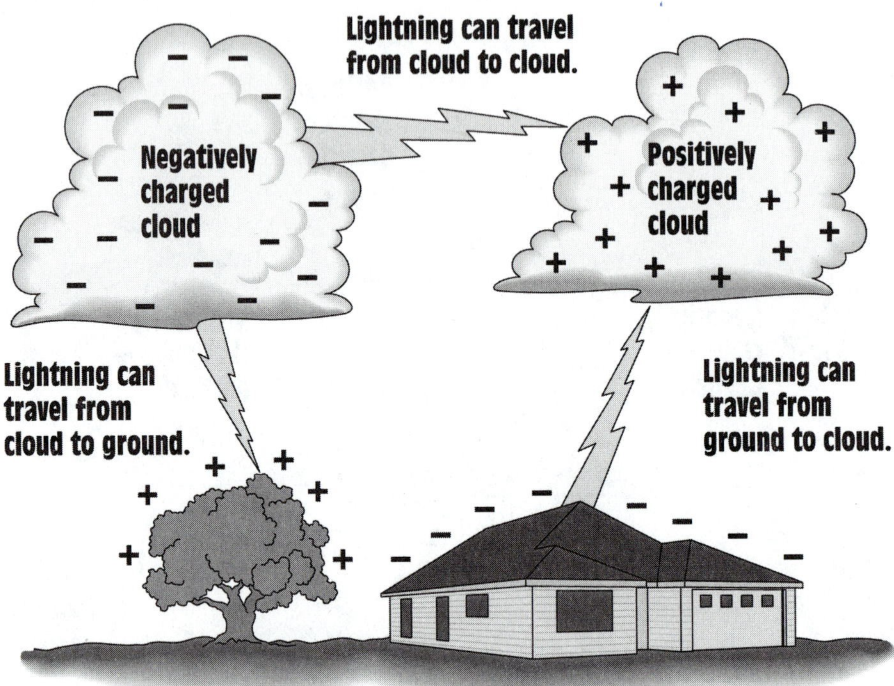

Figure 3-10 Lightning travels from negative to positive.

object is struck, the high voltage of the lightning arcs across to the lightning arrestor and bleeds to ground.

Power lines are often protected by lightning arrestors, which exhibit a very high resistance at the normal voltage of the line. If the power line is struck by lightning, the increase of voltage causes the resistance of the arrestor to decrease and conduct the lightning discharge to ground.

NUISANCE STATIC CHARGES

nuisance static charges

Static charges are sometimes a nuisance. Some examples of **nuisance static charges** are:

1. The static charge that accumulates on automobiles as they move through dry air. These static charges can cause dangerous conditions under certain circumstances. This is the reason why trucks that carry flammable materials, such as gasoline or propane, use a drag chain. One end of the drag chain is attached to the frame of the vehicle and the other end drags the ground. The chain is used to provide a path to ground while the vehicle is moving and to prevent a static charge from accumulating on the body of the vehicle.

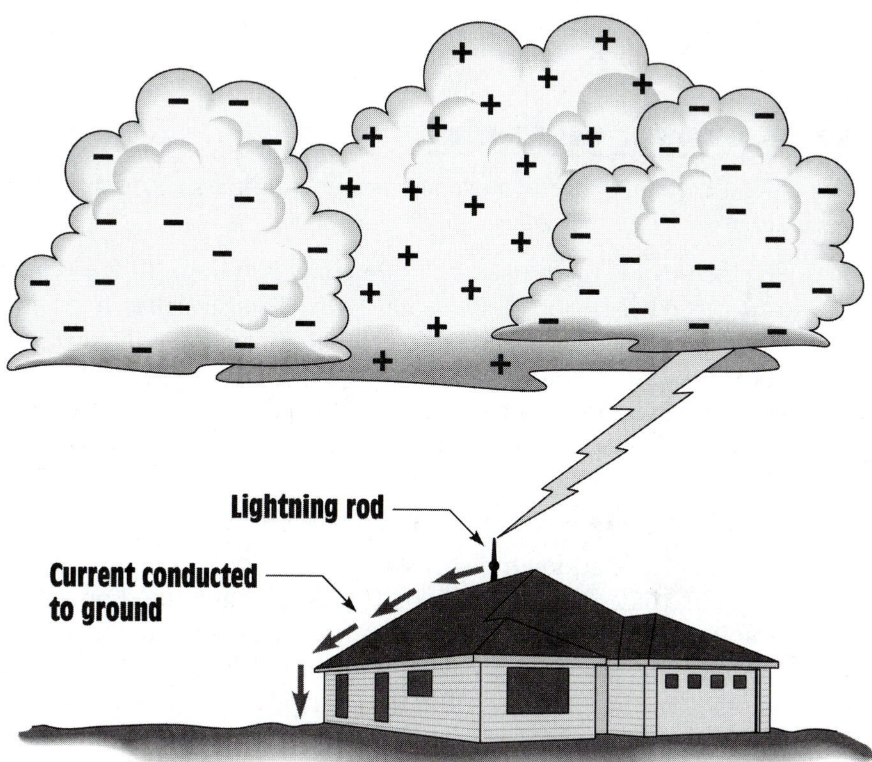

Figure 3-11 A lightning rod provides an easy path to ground.

2. The static charge that accumulates on a person's body when walking across a carpet. This charge can cause a painful shock when a metal object is touched, and it discharges in the form of an electric spark. Most carpets are made from man-made materials that are excellent insulators, such as nylon. In the winter, the heating systems of most dwellings remove the moisture of the air and cause the air to have a low humidity. The dry air, combined with an insulating material, provides an excellent setting for the accumulation of a static charge. This condition can generally be eliminated by the installation of a humidifier. A simple way to prevent the painful shock of a static discharge is to hold a metal object, such as a key or coin, in one hand. Touch the metal object to a grounded surface. The static charge will arc from the metal object ground instead of from your finger to ground.

3. The static charge that accumulates on clothes in a dryer. The static charge is caused by the clothes moving through the dry air. The greatest static charges generally build up on man-made fabrics because

they are the best insulators and retain electrons more readily than natural fabrics, such as cotton or wool.

USEFUL STATIC CHARGES

useful static charges

Not all static charges are a nuisance. Some examples of **useful static charges** are:

1. Static electricity is often used to aid in spray painting. A high-voltage grid is placed in front of the spray gun. This grid has a positive charge. The object to be painted has a negative charge, *Figure 3-12*. As the droplets of paint pass through the grid, the positive charge causes electrons to be removed from the paint droplets. The positively charged droplets are attracted to the negatively charged object to be painted. This static charge helps to prevent waste of the paint and, at the same time, produces a more uniform finish.

selenium

2. Another device that depends on static electricity for operation is the dry copy machine. The copy machine uses an aluminum drum coated with **selenium,** *Figure 3-13*. Selenium is a semiconductor material that changes its conductivity with a change of light intensity. When selenium is in the presence of light, it has a very high conductivity. When it is in darkness, it has a very low conductivity.

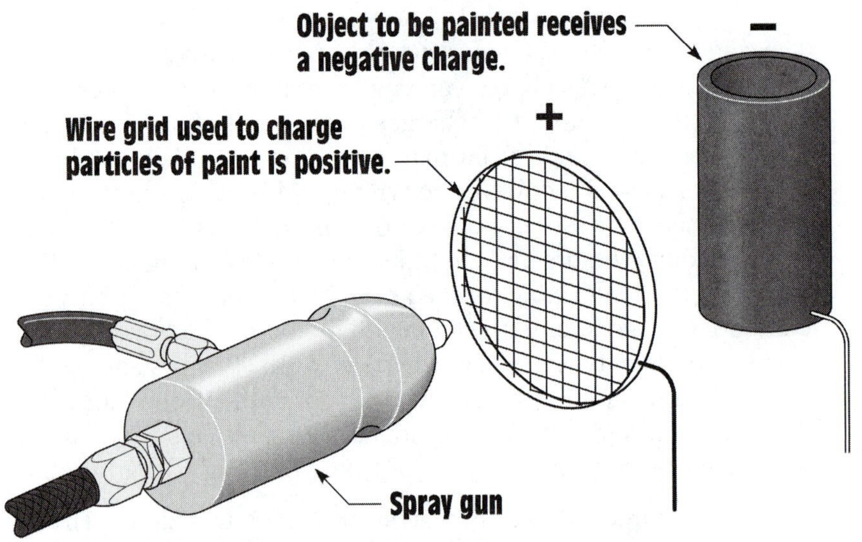

Figure 3-12 Static electric charges are often used in spray painting.

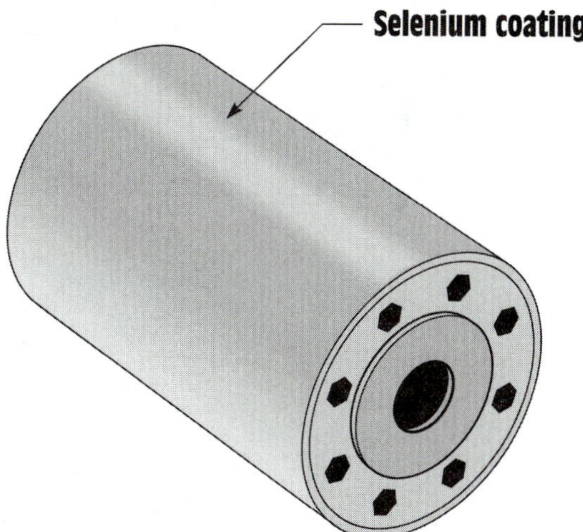

Figure 3-13 The drum of a copy machine is coated with selenium.

A high-voltage wire located near the drum causes the selenium to have a positive charge as it rotates, *Figure 3-14*. The drum is in darkness when it is charged. An image of the material to be copied is reflected on the drum by a system of lenses and mirrors, *Figure 3-15*. The light portions of the

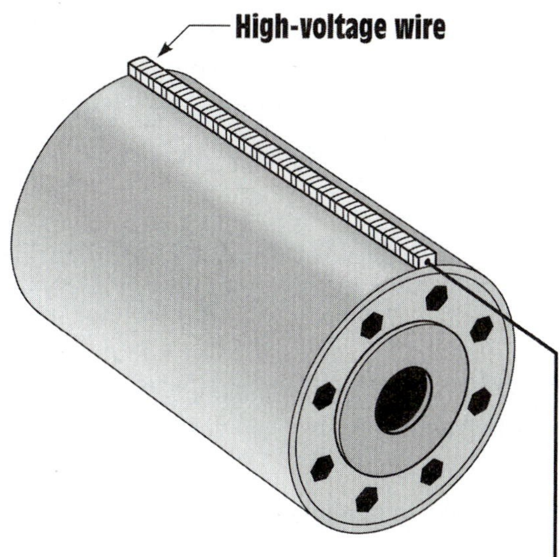

Figure 3-14 The drum receives a positive charge.

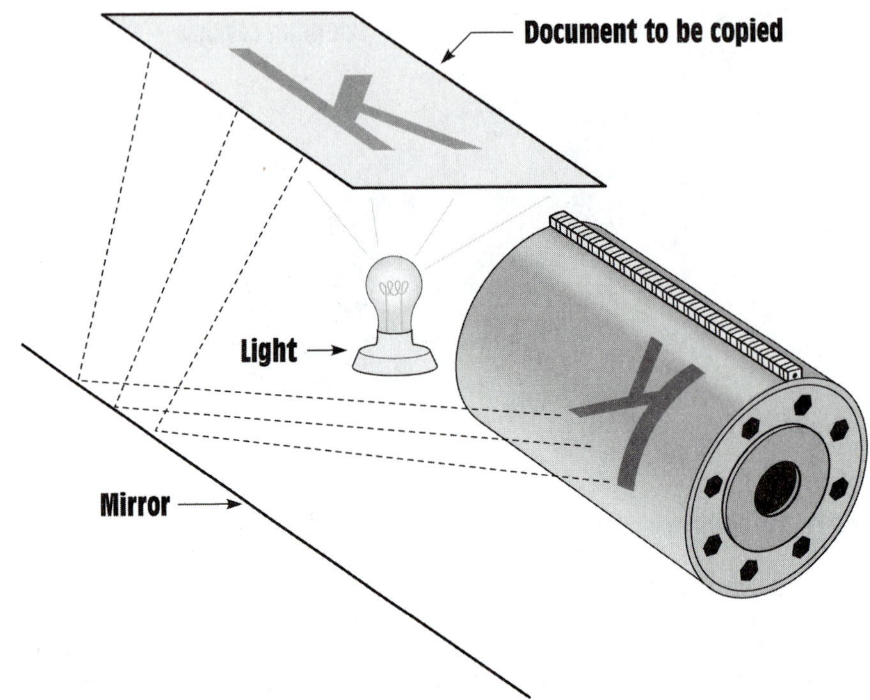

Figure 3-15 The image is transferred to the selenium drum.

paper reflect more light than the dark portions. When the reflected light strikes the drum, the conductivity of the selenium increases greatly and negative electrons from the aluminum drum neutralize the selenium charge at that point. The dark area of the paper causes the drum to retain a positive charge.

A dark powder that has a negative charge is applied to the drum, *Figure 3-16*. The powder is attracted to the positively charged areas on the drum. The powder on the neutral areas of the drum falls away.

A piece of positively charged paper passes under the drum, *Figure 3-17*, and attracts the powder from the drum. The paper then passes under a heating element, which melts the powder into the paper and causes the paper to become a permanent copy of the original.

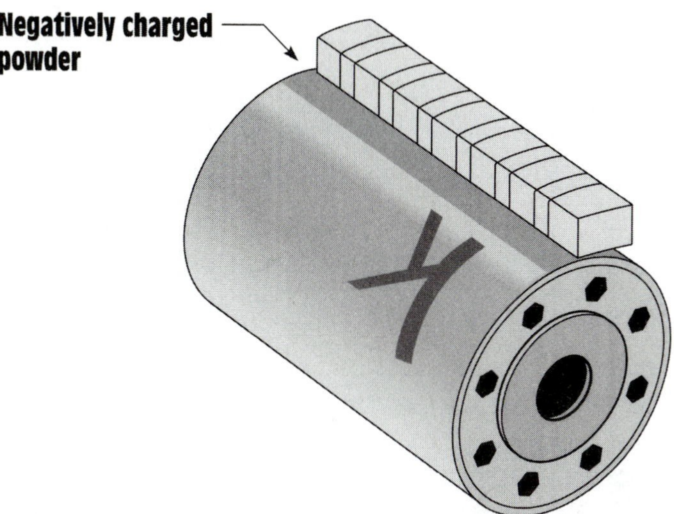

Figure 3-16 Negatively charged powder is applied to the positively charged drum.

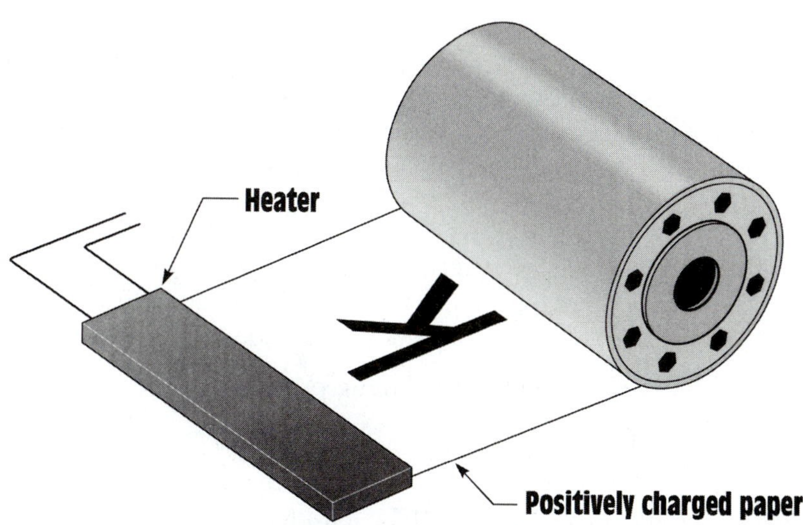

Figure 3-17 The negatively charged powder is attracted to the positively charged paper.

SUMMARY

1. The word *static* means not moving.
2. An object can be positively charged by removing electrons from it.
3. An object can be negatively charged by adding electrons to it.
4. An electroscope is a device used to determine the charge of an object.
5. Static charges accumulate on insulator materials.
6. Lightning is an example of a natural static charge.

REVIEW QUESTIONS

1. Why is static electricity considered to be a charge and not a current?
2. If electrons are removed from an object, is the object now positively or negatively charged?
3. Why do static charges accumulate on insulator materials only?
4. What is an electroscope?
5. An electroscope has been charged with a negative charge. An object with an unknown charge is brought close to the electroscope. The leaves of the electroscope come closer together. Does the object have a positive or negative charge?
6. Can one thundercloud contain both positive and negative charges?
7. A thundercloud has a negative charge, and an object on the ground has a positive charge. Will the lightning discharge go from the cloud to the ground or from the ground to the cloud?
8. Name two devices used for lightning protection.
9. What type of material is used to coat the aluminum drum of a copy machine?
10. What special property does the material in question 9 have that makes it useful in a copy machine?

Unit 4

Magnetism

objectives

*fter studying this unit, you should be able to:*

- Discuss the properties of permanent magnets.
- Discuss the difference between the axis poles of the earth and the magnetic poles of the earth.
- Discuss the operation of electromagnets.
- Determine the polarity of an electromagnet when the current direction is known.
- Discuss the different systems used to measure magnetism.
- Define terms used to describe magnetism and magnetic quantities.

Magnetism is one of the most important phenomena in the electrical field. It is the force used to produce most of the electrical power in the world. The force of magnetism has been known for over two thousand years. It was first discovered by the Greeks when they found that a certain type of stone was attracted to iron. This stone was first found in Magnesia in Asia Minor and was named *magnetite*. The first compass was invented when it was noticed

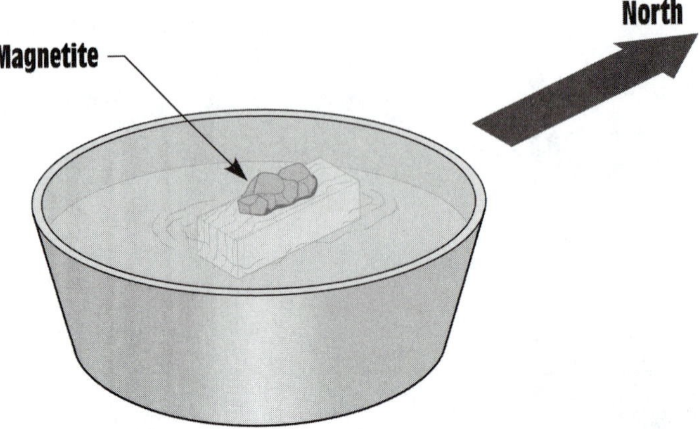

Figure 4-1 The first compass.

that a magnetite placed on a small piece of wood floating in water always aligned itself north and south, *Figure 4-1*. Due to their ability to always align themselves north and south, natural magnets become known as "leading stones" or **lodestones.** In the dark ages, the strange powers of the magnet were believed to be caused by evil spirits or the devil.

lodestones

THE EARTH IS A MAGNET

The reason why the lodestone aligns itself north and south is because the earth itself contains magnetic poles. *Figure 4-2* illustrates the position of the true north and south poles, or axes, of the earth and the position of the magnetic poles. Notice that the magnetic pole is not located at the true north pole of the earth. This is the reason why navigators must distinguish between true north and magnetic north. The angular difference between the two is known as *the angle of declination.* The illustration shows the magnetic lines of force to be only on each side of the earth. It should be understood that the lines actually surround the entire earth like a magnetic shell. Also notice that the magnetic north pole is located near the southern polar axis and the magnetic south pole is located near the northern polar axis. The reason why the *geographic poles* (axes) are called north and south is because the north pole of a compass needle points in the direction of the north geographic pole. Because unlike magnetic poles attract, the north magnetic pole of the compass needle is attracted to the south magnetic pole of the earth.

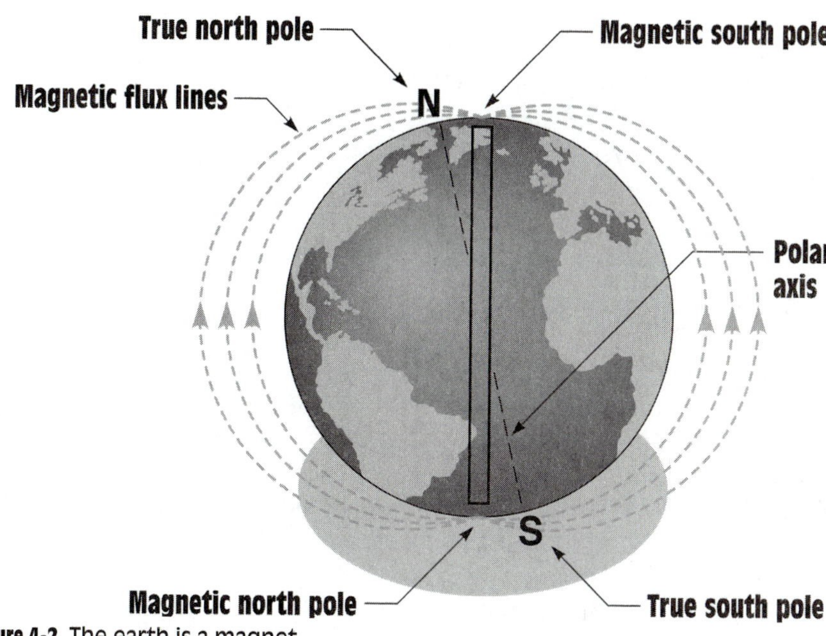

Figure 4-2 The earth is a magnet.

PERMANENT MAGNETS

Permanent magnets are magnets that do not require any power or force to maintain their field. They are an excellent example of one of the basic laws of magnetism, which states that **energy is required to create a magnetic field, but no energy is required to maintain a magnetic field.** Man-made permanent magnets are much stronger and can retain their magnetism longer than natural magnets.

permanent magnets

Energy is required to create a magnetic field, but no energy is required to maintain a magnetic field.

THE ELECTRON THEORY OF MAGNETISM

There are actually only three substances that form natural magnets: iron, nickel, and cobalt. Our understanding of why these materials form magnets is the result of rather complex scientific investigations resulting in an explanation of magnetism based on **electron spin patterns.** It is believed that internal gravity causes electrons to spin on their axes as they orbit around the nucleus of the atom. This spinning motion causes each electron to become a tiny permanent magnet. Although all electrons spin, they do not all spin in the same direction. In most atoms, electrons that spin in opposite directions tend to form pairs, *Figure 4-3*. Since the electron pairs spin in opposite directions, their magnetic effect cancels each other out, as far as having any effect on distant objects. It would be very similar to

electron spin patterns

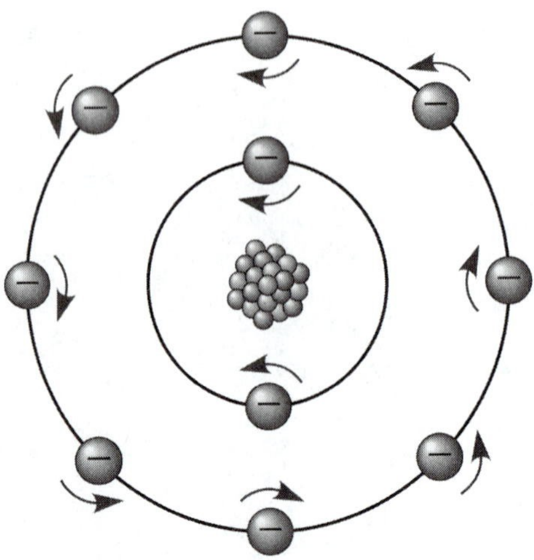

Figure 4-3 Electron pairs generally spin in opposite directions.

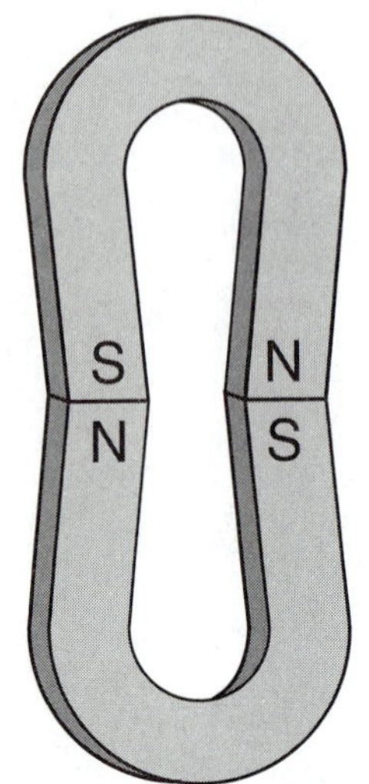

Figure 4-4 Two horseshoe magnets attract each other.

connecting two horseshoe magnets together. They would be strongly attracted to each other but would have little effect on surrounding objects, *Figure 4-4*.

An atom of iron contains twenty-six electrons. Of these twenty-six, twenty-two are paired and spin in opposite directions, canceling each other's magnetic effect. In the next-to-outermost shell, however, there are four electrons that are not paired and thus spin in the same direction. It is these electrons that account for the magnetic properties of iron. At a temperature of 1,420°F or 771.1°C, the electron spin patterns rearrange themselves and iron loses its magnetic properties.

magnetic molecules

When the atoms of most materials combine to form molecules, they arrange themselves in a manner that produces a total of eight valence electrons. The electrons form a spin pattern that cancels the magnetic field of the material. When the atoms of iron, nickel, and cobalt combine, however, this is not the case. Their electrons combine so that they share valence electrons in such a way that their spin patterns are in the same direction, causing their magnetic fields to add instead of cancel. The additive effect forms regions in the molecular structure of the metal called **magnetic domains** or **magnetic molecules.** These magnetic domains act like small permanent magnets.

A piece of nonmagnetized metal has its molecules in a state of disarray, as shown in *Figure 4-5*. When the metal is magnetized, its molecules align themselves in an orderly pattern, as shown in *Figure 4-6*. In theory, each molecule of a magnetic material is itself a small magnet. If a permanent magnet is cut into pieces, each piece is a separate magnet, *Figure 4-7*.

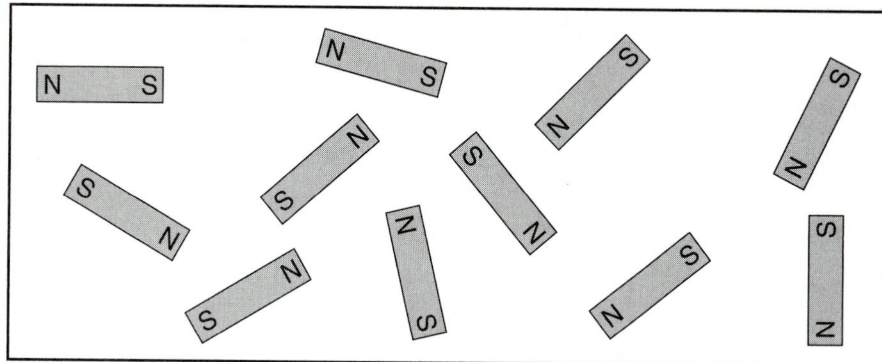

Figure 4-5 The molecules in a piece of nonmagnetized metal are disarrayed.

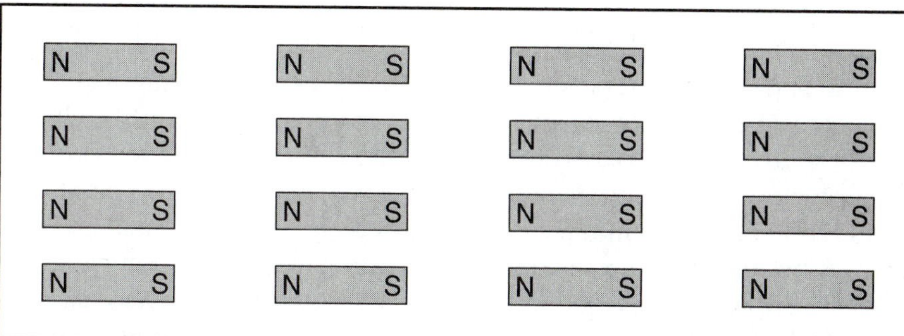

Figure 4-6 The molecules in a piece of magnetized metal are aligned in an orderly fashion.

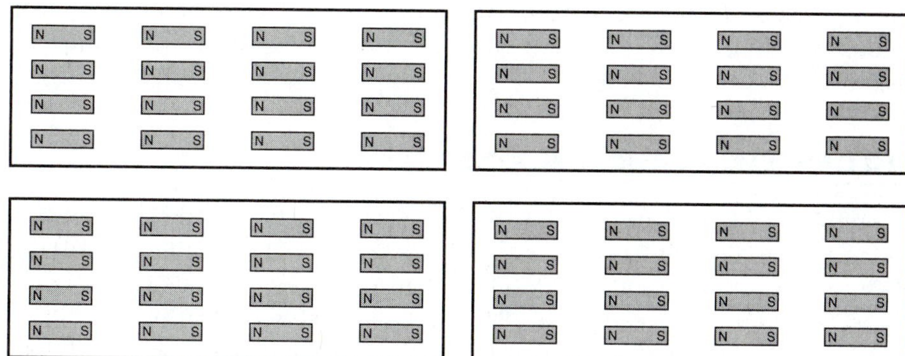

Figure 4-7 When a magnet is cut apart, each piece becomes a separate magnet.

MAGNETIC MATERIALS

Magnetic materials can be divided into three basic classifications. These are:

Ferromagnetic Materials: These materials are metals that are easily magnetized. Examples of these materials are iron, nickel, cobalt, and manganese.

Paramagnetic Materials: These materials can be magnetized, but not as easily as ferromagnetic materials. Some examples of paramagnetic materials are platinum, titanium, and chromium.

Diamagnetic Materials: These are materials that cannot be magnetized. The magnetic lines of force tend to go around them instead of through them. Some examples of these materials are copper, brass, and antimony.

Some of the best materials for the production of permanent magnets are alloys. One of the best permanent magnet materials is Alnico 5, which is made from a combination of aluminum, nickel, cobalt, copper, and iron. Another type of permanent magnet material is made from a combination of barium ferrite and strontium ferrite. Ferrites can have an advantage in some situations because they are insulators and not conductors. They have a resistance of approximately one million ohms per centimeter. These two materials can be powdered. The powder is heated to the melting point and then rolled and heat treated, which changes the grain structure and magnetic properties of the material. This type of material has a property more like stone than metal and is known as a *ceramic magnet.* Ceramic magnets can be powdered and mixed with rubber or plastic or with liquids. Ceramic magnetic materials mixed with liquids can be used to make magnetic ink, which is used on checks. Another frequently used magnetic material is iron oxide, which is used to make magnetic recording tape and computer diskettes.

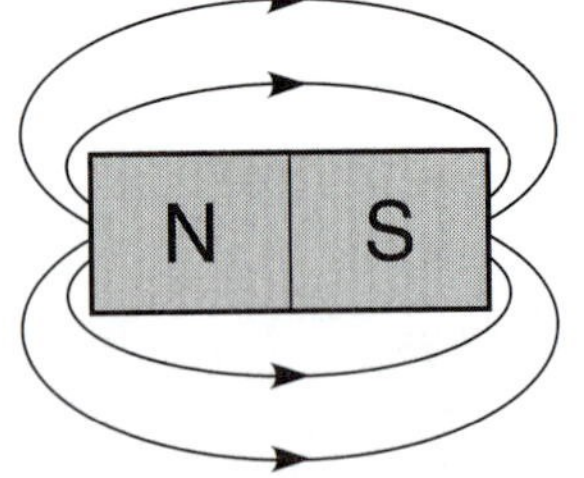

Figure 4-8 Magnetic lines of force are called lines of flux.

flux

MAGNETIC LINES OF FORCE

Magnetic lines of force are called **flux.** The symbol used to represent flux is the Greek letter phi (Φ). Flux lines can be seen by placing a piece of cardboard on a magnet and sprinkling iron filings on the cardboard. The filings will align themselves in a pattern similar to the one shown in *Figure 4-8.* It should be understood that the pattern produced by the iron filings forms a two-dimensional figure. The flux lines actually surround the entire magnet, *Figure 4-9*. Magnetic **lines of flux** repel each other and never cross. Although magnetic lines of force do not flow, it is assumed they are oriented in a direction from north to south.

lines of flux

A basic law of magnetism states that **unlike poles attract and like poles repel.** *Figure 4-10* illustrates what happens if a piece of cardboard is placed over two magnets with their north and south poles facing each other and iron filings are sprinkled on the cardboard. The filings form a pattern showing that the magnetic lines of force are attracted to each other. *Figure 4-11* illustrates the pattern formed by the iron filings when the cardboard is placed over two magnets with like poles facing each other. The filings show that the magnetic lines of force repel each other.

Unlike poles attract and like poles repel.

If the opposite poles of two magnets are brought close to each other, they will be attracted to each other, as shown in *Figure 4-12*. If like poles of the two magnets are brought together, they will repel each other.

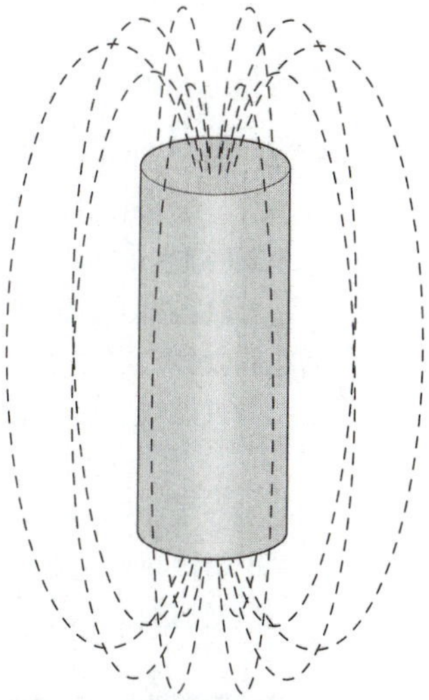

Figure 4-9 Magnetic lines of flux surround the entire magnet.

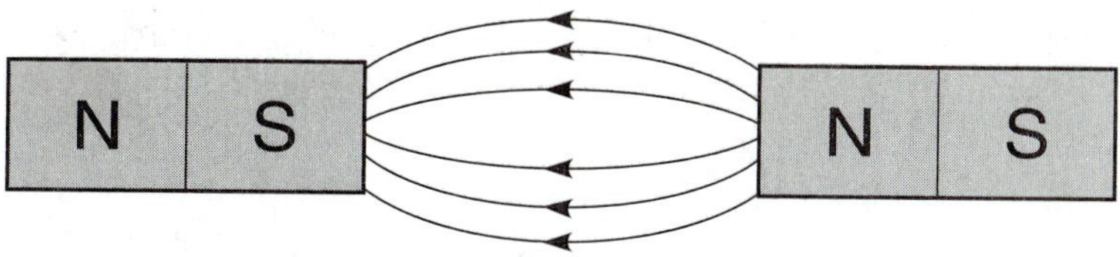

Figure 4-10 Opposite magnetic poles attract each other.

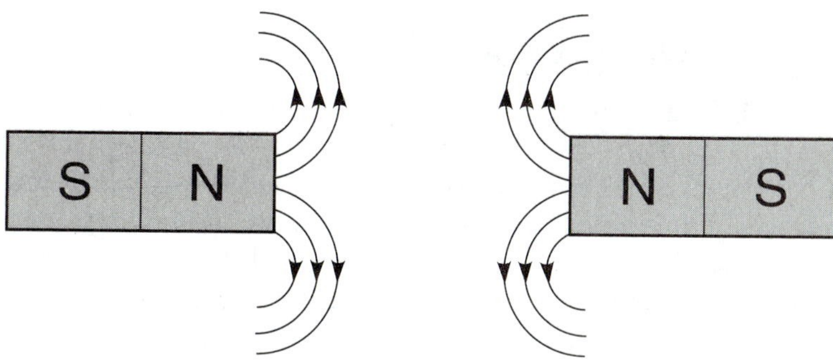

Figure 4-11 Like magnetic poles repel each other.

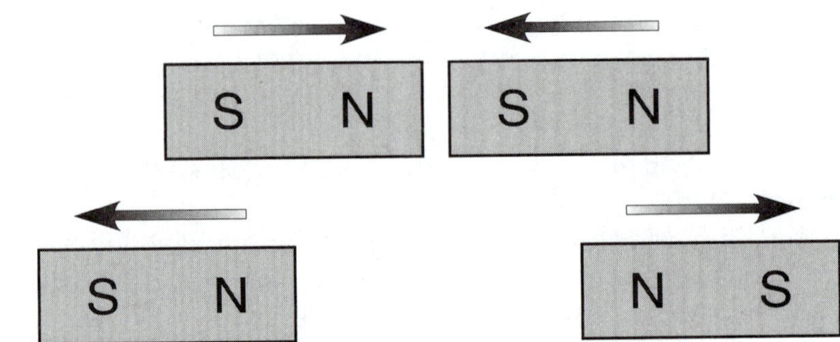

Figure 4-12 Opposite poles of a magnet attract and like poles repel.

ELECTROMAGNETS

Whenever an electric current flows through a conductor, a magnetic field is formed around the conductor.

electromagnets

ampere-turns

A basic law of physics states that **whenever an electric current flows through a conductor, a magnetic field is formed around the conductor. Electromagnets** depend on electric current flow to produce a magnetic field. They are generally designed to produce a magnetic field only as long as the current is flowing and thus do not retain their magnetism when current flow stops. Electromagnets operate on the principle that current flowing through a conductor produces a magnetic field around the conductor, *Figure 4-13.* If the conductor is wound into a coil, as shown in *Figure 4-14,* the magnetic lines of force add to produce a stronger magnetic field. A coil with ten turns of wire will produce a magnetic field that is ten times stronger than the magnetic field around a single conductor.

Another factor that affects the strength of an electromagnetic field is the amount of current flowing through the wire. An increase in current flow will cause an increase in magnetic field strength. The two factors that determine the number of flux lines produced by an electromagnet are the number of turns of wire and the amount of current flow through the wire. The strength of an electromagnet is proportional to its **ampere-turns.**

MAGNETIC INDUCTION

Magnetic induction is one of the most important concepts in the electrical field. It is the basic operating principle underlying alternators, generators, transformers, and most alternating current motors. It is imperative that anyone desiring to work in the electrical field have an understanding of the principles involved.

As stated, one of the basic laws of electricity is that whenever current flows through a conductor, a magnetic field is created around the conductor (see Figure 4-13). The direction of current flow determines the polarity

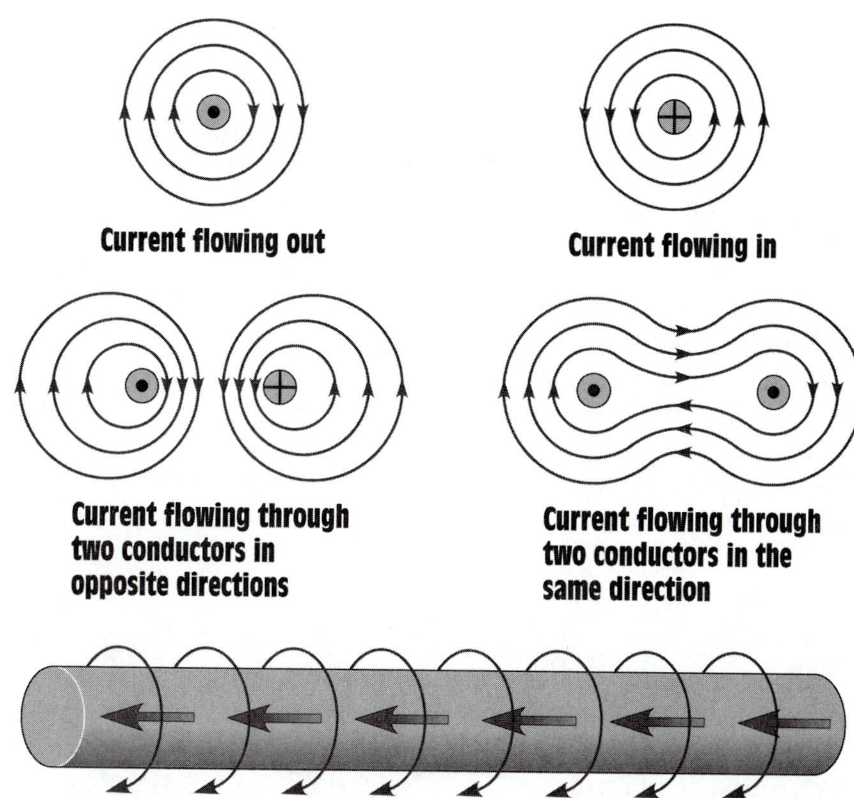

Figure 4-13 Current flowing through a conductor produces a magnetic field around the conductor.

Figure 4-14 Winding the wire into a coil increases the strength of the magnetic field.

The principle of magnetic induction states that whenever a conductor cuts through magnetic lines of flux, a voltage is induced into the conductor.

of the magnetic field, and the amount of current determines the strength of the magnetic field.

That basic law in reverse is the principle of **magnetic induction,** which states that **whenever a conductor cuts through magnetic lines of flux, a voltage is induced into the conductor.** The conductor in *Figure 4-15* is connected to a zero-center microammeter, creating a complete circuit. When the conductor is moved downward through the magnetic lines of flux, the induced voltage will cause electrons to flow in the direction indicated by the arrows. This flow of electrons causes the pointer of the meter to be deflected from the center-zero position.

If the conductor is moved upward, the polarity of inducted voltage will be reversed and the current will flow in the opposite direction (*Figure 4-16*). The pointer will be deflected in the opposite direction.

The polarity of the inducted voltage can also be changed by reversing the polarity of the magnetic field (*Figure 4-17*). In this example, the con-

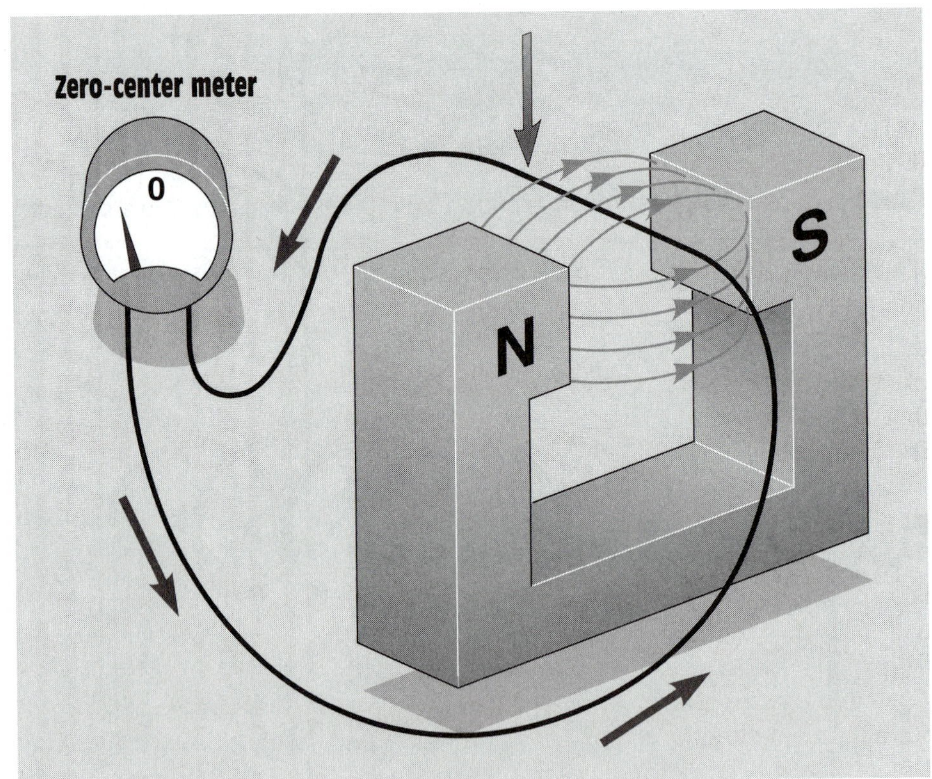

Figure 4-15 A voltage is induced when a conductor cuts magnetic lines of flux.

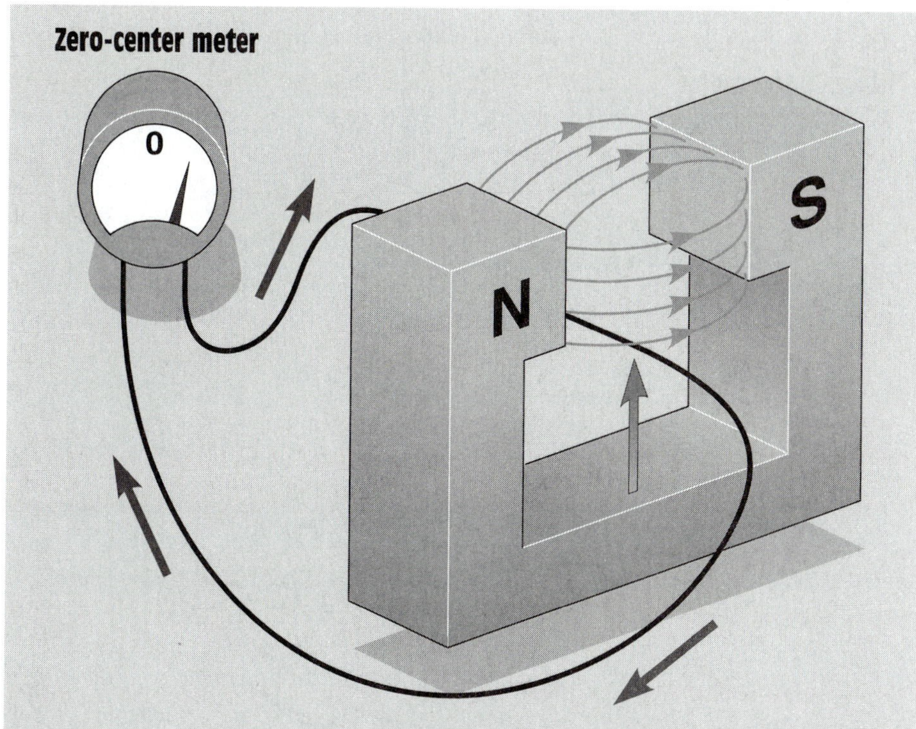

Figure 4-16 Reversing the direction of movement reverses the polarity of the voltage.

The polarity of the induced voltage is determined by the polarity of the magnetic field in relation to the direction of movement.

ductor is again moved downward through the lines of flux, but the polarity of the magnetic field has been reversed. Therefore the polarity of the inducted voltage will be the opposite of that in Figure 4-15, and the pointer of the meter will be deflected in the opposite direction. It can be concluded that **the polarity of the inducted voltage is determined by the polarity of the magnetic field in relation to the direction of movement.**

left-hand generator rule

FLEMING'S LEFT-HAND GENERATOR RULE

Fleming's **left-hand generator rule** can be used to determine the relationship of the motion of the conductor in a magnetic field to the direction of the induced current. To use the left-hand rule, place the thumb, forefinger, and center finger at right angles to each other as shown in *Figure 4-18.* **The forefinger points in the direction of the field flux,** assuming that magnetic lines of force are in a direction of north to south. **The thumb points in the direction of thrust,** or movement of the conductor, **and the center finger shows the direction of the current induced into**

The forefinger points in the direction of the field flux. The thumb points in the direction of thrust, and the center finger shows the direction of the current inducted into the armature.

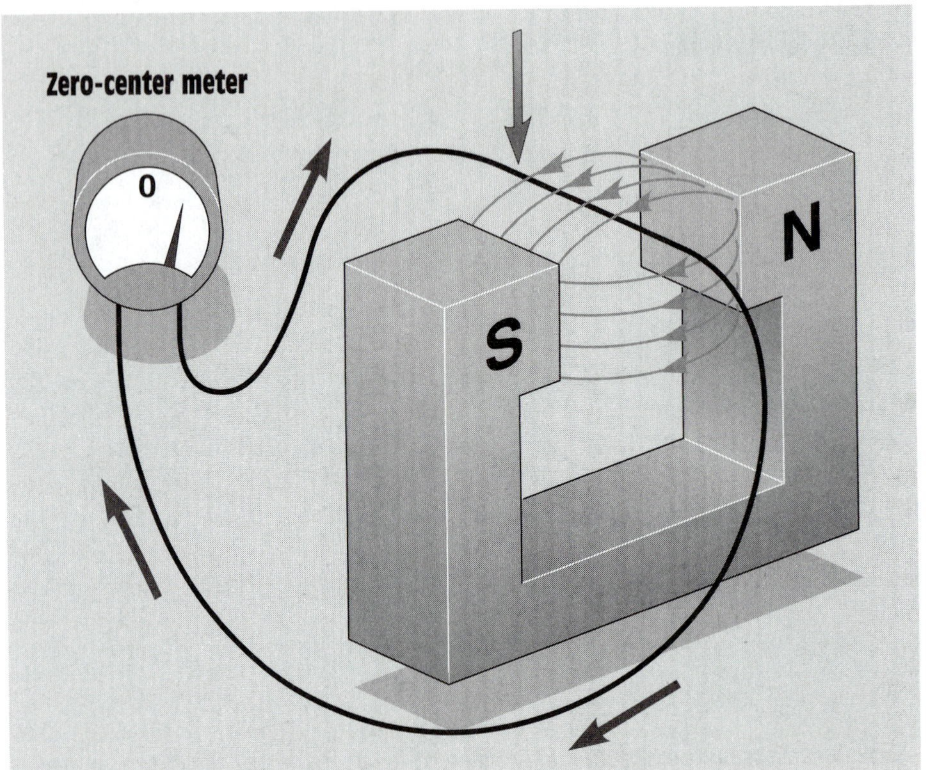

Figure 4-17 Reversing the polarity of the magnetic field reverses the polarity of the voltage.

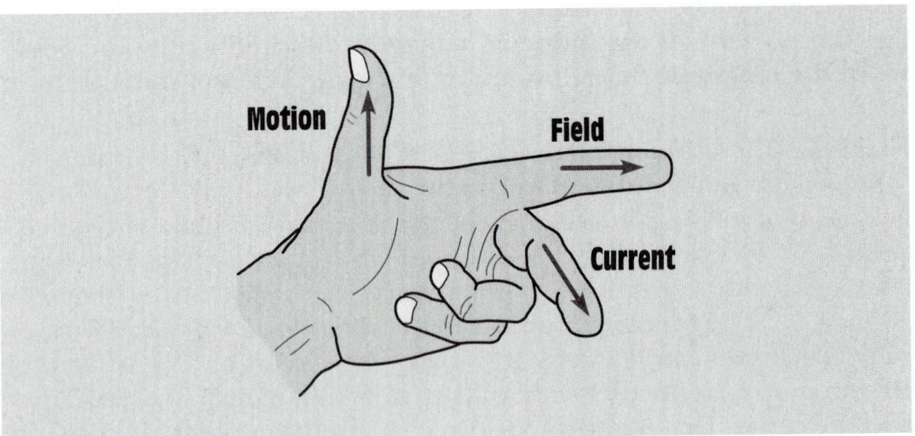

Figure 4-18 Left-hand generator rule.

the armature. An easy method of remembering which finger represents which quantity is shown below.

THumb = **THrust**

Forefinger = **Flux**

Center finger = **Current**

The left-hand rule can be used to clearly illustrate that if the polarity of the magnetic field is changed or if the direction of armature rotation is changed, the direction of induced current will change also.

MOVING MAGNETIC FIELDS

The important factors concerning magnetic induction are conductor, a magnetic field, and relative motion. In practice, it is often desirable to move the magnet instead of the conductor. Most alternating current generators or alternators operate on this principle. In *Figure 4-19,* a coil of wire is held stationary while a magnet is moved through the coil. As the magnet is moved, the lines of flux cut through the windings of the coil and induce a voltage into them.

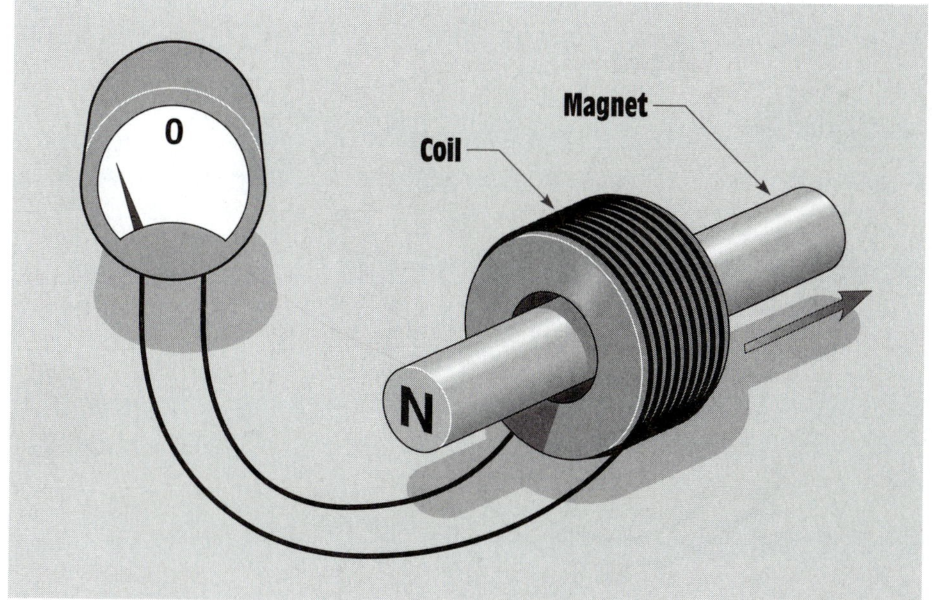

Figure 4-19 Voltage is induced by a moving magnetic field.

turns of wire

strength of magnetic field

speed

weber (Wb)

DETERMINING THE AMOUNT OF INDUCED VOLTAGE

Three factors determine the amount of voltage that will be induced in a conductor:

1. **the number of turns of wire,**
2. **the strength of the magnetic field** (flux density), and
3. **the speed of the cutting action.**

In order to induce 1 V in a conductor, the conductor must cut 100,000,000 lines of magnetic flux in 1 s. In magnetic measurement, 100,000,000 lines of flux are equal to one **weber (Wb).** Therefore, if a conductor cuts magnetic lines of flux at a rate of 1 Wb/s, a voltage of 1 V will be induced. A simple one-loop generator is shown in *Figure 4-20.* The loop is attached to a rod that is free to rotate. This assembly is suspended between the poles of two stationary magnets. If the loop is turned, the conductor cuts through magnetic lines of flux and a voltage is induced into the conductor.

If the speed of rotation is increased, the conductor cuts more lines of flux per second, and the amount of induced voltage increases. If the speed of rotation remains constant and the strength of the magnetic field is increased, there will be more lines of flux per square inch. When there are more lines of flux, the number of lines cut per second increases and the

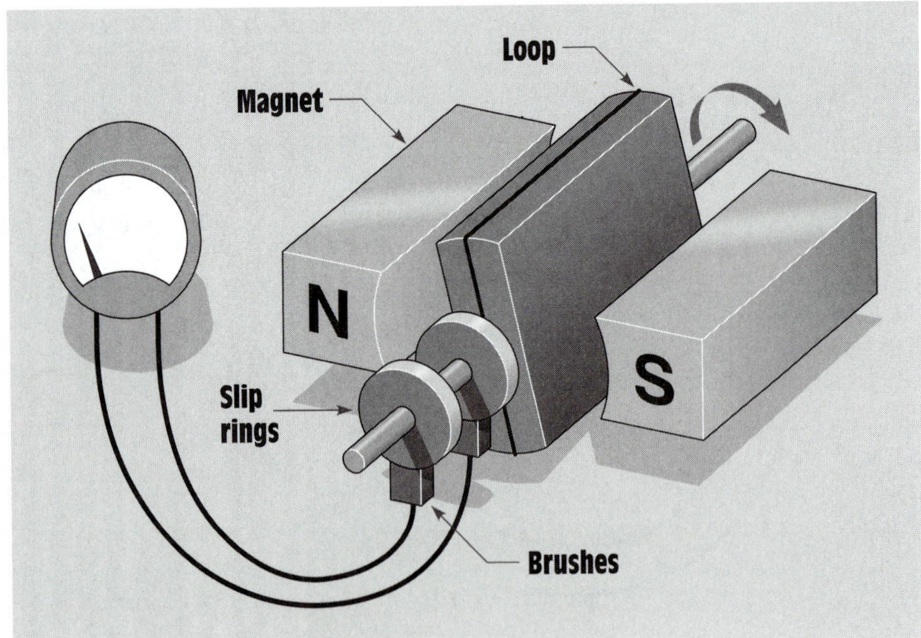

Figure 4-20 A single-loop generator.

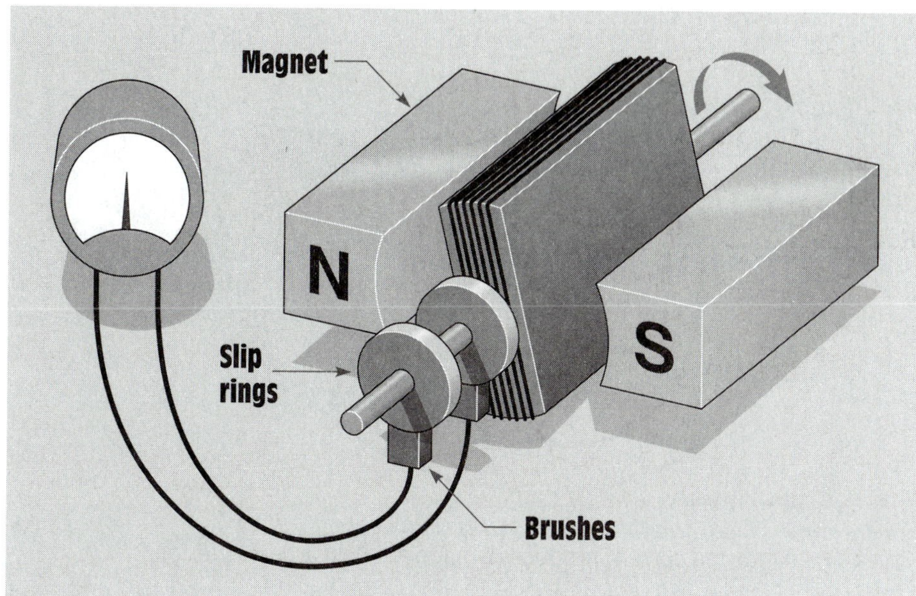

Figure 4-21 Increasing the number of turns increases the induced voltage.

induced voltage increases. If more turns of wire are added to the loop (*Figure 4-21*), more flux lines are cut per second and the amount of induced voltage increases again. Adding more turns has the effect of connecting single conductors in series, and the amount of induced voltage in each conductor adds.

LENZ'S LAW

When a voltage is induced in a coil and there is a complete circuit, current will flow through the coil (*Figure 4-22*). When current flows through the coil, a magnetic field is created around the coil. This magnetic field develops a polarity opposite that of the moving magnet. The magnetic field developed by the induced current acts to attract the moving magnet and pull it back inside the coil.

If the direction of motion is reversed, the polarity of the induced current is reversed, and the magnetic field created by the induced current again opposes the motion of the magnet. This principle was first noticed by Heinrich Lenz many years ago and is summarized in **Lenz's law,** which states that **an induced voltage or current opposes the motion that causes it.** From this basic principle, other laws concerning inductors have been developed. One is that **inductors always oppose a change of current.** The coil in *Figure 4-23*, for example, has no induced voltage and therefore no induced current. If the magnet is moved toward the coil, however, magnetic

Lenz's law

This law stipulates that the polarity of an induced emf will be such that any resulting current will have a magnetic field (flux) that opposes the original action (change in flux) that produced the induced current. In other words, the induced voltage always opposes the original change in current. That is why the induced voltage is known as the counter-emf (cemf), or back-emf.

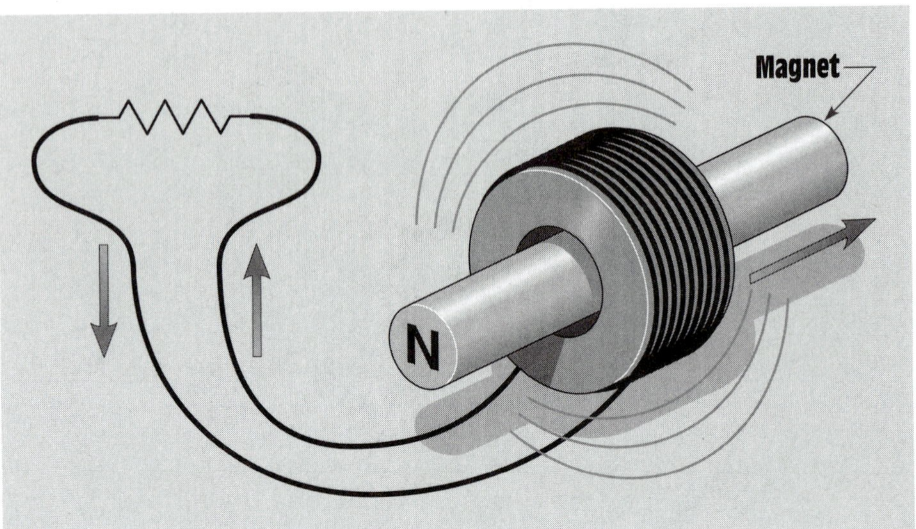

Figure 4-22 An induced current produces a magnetic field around the coil.

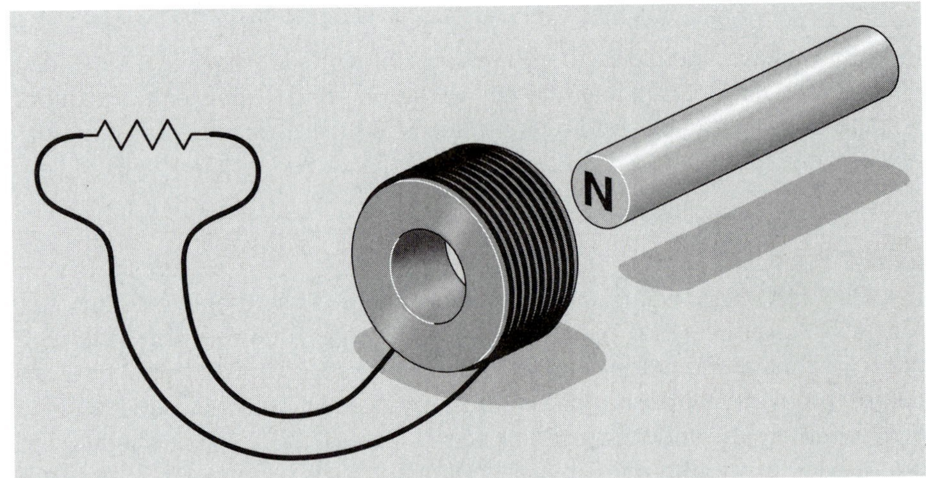

Figure 4-23 No current flows through the coil.

Inductors always oppose a change of current.

lines of flux will begin to cut the conductors of the coil, and a current will be induced in the coil. The induced current causes magnetic lines of flux to expand outward around the coil (*Figure 4-24*). As this expanding magnetic field cuts through the conductors of the coil, a voltage is induced in the coil. The polarity of the voltage is such that it opposes the induced current caused by the moving magnet.

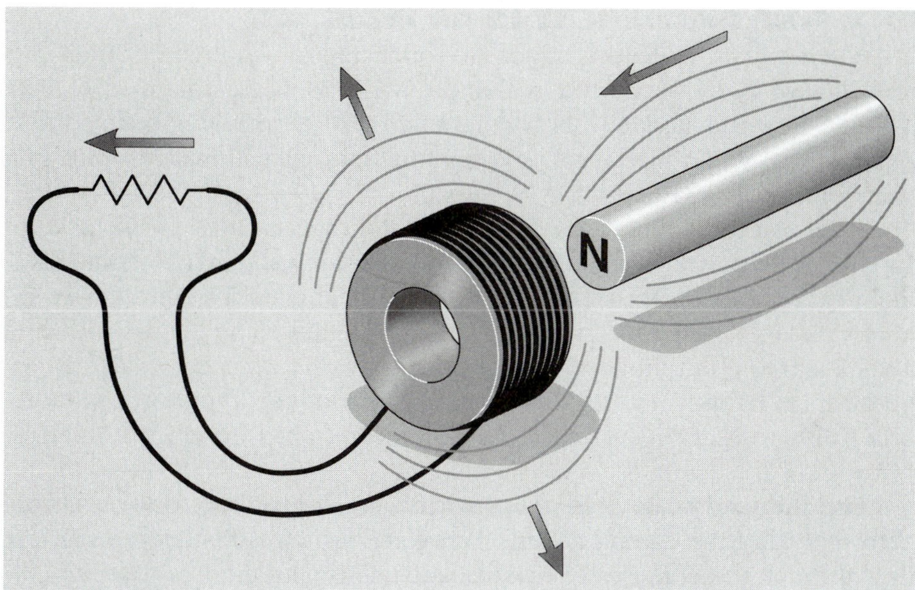

Figure 4-24 Induced current produces a magnetic field around the coil.

If the magnet is moved away, the magnetic field around the coil will collapse and induce a voltage in the coil (*Figure 4-25*). Since the direction of movement of the collapsing field has been reversed, the induced voltage will be opposite in polarity, forcing the current to flow in the same direction.

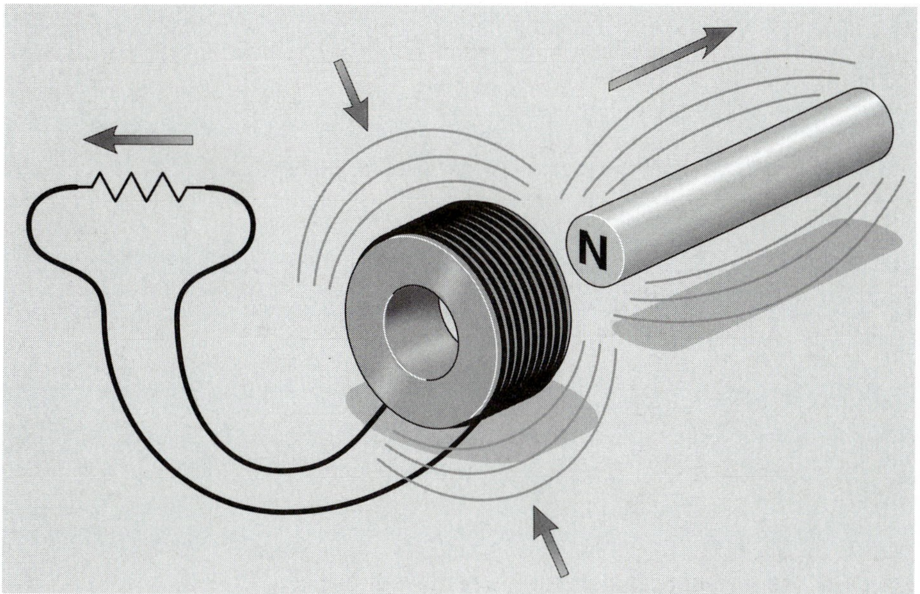

Figure 4-25 The induced voltage forces current to flow in the same direction.

RISE TIME OF CURRENT IN AN INDUCTOR

When a resistive load is suddenly connected to a source of direct current (*Figure 4-26*), the current will instantly rise to its maximum value. The resistor shown in Figure 4-26 has a value of 10 Ω and is connected to a 20-V source. When the switch is closed the current will instantly rise to a value of 2 A (20 V/10 Ω = 2 A).

If the resistor is replaced with an inductor that has a wire resistance of 10 Ω and the switch is closed, the current cannot instantly rise to its maximum value of 2 A (*Figure 4-27*). As current begins to flow through an inductor, the expanding magnetic field cuts through the conductors, inducing a voltage into them. In accord with Lenz's law, the induced voltage is opposite in polarity to the applied voltage. The induced voltage, therefore, acts like a resistance to hinder the flow of current through the inductor (*Figure 4-28*).

The induced voltage is proportional to the rate of change of current.

The induced voltage is proportional to the rate of change of current (speed of the cutting action). When the switch is first closed, current flow through the coil tries to rise instantly. This extremely fast rate of current change induces maximum voltage in the coil. As the current flow approaches its maximum Ohm's law value, in this example 2 A, the rate of change becomes less and the amount of induced voltage decreases.

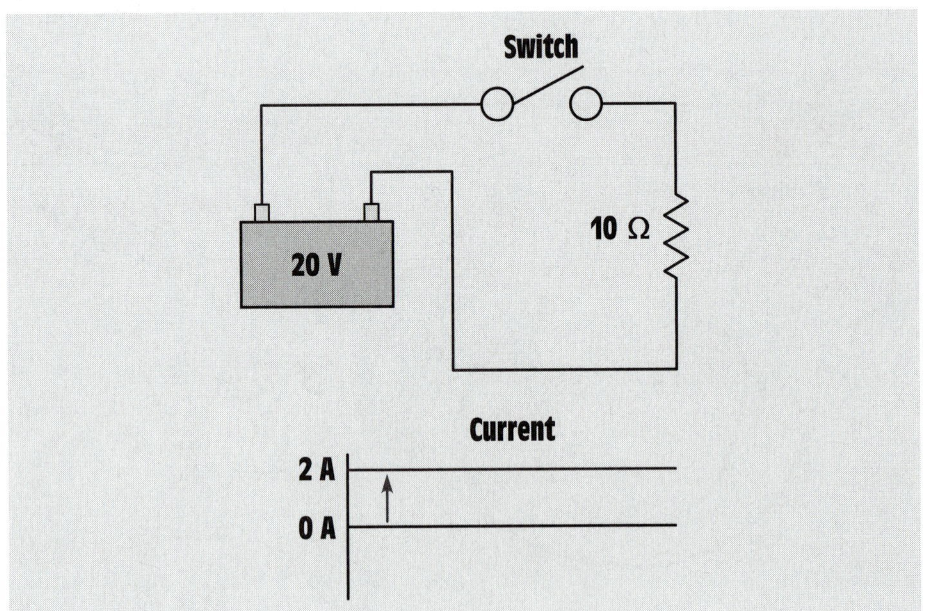

Figure 4-26 The current rises instantly in a resistive circuit.

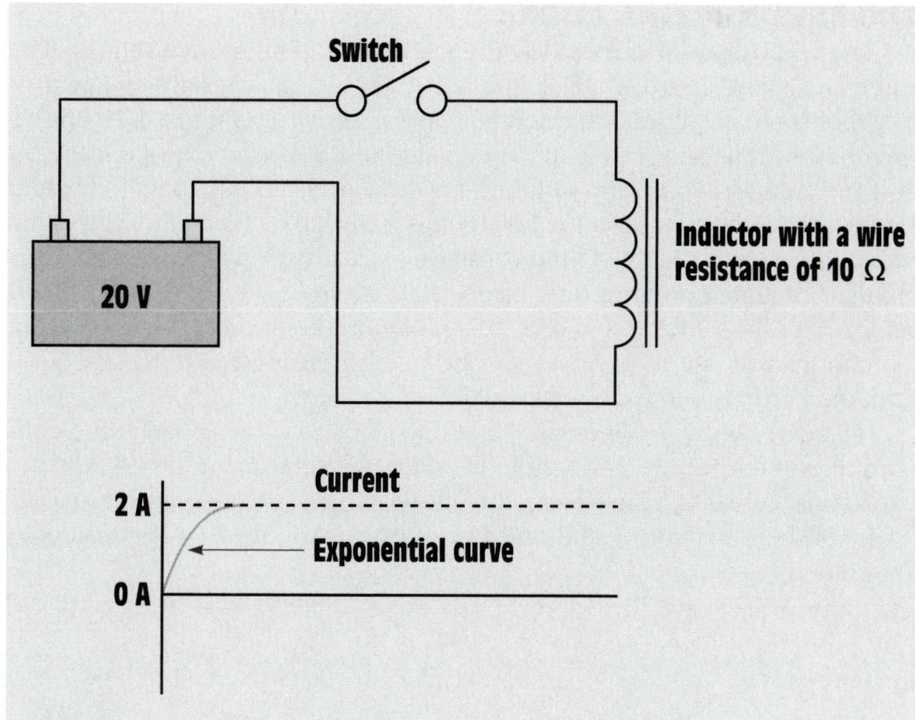

Figure 4-27 Current rises through an indicator at an exponential rate.

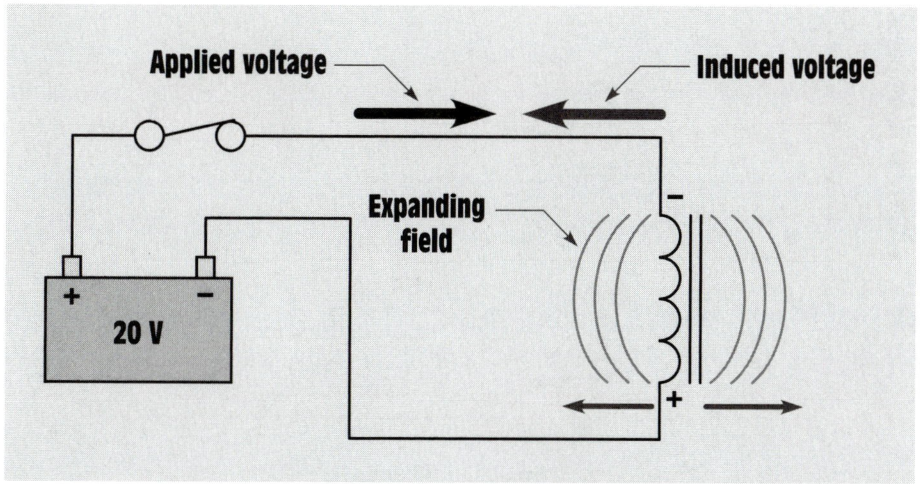

Figure 4-28 The applied voltage is opposite in polarity to the induced voltage.

THE EXPONENTIAL CURVE

exponential curve

The **exponential curve** describes a rate of certain occurrences. The curve is divided into five time constants. Each time constant is equal to 63.2% of some value. An exponential curve is shown in *Figure 4-29*. In this example, current must rise from 0 to a value of 1.5 A at an exponential rate. In this example, 100 ms are required for the current to rise to its full value. Since the current requires a total of 100 ms to rise to its full value, each time constant is 20 ms (100 ms/5 time constants = 20 ms per time constant). During the first time constant, the current will rise from 0 to 63.2% of its total value, or 0.948 A (1.5 × 0.632 = 0.948). During the second time constant the current will rise to a value of 1.297 A, and during the third time constant the current will reach a total value of 1.425 A.

Because the current increases at a rate of 63.2% during each time constant, it is theoretically impossible to reach the total value of 1.5 A. After five times constants, however, the current has reached approximately 99.3% of the maximum value and for all practical purposes is considered to be complete.

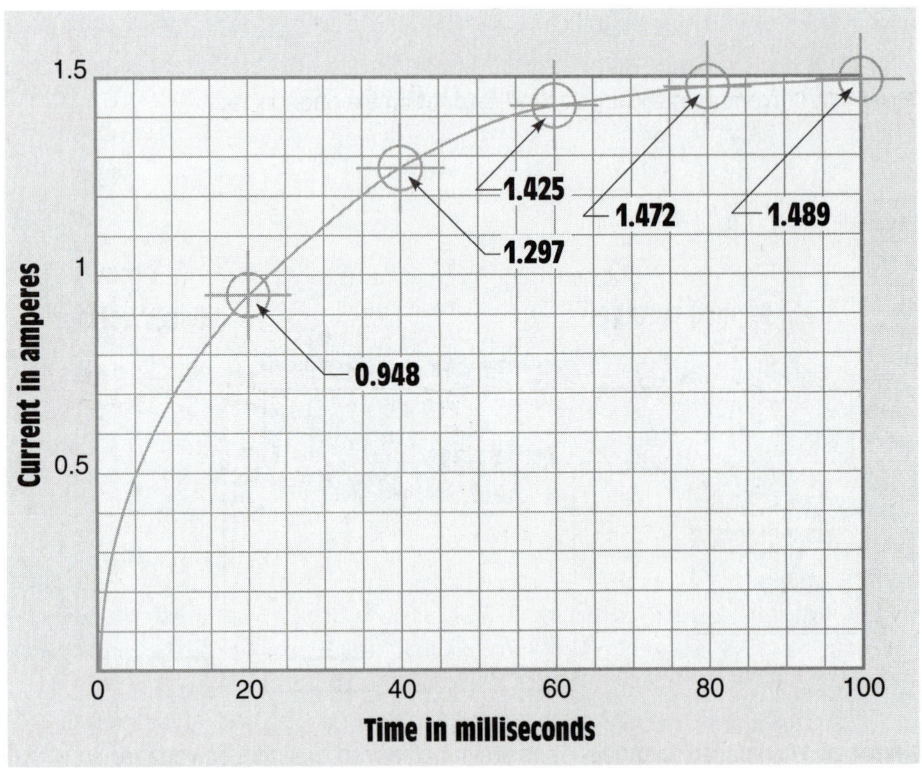

Figure 4-29 An exponential curve.

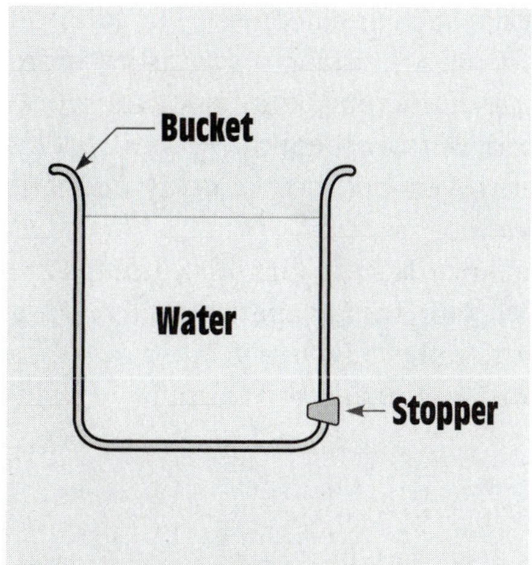

Figure 4-30 Exponential curves can be found in nature.

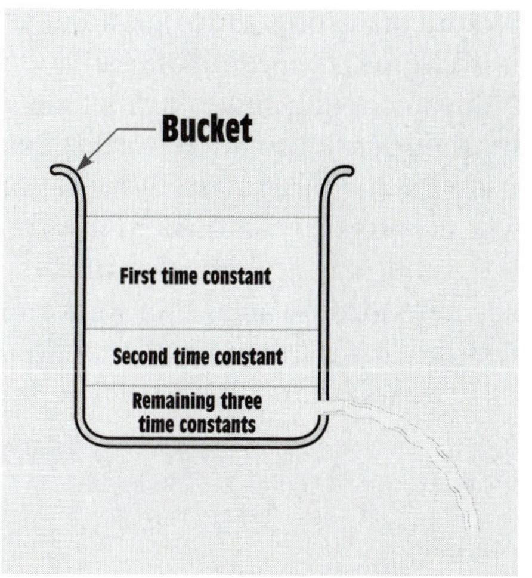

Figure 4-31 Water flows from a bucket at an exponential rate.

The exponential curve can often be found in nature. If clothes are hung on a line to dry, they will dry at an exponential rate. Another example of the exponential curve can be seen in *Figure 4-30.* In this example, a bucket has been filled to a certain mark with water. A hole has been cut at the bottom of the bucket and a stopper placed in the hole. When the stopper is removed from the bucket, water will flow out at an exponential rate. Assume, for example, it takes 5 min for the water to flow out of the bucket. Exponential curves are always divided into five time constants so in this case each time constant has a value of 1 min. In *Figure 4-31,* if the stopper is removed and water is permitted to drain from the bucket for a period of 1 min before the stopper is replaced, during that first time constant 63.2% of the water in the bucket will drain out. If the stopper is again removed for a period of 1 min, 63.2% of the water remaining in the bucket will drain out. Each time the stopper is removed for a period of one time constant, the bucket will lose 63.2% of its remaining water.

INDUCTANCE

Inductance is measured in units called the **henry** (H) and is represented by the letter *L.* **A coil has an inductance of one henry when a current change of one ampere per second results in an induced voltage of one volt.**

The amount of inductance a coil will have is determined by its physical properties and construction. A coil wound on a nonmagnetic core material such as wood or plastic is referred to as an *air core* inductor. If the coil is

henry (H)

A coil has an inductance of one henry when a current change of one ampere per second results in an induced voltage of one volt.

eddy current

hysteresis loss

wound on a core made of magnetic material such as silicon steel or soft iron it is referred to as an *iron core* inductor. Iron core inductors produce more inductance with fewer turns than air core inductors because of the good magnetic path provided by the core material. Iron core inductors cannot be used for high-frequency applications, however, because of **eddy current** loss and **hysteresis loss** in the core material.

Another factor that determines inductance is how far the windings are separated from each other. If the turns of wire are far apart they will have less inductance than turns wound closer together (*Figure 4-32*).

The inductance of a coil can be determined using the formula

$$L = \frac{0.4\pi N^2 \mu A}{l}$$

where

L = inductance in henrys

π = 3.1416

N = number of turns of wire

μ = permeability of the core material

A = cross-sectional area of the core

l = length of the core

The formula indicates that the inductance is proportional to the number of turns of wire, the type of core material used, and the cross-sectional area of the core, but inversely proportional to core length. An inductor is basically an electromagnet that changes its polarity at regular intervals. Because the inductor is an electromagnet, the same factors that affect magnets affect inductors. The permeability of the core material is just as important to an inductor as it is to any other electromagnet. Flux lines pass through a core material with a high permeability (such as silicon, steel, or soft iron) better than through a material with a low permeability (such as brass, copper, or aluminum). Once the core material has become saturated, however, the permeability value becomes approximately 1 and an increase in turns of wire has only a small effect on the value of inductance.

CORE MATERIAL

When a coil is wound around a nonmagnetic material such as wood or plastic, it is known as an *air core* magnet. When a coil is wound around a magnetic material such as iron or soft steel, it is known as an *iron core* magnet. The addition of magnetic material to the center of the coil can greatly

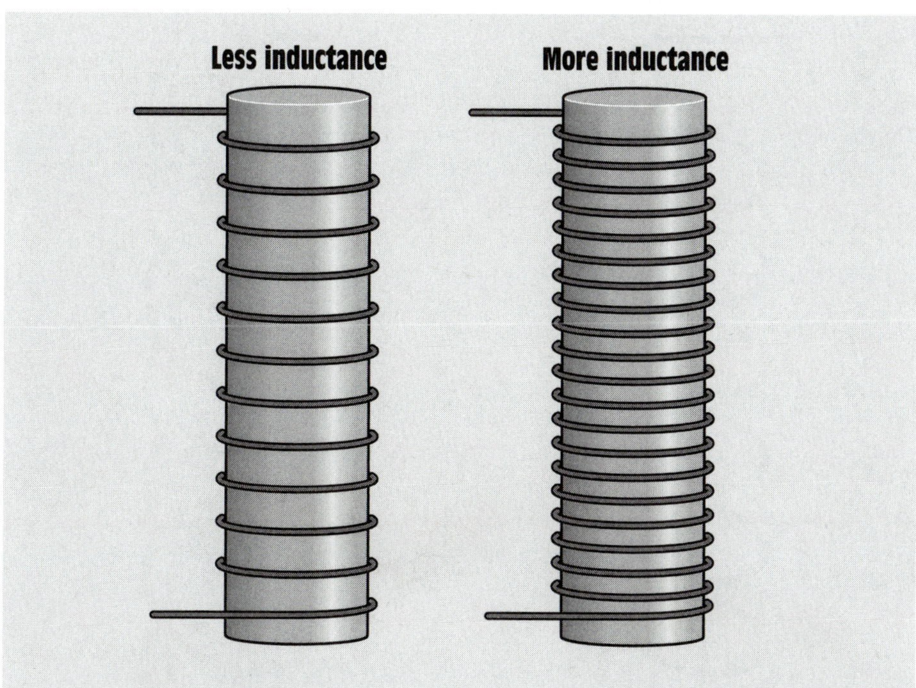

Figure 4-32 Inductance is determined by the physical construction of the coil.

increase the strength of the magnet. If the core material causes the magnetic field to become fifty times stronger, the core material has a **permeability** of 50, *Figure 4-33.* Permeability is a measure of a material's ability to become magnetized. The number of flux lines produced is proportional to the ampere-turns. The magnetic core material provides an easy path for the flow of magnetic lines, in much the same way that a conductor provides an easy path for the flow of electrons. This permits the flux lines to be concentrated in a smaller area, which increases the number of lines per square inch or square centimeter. It is very similar to a person using a garden hose with an adjustable nozzle attached. The nozzle can be adjusted to permit the water to spray in a fine mist, which covers a large area, or spray in a concentrated stream.

perme-ability

Another common magnetic measurement is **reluctance.** Reluctance is resistance to magnetism. A material such as soft iron or steel has a high permeability and low reluctance because it is easily magnetized. A material such as copper has a low permeability and high reluctance.

reluctance

If the current flow in an electromagnet is continually increased, the magnet will eventually reach a point at which its strength will increase only slightly with an increase in current. When this condition occurs, the

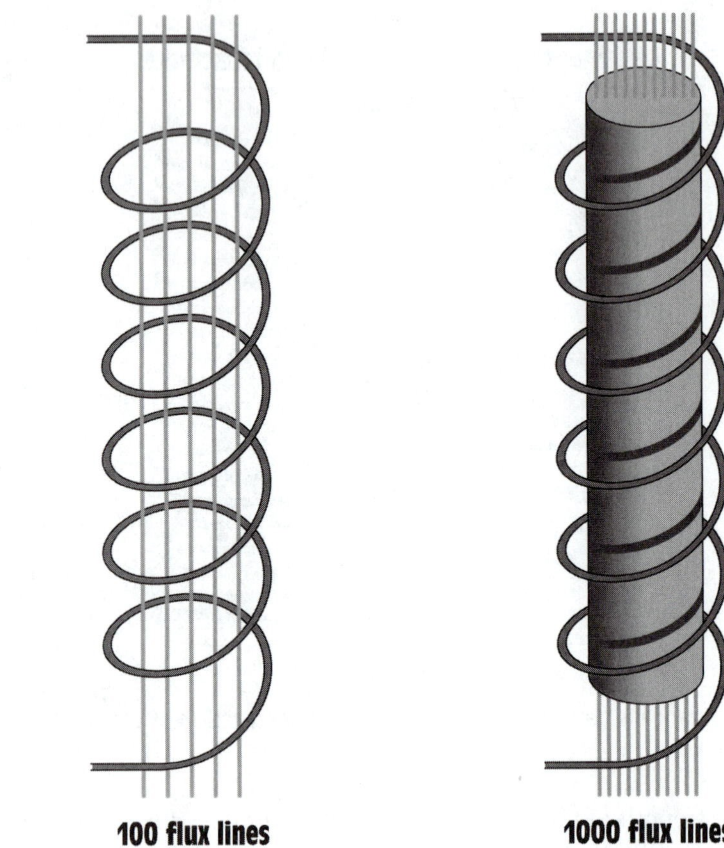

Figure 4-33 An iron core increases the number of flux lines per square inch.

saturation

magnetic material is at a point of **saturation.** Saturation occurs when all the molecules of the magnetic material are lined up. Saturation is similar to pouring five gallons of water into a five-gallon bucket. Once the bucket is full, it cannot hold any more water. If it became necessary to construct a stronger magnet, a larger piece of core material would be required.

residual magnetism

When the current flow through the coil of a magnet is stopped, there may be some magnetism left in the core material. The amount of magnetism left in a material after the magnetizing force has stopped is called **residual magnetism.** If the residual magnetism of a piece of core material is hard to remove, the material has a high *coercive force.* Coercive force is a measure of a material's ability to retain magnetism. A high coercive force is desirable in materials that are intended to be used as permanent magnets. A low coercive force is generally desirable for materials intended to be used as electromagnets. Coercive force is measured by determining the amount of current flow through the coil in the reverse direction that is required to remove the residual magnetism. Another term that is used to describe a material's ability to retain magnetism is *retentivity*.

MAGNETIC MEASUREMENT

The terms used to measure the strength of a magnetic field are determined by the system that is being used. There are three different systems used to measure magnetism: the *English* system, the *CGS* system, and the *MKS* system.

THE ENGLISH SYSTEM

In the English system of measure, magnetic strength is measured in a term called *flux density*. Flux density is measured in lines per square inch. The Greek letter phi (Φ) is used to measure flux. The letter B is used to represent flux density. The following formula is used to determine flux density.

$$\text{B (flux density)} = \frac{\Phi \text{ (flux lines)}}{\text{A (area)}}$$

In the English system, the term used to describe the total force producing a magnetic field, or flux, is **magnetomotive force** (mmf). Magnetomotive force can be computed using the formula:

magneto-motive force (mmf)

$$\text{mmf} = \Phi \times \text{rel (reluctance)}$$

The following formula can be used to determine the strength of the magnet.

$$\text{Pull (in pounds)} = \frac{\text{B} \times \text{A}}{72{,}000{,}000}$$

where

B = flux density in lines per square inch
A = area of the magnet

THE CGS SYSTEM (CENTIMETER-GRAM-SECOND)

In the CGS system of measurement, one magnetic line of force is known as a *maxwell*. A *gauss* represents a magnetic force of one maxwell per square centimeter. In the English system, magnetomotive force is measured in ampere-turns. In the CGS system, *gilberts* are used to represent the same measurement. Since the main difference between these two systems of measurement is that one uses English units of measure and the other uses metric units of measure, a conversion factor can be used to help convert one set of units to the other.

$$1 \text{ gilbert} = 1.256 \text{ ampere-turns}$$

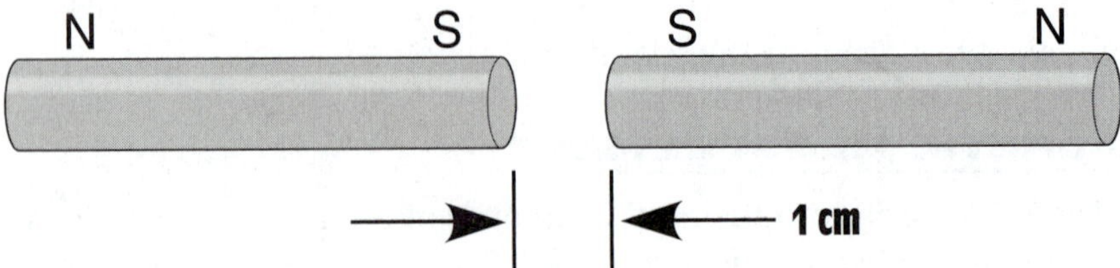

Figure 4-34 A unit magnetic pole produces a force of one dyne.

THE MKS SYSTEM (METER-KILOGRAM-SECOND)

The MKS system uses metric units of measure also. In this system, the main unit of magnetic measurement is the *dyne.* The dyne is a very weak amount of force. One dyne is equal to 1/27,800 of an ounce, or it requires 27,800 dynes to equal a force of one ounce. In the MKS system, a standard called the *unit magnetic pole* is used. In *Figure 4-34,* two magnets are separated by a distance of 1 centimeter (cm). These magnets repel each other with a force of 1 dyne. When two magnets separated by a distance of 1 cm exert a force on each other of 1 dyne, they are considered to be a unit magnetic pole. Magnetic force can then be determined using the following formula:

$$\text{Force (in dynes)} = \frac{M_1 \times M_2}{D}$$

where

M_1 = strength of first magnet, in unit magnetic poles
M_2 = strength of second magnet, in unit magnetic poles
D = distance between the poles, in centimeters

MAGNETIC POLARITY

left-hand rule

The polarity of an electromagnet can be determined using the **left-hand rule.** When the fingers of the left hand are placed around the windings in the direction of electron current flow, the thumb will point to the north magnetic pole, *Figure 4-35.* If the direction of current flow is reversed, the polarity of the magnetic field will reverse also.

DEMAGNETIZING

demagnetized

When an object is to be **demagnetized,** its molecules must be disarranged as they are in a nonmagnetized material. This can be done by placing the object in the field of a strong electromagnet connected to an alternating current line. Since the magnet is connected to AC current, the

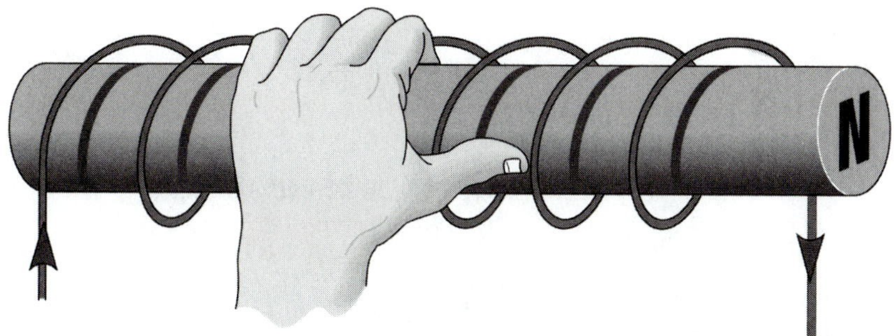

Figure 4-35 The left-hand rule can be used to determine the polarity of an electromagnet.

polarity of the magnetic field reverses each time the current changes direction. The molecules of the object to be demagnetized are, therefore, aligned first in one direction and then in the other. If the object is pulled away from the AC magnetic field, the effect of the field becomes weaker as the object is moved farther away, *Figure 4-36*. This causes the molecules of the object to be left in a state of disarray. The ease or difficulty with which an object can be demagnetized depends on the strength of the AC magnetic field and the coercive force of the object.

There are two other ways in which an object can be demagnetized, *Figure 4-37*. If a magnetized object is struck, the vibration will often cause

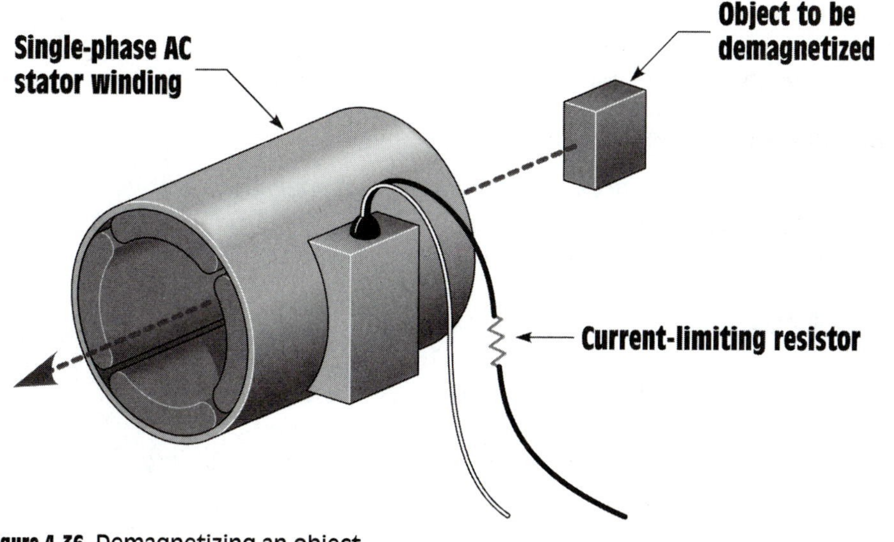

Figure 4-36 Demagnetizing an object.

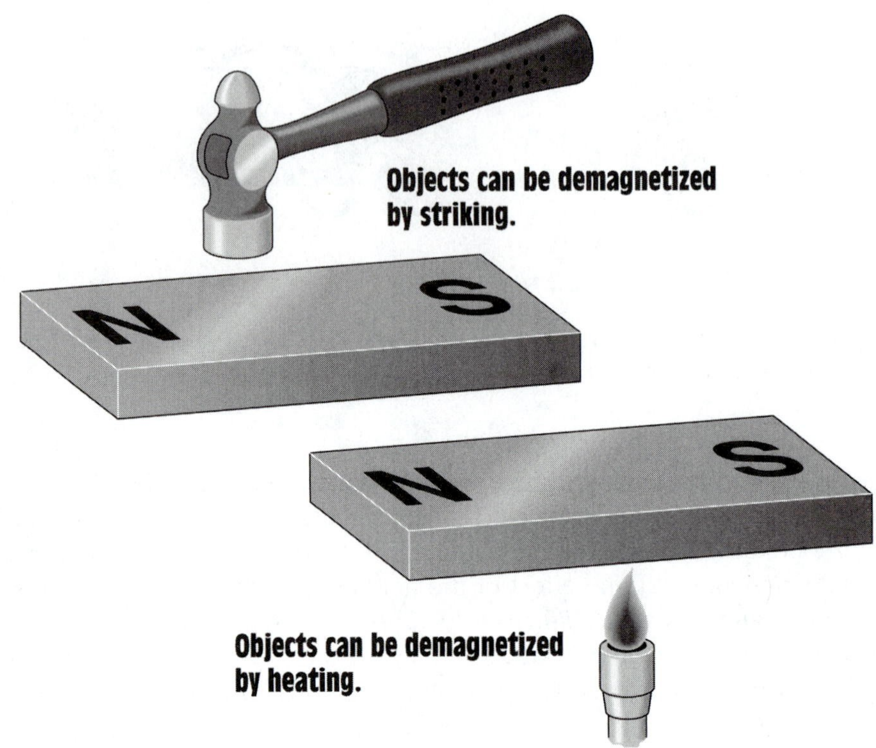

Figure 4-37 Other methods for demagnetizing objects.

the molecules to rearrange themselves in a disordered fashion. It may be necessary to strike the object several times. Heating an object also will demagnetize it. When the temperature becomes high enough, the molecules will rearrange themselves in a disordered fashion.

MAGNETIC DEVICES

A list of all the devices that operate on magnetism would be very long indeed. Some of the more common devices are electromagnets, measuring instruments, inductors, transformers, and motors.

THE SPEAKER

The speaker is a common device that operates on the principle of magnetism, *Figure 4-38*. The speaker produces sound by moving a cone, which causes a displacement of air. The tone is determined by how fast the cone vibrates. Low or bass sounds are produced by vibrations in the range of 20

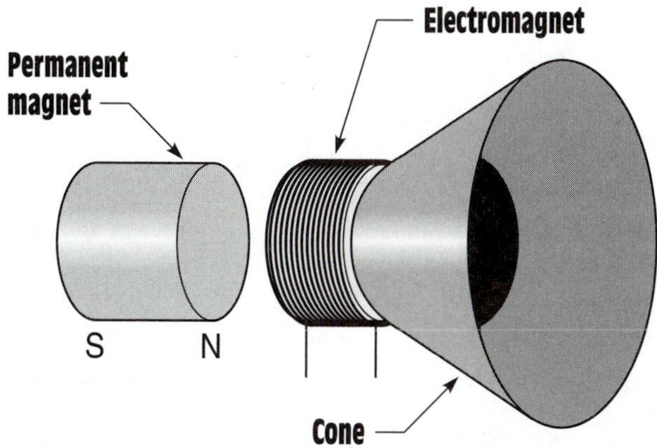

Figure 4-38 A speaker uses both an electromagnet and a permanent magnet.

cycles per second. High sounds are produced when the speaker vibrates in the range of 20,000 cycles per second.

The speaker uses two separate magnets. One is a permanent magnet, and the other is an electromagnet. The permanent magnet is held stationary, and the electromagnet is attached to the speaker cone. When current flows through the coil of the electromagnet, a magnetic field is produced. The polarity of the field is determined by the direction of current flow. When the electromagnet becomes a north polarity, it is repelled away from the permanent magnet. This causes the speaker cone to move outward, which causes a displacement of air. When the current flow reverses through the coil, the electromagnet becomes a south polarity and is attracted to the permanent magnet. This causes the speaker cone to move inward and again displace air. The tone of the speaker determines the number of times per second the current through the coil reverses.

SUMMARY

1. Early natural magnets were known as lodestones.
2. The earth has a north and a south magnetic pole.
3. The magnetic poles of the earth and the axis poles are not the same.
4. Like poles of a magnet repel each other and unlike poles attract each other.
5. Some materials have the ability to become better magnets than others.

6. Three basic types of magnetic material are:

   A. ferromagnetic
   B. paramagnetic
   C. diamagnetic

7. When current flows through a wire, a magnetic field is created around the wire.

8. The direction of current flow through the wire determines the polarity of the magnetic field.

9. The strength of an electromagnet is determined by the ampere-turns.

10. The type of core material used in an electromagnet can determine its strength.

11. Three different systems used to measure magnetic values are:

    A. the English System
    B. the CGS System
    C. the MKS System

12. An object can be demagnetized by placing it in an AC magnetic field and pulling it away, by striking, and by heating.

13. When a conductor is cut by a magnetic field a voltage is induced into the conductor.

14. Three factors that determine the amount of induced voltage are:

    A. strength of the magnetic field
    B. number of turns of wire (sometimes stated as length of conductor)
    C. speed of the cutting action

15. The polarity of the induced voltage is determined by the polarity of the magnetic field and the direction of conductor movement.

16. Lenz's law states that an induced voltage or current opposes the motion that causes it.

REVIEW QUESTIONS

1. Is the north magnetic pole of the earth a north polarity or a south polarity?

2. What were early natural magnets known as?

3. The south pole of one magnet is brought close to the south pole of another magnet. Will the magnets repel or attract each other?

4. How can the polarity of an electromagnet be determined if the direction of current flow is known?
5. Define the following terms.
   A. flux density
   B. permeability
   C. reluctance
   D. saturation
   E. coercive force
   F. residual magnetism
6. A force of one ounce is equal to how many dynes?
7. A weber is defined as __________ lines of magnetic flux.
8. A coil contains 50 loops of wire and cuts through magnetic lines of flux at a rate of 20 webers per second. How much voltage is induced into the coil?

5

Unit

Series Circuits

objectives

fter studying this unit, you should be able to:

- Discuss the properties of series circuits.
- Compute values of voltage, current, and resistance for series circuits.
- List rules for solving electrical values of series circuits.

series circuit

Electrical circuits can be divided into three major types: the series, the parallel, and the combination. Combination circuits are circuits that contain both series and parallel paths. The first type to be discussed is the **series circuit.**

SERIES CIRCUIT

A series circuit is a circuit that has only one path for current flow.

A series circuit is a circuit that has only one path for current flow, *Figure 5-1.* Since there is only one path for current flow, the current is the same at any point in the circuit. Imagine that an electron leaves the negative terminal of the battery. This electron must flow through each resistor before it can complete the circuit to the positive battery terminal.

fuses

circuit breakers

One of the best examples of a series-connected device is a fuse or circuit breaker, *Figure 5-2.* Since **fuses** and **circuit breakers** are connected in series with the rest of the circuit, all the circuit current must flow through them. If the current becomes excessive, the fuse or circuit breaker will open and disconnect the rest of the circuit from the power source.

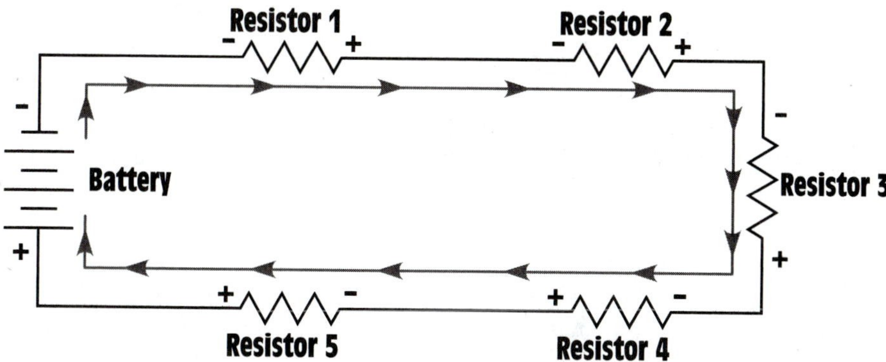

Figure 5-1 In a series circuit, there is only one path for current flow.

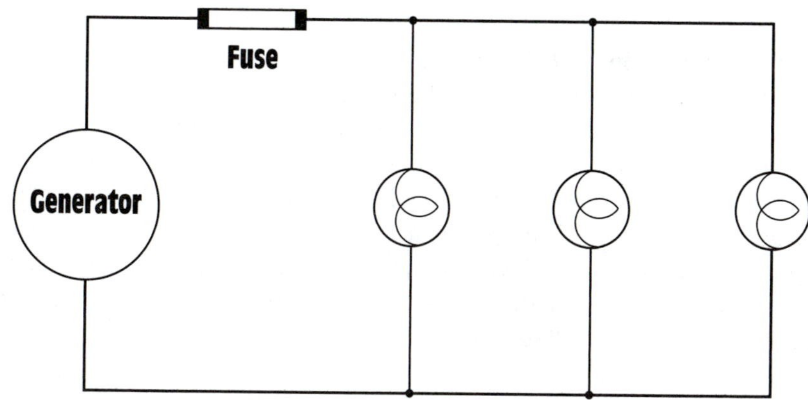

Figure 5-2 All the current must flow through the fuse.

voltage drop

In a series circuit, the sum of all the voltage drops across all the resistors must equal the voltage applied to the circuit.

VOLTAGE DROPS IN A SERIES CIRCUIT

Voltage is the force that pushes the electrons through a resistance. The amount of voltage required is determined by the amount of current flow and resistance. If a voltmeter is connected across a resistor, *Figure 5-3,* the amount of voltage necessary to push the current through that resistor will be indicated by the meter. This is known as **voltage drop.** It is similar to pressure drop in a water system. **In a series circuit, the sum of all the voltage drops across all resistors must equal the voltage applied to the circuit.** The amount of voltage drop across each resistor will be proportional to its resistance and the circuit current.

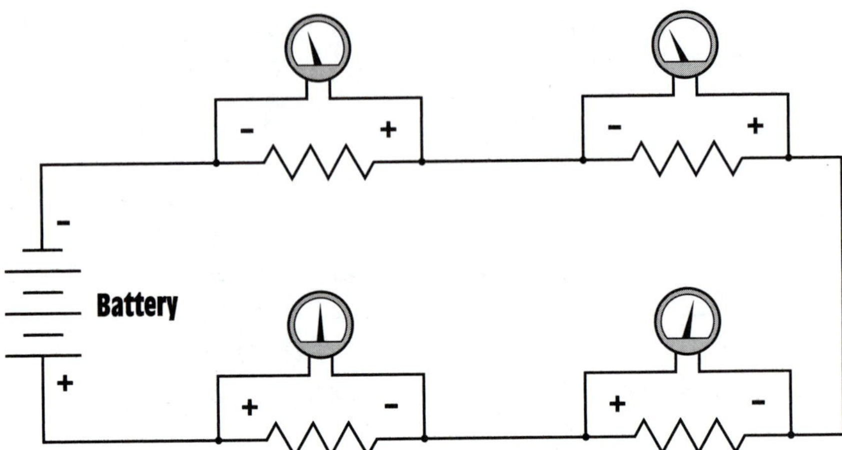

Figure 5-3 The voltage drops in a series circuit must equal the applied voltage.

In the circuit shown in *Figure 5-4,* four resistors are connected in series. It is assumed that each resistor has the same value. The circuit is connected to a 24-V battery. Since each resistor has the same value, the voltage drop across each will be 6 V (24 V/4 resistors = 6 V). Note that each resistor will have the same voltage drop only if each resistor is the same value. The circuit shown in *Figure 5-5* illustrates a series circuit comprised of resistors having different values. Notice that the voltage drop across each resistor is proportional to its resistance. Also notice that the sum of the voltage drops will equal the applied voltage of 24 V.

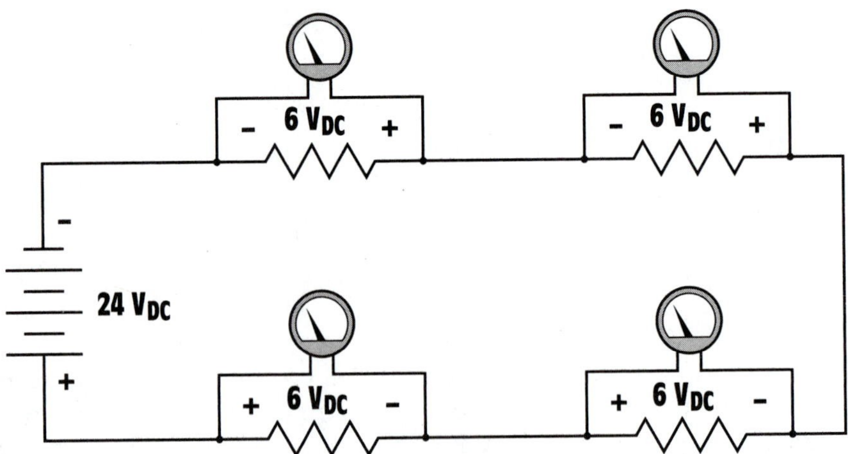

Figure 5-4 The voltage drop across each resistor is proportional to its resistance.

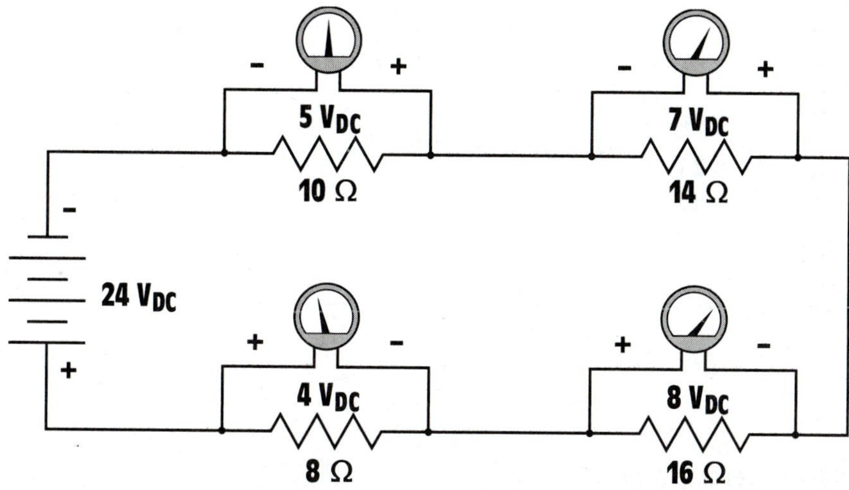

Figure 5-5 Series circuit with four resistors having different voltage drops.

RESISTANCE IN A SERIES CIRCUIT

Since there is only one path for the current to flow through a series circuit, it must flow through each resistor in the circuit (Figure 5-1). Each resistor limits or impedes the flow or current in the circuit. Therefore, the total amount of **series resistance** offered to the flow of current is proportional to the sum of all the resistors in the circuit.

series resistance

CALCULATING SERIES CIRCUIT VALUES

There are three rules that can be used with Ohm's Law for finding values of voltage, current, resistance, and power in any series circuit.

1. The current is the same at any point in the circuit.
2. The total resistance is the sum of the individual resistors.
3. The applied voltage is equal to the sum of the voltage drops across all resistors.

The circuit shown in *Figure 5-6* shows the values of current flow, voltage drop, and resistance for each of the resistors. Notice that the total resistance (R_T) of the circuit can be found by adding the values of the resistors **(resistance adds).**

resistance adds

$$R_{T(total)} = R_1 + R_2 + R_3$$

$$R_T = 20\ \Omega + 10\ \Omega + 30\ \Omega$$

$$R_T = 60\ \Omega$$

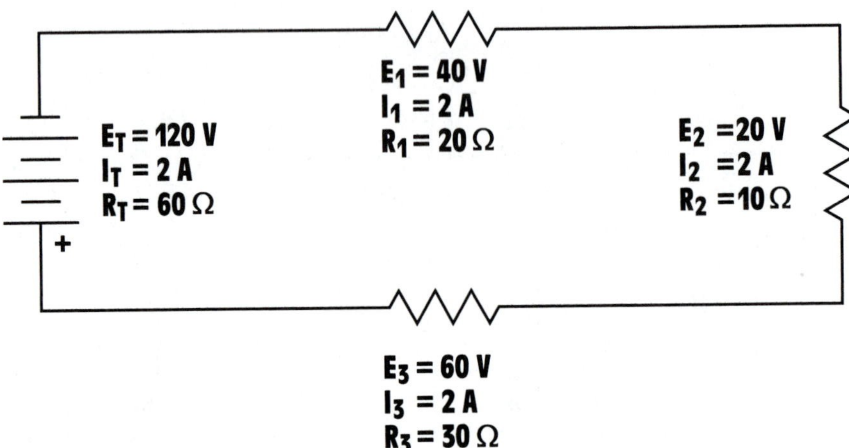

Figure 5-6 Series circuit values.

Ohm's Law

The amount of current flow in the circuit can be found by using **Ohm's Law.**

$$I = \frac{E}{R}$$

$$I = \frac{120}{60}$$

$$I = 2\text{ A}$$

A current of 2 A flows through each resistor in the circuit.

$$(I_{T(total)} = I_1 = I_2 = I_3)$$

Since the amount of current flowing through resistor R_1 is known, the voltage drop across the resistor can be found using Ohm's Law.

$$E_1 = I_1 \times R_1$$

$$E_1 = 2\text{ A} \times 20\ \Omega$$

$$E_1 = 40\text{ V}$$

This means that it takes 40 V to push 2 A of current through 20 Ω of resistance. If a voltmeter is connected across resistor R_1, it would indicate a value of 40 V, *Figure 5-7.* This is the amount of voltage drop across

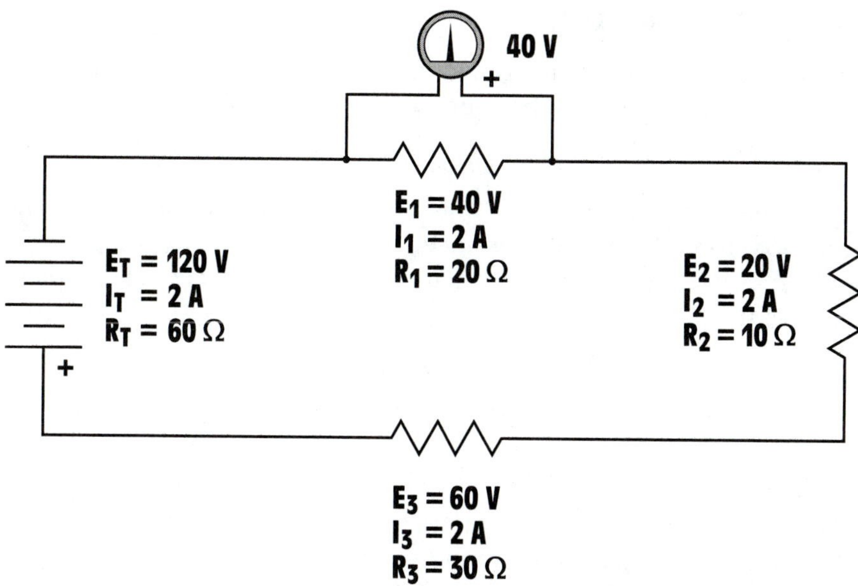

Figure 5-7 The voltmeter indicates a voltage drop of 40 V.

resistor R_1. The voltage drop across resistors R_2 and R_3 can be found in the same way.

$$E_2 = I_2 \times R_2$$

$$E_2 = 2\ \text{A} \times 10\ \Omega$$

$$E_2 = 20\ \text{V}$$

$$E_3 = I_3 \times R_3$$

$$E_3 = 2\ \text{A} \times 30\ \Omega$$

$$E_3 = 60\ \text{V}$$

If the voltage drop across all the resistors is added, it equals the total applied voltage (E_T).

$$(E_{T(total)} = E_1 + E_2 + E_3)$$

$$E_T = 40\ \text{V} + 20\ \text{V} + 60\ \text{V}$$

$$E_T = 120\ \text{V}$$

SOLVING CIRCUITS

In the following problems, circuits will be shown that have missing values. The missing values can be found by using the rules for series circuits and Ohm's Law.

Example 1

The first step in finding the missing values in the circuit shown in *Figure 5-8* is to find the total resistance. This can be done by using the third rule of series circuits, which states that resistances add to equal the total resistance of the circuit.

$$R_T = R_1 + R_2 + R_3 + R_4$$

$$R_T = 100\ \Omega + 250\ \Omega + 150\ \Omega + 300\ \Omega$$

$$R_T = 800\ \Omega$$

Now that the total voltage and total resistance are known, the current flow through the circuit can be found using Ohm's Law.

$$I = \frac{E}{R}$$

$$I = \frac{40}{800}$$

$$I = 0.050\ A$$

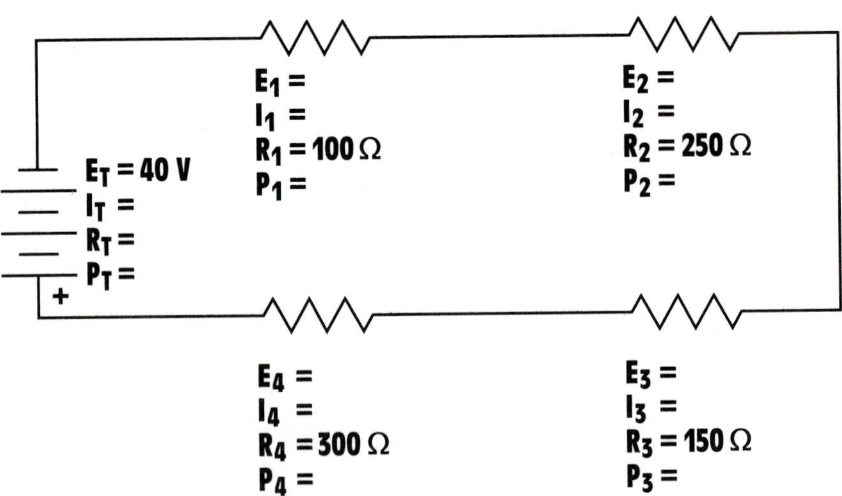

Figure 5-8 Series circuit, Example 1.

The second rule of series circuits states that current remains the same at any point in the circuit. Therefore, 0.050 A flows through each resistor in the circuit, *Figure 5-9*. The voltage drop across each resistor can now be found using Ohm's Law, *Figure 5-10*.

$$E_1 = I_1 \times R_1$$
$$E_1 = 0.050 \times 100$$
$$E_1 = 5\ V$$

$$E_2 = I_2 \times R_2$$
$$E_2 = 0.050 \times 250$$
$$E_2 = 12.5\ V$$

$$E_3 = I_3 \times R_3$$
$$E_3 = 0.050 \times 150$$
$$E_3 = 7.5\ V$$

$$E_4 = I_4 \times R_4$$
$$E_4 = 0.050 \times 300$$
$$E_4 = 15\ V$$

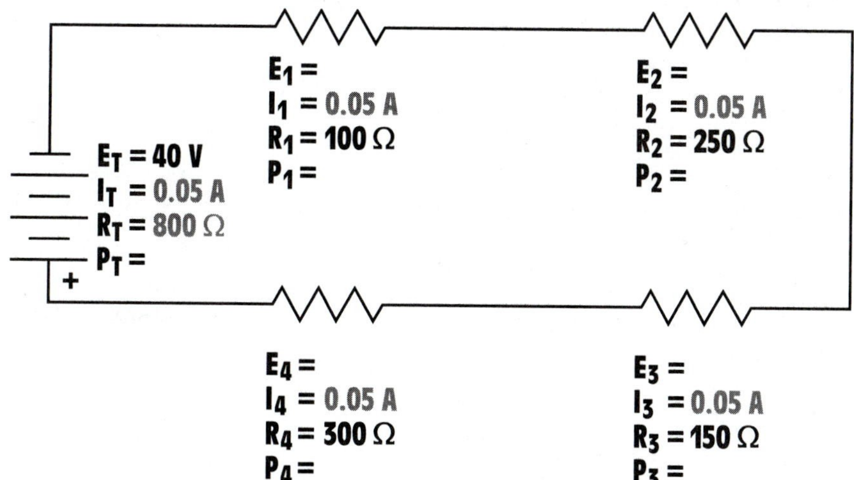

Figure 5-9 In a series circuit, the current is the same at any point.

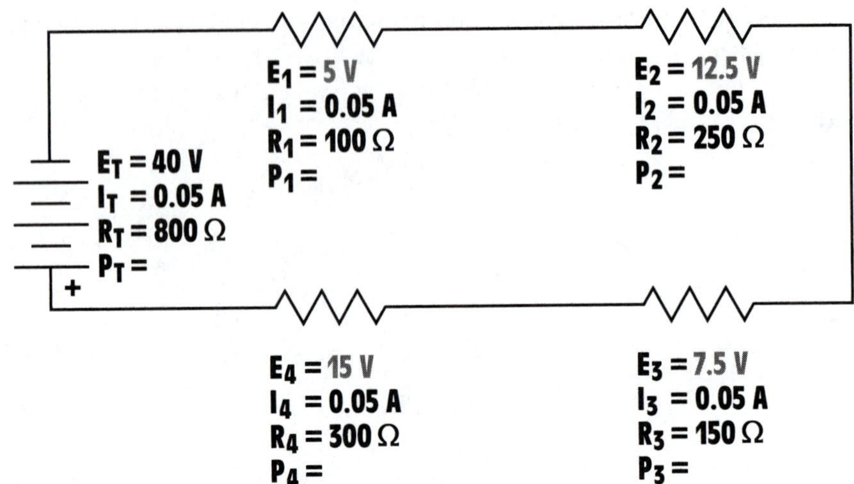

Figure 5-10 The voltage drop across each resistor can be found using Ohm's Law.

Several formulas can be used to determine the amount of power dissipated (changed into heat energy) by each resistor. The power dissipation of resistor R_1 will be found using the formula:

$$P_1 = E_1 \times I_1$$

$$P_1 = 5 \times 0.05$$

$$P_1 = 0.25 \text{ W}$$

The amount of power dissipation for resistor R_2 will be computed using the formula:

$$P_2 = \frac{E_2^{\,2}}{R_2}$$

$$P_2 = \frac{156.25}{250}$$

$$P_2 = 0.625 \text{ W}$$

The amount of power dissipation for resistor R_3 will be computed using the formula:

$$P_3 = I_3^{\,2} \times R_3$$

$$P_3 = 0.0025 \times 150$$

$$P_3 = 0.375 \text{ W}$$

The amount of power dissipation for resistor R_4 will be found using the formula:

$$P_4 = E_4 \times I_4$$

$$P_4 = 15 \times 0.05$$

$$P_4 = 0.75 \text{ W}$$

A good rule to remember when calculating values of electrical circuits is that **the total power used in a circuit is equal to the sum of the power used by all parts.** This means that the total power can be found in any kind of a circuit (series, parallel, or combination) by adding the power dissipation of each part. The total power for this circuit can be found using the formula:

The total power used in a circuit is equal to the sum of the power used by all parts.

$$P_T = P_1 + P_2 + P_3 + P_4$$

$$P_T = 0.25 + 0.625 + 0.375 + 0.75$$

$$P_T = 2 \text{ W}$$

Now that all the missing values have been found, *Figure 5-11,* the circuit can be checked by using the first rule of series circuits, which states that voltage drops add to equal the applied voltage.

$$E_T = E_1 + E_2 + E_3 + E_4$$

$$E_T = 5 + 12.5 + 7.5 + 15$$

$$E_T = 40 \text{ V}$$

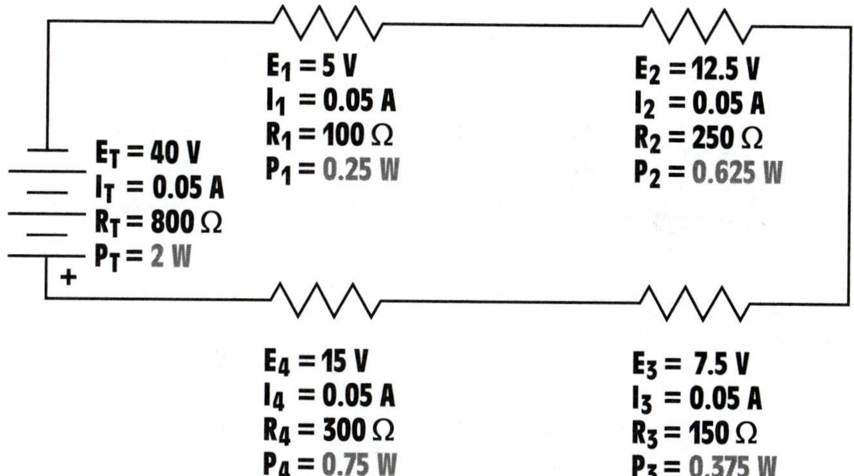

Figure 5-11 The final values for the circuit in Example 1.

Example 2

The second circuit to be solved is shown in *Figure 5-12*. In this circuit the total resistance is known, but the value of resistor R_2 is not. The third rule of series circuits states that resistance adds to equal the total resistance of the circuit. Since the total resistance is known, the missing resistance of R_2 can be found by adding the value of the other resistors and subtracting their sum from the total resistance of the circuit, *Figure 5-13*.

$$R_2 = R_T - (R_1 + R_3 + R_4)$$

$$R_2 = 6000 - (1000 + 2000 + 1200)$$

$E_1 =$
$I_1 =$
$R_1 = 1000\ \Omega$
$P_1 =$

$E_2 =$
$I_2 =$
$R_2 =$
$P_2 =$

$E_T = 120$ V
$I_T =$
$R_T = 6000\ \Omega$
$P_T =$
+

$E_4 =$
$I_4 =$
$R_4 = 1200\ \Omega$
$P_4 =$

$E_3 =$
$I_3 =$
$R_3 = 2000\ \Omega$
$P_3 =$

Figure 5-12 Series circuit, Example 2.

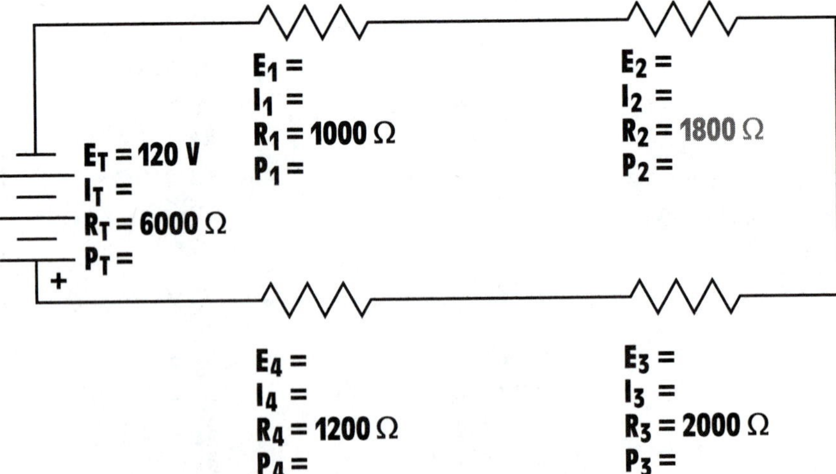

Figure 5-13 The missing resistor value.

$$R_2 = 6000 - 4200$$

$$R_2 = 1800\ \Omega$$

The amount of current flow in the circuit can be found using Ohm's Law.

$$I = \frac{E}{R}$$

$$I = \frac{120}{6000}$$

$$I = 0.020\ A$$

Since the amount of current flow is the same through all elements of a series circuit, *Figure 5-14,* the voltage drop across each resistor can be found using Ohm's Law, *Figure 5-15.*

$$E_1 = I_1 \times R_1$$

$$E_1 = 0.020 \times 1000$$

$$E_1 = 20\ V$$

$$E_2 = I_2 \times R_2$$

$$E_2 = 0.020 \times 1800$$

$$E_2 = 36\ V$$

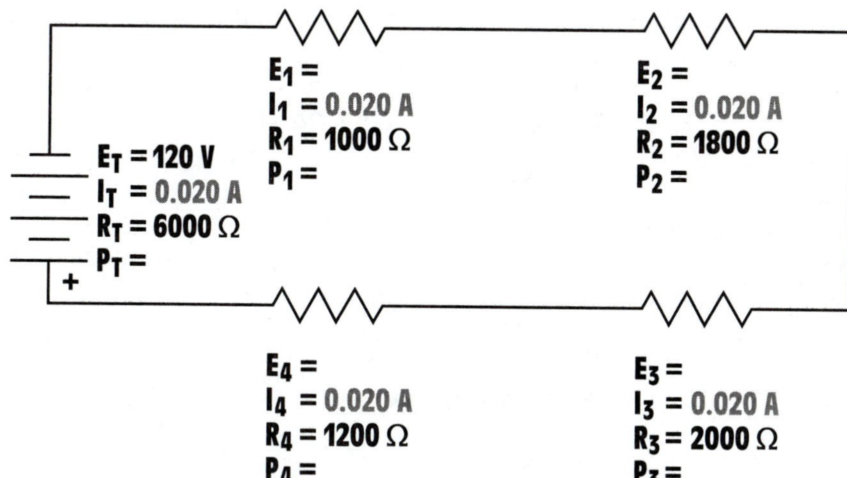

Figure 5-14 The current is the same through each circuit element.

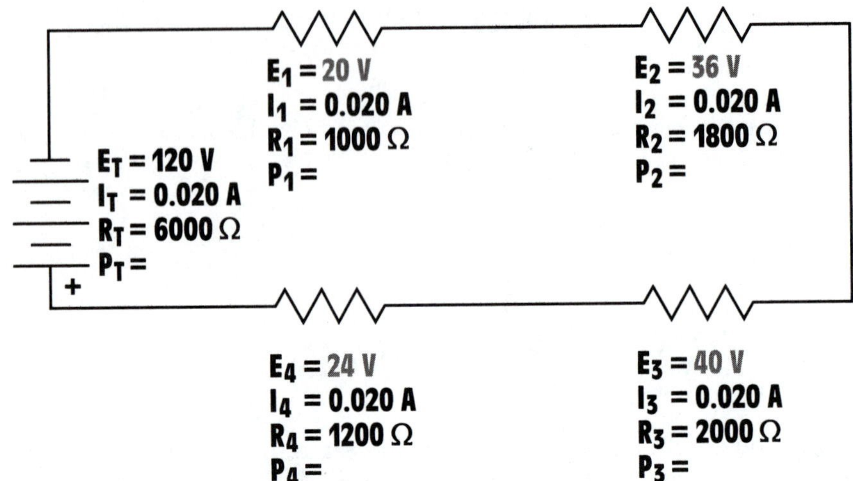

Figure 5-15 The voltage drops across each resistor.

$$E_3 = I_3 \times R_3$$

$$E_3 = 0.020 \times 2000$$

$$E_3 = 40 \text{ V}$$

$$E_4 = I_4 \times R_4$$

$$E_4 = 0.020 \times 1200$$

$$E_4 = 24 \text{ V}$$

The first rule of series circuits can be used to check the answers.

$$E_T = E_1 + E_2 + E_3 + E_4$$

$$E_T = 20 + 36 + 40 + 24$$

$$E_T = 120 \text{ V}$$

The amount of power dissipation for each resistor in the circuit can be computed in the same manner as in circuit one. The power dissipated by resistor R_1 will be computed using the formula:

$$P_1 = E_1 \times I_1$$

$$P_1 = 20 \times 0.02$$

$$P_1 = 0.4 \text{ W}$$

The amount of power dissipation for resistor R_2 will be found by using the formula:

$$P_2 = \frac{E_2^2}{R_2}$$

$$P_2 = \frac{1296}{1800}$$

$$P_2 = 0.72\ W$$

The power dissipation of resistor R_3 will be found using the formula:

$$P_3 = I_3^2 \times R_3$$

$$P_3 = 0.0004 \times 2000$$

$$P_3 = 0.8\ W$$

The power dissipation of resistor R_4 will be computed using the formula:

$$P_4 = E_4 \times I_4$$

$$P_4 = 24 \times 0.02$$

$$P_4 = 0.48\ W$$

The total power will be computed using the formula:

$$P_T = E_T \times I_T$$

$$P_T = 120 \times 0.02$$

$$P_T = 2.4\ W$$

The circuit with all computed values is shown in *Figure 5-16.*

Example 3

In the circuit shown in *Figure 5-17,* resistor R_1 has a voltage drop of 6.4 V, resistor R_2 has a power dissipation of 0.102 W, resistor R_3 has a power dissipation of 0.154 W, resistor R_4 has a power dissipation of 0.307 W, and the total power consumed by the circuit is 0.768 W.

The only value that can be found with the given quantities is the amount of power dissipated by resistor R_1. Since the total power is known and the power dissipated by all other resistors is known, the power dissipated by resistor R_1 can be found by subtracting the power dissipated by resistors R_2, R_3, and R_4 from the total power used in the circuit.

$$P_1 = P_T - (P_2 + P_3 + P_4)$$

or

$$P_1 = P_T - P_2 - P_3 - P_4$$

$$P_1 = 0.768 - 0.102 - 0.154 - 0.307$$

$$P_1 = 0.205\ W$$

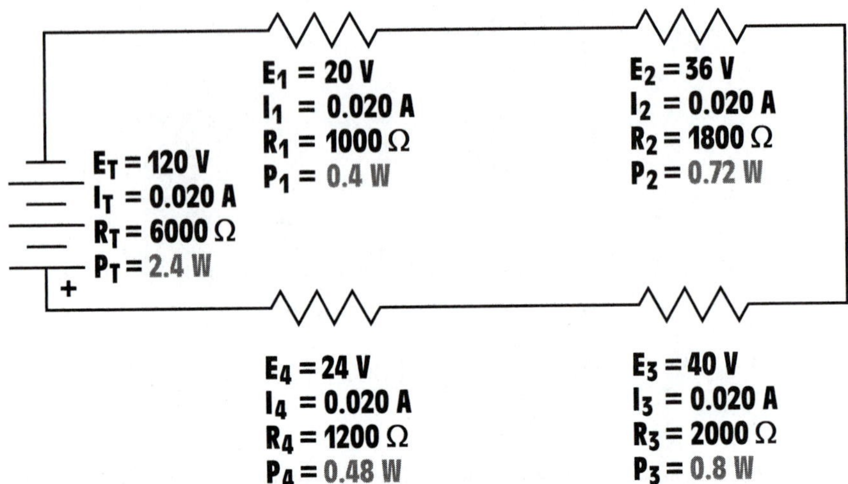

Figure 5-16 The remaining unknown values for the circuit in Example 2.

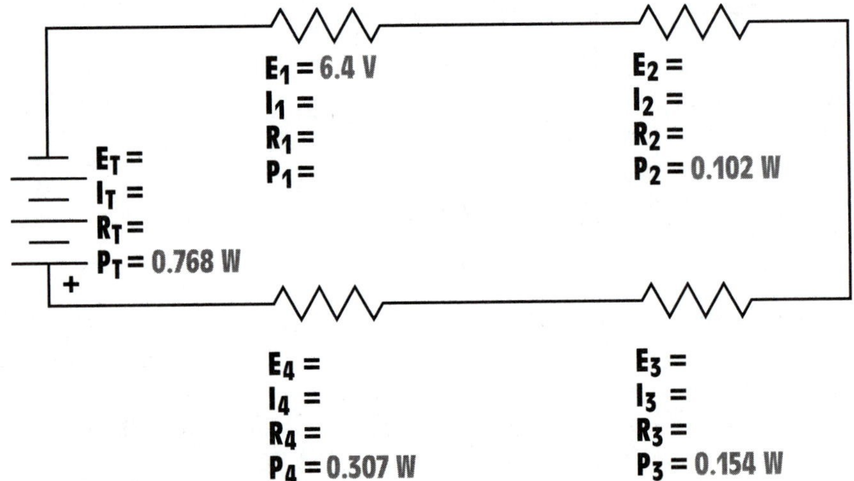

Figure 5-17 Series circuit, Example 3.

Now that the amount of power dissipated by resistor R_1 and the amount of voltage dropped across R_1 are known, the current flow through resistor R_1 can be found using the formula:

$$I = \frac{P}{E}$$

$$I = \frac{0.205}{6.4}$$

$$I = 0.032 \text{ A}$$

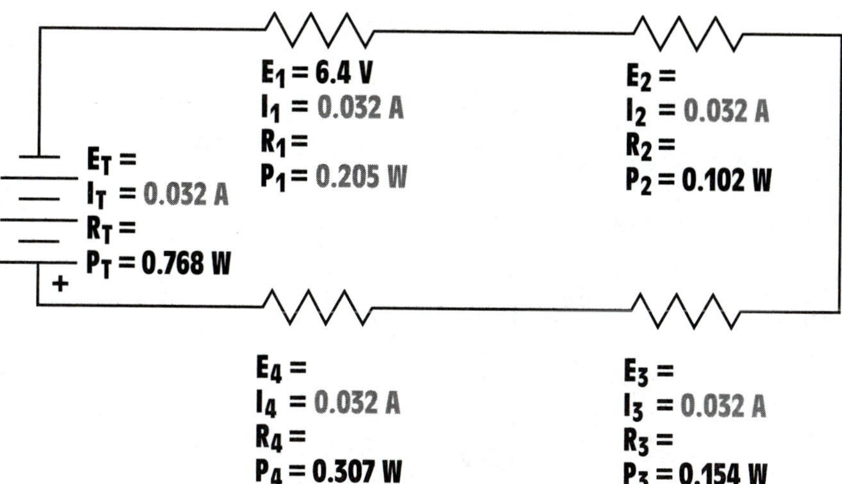

Figure 5-18 The current flow in the circuit in Example 3.

Since the current in a series circuit must be the same at any point in the circuit, it must be the same through all circuit components, *Figure 5-18.*

Now that the power dissipation of each resistor and the amount of current flowing through each resistor are known, the voltage drop of each resistor can be computed, *Figure 5-19.*

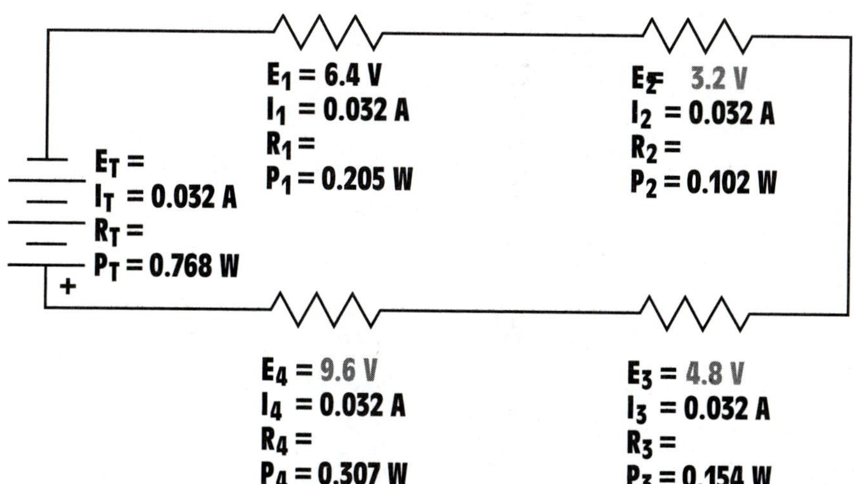

Figure 5-19 The voltage drops across each resistor.

$$E_2 = \frac{P_2}{I_2}$$

$$E_2 = \frac{0.102}{0.032}$$

$$E_2 = 3.2 \text{ V}$$

$$E_3 = \frac{P_3}{I_3}$$

$$E_3 = \frac{0.154}{0.032}$$

$$E_3 = 4.8 \text{ V}$$

$$E_4 = \frac{P_4}{I_4}$$

$$E_4 = \frac{0.307}{0.032}$$

$$E_4 = 9.6 \text{ V}$$

Ohm's Law can now be used to find the ohmic value of each resistor in the circuit, *Figure 5-20*.

$$R_1 = \frac{E_1}{I_1}$$

$$R_1 = \frac{6.4}{0.032}$$

$$R_1 = 200 \ \Omega$$

$$R_2 = \frac{E_2}{I_2}$$

$$R_2 = \frac{3.2}{0.032}$$

$$R_2 = 100 \ \Omega$$

$$R_3 = \frac{E_3}{I_3}$$

$$R_3 = \frac{4.8}{0.032}$$

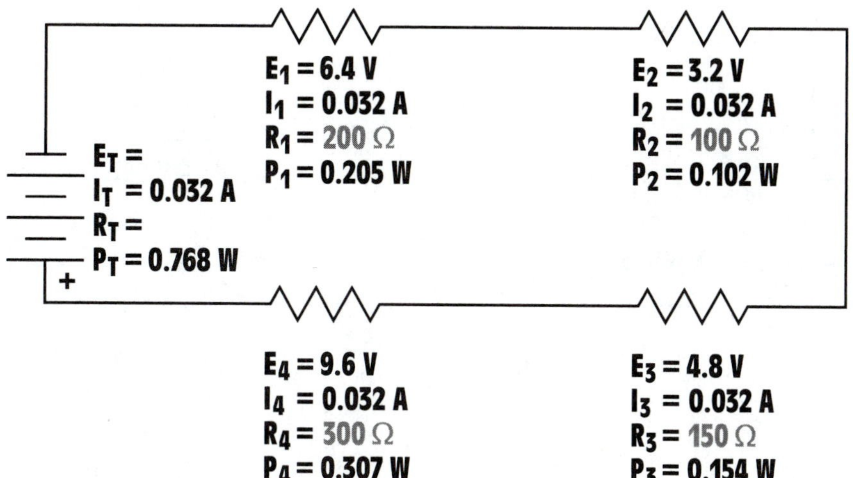

Figure 5-20 The ohmic value of each resistor.

$$R_3 = 150\ \Omega$$

$$R_4 = \frac{E_4}{I_4}$$

$$R_4 = \frac{9.6}{0.032}$$

$$R_4 = 300\ \Omega$$

The voltage applied to the circuit can be found by adding the voltage drops across each resistor, *Figure 5-21*.

$$E_T = E_1 + E_2 + E_3 + E_4$$

$$E_T = 6.4 + 3.2 + 4.8 + 9.6$$

$$E_T = 24\ \text{V}$$

The total resistance of the circuit can be found in a similar manner. The total resistance is equal to the sum of all the resistive elements in the circuit.

$$R_T = R_1 + R_2 + R_3 + R_4$$

$$R_T = 200 + 100 + 150 + 300$$

$$R_T = 750\ \Omega$$

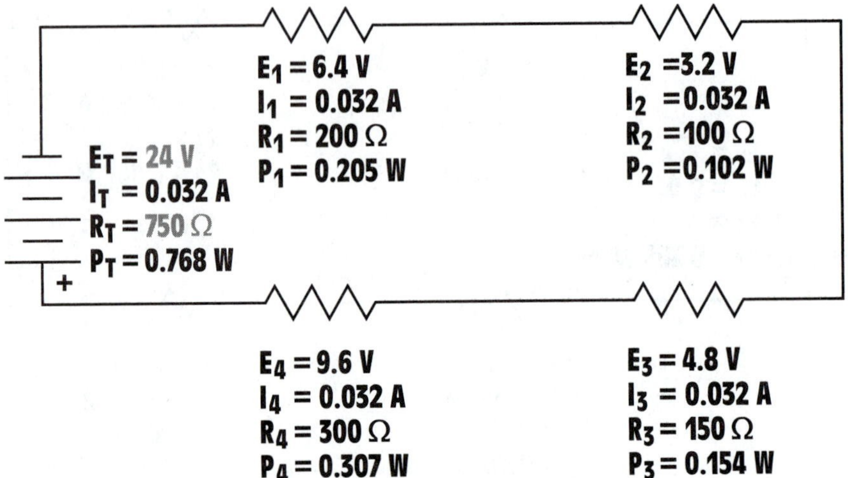

Figure 5-21 The applied voltage and the total resistance.

VOLTAGE DIVIDERS

voltage divider

One common use for series circuits is the construction of voltage dividers. A **voltage divider** works on the principle that the sum of the voltage drops across a series circuit must equal the applied voltage. Voltage dividers are used to provide different voltages across certain points, *Figure 5-22*. If a voltmeter is connected between points A and B, a voltage of 20 V will be seen. If the voltmeter is connected between points B and D, a voltage of 80 V will be seen.

Voltage dividers can be constructed to produce any voltage desired. For example, assume that a voltage divider is connected to a source of 120 V and is to provide voltage drops of 36 V, 18 V, and 66 V. Notice that the sum of the voltage drops equals the applied voltage. The next step is to decide how much current is to flow through the circuit. Since there is only one path for current flow, the current will be the same through each resistor. In this circuit, a current flow of 15 mA (0.015 A) will be used. The resistance value of each resistor can now be determined.

$$R = \frac{E}{I}$$

$$R_1 = \frac{36}{0.015}$$

$$R_1 = 2.4\ \text{k}\Omega\ (2400\ \Omega)$$

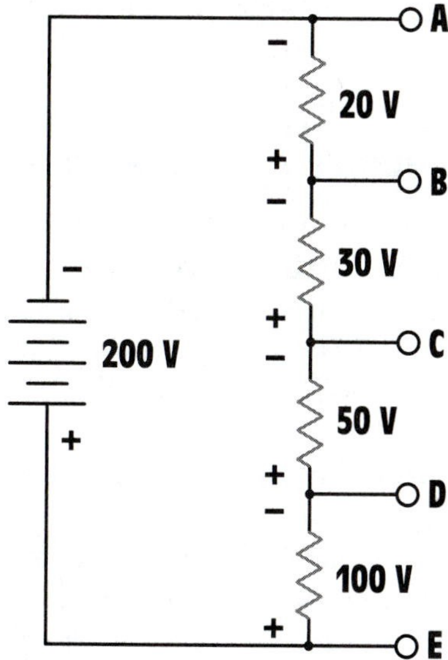

Figure 5-22 Series circuit used as a voltage divider.

$$R_2 = \frac{18}{0.015}$$

$$R_2 = 1.2\ \text{k}\Omega\ (1200\ \Omega)$$

$$R_3 = \frac{66}{0.015}$$

$$R_3 = 4.4\ \text{k}\Omega\ (4400\ \Omega)$$

VOLTAGE POLARITY

voltage polarity

It is often necessary to know the polarity of the voltage developed across a resistor. **Voltage polarity** can be determined by observing the direction of current flow through the circuit. In the circuit shown in Figure 5-22 it will be assumed that the current flows from the negative terminal of the battery to the positive terminal. Point A is connected to the negative battery terminal and point E is connected to the positive terminal. If a voltmeter is connected across terminals A and B, terminal B will be positive with respect to A. If a voltmeter is connected across terminals B and C, however, terminal B will be negative with respect to terminal C. Notice that terminal B is located closer to the negative terminal of the battery. This means that electrons flow through the resistor in a direction that makes terminal B more negative than C. Terminal C would be negative with respect to terminal D for the same reason.

USING GROUND AS A REFERENCE

earth ground

ground point

chassis ground

Two symbols are used to represent ground, *Figure 5-23*. The symbol shown in Figure 5-23A is an **earth ground** symbol. It symbolizes a **ground point** that is made by physically driving an object, such as a rod or pipe, into the ground. The symbol shown in Figure 5-23B symbolizes a **chassis ground.** This is a point that is used as a common connection for other parts of a circuit but is not actually driven into the ground. Although the symbol shown in Figure 5-23B is the accepted symbol for a chassis ground, the symbol shown in Figure 5-23A is also often used to represent a chassis ground.

An excellent example of using ground as a common connection can be found in the electrical system of an automobile. The negative terminal of the battery is grounded to the frame or chassis of the vehicle. This does not mean that the frame of the automobile is connected directly to earth ground. The chassis is insulated from the ground by rubber tires. In the case of an automobile electrical system, the chassis of the vehicle is the negative side of the circuit. An electrical circuit using ground as a common connection point is shown in *Figure 5-24*. This circuit is an electronic burglar alarm. Notice the numerous ground points in the schematic. In practice, when the circuit is connected, all the ground points will be connected.

In voltage divider circuits, ground is often used to provide a common reference point to produce voltages that are above and below ground, *Figure 5-25*. An above-ground voltage is a voltage that is positive with respect to ground. A below-ground voltage is negative with respect to ground. In Figure 5-25, one terminal of a zero-center voltmeter is connected to ground. If the probe is connected to point A, the pointer of the voltmeter will give a negative indication for voltage. If the probe is connected to point B, the pointer will indicate a positive voltage.

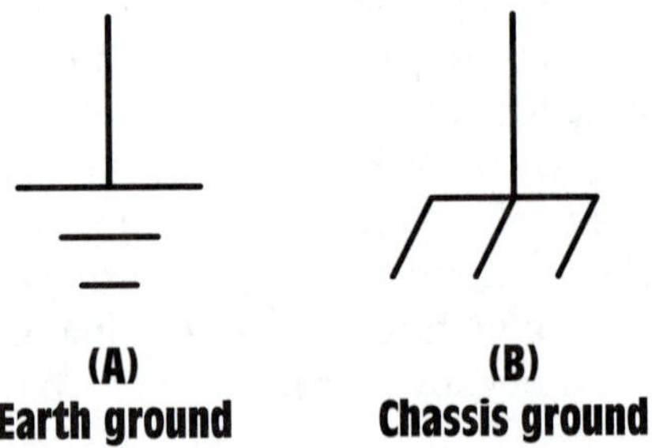

Figure 5-23 Ground symbols.

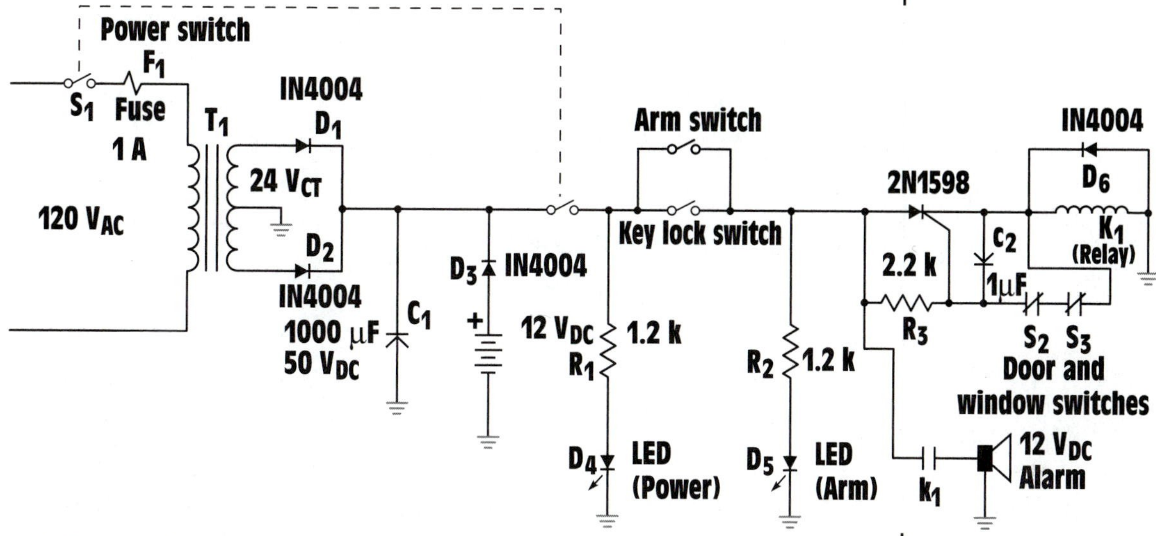

Figure 5-24 Burglar alarm with battery back-up.

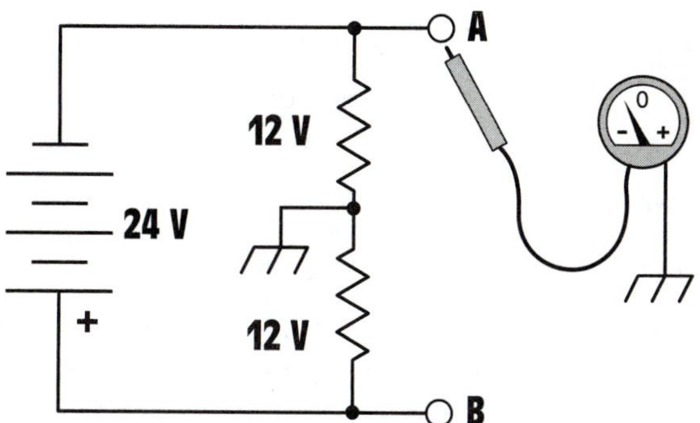

Figure 5-25 A common ground used to produce above- and below-ground voltage.

SUMMARY

1. Series circuits have only one path for current flow.
2. The individual voltage drops in a series circuit can be added to equal the applied voltage.
3. The current is the same at any point in a series circuit.
4. The individual resistors can be added to equal the total resistance of the circuit.

5. Fuses and circuit breakers are connected in series with the devices they are intended to protect.
6. The total power in any circuit is equal to the sum of the power dissipated by all parts of the circuit.

REVIEW QUESTIONS

1. A series circuit has individual resistor values of 200 Ω, 86 Ω, 91 Ω, 180 Ω, and 150 Ω. What is the total resistance of the circuit?
2. A series circuit contains four resistors. The total resistance of the circuit is 360 Ω. Three of the resistors have values of 56 Ω, 110 Ω, and 75 Ω. What is the value of the fourth resistor?
3. A series circuit contains five resistors. The total voltage applied to the circuit is 120 V. Four resistors have voltage drops of 35 V, 28 V, 22 V and 15 V. What is the voltage drop of the fifth resistor?
4. A circuit has three resistors connected in series. The resistance of resistor R_2 is 220 Ω, and it has a voltage drop of 44 V. What is the current flow through resistor R_3?
5. A circuit has four resistors connected in series. If each resistor has a voltage drop of 60 V, what is the voltage applied to the circuit?
6. Define a series circuit.
7. State the three rules for series circuits.
8. A series circuit has resistance values of 160 Ω, 100 Ω, 82 Ω, and 120 Ω. What is the total resistance of this circuit?
9. If a voltage of 24 V is applied to the circuit in question 8, what will be the total amount of current flow in the circuit?
10. What will be the voltage drop across each of the resistors?

    A. 160 Ω____________V

    B. 100 Ω____________V

    C. 82 Ω____________V

    D. 120 Ω____________V

Practice Problems

CIRCUITS

1. Find the missing values in the circuit shown in *Figure 5-26.*
2. Find the missing values in the circuit shown in *Figure 5-27.*

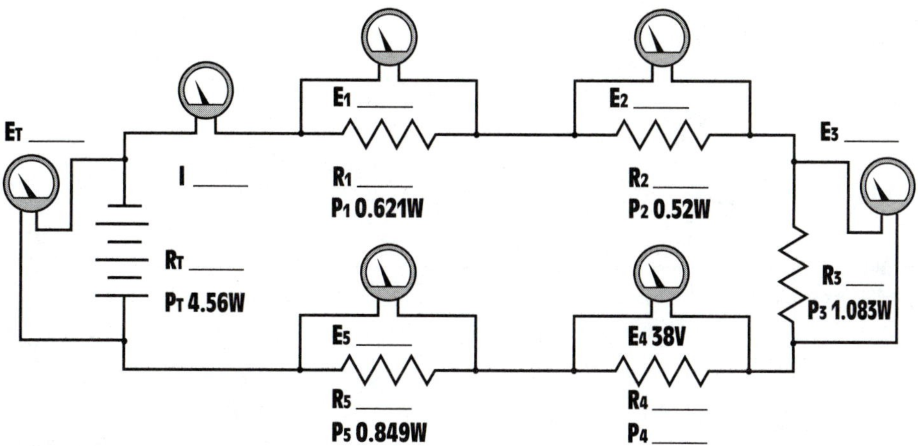

Figure 5-26 Practice Problem 1.

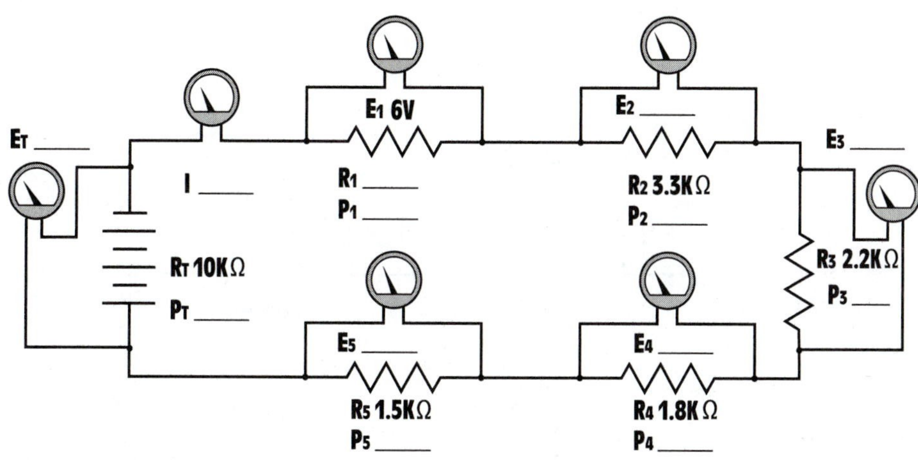

Figure 5-27 Practice Problem 2.

3. Find the missing values in the circuit shown in *Figure 5-28.*
4. Find the missing values in the circuit shown in *Figure 5-29.*
5. Find the missing values in the circuit shown in *Figure 5-30.*

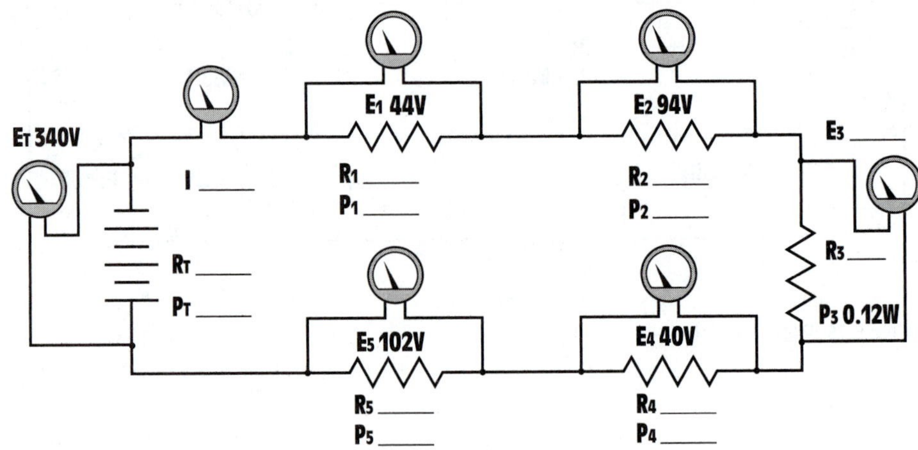

Figure 5-28 Practice Problem 3.

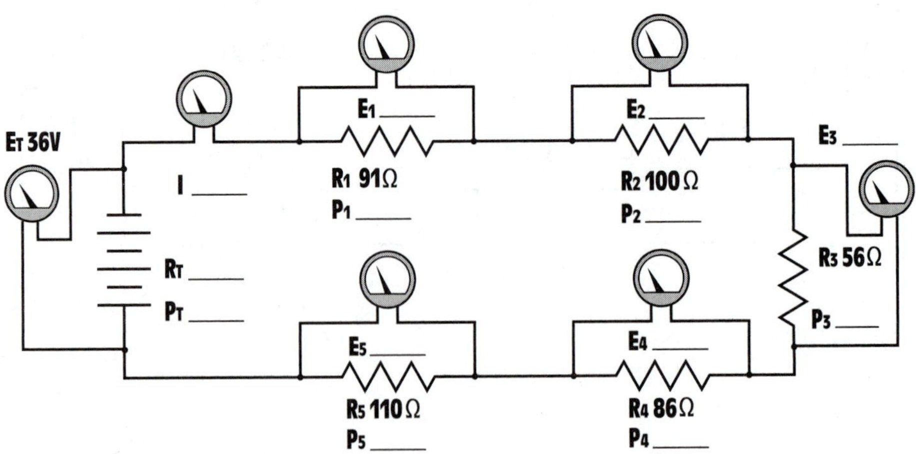

Figure 5-29 Practice Problem 4.

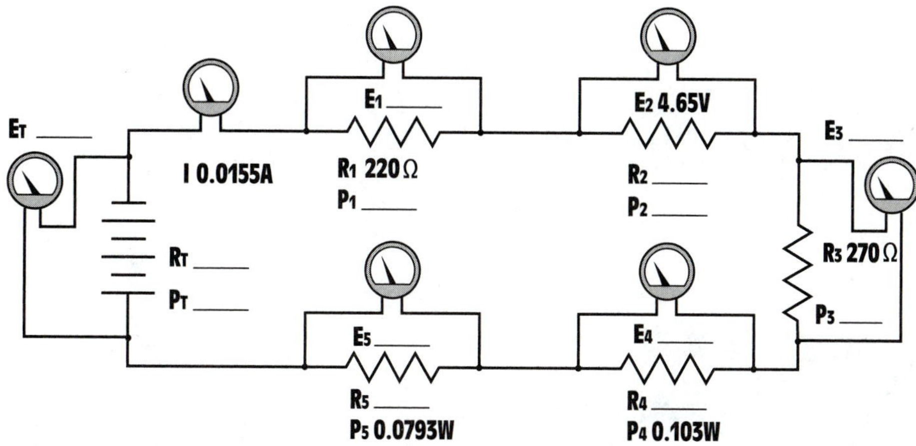

Figure 5-30 Practice Problem 5.

6 Unit

Parallel Circuits

objectives

fter studying this unit, you should be able to:

- Discuss the characteristics of a parallel circuit.
- State three rules for parallel circuits.
- Solve the missing values in a parallel circuit using the three rules and Ohm's Law.

parallel circuits

Parallel circuits are probably the type of circuit with which most people are familiar. Most devices such as lights and receptacles in homes and office buildings are connected in parallel. Imagine if the lights in your home were wired in series. All the lights in the home would have to be turned on in order for any light to operate, and if one were to burn out, all the lights would go out. The same is true for receptacles. If receptacles were connected in series, some device would have to be connected into each receptacle before power could be supplied to any other device.

Parallel circuits are circuits that have more than one path for current flow.

PARALLEL CIRCUIT VALUES

TOTAL CURRENT

Parallel circuits are circuits that have more than one path for current flow, *Figure 6-1*. If it is assumed that current leaves terminal A and

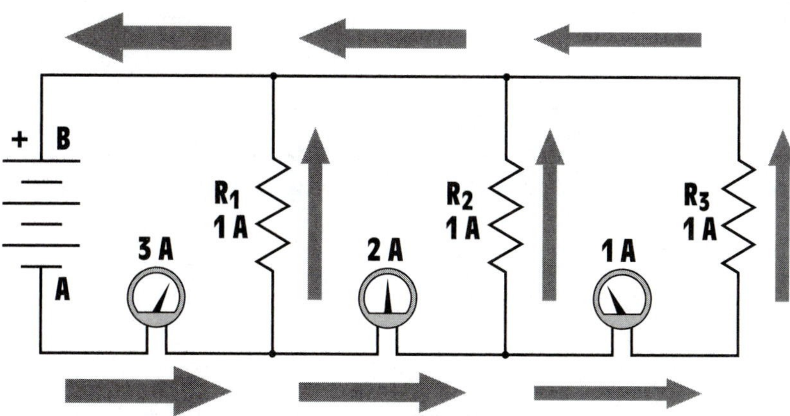

Figure 6-1 Parallel circuits provide more than one path for current flow.

returns to terminal B, it can be seen that there are three separate paths that can be taken by the electrons. In Figure 6-1, 3 A of current leaves terminal A. One amp flows through resistor R_1 and 2 A flows to resistors R_2 and R_3. At the junction of resistors R_2 and R_3, 1 A flows through resistor R_2 and 1 A flows to resistor R_3. Notice that the power supply, terminals A and B, must furnish all the current that flows through each individual resistor or **circuit branch**. One of the rules for parallel circuits states that **the total current flow in the circuit is equal to the sum of the currents through all of the branches. This is known as current adds.**

circuit branch

The total current flow in the circuit is equal to the sum of the currents through all of the branches; this is known as current adds.

VOLTAGE DROP

Figure 6-2 shows another parallel circuit and gives the values of voltage and current for each individual resistor or branch. Notice that the voltage drop across each resistor is the same. If the circuit is traced, it can be seen that each resistor is connected directly to the power source. A second rule for parallel circuits states that **the voltage drop across any branch of a parallel circuit is the same as the applied voltage.** For this reason, most electrical circuits in homes are connected in parallel. Each lamp and receptacle is supplied with 120 V, *Figure 6-3*.

The voltage drop across any branch of a parallel circuit is the same as the applied voltage.

TOTAL RESISTANCE

In the circuit shown in *Figure 6-4,* three separate resistors have values of 15 Ω, 10 Ω, and 30 Ω. The total resistance of the circuit, however, is 5 Ω. **The total resistance of a parallel circuit is always less than the resistance of the lowest-value resistor, or branch, in the circuit.** This is because there is more than one path for current flow. Each time another

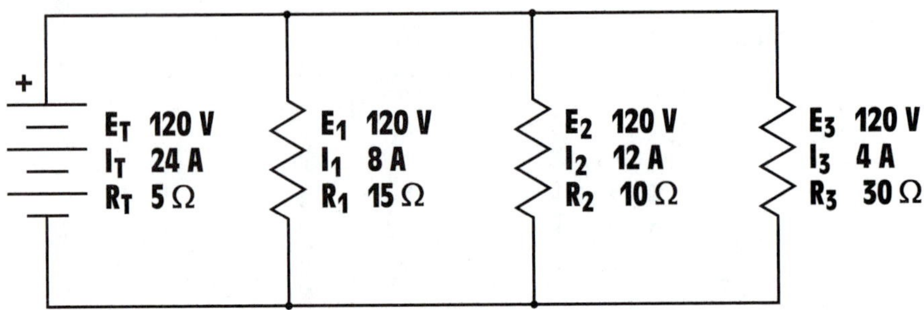

Figure 6-2 Parallel circuit values.

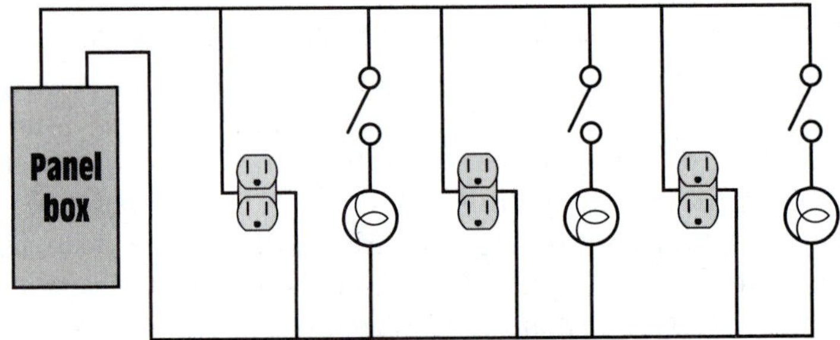

Figure 6-3 Lights and receptacles are connected in parallel.

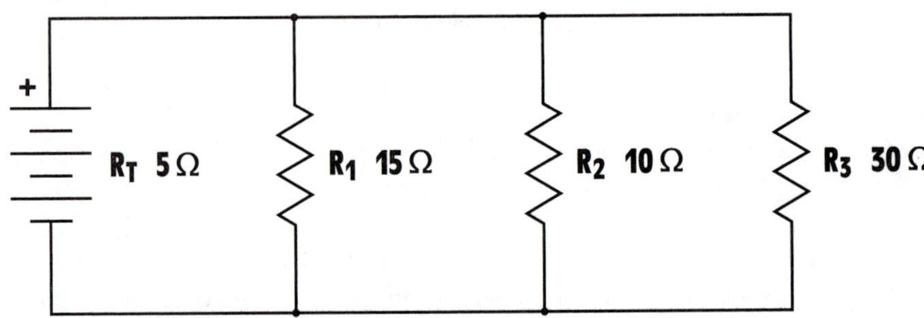

Figure 6-4 Total resistance is always less than the resistance of any single branch.

The total resistance of a parallel circuit is always less than the resistance of the lowest-value resistor, or branch, in the circuit.

element is connected in parallel, there is less opposition to the flow of current through the entire circuit. Imagine a water system consisting of a holding tank, a pump, and return lines to the tank, *Figure 6-5.* Although large return pipes have less resistance to the flow of water than small pipes, the small pipes do provide a return path to the holding tank. Each time another return path is added, regardless of size, there is less resistance to flow and the rate of flow increases.

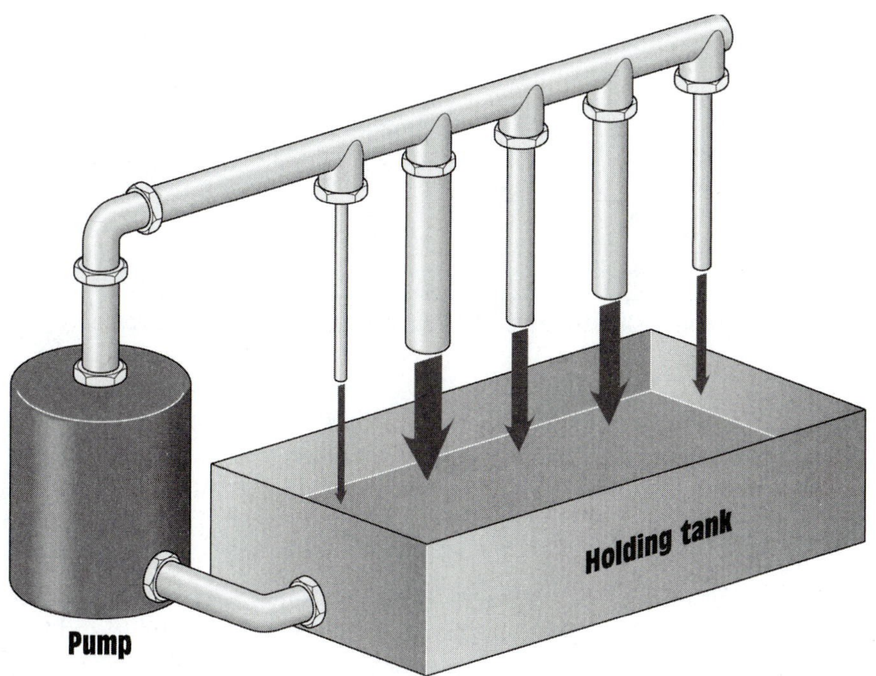

Figure 6-5 Each new path reduces the total resistance to the flow of water.

This concept often causes confusion concerning the definition of **load** among students of electricity. Students often think that an increase of resistance constitutes an increase of load. The reason for this is that in a laboratory exercise, students often see the circuit current increase each time a resistive element is connected to the circuit. They conclude that an increase of resistance must, therefore, cause an increase of current. This conclusion is, of course, completely contrary to Ohm's Law, which states that an increase of resistance must cause a proportional decrease of current. The false concept that an increase of resistance causes an increase of current can be overcome once the student understands that if the resistive elements are being connected in parallel, the circuit resistance is actually being decreased and not increased.

load

When all resistors are of equal value, the total resistance is equal to the value of one individual resistor, or branch, divided by the number (N) of resistors, or branches.

PARALLEL RESISTANCE FORMULAS

RESISTORS OF EQUAL VALUE

Three formulas can be used to determine the total resistance of a parallel circuit. The first formula shown can be used only when all the resistors in the circuit are of equal value. This formula states that **when all resistors are of equal value, the total resistance is equal to the value**

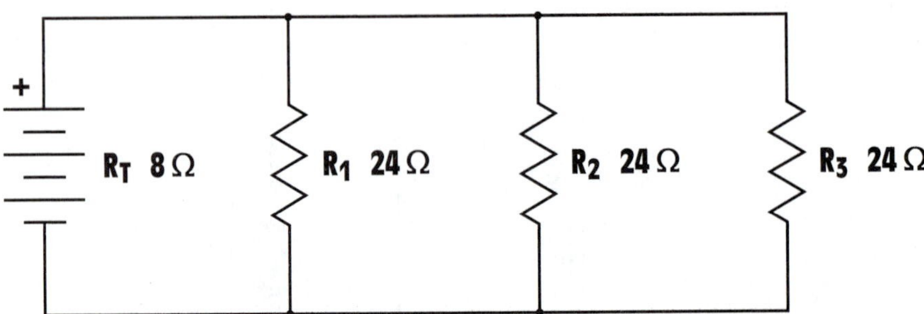

Figure 6-6 Finding the total resistance when all resistors have the same value.

of one individual resistor, or branch, divided by the number (N) of resistors, or branches.

$$R_T = \frac{R}{N}$$

Assume that three resistors, each having a value of 24 Ω are connected in parallel, *Figure 6-6*. The total resistance of this circuit can be found by dividing the resistance of one single resistor by the total number of resistors.

$$R_T = \frac{R}{N}$$

$$R_T = \frac{24}{3}$$

$$R_T = 8\ \Omega$$

Product over Sum

The second formula used to determine the total resistance in a parallel circuit divides the product of two resistors by their sum. This is commonly referred to as the **product over sum method.**

product over sum method

$$R_T = \frac{R_1 \times R_2}{R_1 + R_2}$$

In the circuit shown in *Figure 6-7,* three branches having single resistors with values of 20 Ω, 30 Ω, and 60 Ω are connected in parallel. To find the total resistance of the circuit using the previous formula, find the total resistance of the first two branches in the circuit, *Figure 6-8.*

$$R_T = \frac{R_2 \times R_3}{R_2 + R_3}$$

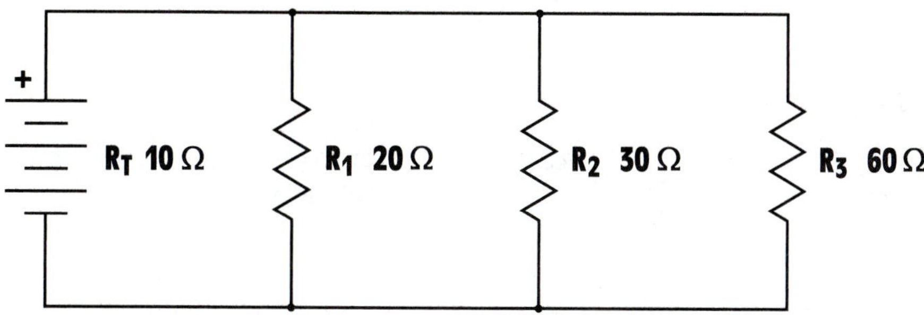

Figure 6-7 Finding the total resistance of a parallel circuit by dividing the product of two resistors by their sum.

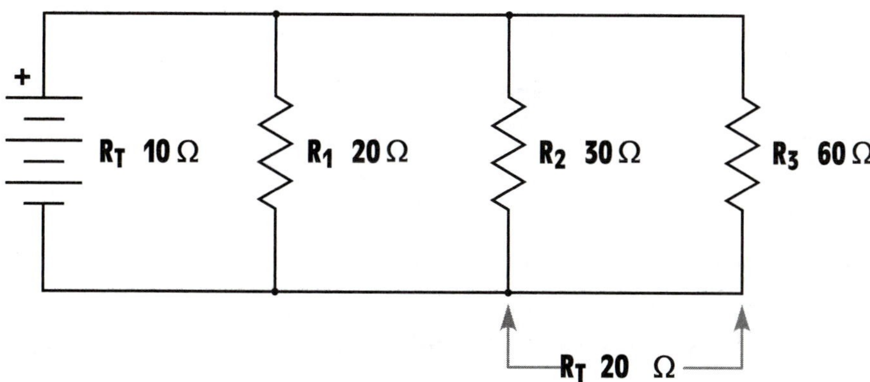

Figure 6-8 Finding the total resistance of the first two branches.

$$R_T = \frac{30 \times 60}{30 + 60}$$

$$R_T = \frac{1800}{90}$$

$$R_T = 20\ \Omega$$

The total resistance of the last two resistors in the circuit is 20 Ω. This 20 Ω, however, is connected in parallel with a 20-Ω resistor. The total resistance of the last two resistors is now used to substitute for the value of R_2 in the formula, and the value of the next resistor is used to substitute for the value of R_1, *Figure 6-9*.

$$R_T = \frac{R_1 \times R_2}{R_1 + R_2}$$

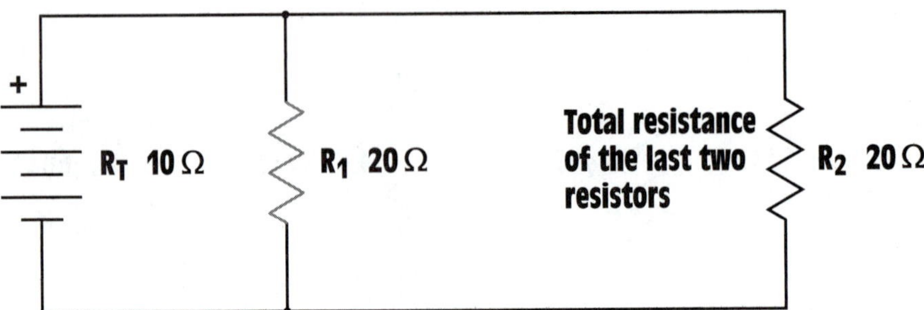

Figure 6-9 The total value of the last two resistors is used as resistor 2.

$$R_T = \frac{20 \times 20}{20 + 20}$$

$$R_T = \frac{400}{40}$$

$$R_T = 10\ \Omega$$

RECIPROCAL FORMULA

reciprocal formula

The total resistance of a parallel circuit is the reciprocal of the sum of the reciprocals of the individual branches.

The third formula used to find the total resistance of a parallel circuit is often referred to as the **reciprocal formula.**

$$\frac{1}{R_T} = \frac{1}{R_1} + \frac{1}{R_2} + \frac{1}{R_3} + \frac{1}{R_N}$$

Notice that this formula actually finds the reciprocal of the total resistance, instead of the total resistance. To make the formula equal to the total resistance it can be rewritten as follows:

$$R_T = \frac{1}{\frac{1}{R_1} + \frac{1}{R_2} + \frac{1}{R_3} + \frac{1}{R_N}}$$

The value R_N stands for R number and means the number of resistors in the circuit. If the circuit has twenty-five resistors connected in parallel, for example, the last resistor in the formula would be R_{25}.

This formula is referred to as the reciprocal formula because the reciprocal of any number is that number divided into 1. The reciprocal of 4, for example, is 0.25 because 1 / 4 = 0.25. A third rule of parallel circuits is **the total resistance of a parallel circuit is the reciprocal of the sum of the reciprocals of the individual branches.**

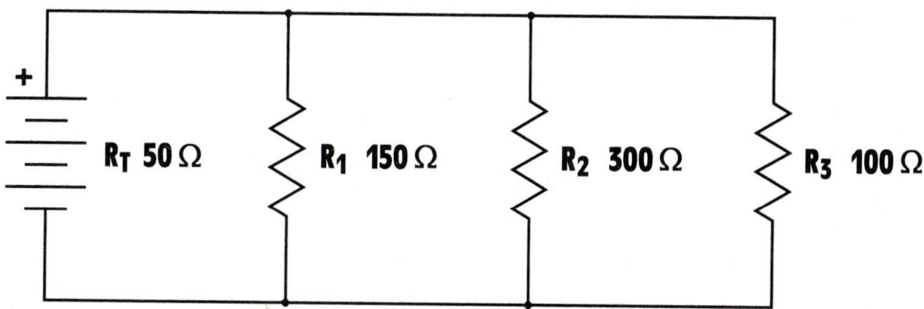

Figure 6-10 Finding the total resistance using the reciprocal method.

Before the invention of hand-held calculators, the slide rule was often employed to help with the mathematical calculations in electrical work. At that time, the product over sum method of finding total resistance was the most popular. Since the invention of calculators, however, the reciprocal formula has become the most popular because scientific calculators have a reciprocal key (1/X), which makes computing total resistance using the reciprocal method very easy.

In *Figure 6-10,* three resistors having values of 150 Ω, 300 Ω, and 100 Ω are connected in parallel. The total resistance will be found using the reciprocal formula.

$$R_T = \frac{1}{\frac{1}{R_1} + \frac{1}{R_2} + \frac{1}{R_3} + \frac{1}{R_N}}$$

$$R_T = \frac{1}{\frac{1}{150} + \frac{1}{300} + \frac{1}{100}}$$

$$R_T = \frac{1}{0.006667 + 0.003333 + 0.01}$$

$$R_T = \frac{1}{0.02}$$

$$R_T = 50\ \Omega$$

SOLVING CIRCUIT VALUES

Example 1

In the circuit shown in *Figure 6-11,* three resistors having values of 300 Ω, 200 Ω, and 600 Ω are connected in parallel. Total current flow through the circuit is 0.6 A. Find all of the missing values in the circuit.

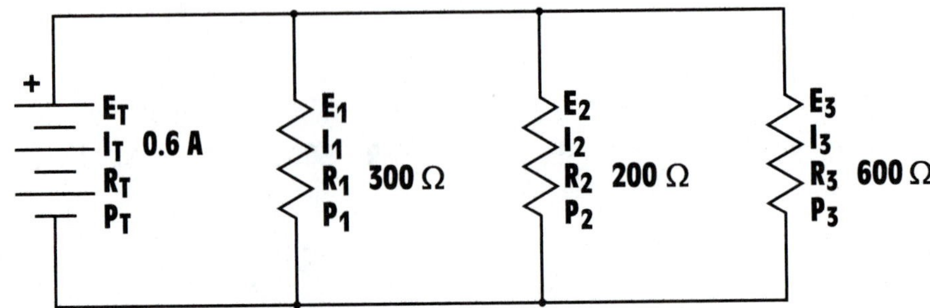

Figure 6-11 Parallel circuit, Example 1.

Solution:

The first step is to find the total resistance of the circuit. The reciprocal formula will be used.

$$R_T = \frac{1}{\frac{1}{R_1} + \frac{1}{R_2} + \frac{1}{R_3} + \frac{1}{R_N}}$$

$$R_T = \frac{1}{\frac{1}{300} + \frac{1}{200} + \frac{1}{600}}$$

$$R_T = \frac{1}{0.003333 + 0.005 + 0.001667}$$

$$R_T = \frac{1}{0.01}$$

$$R_T = 100\ \Omega$$

Now that the total resistance of the circuit is known, the voltage applied to the circuit can be found by using the total current value and Ohm's Law.

$$E_T = I_T \times R_T$$

$$E_T = 0.6 \times 100$$

$$E_T = 60\ \text{V}$$

One of the rules for parallel circuits states that the voltage across all of the parts of a parallel circuit are the same as the total voltage. Therefore, the voltage drop across each resistor is 60 V, *Figure 6-12*.

$$E_T = E_1 = E_2 = E_3$$

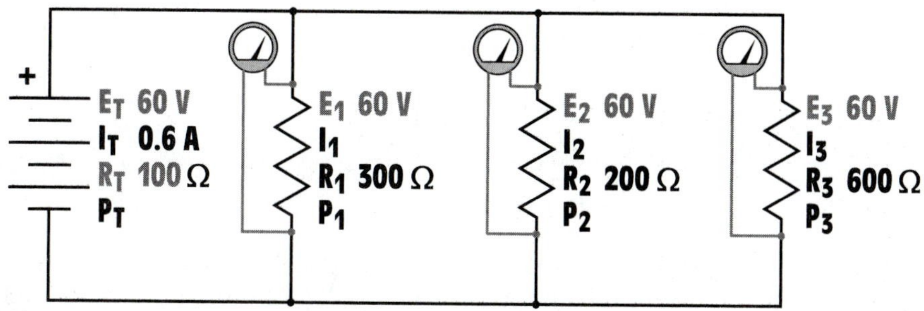

Figure 6-12 The voltage is the same across all branches of a parallel circuit.

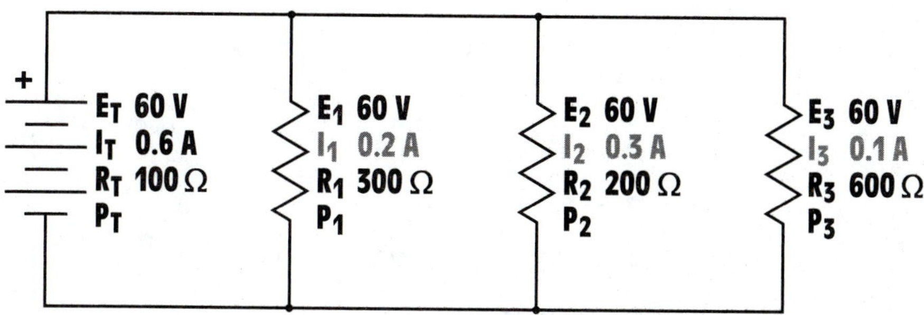

Figure 6-13 Ohm's Law is used to compute the amount of current through each branch.

Since the voltage drop and resistance of each resistor is known, Ohm's Law can be used to determine the amount of current flow through each resistor, *Figure 6-13*.

$$I_1 = \frac{E_1}{R_1}$$

$$I_1 = \frac{60}{300}$$

$$I_1 = 0.2\text{ A}$$

$$I_2 = \frac{E_2}{R_2}$$

$$I_2 = \frac{60}{200}$$

$$I_2 = 0.3\text{ A}$$

$$I_3 = \frac{E_3}{R_3}$$

$$I_3 = \frac{60}{600}$$

$$I_3 = 0.1\ A$$

The amount of power (watts) used by each resistor can be found by using Ohm's Law. A different formula will be used to find the amount of electrical energy converted into heat by each of the resistors.

$$P_1 = \frac{E_1^2}{R_1}$$

$$P_1 = \frac{60 \times 60}{300}$$

$$P_1 = 12\ W$$

$$P_2 = I_2^2 \times R_2$$

$$P_2 = 0.3 \times 0.3 \times 200$$

$$P_2 = 18\ W$$

$$P_3 = E_2 \times I_3$$

$$P_3 = 60 \times 0.1$$

$$P_3 = 6\ W$$

In Unit 5, it was stated that the total amount of power in a circuit is equal to the sum of the power used by all the parts. This is true for any type of circuit. Therefore, the total amount of power used by this circuit can be found by taking the sum of the power used by all resistors, *Figure 6-14.*

$$P_T = P_1 + P_2 + P_3$$

$$P_T = 12 + 18 + 6$$

$$P_T = 36\ W$$

Example 2

In the circuit shown in *Figure 6-15*, three resistors are connected in parallel. Two of the resistors have a value of 900 Ω and 1800 Ω. The value of resistor R_2 is unknown. The total resistance of the circuit is 300 Ω. Resistor R_2 has a current flow through it of 0.2 A. Find the missing circuit values.

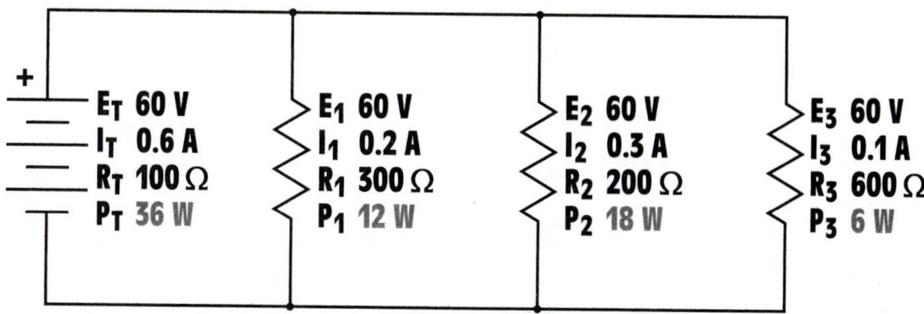

Figure 6-14 Finding the amount of power used by the circuit.

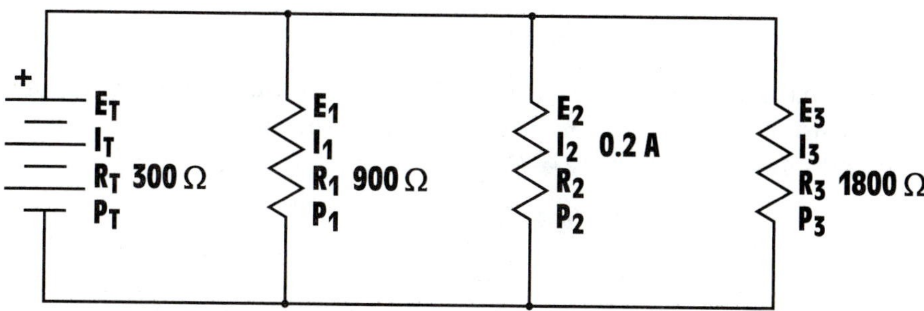

Figure 6-15 Parallel circuit, Example 2.

Solution:

The first step in solving this problem is to find the missing resistor value. This can be done by changing the reciprocal formula as shown:

$$\frac{1}{R_2} = \frac{1}{R_T} - \frac{1}{R_1} - \frac{1}{R_3}$$

or

$$R_2 = \frac{1}{\frac{1}{R_T} - \frac{1}{R_1} - \frac{1}{R_3}}$$

One of the rules for parallel circuits states that the reciprocal of the total resistance is equal to the sum of the reciprocals of the individual resistors. Therefore, the reciprocal of any individual resistor is equal to the difference between the reciprocal of the total resistance and the sum of the reciprocals of the other resistors in the circuit.

$$R_2 = \frac{1}{\frac{1}{R_T} - \frac{1}{R_1} - \frac{1}{R_3}}$$

$$R_2 = \frac{1}{\frac{1}{300} - \frac{1}{900} - \frac{1}{1800}}$$

$$R_2 = \frac{1}{0.003333 - 0.001111 - 0.0005556}$$

$$R_2 = \frac{1}{0.001666}$$

$$R_2 = 600\ \Omega$$

Now that the resistance of resistor R_2 has been found, the voltage drop across resistor R_2 can be determined using the current flow through the resistor and Ohm's Law, *Figure 6-16.*

$$E_2 = I_2 \times R_2$$

$$E_2 = 0.2 \times 600$$

$$E_2 = 120\ \text{V}$$

If 120 volts is dropped across resistor R_2, the same voltage is dropped across each component of the circuit.

$$E_2 = E_T = E_1 = E_3$$

Now that the voltage drop across each part of the circuit is known and the resistance is known, the current flow through each branch can be determined using Ohm's Law, *Figure 6-17.*

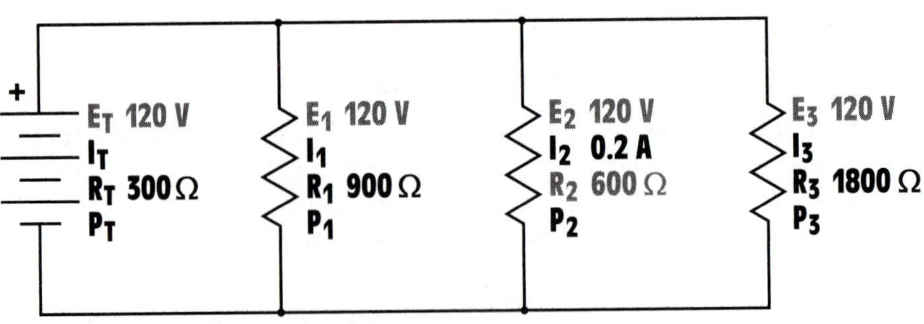

Figure 6-16 Finding the missing resistor value.

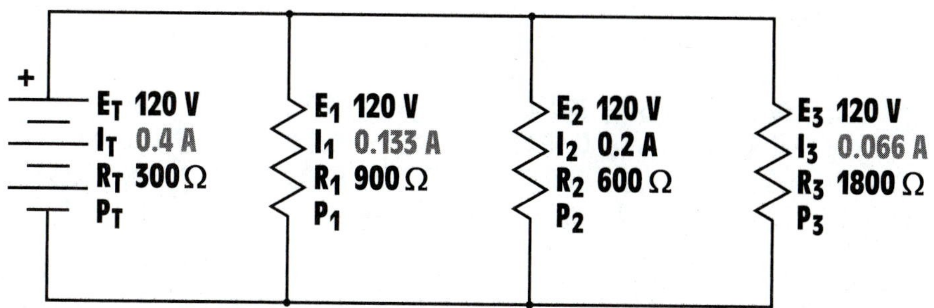

Figure 6-17 Determining the current using Ohm's Law.

$$I_T = \frac{E_T}{R_T}$$

$$I_T = \frac{120}{300}$$

$$I_T = 0.4 \text{ A}$$

$$I_1 = \frac{E_1}{R_1}$$

$$I_1 = \frac{120}{900}$$

$$I_1 = 0.1333 \text{ A}$$

$$I_3 = \frac{E_3}{R_3}$$

$$I_3 = \frac{120}{1800}$$

$$I_3 = 0.0666 \text{ A}$$

The amount of power used by each resistor can be found using Ohm's Law, *Figure 6-18*.

$$P_1 = \frac{{E_1}^2}{R_1}$$

$$P_1 = \frac{120 \times 120}{900}$$

$$P_1 = 16 \text{ W}$$

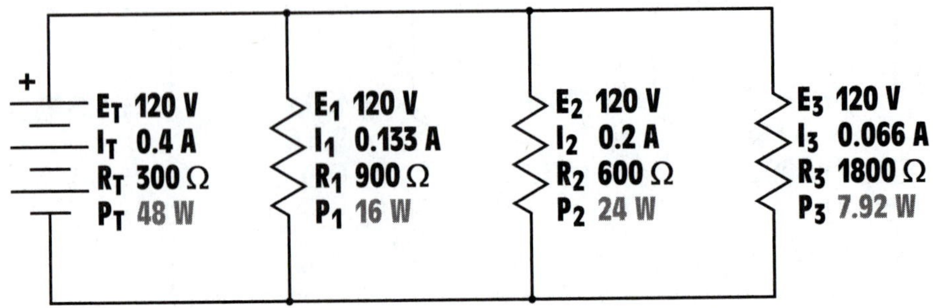

Figure 6-18 Computing values of power for the circuit in Example 2.

$$P_2 = I_2^2 \times R_2$$
$$P_2 = 0.2 \times 0.2 \times 600$$
$$P_2 = 24 \text{ W}$$

$$P_3 = E_3 \times I_3$$
$$P_3 = 120 \times 0.066$$
$$P_3 = 7.92 \text{ W}$$

$$P_T = E_T \times I_T$$
$$P_T = 120 \times 0.4$$
$$P_T = 48 \text{ W}$$

If the wattage values of the three resistors are added to compute total power for the circuit, it will be seen that their total is 47.92 W instead of the computed 48 W. The small difference in answers is caused by the rounding off of other values. In this instance, the current of resistor R_3 was rounded from 0.066666666 to 0.066.

Example 3

In the circuit shown in *Figure 6-19,* three resistors are connected in parallel. Resistor R_1 is producing 0.075 W of heat, R_2 is producing 0.45 W of heat, and R_3 is producing 0.225 W of heat. The circuit has a total current of 0.05 A.

Solution:

Since the amount of power dissipated by each resistor is known, the total power for the circuit can be found by taking the sum of the power used by each component.

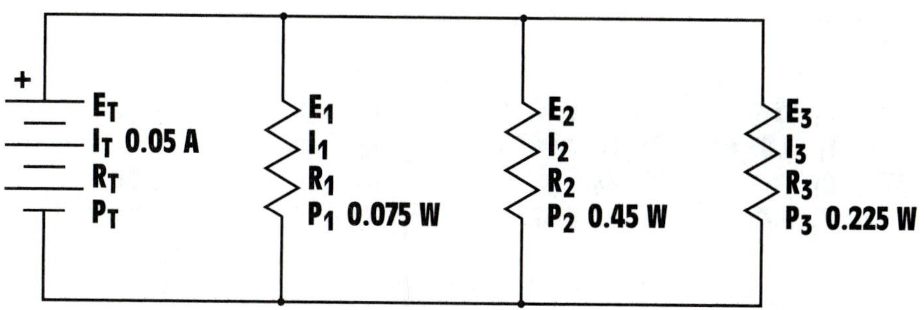

Figure 6-19 Parallel circuit, Example 3.

$$P_T = P_1 + P_2 + P_3$$

$$P_T = 0.075 + 0.45 + 0.225$$

$$P_T = 0.75 \text{ W}$$

Now that the amount of total current and total power for the circuit is known, the applied voltage can be found using Ohm's Law, *Figure 6-20*.

$$E_T = \frac{P_T}{I_T}$$

$$E_T = \frac{0.75}{0.05}$$

$$E_T = 15 \text{ V}$$

The amount of current flow through each resistor can now be found using Ohm's Law, *Figure 6-21*.

$$I_1 = \frac{P_1}{E_1}$$

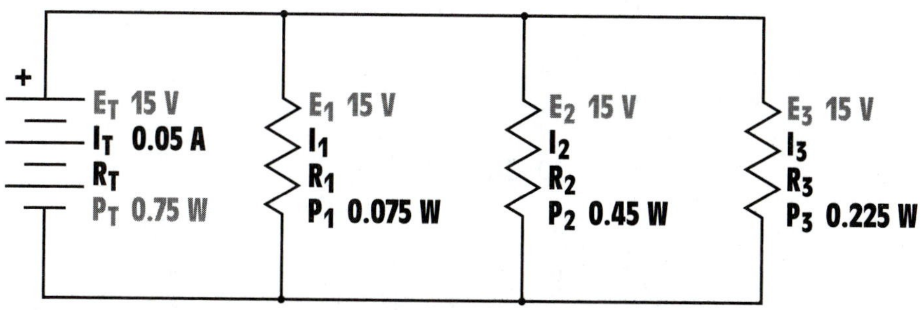

Figure 6-20 Computing the applied voltage for the circuit, Example 3.

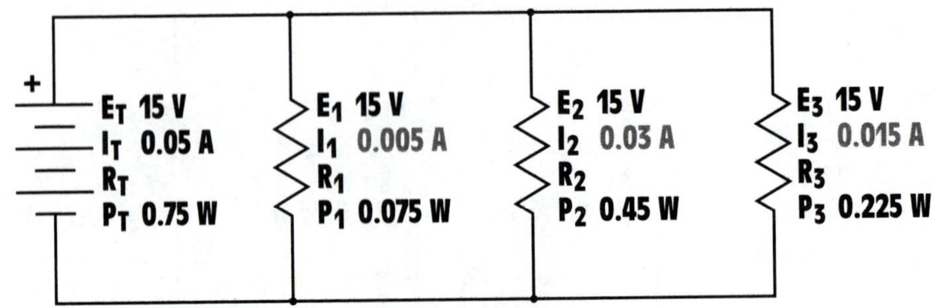

Figure 6-21 Finding the current through each branch.

$$I_1 = \frac{0.075}{15}$$

$$1_1 = 0.005 \text{ A}$$

$$I_2 = \frac{P_2}{E_2}$$

$$I_2 = \frac{0.45}{15}$$

$$I_2 = 0.03 \text{ A}$$

$$I_3 = \frac{P_3}{E_3}$$

$$I_3 = \frac{0.225}{15}$$

$$I_3 = 0.015 \text{ A}$$

All resistance values for the circuit can now be found using Ohm's Law, *Figure 6-22.*

$$R_1 = \frac{E_1}{I_1}$$

$$R_1 = \frac{15}{0.005}$$

$$R_1 = 3000\ \Omega$$

$$R_2 = \frac{E_2}{I_2}$$

$$R_2 = \frac{15}{0.03}$$

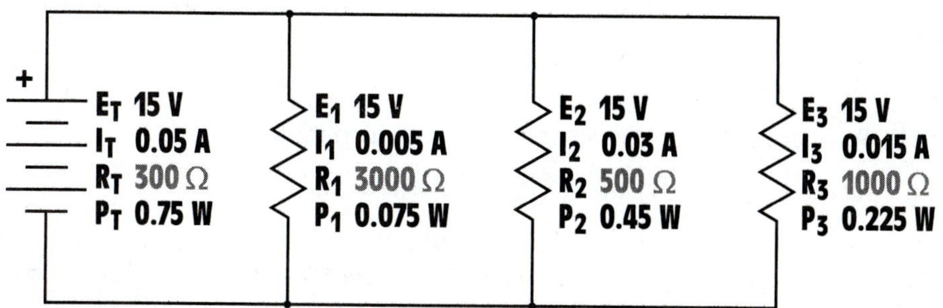

Figure 6-22 Computing the remaining values for the circuit, Example 3.

$$R_2 = 500\ \Omega$$

$$R_3 = \frac{E_3}{I_3}$$

$$R_3 = \frac{15}{0.015}$$

$$R_3 = 1000\ \Omega$$

$$R_T = \frac{E_T}{I_T}$$

$$R_T = \frac{15}{0.05}$$

$$R_T = 300\ \Omega$$

Summary

1. A parallel circuit is characterized by the fact that it has more than one path for current flow.
2. Three rules for solving parallel circuits are:
   A. The total current is the sum of the currents through each individual branch of the circuit.
   B. The voltage is the same across any part of the circuit.
   C. The reciprocal of total resistance is the sum of the reciprocals of each individual branch.
3. Circuits in homes are connected in parallel.
4. The total power in a parallel circuit is equal to the sum of the power dissipation by each component.

Review Questions

1. What characterizes a parallel circuit?
2. Why are circuits in homes connected in parallel?
3. State three rules concerning parallel circuits.
4. A parallel circuit contains four branches. One branch has a current flow of 0.8 A, another has a current flow of 1.2 A, the third has a current flow of 0.25 A, and the fourth has a current flow of 1.5 A. What is the total current flow in the circuit?
5. Four resistors having a value of 100 Ω each are connected in parallel. What is the total resistance of the circuit?
6. A parallel circuit has three branches. An ammeter is connected in series with the output of the power supply and indicates a total current flow of 2.8 A. If branch 1 has a current flow of 0.9 A and branch 2 has a current flow of 1.05 A, what is the current flow through branch 3?
7. Four resistors having values of 270 Ω, 330 Ω, 510 Ω, and 430 Ω are connected in parallel. What is the total resistance of the circuit?
8. A parallel circuit contains four resistors. The total resistance of the circuit is 120 Ω. Three of the resistors have values of 820 Ω, 750 Ω, and 470 Ω. What is the value of the fourth resistor?

Practice Problems

PARALLEL CIRCUITS

1. Find the missing values in the circuit shown in *Figure 6-23.*

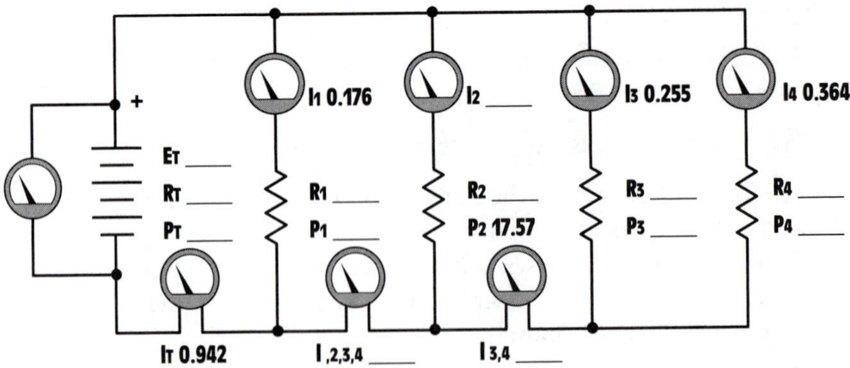

Figure 6-23 Practice Problem 1.

2. Find the missing values in the circuit shown in *Figure 6-24*.

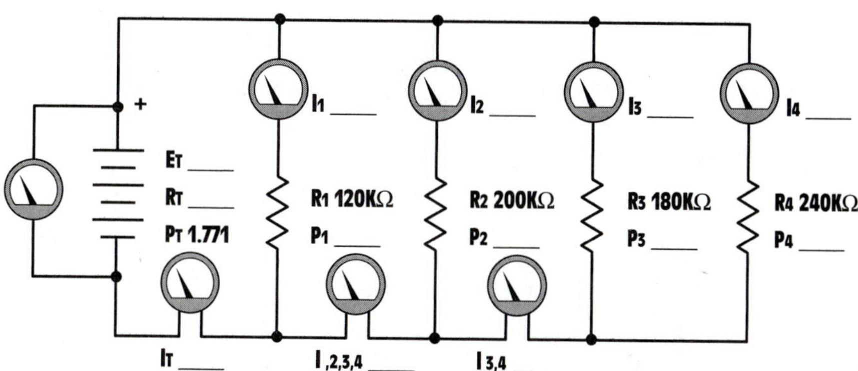

Figure 6-24 Practice Problem 2.

3. Find the missing values in the circuit shown in *Figure 6-25*.

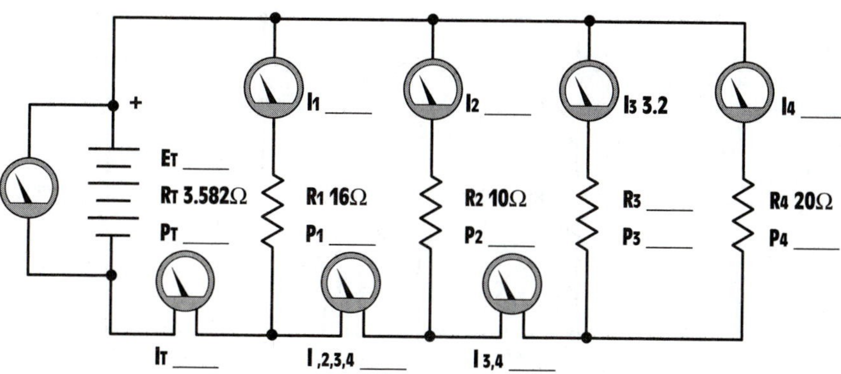

Figure 6-25 Practice Problem 3.

4. Find the missing values in the circuit shown in *Figure 6-26*.

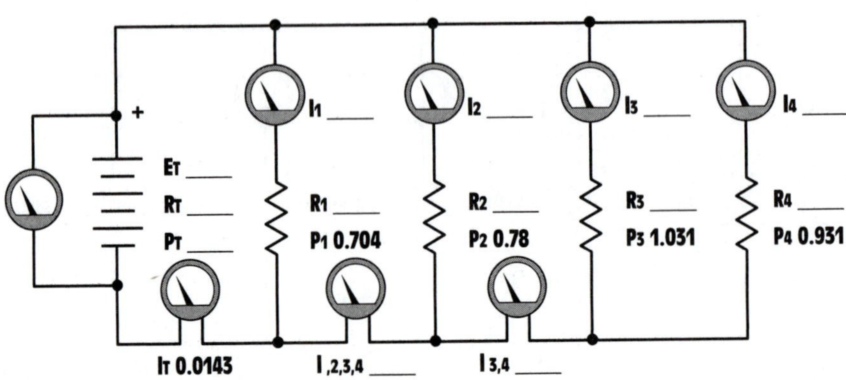

Figure 6-26 Practice Problem 4.

5. Find the missing values in the circuit shown in *Figure 6-27*.

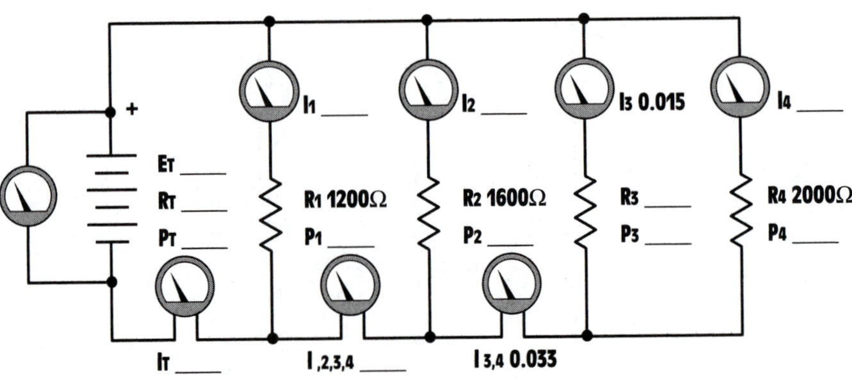

Figure 6-27 Practice Problem 5.

6. Find the missing values in the circuit shown in *Figure 6-28*.

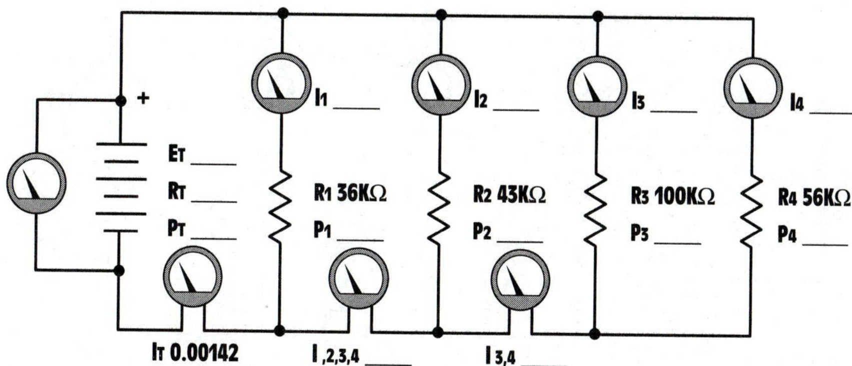

Figure 6-28 Practice Problem 6.

7. Find the missing values in the circuit shown in *Figure 6-29*.

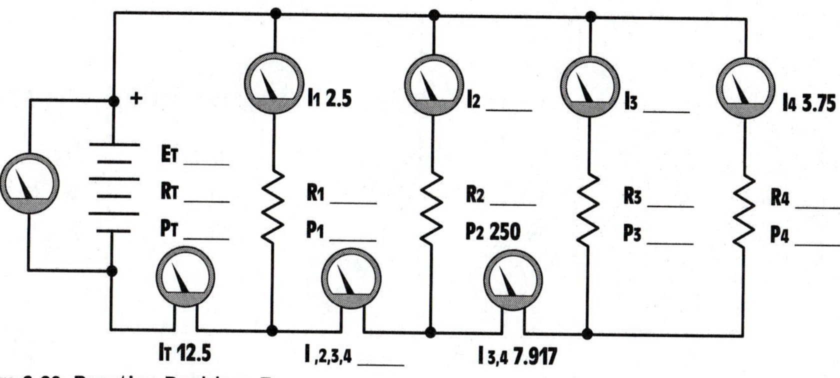

Figure 6-29 Practice Problem 7.

8. Find the missing values in the circuit shown in *Figure 6-30*.

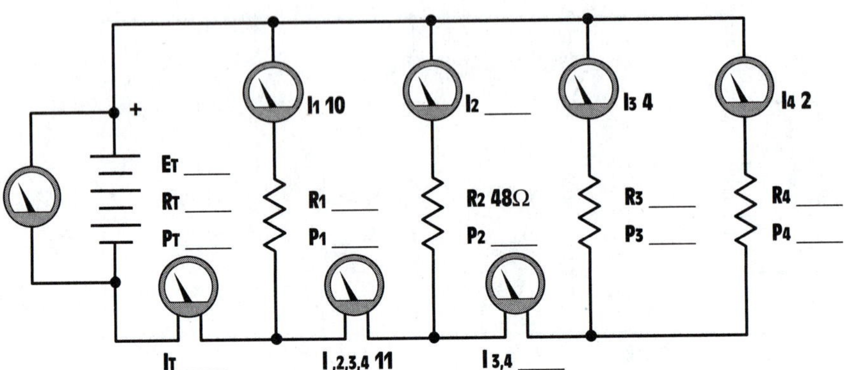

Figure 6-30 Practice Problem 8.

9. Find the missing values in the circuit shown in *Figure 6-31*.

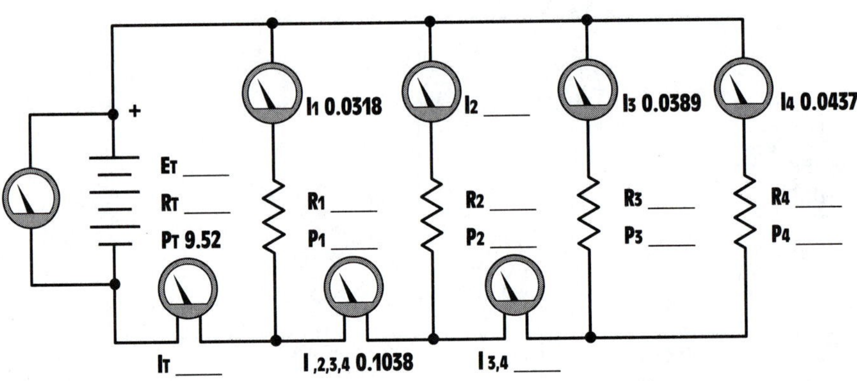

Figure 6-31 Practice Problem 9.

10. Find the missing values in the circuit shown in *Figure 6-32*.

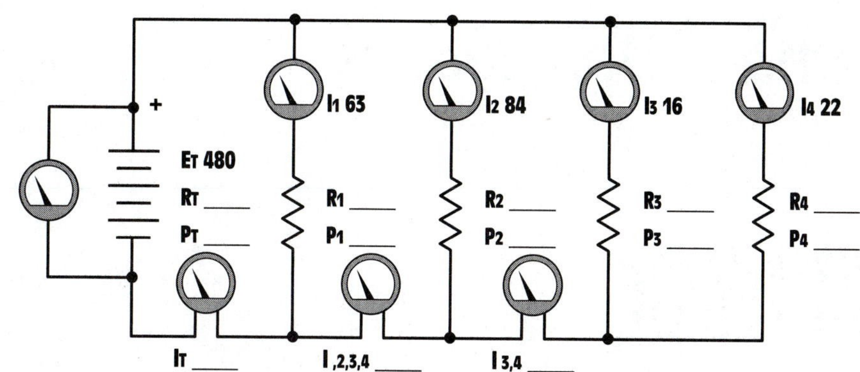

Figure 6-32 Practice Problem 10.

7 Unit

Combination Circuits

objectives

fter studying this unit, you should be able to:

- Define a combination circuit.
- List the rules for parallel circuits.
- List the rules for series circuits.
- Solve combination circuits using the rules for parallel circuits, the rules for series circuits, and Ohm's Law.

Combination circuits contain a combination of both series and parallel elements. A simple combination circuit is shown in *Figure 7-1*. To determine which components are in parallel and which are in series, trace the flow of current through the circuit. Remember, a series circuit is one that has only one path for current flow, and a parallel circuit has more than one path for current flow. Series elements can be identified by the fact that all the circuit current must pass through them. In parallel elements the circuit current will divide, and part will flow through each element.

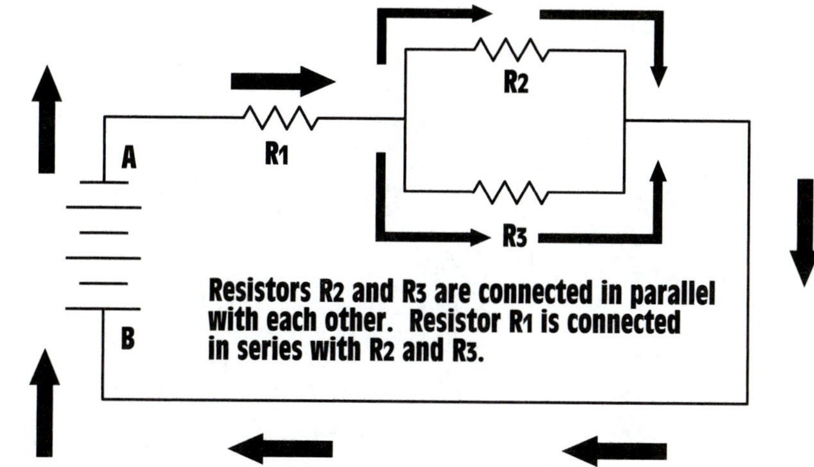

Figure 7-1 A simple combination circuit.

Example 1

COMBINATION CIRCUITS

trace the current path

node

combination circuit

In Figure 7-1, it will be assumed that the current will flow from point A to point B. To identify the series and parallel elements, **trace the current path.** All the current in the circuit must flow through resistor R_1. Resistor R_1 is, therefore, in series with the rest of the circuit. When the current reaches the junction point of resistors R_2 and R_3, however, it splits. A junction point where two or more branches are connected is often referred to as a **node.** Part of the current flows through resistor R_2 and part flows through resistor R_3. These two resistors are in parallel because this circuit contains both series and parallel elements, it is a **combination circuit.**

SOLVING COMBINATION CIRCUITS

The circuit shown in *Figure 7-2* contains four resistors with values of 325 Ω, 275 Ω, 150 Ω, and 250 Ω. The circuit has a total current flow of 1 A. To determine which resistors are in series or parallel, the path for current flow will be traced through the circuit. When the path of current flow is traced, it can be seen that there are two separate paths by which current can flow from the negative terminal to the positive terminal. One path is through resistors R_1 and R_2, and the other path is through resistors R_3 and R_4. These two paths are, therefore, in parallel. However, the same current must flow through resistors R_1 and R_2. These two resistors are in series. The same is true for resistors R_3 and R_4.

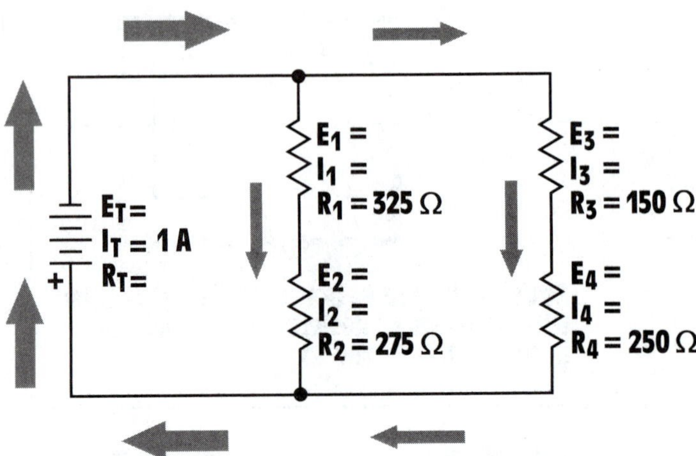

Figure 7-2 Tracing the current paths through combination circuit 1.

When solving the unknown values in a combination circuit, use series circuit rules for those sections of the circuit that are connected in series and parallel circuit rules for those sections connected in parallel. The circuit rules are as follows:

SERIES CIRCUITS

1. The current is the same at any point in the circuit.
2. The total resistance is the sum of the individual resistors.
3. The sum of the voltage drops of the individual resistors must equal the applied voltage.

PARALLEL CIRCUITS

1. The voltage is the same across any circuit branch.
2. The total current is the sum of the current flowing though the individual circuit paths.
3. The reciprocal of the total resistance is equal to the sum of the reciprocals of the branch resistances.

SIMPLIFYING THE CIRCUIT

simple parallel circuit

The circuit shown in Figure 7-2 can be reduced or simplified to a **simple parallel circuit** as shown in *Figure 7-3*. Since resistors R_1 and R_2 are connected in series, their values can be added to form one equivalent resistor, [$R_{C(1\&2)}$], which stands for combination of resistors 1 and 2. The same is true for resistors R_3 and R_4. Their values are added to form resistor

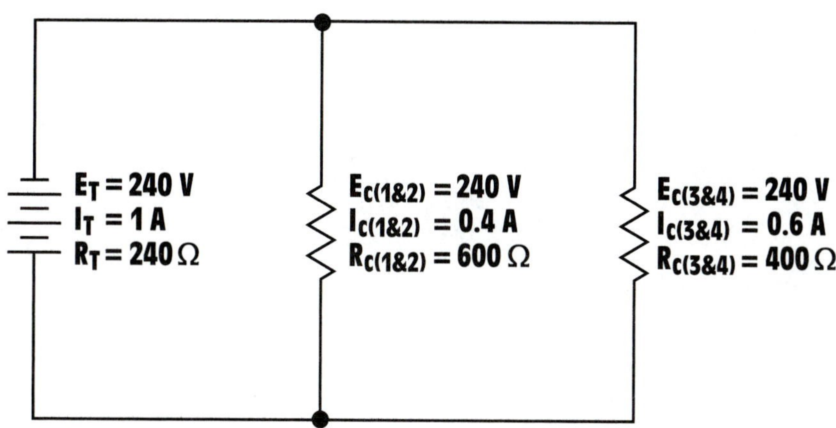

Figure 7-3 Simplifying combination circuit 1.

$R_{C(3\&4)}$. Now that the circuit has been reduced to a simple parallel circuit, the total resistance can be found.

$$R_T = \frac{1}{\frac{1}{R_{C(1\&2)}} + \frac{1}{R_{C(3\&4)}}}$$

$$R_T = \frac{1}{\frac{1}{600} + \frac{1}{400}}$$

$$R_T = \frac{1}{0.0016667 + 0.0025}$$

$$R_T = \frac{1}{0.0041667}$$

$$R_T = 240\ \Omega$$

Now that the total resistance has been found, the other circuit values can be computed. The applied voltage can be found using Ohm's Law.

$$E_T = I_T \times R_T$$

$$E_T = 1 \times 240$$

$$E_T = 240\text{ V}$$

One of the rules for parallel circuits states that the voltage is the same across each branch of the circuit. For this reason, the voltage drops across resistors $R_{C(1\&2)}$ and $R_{C(3\&4)}$ are the same. Since the voltage drop and the resistance are known, Ohm's Law can be used to find the current flow through each branch.

$$I_{C(1\&2)} = \frac{E_{C(1\&2)}}{R_{C(1\&2)}}$$

$$I_{C(1\&2)} = \frac{240}{600}$$

$$I_{C(1\&2)} = 0.4\ \text{A}$$

$$I_{C(3\&4)} = \frac{E_{C(3\&4)}}{R_{C(3\&4)}}$$

$$I_{C(3\&4)} = \frac{240}{400}$$

$$I_{C(3\&4)} = 0.6\ \text{A}$$

These values can now be used to solve the missing values in the original circuit. Resistor $R_{C(1\&2)}$ is actually a combination of resistors R_1 and R_2. The values of voltage and current that apply to $R_{C(1\&2)}$ therefore apply to resistors R_1 and R_2. Resistors R_1 and R_2 are connected in series. One of the rules for a series circuit states that the current is the same at any point in the circuit. Since 0.4 A of current flows through resistor $R_{C(1\&2)}$, the same amount of current flows through resistors R_1 and R_2. Now that the current flow through these two resistors is known, the voltage drop across each can be computed using Ohm's Law.

$$E_1 = I_1 \times R_1$$

$$E_1 = 0.4 \times 325$$

$$E_1 = 130\ \text{V}$$

$$E_2 = I_2 \times R_2$$

$$E_2 = 0.4 \times 275$$

$$E_2 = 110\ \text{V}$$

These values of voltage and current can now be added to the circuit in Figure 7-2 to produce the circuit shown in *Figure 7-4*.

The values of voltage and current for resistor $R_{C(3\&4)}$ apply to resistors R_3 and R_4. The same amount of current that flows through resistor $R_{C(3\&4)}$ flows through resistors R_3 and R_4. The voltage across these two resistors can now be computed using Ohm's Law.

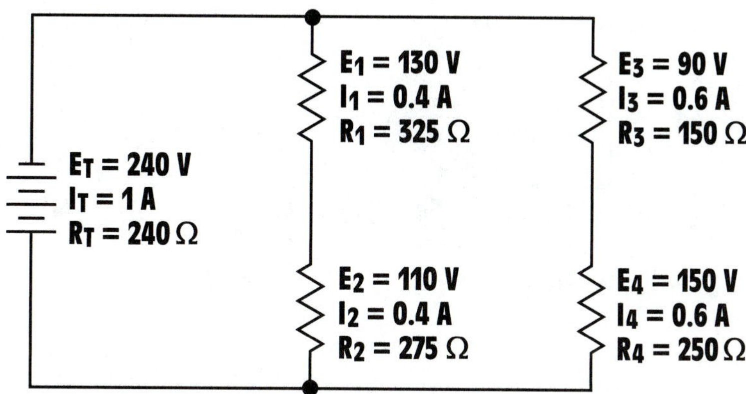

Figure 7-4 Finding all the missing values for combination circuit 1.

$$E_3 = I_3 \times R_3$$

$$E_3 = 0.6 \times 150$$

$$E_3 = 90 \text{ V}$$

$$E_4 = I_4 \times R_4$$

$$E_4 = 0.6 \times 250$$

$$E_4 = 150 \text{ V}$$

Example 2

The second circuit to be discussed is shown in *Figure 7-5*. The first step in finding the missing values is to trace the current path through the circuit to determine which resistors are in series and which are connected in parallel. All the current must flow through Resistor R_1. Resistor R_1 is, therefore, in series with the rest of the circuit. When the current reaches the junction of resistors R_2 and R_3, it divides, and part flows through each resistor. Resistors R_2 and R_3 are in parallel. All the current must then flow through resistor R_4 to the junction of resistors R_5 and R_6. The current path is divided between these two resistors. Resistors R_5 and R_6 are connected in parallel. All the circuit current must then flow through resistor R_7.

reduce

parallel block

SIMPLIFYING THE CIRCUIT

The next step in solving this circuit is to **reduce** it to a simpler circuit. If the total resistance of the first **parallel block** formed by resistors R_2 and R_3 and R_3 is found, this block can be replaced by a single resistor.

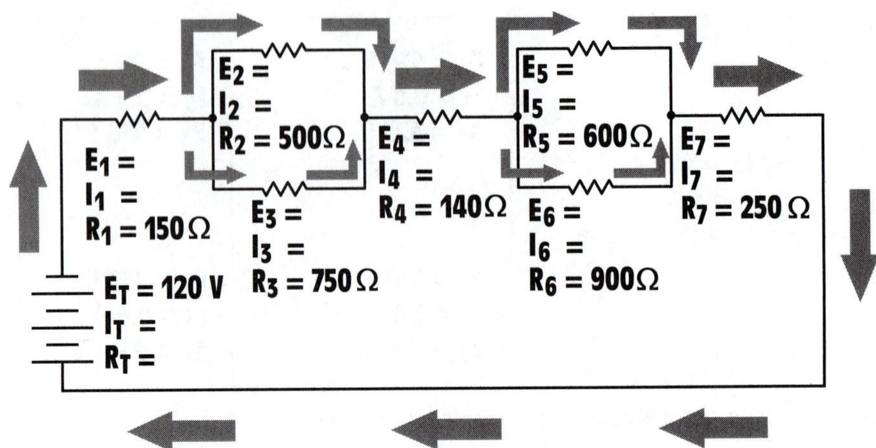

Figure 7-5 Tracing the flow of current through combination circuit 2.

$$R_T = \frac{1}{\frac{1}{R_2} + \frac{1}{R_3}}$$

$$R_T = \frac{1}{0.002 + 0.0013333}$$

$$R_T = \frac{1}{0.0033333}$$

$$R_T = 300\ \Omega$$

The equivalent resistance of the second parallel block can be computed in the same way.

$$R_T = \frac{1}{\frac{1}{R_5} + \frac{1}{R_6}}$$

$$R_T = \frac{1}{\frac{1}{600} + \frac{1}{900}}$$

$$R_T = \frac{1}{0.0016667 + 0.0011111}$$

$$R_T = \frac{1}{0.00277778}$$

$$R_T = 360\ \Omega$$

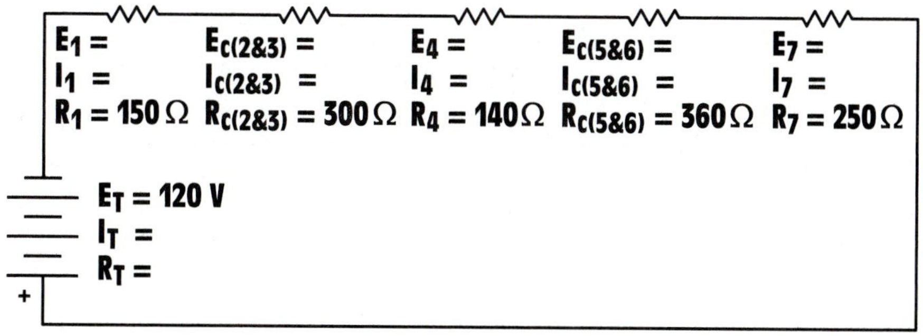

Figure 7-6 Simplifying combination circuit 2.

Now that the total resistance of the second parallel block is known, you can **redraw** the circuit as a simple series circuit as shown in *Figure 7-6*. The first parallel block has been replaced with a single resistor of 300 Ω labeled $R_{C(2\&3)}$, and the second parallel block has been replaced with a single 360 Ω resistor labeled $R_{C(5\&6)}$. Ohm's Law can be used to find the missing values in this series circuit.

redraw

One of the rules for series circuits states that the total resistance of a series circuit is equal to the sum of the individual resistors. R_T (total resistance of the circuit) can be computed by adding the resistance of all resistors.

$$R_T = R_1 + R_{C(2\&3)} + R_4 + R_{C(5\&6)} + R_7$$

$$R_T = 150 + 300 + 140 + 360 + 250$$

$$R_T = 1200\ \Omega$$

Since the total voltage and total resistance are known, the total current flow through the circuit can be computed.

$$I_T = \frac{E_T}{R_T}$$

$$I_T = \frac{120}{1200}$$

$$I_T = 0.1\ \text{A}$$

The first rule of series circuits states that the current is the same at any point in the circuit. The current flow through each resistor is, therefore, 0.1 A. The voltage drop across each resistor can now be computed using Ohm's Law.

$$E_1 = I_1 \times R_1$$

$$E_1 = 0.1 \times 150$$

$$E_1 = 15\ \text{V}$$

$$E_{C(2\&3)} = I_{C(2\&3)} \times R_{C(2\&3)}$$

$$E_{C(2\&3)} = 0.1 \times 300$$

$$E_{C(2\&3)} = 30 \text{ V}$$

$$E_4 = I_4 \times R_4$$

$$E_4 = 0.1 \times 140$$

$$E_4 = 14 \text{ V}$$

$$E_{C(5\&6)} = I_{C(5\&6)} \times R_{C(5\&6)}$$

$$E_{C(5\&6)} = 0.1 \times 360$$

$$E_{C(5\&6)} = 36 \text{ V}$$

$$E_7 = I_7 \times R_7$$

$$E_7 = 0.1 \times 250$$

$$E_7 = 25 \text{ V}$$

The series circuit with all solved values is shown in *Figure 7-7*. These values can now be used to solve missing parts in the original circuit.

Resistor $R_{C(2\&3)}$ is actually the parallel block containing resistors R_2 and R_3. The values for $R_{C(2\&3)}$, therefore, apply to this parallel block. One of the rules for a parallel circuit states that the voltage drop of a parallel circuit is the same at any point in the circuit. Since 30 V is dropped across resistor $R_{C(2\&3)}$, the same 30 V is dropped across resistors R_2 and R_3, *Figure 7-8*. The current flow through these resistors can now be computed using Ohm's Law.

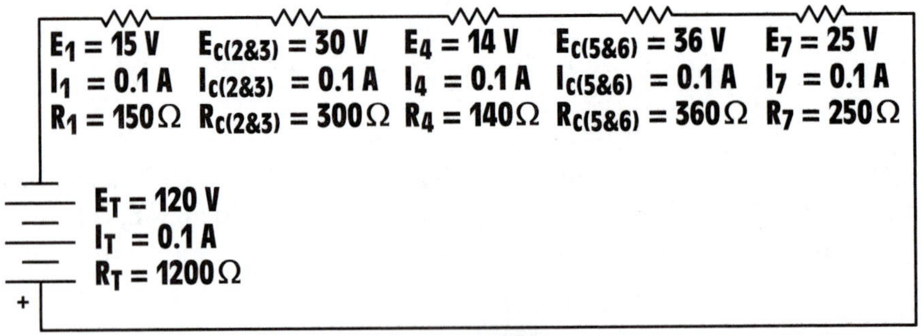

Figure 7-7 The simplified circuit with all values solved.

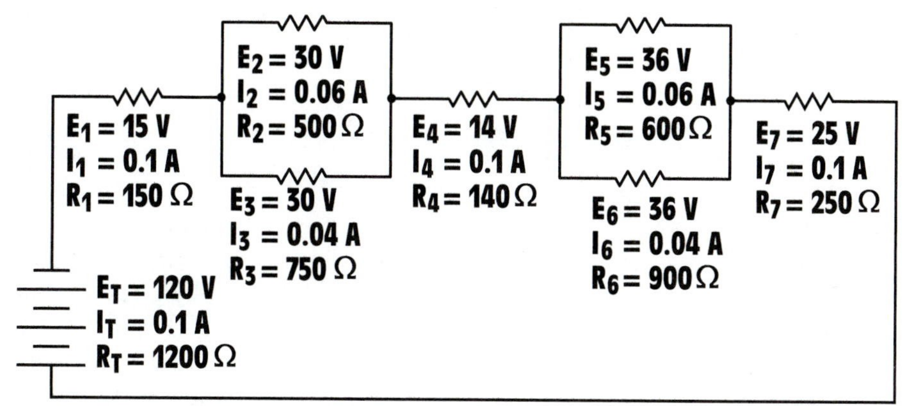

Figure 7-8 All values solved for combination circuit 2.

$$I_2 = \frac{E_2}{R_2}$$

$$I_2 = \frac{30}{500}$$

$$I_2 = 0.06 \text{ A}$$

$$I_3 = \frac{E_3}{R_3}$$

$$I_3 = \frac{30}{750}$$

$$I_3 = 0.04 \text{ A}$$

The values of resistor $R_{C(5\&6)}$ can be applied to the parallel block composed of resistors R_5 and R_6. $E_{C(5\&6)}$ is 36 V. This is the voltage drop across resistors R_5 and R_6. The current flow through these two resistors can be computed using Ohm's Law.

$$I_5 = \frac{E_5}{R_5}$$

$$I_5 = \frac{36}{600}$$

$$I_5 = 0.06 \text{ A}$$

$$I_6 = \frac{E_6}{R_6}$$

$$I_6 = \frac{36}{900}$$

$$I_6 = 0.04\text{ A}$$

Example 3

Both of the preceding circuits were solved by first determining which parts of the circuit were in series and which were in parallel. The circuits were then reduced to a simple series or parallel circuit. This same procedure can be used for any combination circuit. The circuit shown in *Figure 7-9* will be reduced to a simpler circuit first. Once the values of the simple circuit are found, they can be placed back in the original circuit to find other values.

Solution:

The first step will be to reduce the top part of the circuit to a single resistor. This part consists of resistors R_3 and R_4. Since these two resistors are connected in series, their values can be added to form one single resistor. This combination will form R_{C1}, *Figure 7-10*.

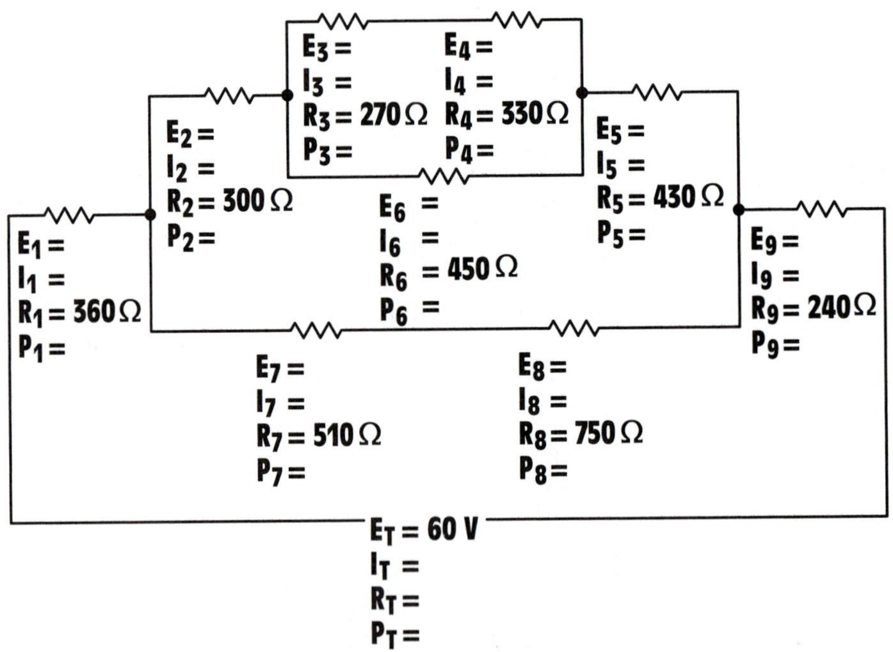

Figure 7-9 Circuit 3.

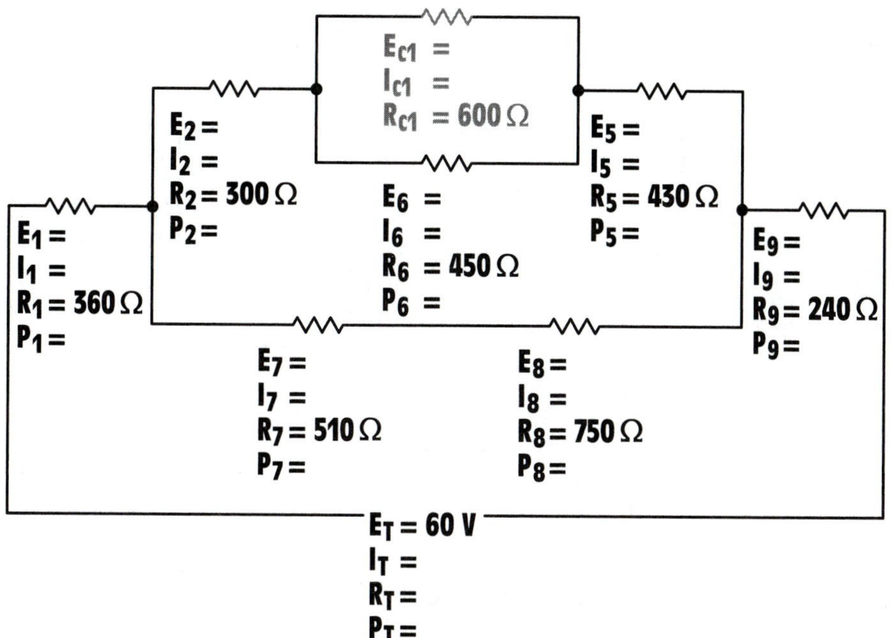

Figure 7-10 Resistors R_1 and R_2 are combined to form R_{C1}.

$$R_{C1} = R_3 + R_4$$

$$R_{C1} = 270 + 330$$

$$R_{C1} = 600\ \Omega$$

The top part of the circuit is now formed by resistors R_{C1} and R_6. These two resistors are in parallel. If their total resistance is computed, they can be changed into one single resistor with a value of 257.143 Ω. This combination will become resistor R_{C2}, *Figure 7-11*.

$$R_{C2} = \frac{1}{\frac{1}{R_{C1}} + \frac{1}{R_6}}$$

$$R_{C2} = \frac{1}{\frac{1}{600} + \frac{1}{450}}$$

$$R_T = 257.143\ \Omega$$

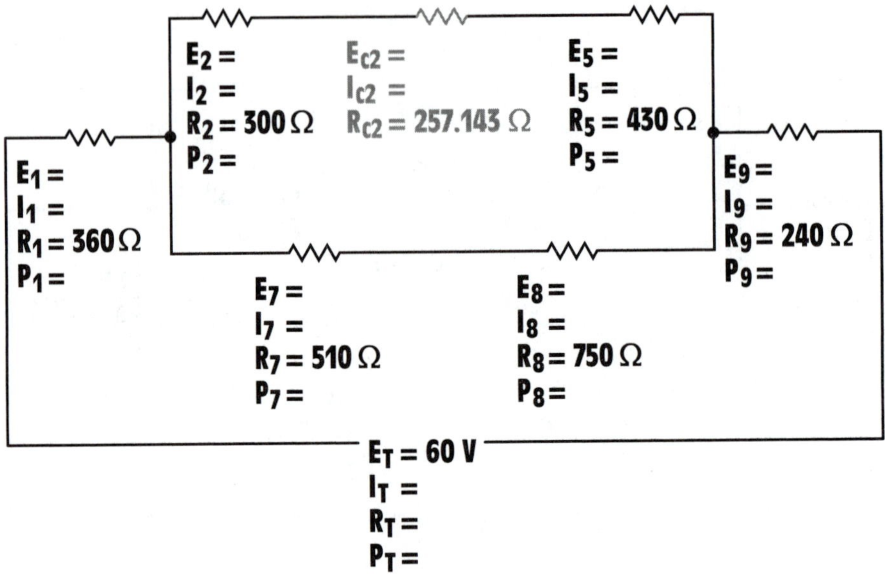

Figure 7-11 Resistors R_{C1} and R_6 are combined to form R_{C2}.

The top of the circuit now consists of resistors R_2, R_{C2}, and R_5. These three resistors are connected in series with each other. They can be combined to form resistor R_{C3} by adding their resistances together, *Figure 7-12*.

$$R_{C3} = R_2 + R_{C2} + R_5$$

$$R_{C3} = 300 + 257.143 + 430$$

$$R_{C3} = 987.143\ \Omega$$

Resistors R_7 and R_8 are connected in series also. These two resistors will be added to form resistor R_{C4}, *Figure 7-13*.

$$R_{C4} = R_7 + R_8$$

$$R_{C4} = 510 + 750$$

$$R_{C4} = 1260\ \Omega$$

Resistors R_{C3} and R_{C4} are connected in parallel. Their total resistance can be computed to form resistor R_{C5}, *Figure 7-14*.

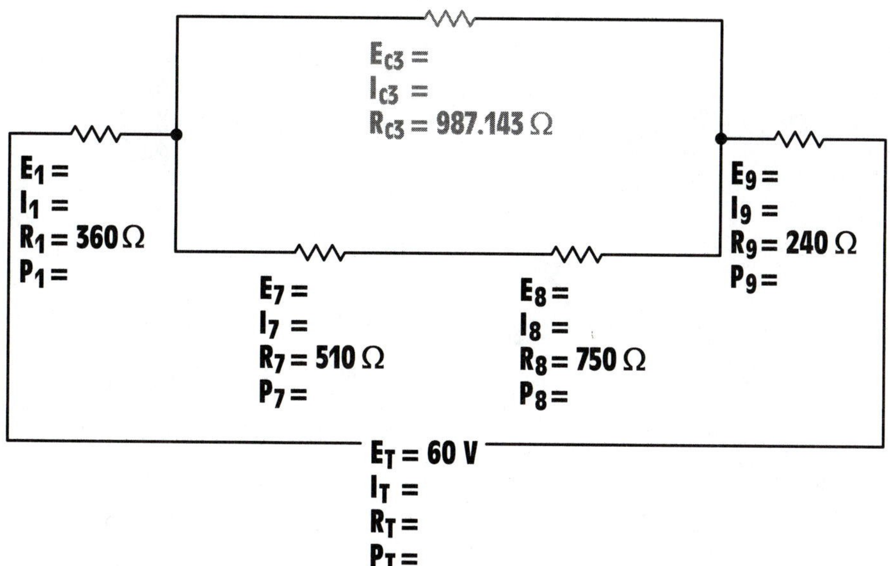

Figure 7-12 Resistors R_2, R_{c2}, and R_5 are combined to form R_{c3}.

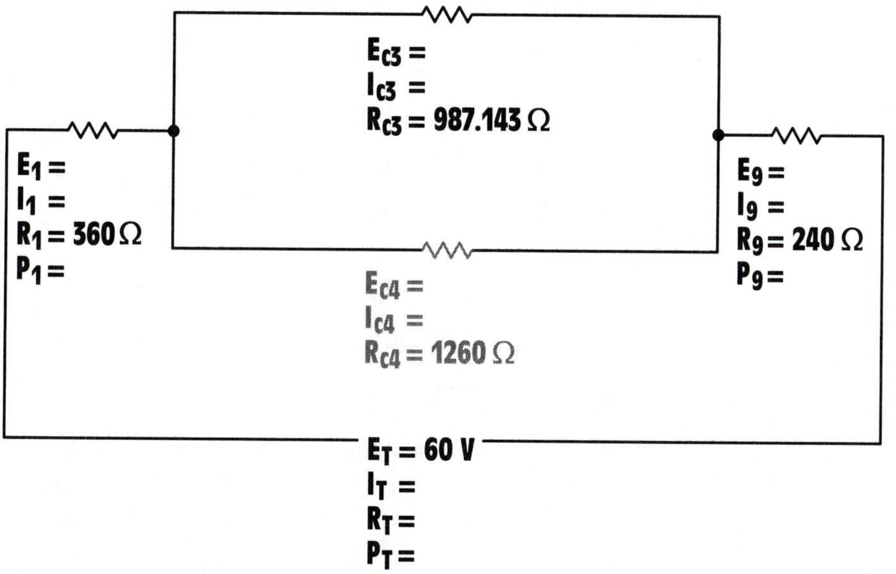

Figure 7-13 Resistors R_7 and R_8 are combined to form R_{c4}.

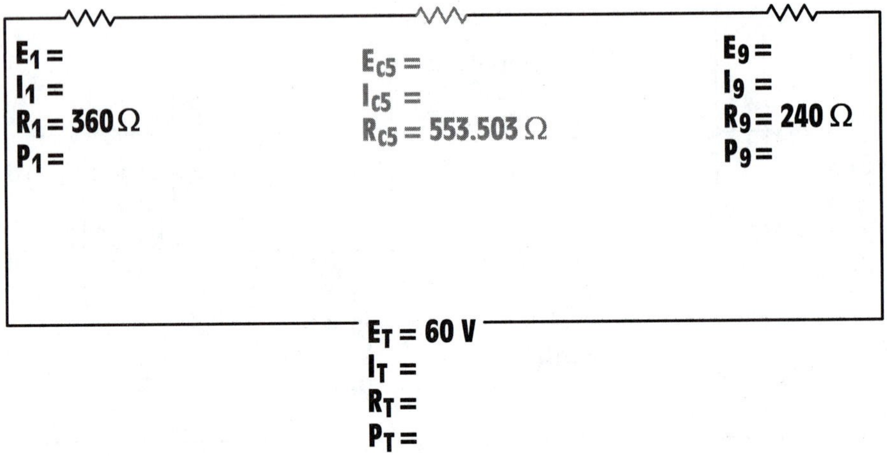

$E_1 =$
$I_1 =$
$R_1 = 360\,\Omega$
$P_1 =$

$E_{C5} =$
$I_{C5} =$
$R_{C5} = 553.503\,\Omega$

$E_9 =$
$I_9 =$
$R_9 = 240\,\Omega$
$P_9 =$

$E_T = 60\text{ V}$
$I_T =$
$R_T =$
$P_T =$

Figure 7-14 Resistors R_{C3} and R_{C4} are combined to form R_{C5}.

$$R_{C5} = \frac{1}{\frac{1}{R_{C3}} + \frac{1}{R_{C4}}}$$

$$R_{C5} = \frac{1}{\frac{1}{987.143} + \frac{1}{1260}}$$

$$R_{C5} = 553.503\ \Omega$$

The circuit has now been reduced to a simple series circuit containing three resistors. The total resistance of the circuit can be computed by adding resistors R_1, R_{C5}, and R_9.

$$R_T = R_1 + R_{C5} + R_9$$

$$R_T = 360 + 553.503 + 240$$

$$R_T = 1153.503\ \Omega$$

Now that the total resistance and total voltage are known, the total circuit current and total circuit power can be computed using Ohm's Law.

$$I_T = \frac{E_T}{R_T}$$

$$I_T = \frac{60}{1153.503}$$

$$I_T = 0.052\text{ A}$$

$$P_T = E_T \times I_T$$
$$P_T = 60 \times 0.052$$
$$P_T = 3.12\ W$$

Ohm's Law can now be used to find the missing values for resistors R_1, R_{C5}, and R_9, *Figure 7-15*.

$$E_1 = I_1 \times R_1$$
$$E_1 = 0.052 \times 360$$
$$E_1 = 18.72\ V$$

$$P_1 = E_1 \times I_1$$
$$P_1 = 18.72 \times 0.052$$
$$P_1 = 0.973\ W$$

$$E_{C5} = I_{C5} \times R_{C5}$$
$$E_{C5} = 0.052 \times 553.143$$
$$E_{C5} = 28.763\ V$$

$$E_9 = I_9 \times R_9$$
$$E_9 = 0.052 \times 240$$
$$E_9 = 12.48\ V$$

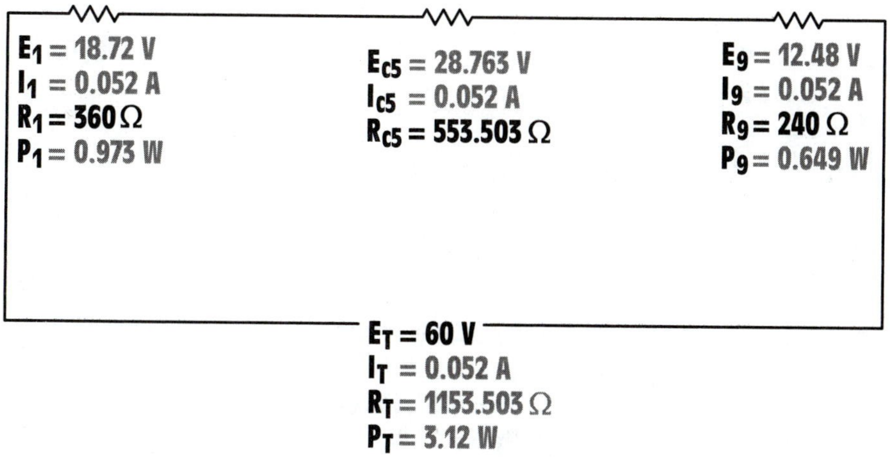

$E_1 = 18.72\ V$
$I_1 = 0.052\ A$
$R_1 = 360\ \Omega$
$P_1 = 0.973\ W$

$E_{C5} = 28.763\ V$
$I_{C5} = 0.052\ A$
$R_{C5} = 553.503\ \Omega$

$E_9 = 12.48\ V$
$I_9 = 0.052\ A$
$R_9 = 240\ \Omega$
$P_9 = 0.649\ W$

$E_T = 60\ V$
$I_T = 0.052\ A$
$R_T = 1153.503\ \Omega$
$P_T = 3.12\ W$

Figure 7-15 Missing values are found for the first part of the circuit.

Resistor R_{C5} is actually the combination of resistors R_{C3} and R_{C4}. The values of R_{C5}, therefore, apply to resistors R_{C3} and R_{C4}. Since these two resistors are connected in parallel with each other, the voltage drop across them will be the same. Each will have the same voltage drop as resistor R_{C5}, *Figure 7-16*. Ohm's Law can now be used to find the remaining values of R_{C3} and R_{C4}.

$$I_{C4} = \frac{E_{C4}}{R_{C4}}$$

$$I_{C4} = \frac{28.763}{1260}$$

$$I_{C4} = 0.0228 \text{ A}$$

$$I_{C3} = \frac{E_{C3}}{R_{C3}}$$

$$I_{C3} = \frac{28.763}{987.143}$$

$$I_{C3} = 0.0291 \text{ A}$$

Resistor R_{C4} is the combination of resistors R_7 and R_8. The values of resistor R_{C4} apply to resistors R_7 and R_8. Since resistors R_7 and R_8 are connected

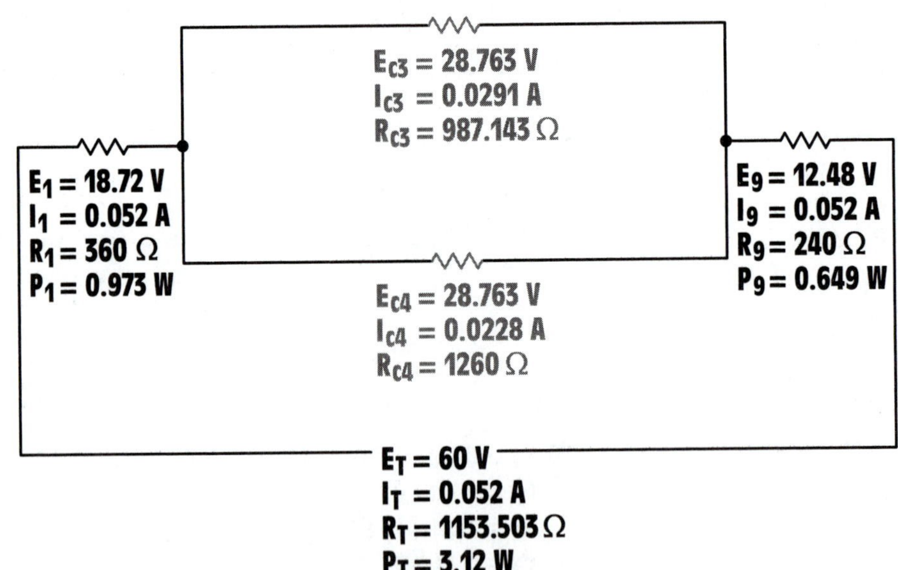

Figure 7-16 Solving the values for resistors R_{C3} and R_{C4}.

in series, the current flow will be the same through both, *Figure 7-17.* Ohm's Law can now be used to compute the remaining values for these two resistors.

$$E_7 = I_7 \times R_7$$
$$E_7 = 0.0228 \times 510$$
$$E_7 = 11.268 \text{ V}$$

$$P_7 = E_7 \times I_7$$
$$P_7 = 11.268 \times 0.0228$$
$$P_7 = 0.265 \text{ W}$$

$$E_8 = I_8 \times R_8$$
$$E_8 = 0.0228 \times 750$$
$$E_8 = 17.1 \text{ V}$$

$$P_8 = E_8 \times I_8$$
$$P_8 = 17.1 \times 0.0228$$
$$P_8 = 0.390 \text{ W}$$

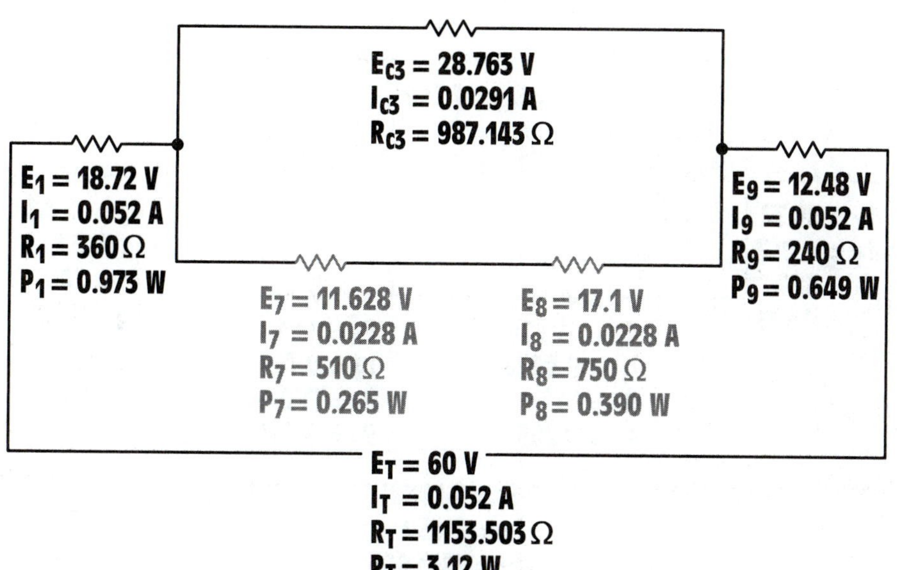

Figure 7-17 Solving the values for resistors R_7 and R_8.

Resistor R_{C3} is the combination of resistors R_2, R_{C2} and R_5. Since these resistors are connected in series, the current flow through each will be the same as the current flow through R_{C3}. The remaining values can now be computed using Ohm's Law, *Figure 7-18.*

$$E_{C2} = I_{C2} \times R_{C2}$$

$$E_{C2} = 0.0292 \times 257.143$$

$$E_{C2} = 7.509 \text{ V}$$

$$E_2 = I_2 \times R_2$$

$$E_2 = 0.0292 \times 300$$

$$E_2 = 8.76 \text{ V}$$

$$P_2 = E_2 \times I_2$$

$$P_2 = 8.76 \times 0.292$$

$$P_2 = 0.256 \text{ W}$$

$$E_5 = I_5 \times R_5$$

$$E_5 = 0.0292 \times 430$$

$$E_5 = 12.556 \text{ V}$$

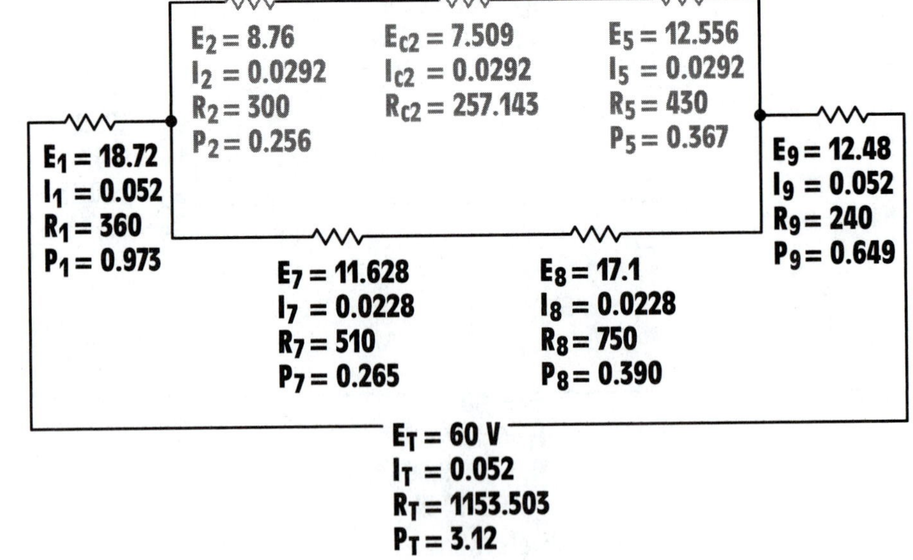

Figure 7-18 Determining values for R_2, R_{C2}, and R_5.

$$P_5 = E_5 \times I_5$$

$$P_5 = 12.556 \times 0.0292$$

$$P_5 = 0.367 \text{ W}$$

Resistor R_{C2} is the combination of resistors R_{C1} and R_6. Resistors R_{C1} and R_6 are connected in parallel and will, therefore, have the same voltage drop as resistor R_{C2}. Ohm's Law can be used to compute the remaining values for R_{C2} and R_6, *Figure 7-19.*

$$I_{c1} = \frac{E_{c1}}{R_{c1}}$$

$$I_{c1} = \frac{7.509}{600}$$

$$I_{c1} = 0.0125 \text{ A}$$

$$I_6 = \frac{E_6}{R_6}$$

$$I_6 = \frac{7.509}{450}$$

$$I_6 = 0.0167 \text{ A}$$

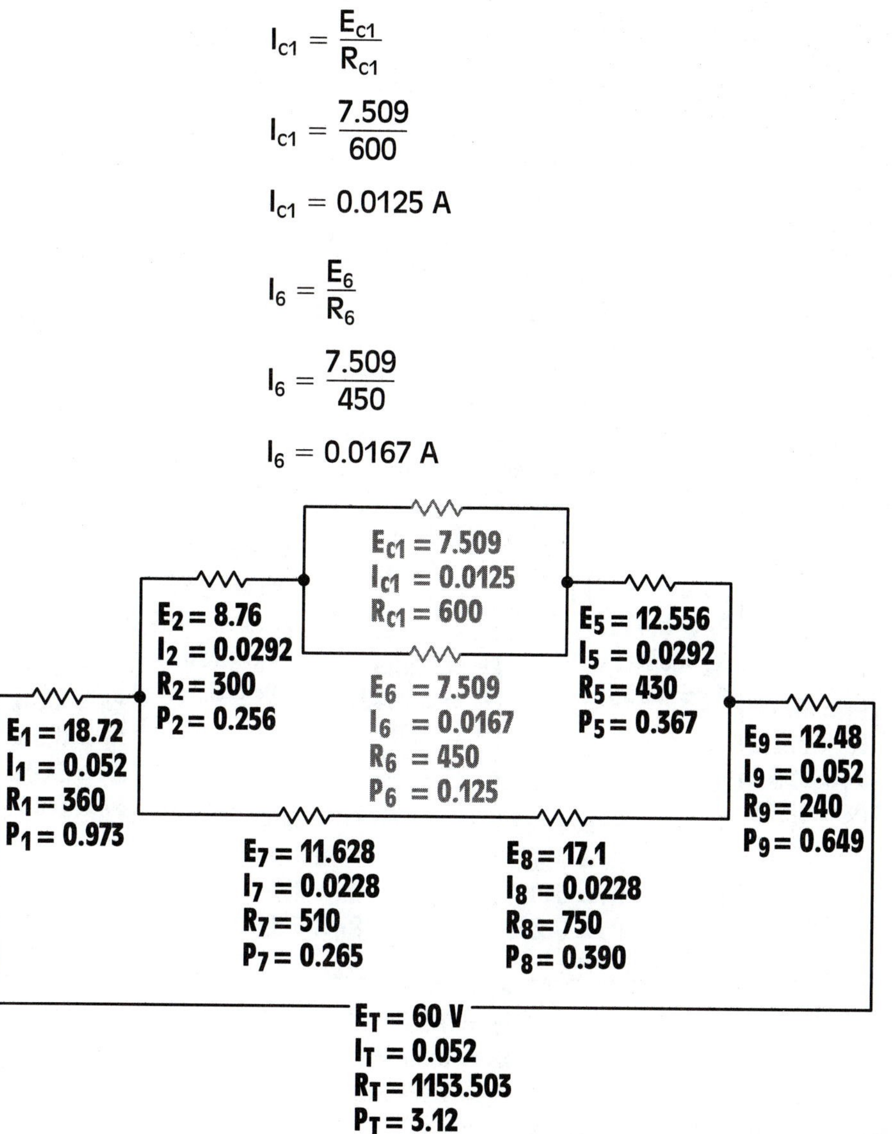

Figure 7-19 Computing the missing values for R_{c1} and R_6.

$$P_6 = E_6 \times I_6$$

$$P_6 = 7.509 \times 0.0167$$

$$P_6 = 0.125 \text{ W}$$

Resistor R_{C1} is the combination of resistors R_3 and R_4. Since these two resistors are connected in series, the amount of current flow through resistor R_{C1} will be the same as the flow through R_3 and R_4. The remaining values of the circuit can now be found using Ohm's Law, *Figure 7-20*.

$$E_3 = I_3 \times R_3$$

$$E_3 = 0.0125 \times 270$$

$$E_3 = 3.375 \text{ V}$$

$$P_3 = E_3 \times I_3$$

$$P_3 = 3.375 \times 0.0125$$

$$P_3 = 0.0423 \text{ W}$$

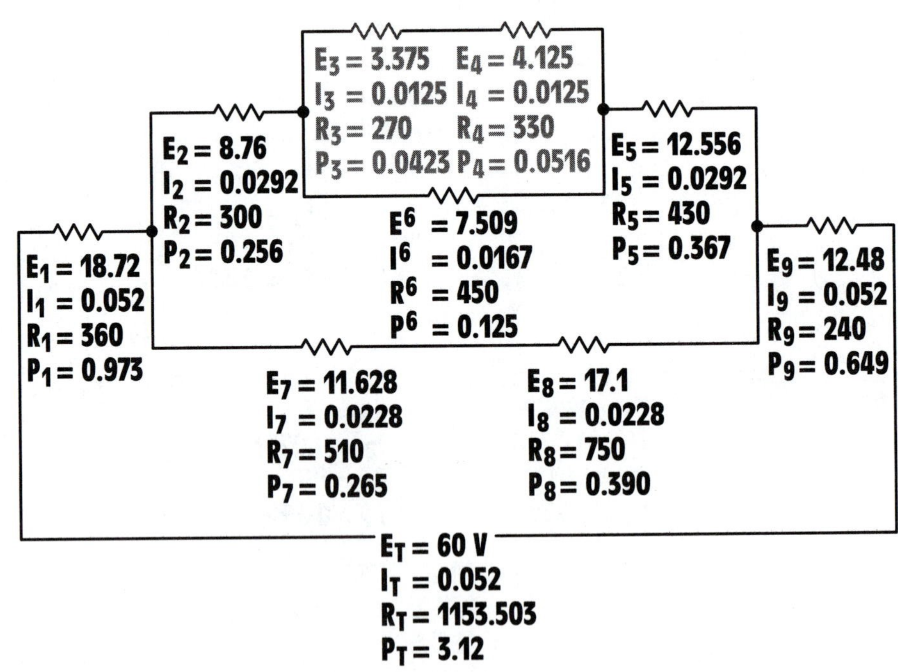

Figure 7-20 Computing the values for R_3 and R_4.

$$E_4 = I_4 \times R_4$$

$$E_4 = 0.0125 \times 330$$

$$E_4 = 4.125 \text{ V}$$

$$P_4 = E_4 \times I_4$$

$$P_4 = 4.125 \times 0.0125$$

$$P_4 = 0.0516 \text{ W}$$

SUMMARY

1. Combination circuits are circuits that contain both series and parallel branches.
2. The three rules for series circuits are:
   A. The current is the same at any point in a series circuit.
   B. The total resistance is the sum of each individual resistor.
   C. The applied voltage is equal to the sum of the voltage drops across each individual component.
3. The three rules for parallel circuits are:
   A. The voltage is the same across any branch of a parallel circuit.
   B. The total current is the sum of the individual currents through each path in the circuit.
   C. The reciprocal of the total resistance is the sum of the reciprocals of the branch resistances.
4. Solving combination circuits is generally easier if the circuit is reduced to simpler circuits.

REVIEW QUESTIONS

1. Refer to Figure 7-2. Replace the values shown with the listed values. Solve for all of the unknown values.

   $I_T = 0.6$ A
   $R_1 = 470\ \Omega$
   $R_2 = 360\ \Omega$
   $R_3 = 510\ \Omega$
   $R_4 = 430\ \Omega$

2. Refer to Figure 7-5. Replace the values shown with the listed values. Solve for all of the unknown values.

$E_T = 63\text{ V}$
$R_1 = 1000\ \Omega$
$R_2 = 2200\ \Omega$
$R_3 = 1800\ \Omega$
$R_4 = 910\ \Omega$
$R_5 = 3300\ \Omega$
$R_6 = 4300\ \Omega$
$R_7 = 860\ \Omega$

Practice Problems

1. Refer to the circuit shown in *Figure 7-21*. Determine the amount of resistance that would be measured across the points indicated.

A-B ________ Ω A-C ________ Ω A-D ________ Ω

B-C ________ Ω B-D ________ Ω C-D ________ Ω

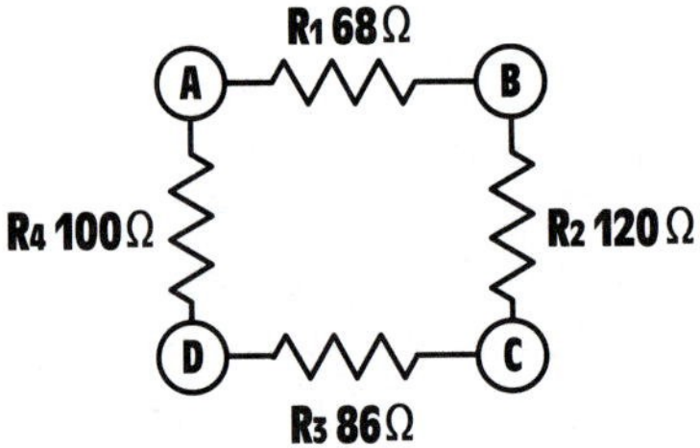

Figure 7-21 Practice Problem 1.

2. Refer to the circuit shown in *Figure 7-22*. Determine the amount of resistance that would be measured across the points indicated.

A-B ________ Ω A-C ________ Ω A-D ________ Ω

B-C ________ Ω B-D ________ Ω C-D ________ Ω

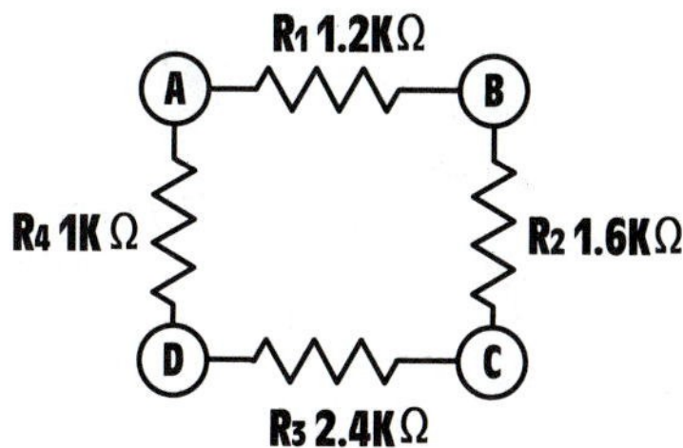

Figure 7-22 Practice Problem 2.

3. Refer to the circuit shown in *Figure 7-23*. Determine the amount of resistance that would be measured across the points indicated.

A-B ________ Ω A-C ________ Ω A-D ________ Ω

B-C ________ Ω B-D ________ Ω C-D ________ Ω

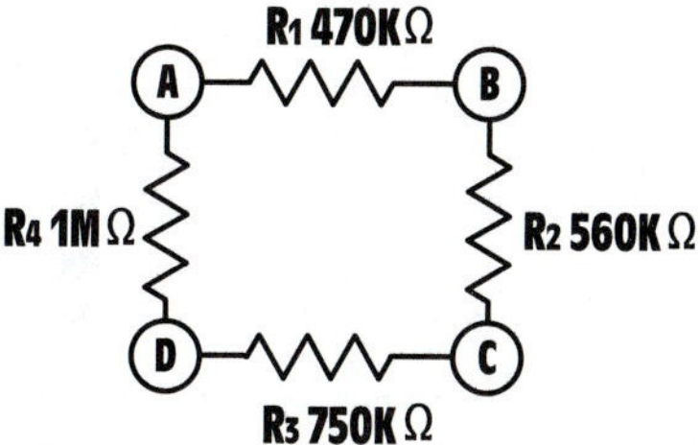

Figure 7-23 Practice Problem 3.

4. Refer to the circuit shown in *Figure 7-24*. Determine the amount of resistance that would be measured across the points indicated.

A-B ________ Ω A-C ________ Ω A-D ________ Ω

B-C ________ Ω B-D ________ Ω C-D ________ Ω

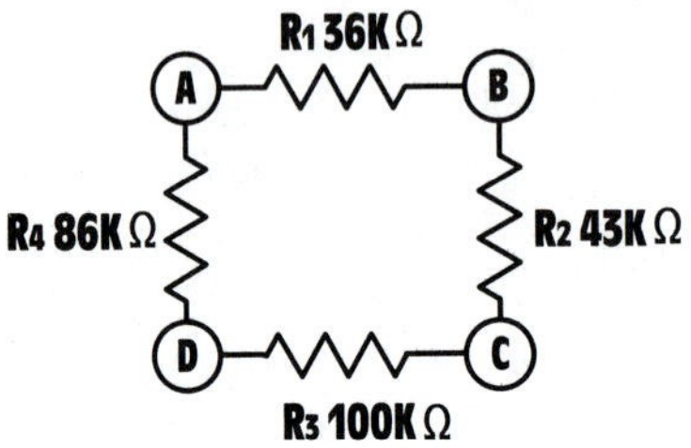

Figure 7-24 Practice Problem 4.

5. Find all the missing values in the circuit shown in *Figure 7-25*.

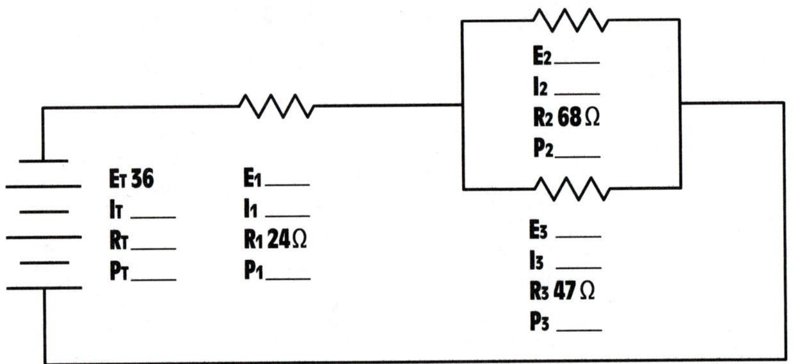

Figure 7-25 Practice Problem 5.

6. Find all the missing values in the circuit shown in *Figure 7-26*.

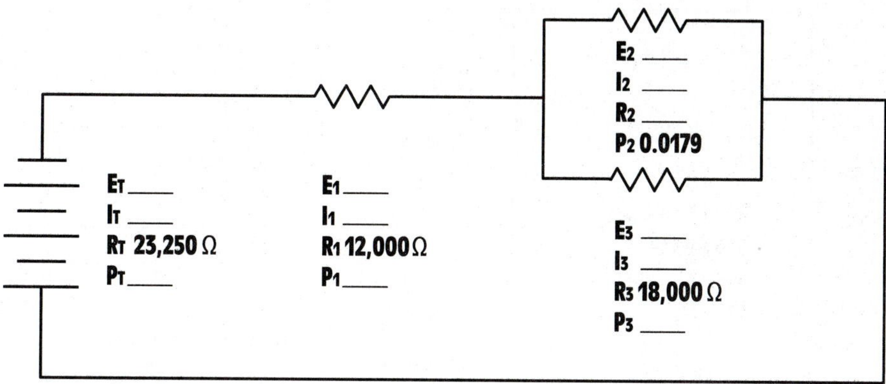

Figure 7-26 Practice Problem 6.

7. Find all the missing values in the circuit shown in *Figure 7-27*.

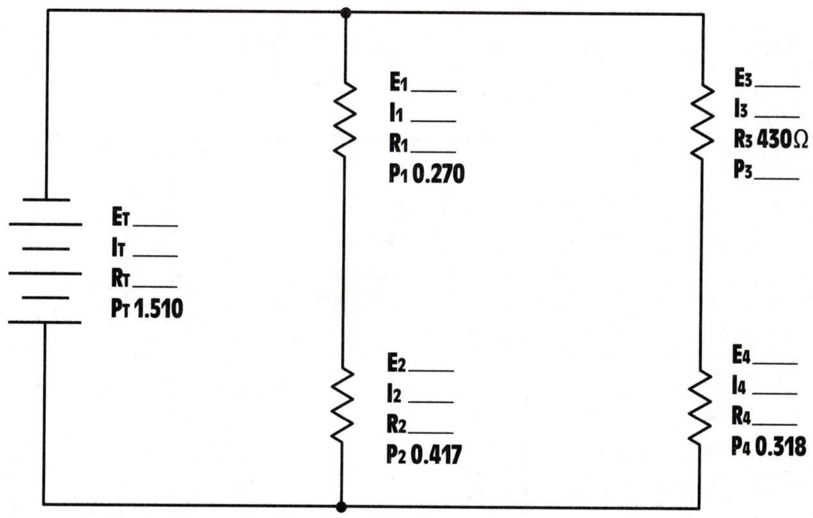

Figure 7-27 Practice Problem 7.

8. Find all the missing values in the circuit shown in *Figure 7-28*.

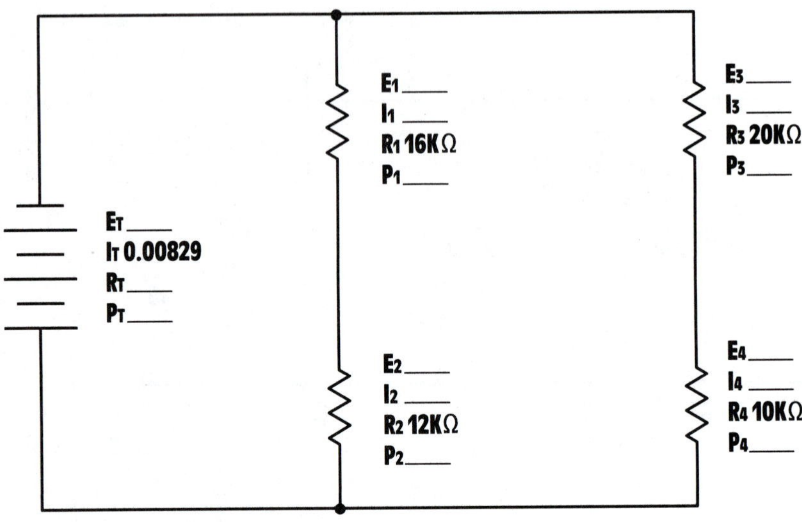

Figure 7-28 Practice Problem 8.

9. Find all the missing values in the circuit shown in *Figure 7-29*.

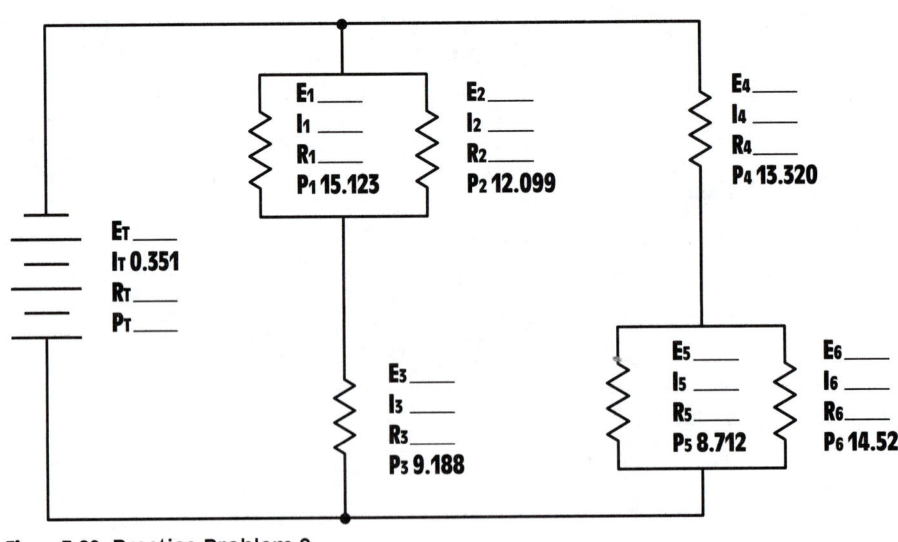

Figure 7-29 Practice Problem 9.

10. Find all the missing values in the circuit shown in *Figure 7-30.*

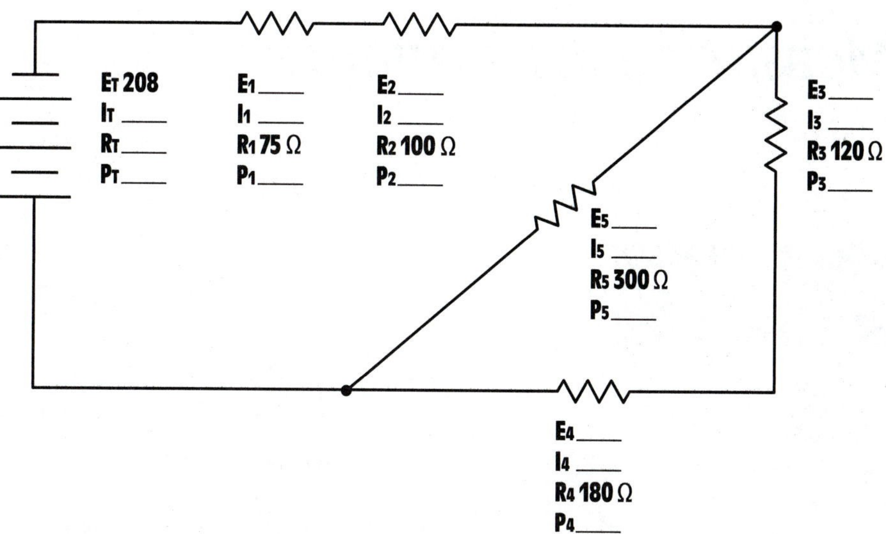

Figure 7-30 Practice Problem 10.

8 **Unit**

Measuring Instruments

objectives

fter studying this unit, you should be able to:

- Discuss the operation of a d'Arsonival meter movement.
- Discuss the operation of a moving iron type of movement.
- Connect a voltmeter to a circuit.
- Connect and read an analog multimeter.
- Connect an ammeter.
- Measure resistance using an ohmmeter.

Anyone desiring to work in the electrical and electronics field must become proficient with the common instruments used to measure electrical quantities. These instruments are the voltmeter, ammeter, and ohmmeter. Without meters it would be impossible to make meaningful interpretations of what is happening in a circuit. Meters can be divided into two general types, analog and digital.

ANALOG METERS

Analog meters are characterized by the fact that they use a pointer and scale to indicate their value, *Figure 8-1*. There are different types of analog meter movements. One of the most common is the **d'Arsonival movement** shown in *Figure 8-2*. This type of movement is often referred to as a **moving coil** type meter. A coil of wire is suspended between the poles of a permanent magnet. The coil is suspended either by jeweled movements similar to those used in watches or by taut bands. The taut band type offers less turning friction than the jeweled movement. These meters can be made to operate on very small amounts of current and are often referred to as **galvanometers.**

analog meters

d'Arsonival movement

moving coil meter

galvanometers

PRINCIPLE OF OPERATION

Analog meters operate on the principle that like magnetic poles repel each other. As current passes through the coil, a magnetic field is created around the coil. The direction of current flow through the meter is such that the same polarity of magnetic pole is created around the coil as that of the permanent magnet. This causes the coil to be deflected away from the pole of the magnet. A spring is used to retard the turning of the coil. The distance the coil turns against the spring is proportional to the strength of the magnetic field developed in the coil. If a pointer is added to the coil and a scale placed behind the pointer, a meter movement is created.

Figure 8-1 An analog meter.

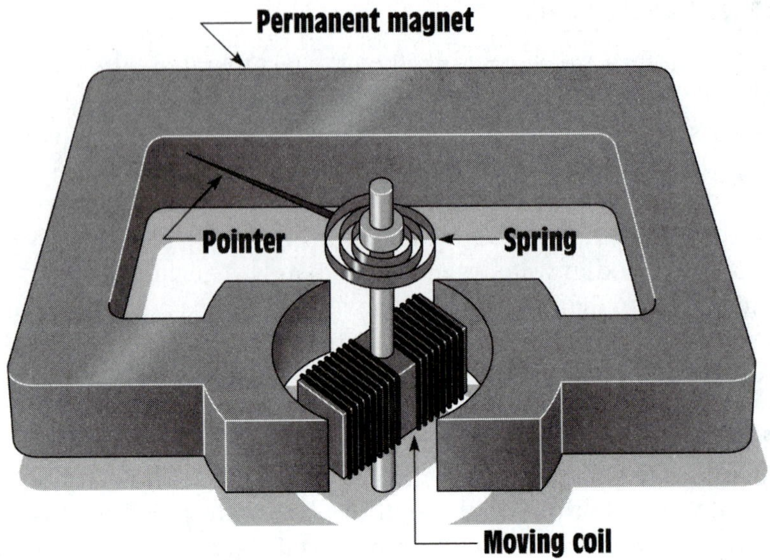

Figure 8-2 Basic d'Arsonival meter movement.

Since the turning force of this meter depends on the repulsion of magnetic fields, it will operate on DC current only. If a 60-Hz AC current is connected to the moving coil, the magnetic polarity will change 60 times per second and the net turning force will be zero. For this reason, a DC voltmeter will indicate zero if connected to an AC line. When this type of movement is to be used to measure AC values, the current must be rectified or changed into DC before it is applied to the meter, *Figure 8-3*.

ANALOG VOLTMETERS

voltmeter

The **voltmeter** is designed to be connected directly across the source of power. *Figure 8-4* shows an analog voltmeter being used to test the voltage of a battery. Notice that the leads of the meter are connected directly across the source of voltage. A voltmeter can be connected directly across the power source because it has a very high resistance connected in series with the meter movement, *Figure 8-5*. The industrial standard for an analog type voltmeter is 20,000 Ω per volt for DC and 5,000 Ω per volt for AC. Assume the voltmeter shown in Figure 8-5 is an AC meter and has a full scale range of 300 V. The meter circuit (meter plus resistor) would, therefore, have a resistance of 1,500,000 Ω (300 V × 5,000 Ω per volt = 1,500,000 Ω).

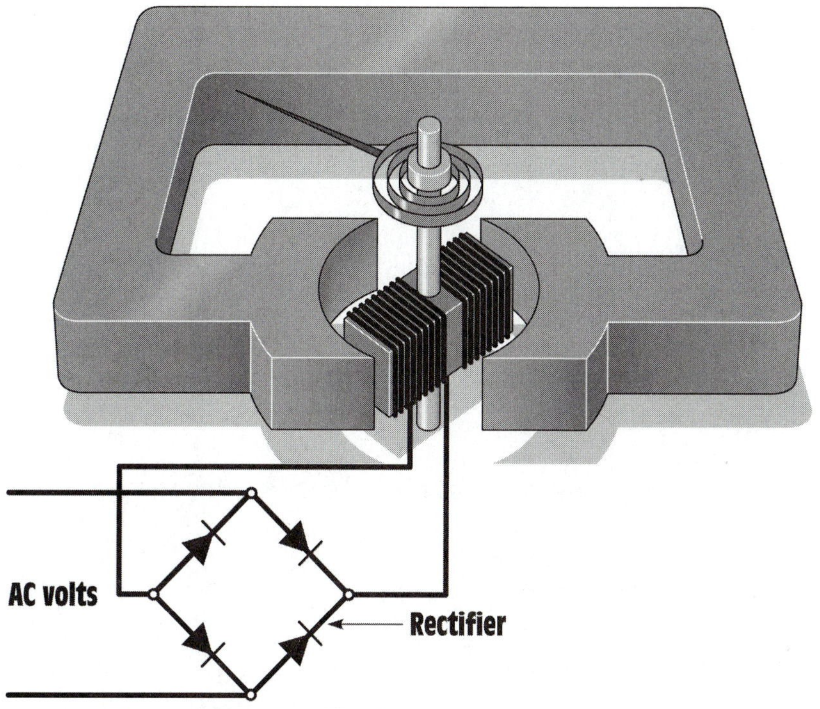

Figure 8-3 A rectifier changes AC voltage into DC voltage.

Figure 8-4 A voltmeter connects directly across the power source.

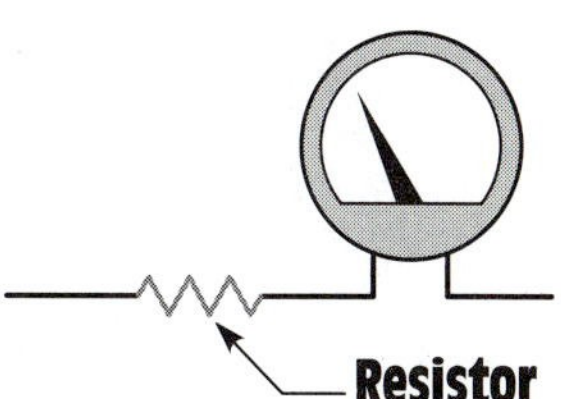

Figure 8-5 A resistor connects in series with the meter.

MULTIRANGE VOLTMETERS

multirange voltmeters

Most voltmeters are **multirange voltmeters,** which means that they are designed to use one meter movement to measure several ranges of voltage. For example, one meter may have a selector switch that permits full scale ranges to be selected. These ranges may be 3-V full scale, 12-V full scale, 30-V full scale, 60-V full scale, 120-V full scale, 300-V full scale, and 600-V full scale. The reason for making a meter with this many scales is to make the device as versatile as possible. If it is necessary to check for a voltage of 480 V, the meter can be set on the 600-V range. If it became necessary to check a 24-V system, this would be very difficult to do on the 600-V range. If the meter is set on the 30-V range, it becomes a simple matter to test for a voltage of 24 V. The meter shown in Figure 8-1 has multirange selection for voltage.

When the selector switch of this meter is turned, steps of resistance are inserted in the circuit to increase the range, or removed from the circuit to decrease the range. The meter shown in *Figure 8-6* has four range settings for full scale voltage: 30 V, 60 V, 300 V and 600 V. Notice that when the higher voltage settings are selected, more resistance is inserted in the circuit.

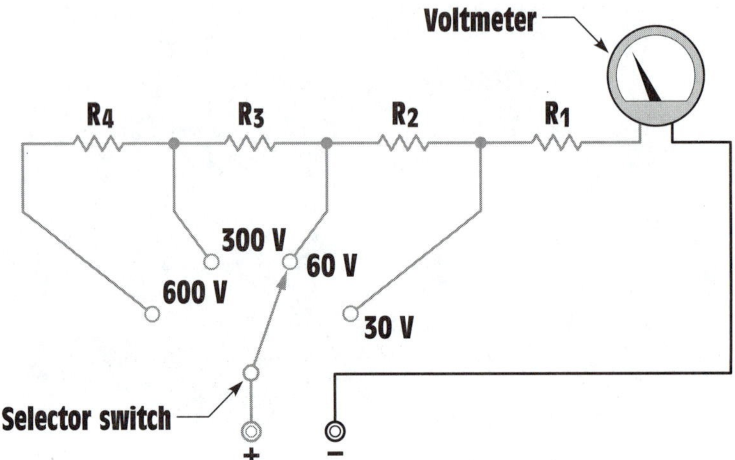

Figure 8-6 A rotary selector switch is used to change the full range setting.

READING A METER

Learning to read the scale of a multimeter takes time and practice. Most people use meters every day without thinking about it. A very common type of meter used daily by most people is shown in *Figure 8-7*. The meter illustrated is a speedometer similar to those seen in automobiles. This meter is designed to measure speed. It is calibrated in miles per hour. The speedometer shown has a full scale value of 80 mph. If the pointer is positioned as shown in Figure 8-7, most people will know instantly that the speed of the automobile is 55 mph.

Figure 8-7 A speedometer.

Figure 8-8 illustrates another common meter used by most people. This meter is used to measure the amount of fuel in the tank of the automobile. Most people can glance at the pointer of the meter and know that the meter is indicating that there is one-quarter of a tank of fuel remaining. Now assume that the tank has a capacity of 20 gallons. The meter is indicating that there are a total of 5 gallons of fuel remaining in the tank.

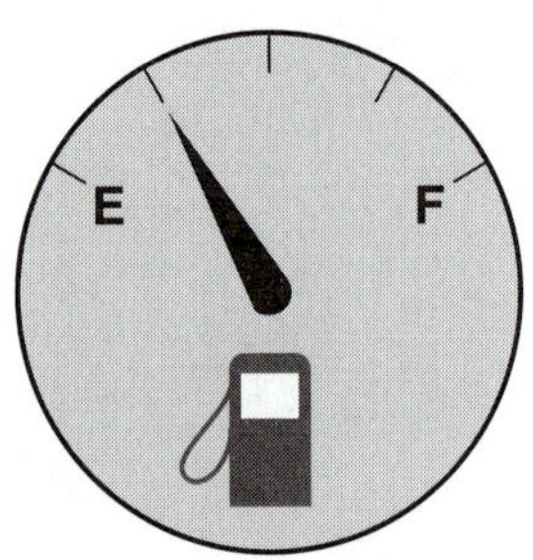

Figure 8-8 A fuel gauge.

Learning to read the scale of a multimeter is similar to learning to read a speedometer or fuel gauge. The meter scale shown in *Figure 8-9* has several scales used to measure different quantities and values. The top of the scale is used to measure resistance or ohms. Notice that the scale begins with infinity at the left and ends with zero at the right. Ohmmeters will be covered later in this unit. The second scale is labeled AC-DC and is used to measure voltage. Notice that this scale has three different full scale values. The top scale is 0–300, the second scale is 0–60, and the third scale is 0–12. The scale used is determined by the setting of the range control switch. The third set of scales is labeled AC amps. This scale is used with a clamp-on

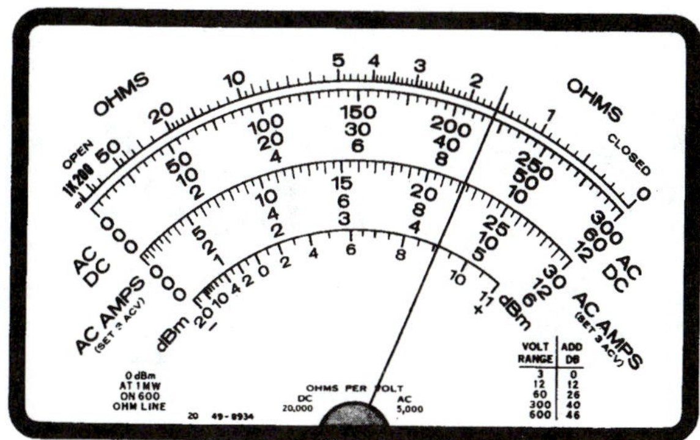

Figure 8-9 A typical multimeter.

ammeter attachment that can be used with some meters. The last scale is labeled dBm and is used to measure decibels.

READING A VOLTMETER

Notice that the three voltmeter scales use the primary numbers 3, 6, and 12 and are in multiples of 10 of these numbers (300 is a multiple of 3 and 60 is a multiple of 6). Since the numbers are in multiples of 10, it is easy to multiply or divide the readings in your head by moving a decimal point. Remember that any number can be multiplied by 10 by moving the decimal point one place to the right, and any number can be divided by 10 by moving the decimal point one place to the left. For example, if the selector switch is set to permit the meter to indicate a voltage of 3-V full scale, the 300-V scale would be used and the reading would be divided by 100. The reading can be divided by 100 by moving the decimal point two places to the left. In *Figure 8-10,* the meter is indicating a voltage of 2.5 V

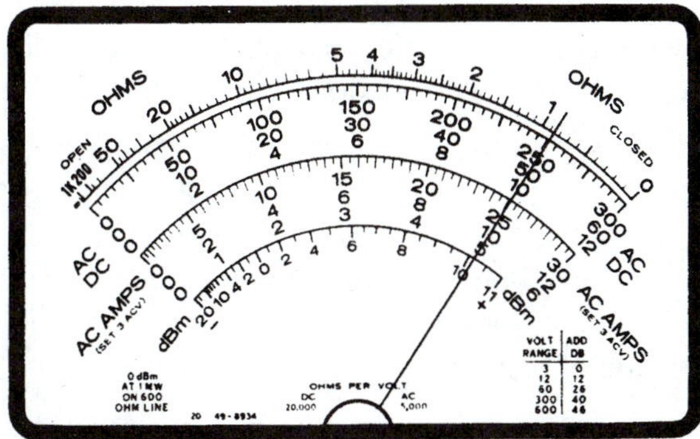

Figure 8-10 Reading the meter.

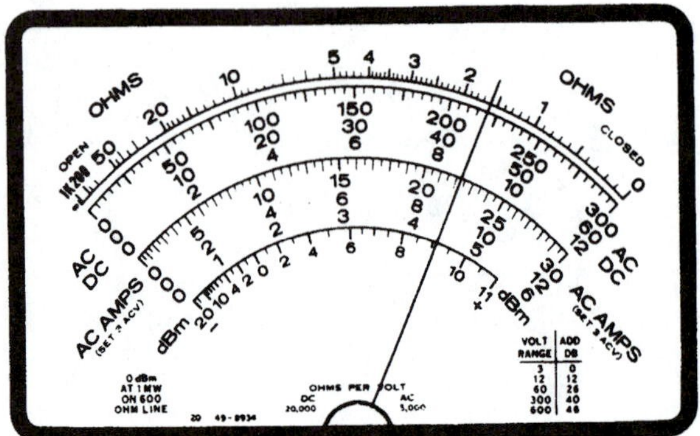

Figure 8-11 Reading the meter.

if the selector switch is set for 3-V full scale. The pointer is indicating a value of 250. Moving the decimal point two places to the left will give a reading of 2.5 V. If the selector switch is set for a full scale value of 30 V, the meter shown in Figure 8-10 would be indicating a value of 25 V. This reading is obtained by dividing the scale by 10 and moving the decimal point one place to the left.

Now assume that the meter has been set to have a full scale value of 600 V. The meter shown in *Figure 8-11* is indicating a voltage of 440 V. Since the full scale value of the meter is set for 600 V, use the 60-V range and multiply the reading on the meter by 10. Do this by moving the decimal point one place to the right. The pointer in Figure 8-11 is indicating a value of 44. If this value is multiplied by 10, the correct reading becomes 440 V.

Three distinct steps should be followed when reading a meter. This is especially true for someone who has not had a great deal of experience reading a multimeter. These steps are:

1. *Determine what the meter indicates.* Is the meter set to read a value of DC voltage, DC current, AC voltage, AC current, or ohms? It is impossible to read a meter if you don't know what it is used to measure.
2. *Determine the full scale value of the meter.* The advantage of a multimeter is that it has the ability to measure a wide range of values and quantities. After it has been determined what quantity the meter is set to measure, it must then be determined what the range of the meter is. There is a great deal of difference in reading when the meter is set to indicate a value of 600-V full scale and when it is set for 30-V full scale.
3. *Read the meter.* The last step is to determine what the meter is indicating. It may be necessary to determine the value of the hash marks

on the meter face for the range for which the selector switch is set. If the meter in Figure 8-9 is set for 300-V full scale, each hash mark has a value of 5 V. If the full scale value of the meter is 60-V, however, each hash mark has a value of 1 V.

THE AMMETER

ammeter

The **ammeter,** unlike the voltmeter, is a very low-impedance device. The ammeter must be connected in series with the load to permit the load to limit the current flow, *Figure 8-12.* An ammeter has a typical impedance of less than 0.1 Ω. If this meter is connected in parallel with the power supply, the impedance of the ammeter is the only thing to limit the amount of current flow in the circuit. Assume that an ammeter with a resistance of 0.1 Ω is connected across a 240-V AC line. The current flow in this circuit would be 2400 A (240/.1 = 2400). The blinding flash of light would be followed by the destruction of the ammeter. Ammeters connected directly into the circuit, as shown in Figure 8-12, are referred to as in-line ammeters. *Figure 8-13* shows an ammeter of this type.

AMMETER SHUNTS

ammeter shunt

DC ammeters are constructed by connecting a common moving coil type of meter across a shunt. An **ammeter shunt** is a low-resistance device used to conduct most of the circuit current away from the meter movement. Since the meter movement is connected in parallel with the shunt, the voltage drop across the shunt is the voltage applied to the meter. Most

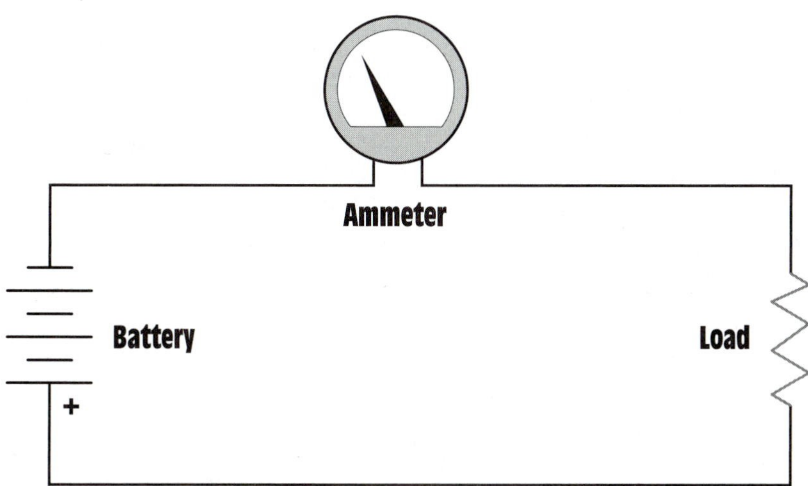

Figure 8-12 An ammeter connects in series with the load.

Figure 8-13 In-line ammeter.

ammeter shunts are manufactured to have a voltage drop of 50 millivolts (mV). If a 50-mV meter movement is connected across the shunt as shown in *Figure 8-14,* the pointer will move to the full scale value when the rated current of the shunt is flowing. In the example shown, the ammeter shunt is rated to have a 50-mV drop when 10 A of current are flowing in the circuit. Since the meter movement has a full scale voltage of 50 mV, it will indicate the full scale value when 10 A of current are flowing through the shunt. An ammeter shunt is shown in *Figure 8-15.*

Ammeter shunts can be purchased to indicate different values. If the same 50-mV movement is connected across a shunt designed to drop 50 mV when 100 A of current flow through it, the meter will now have a full scale value of 100 A.

The resistance of an ammeter shunt can be computed using Ohm's Law. The resistance of a shunt designed to have a voltage drop of 50 mV when 100 A of current flows through it is:

$$R = \frac{E}{I}$$

$$R = \frac{0.050}{100}$$

$$R = 0.0005\ \Omega\text{, or } 0.5\ \text{m}\Omega$$

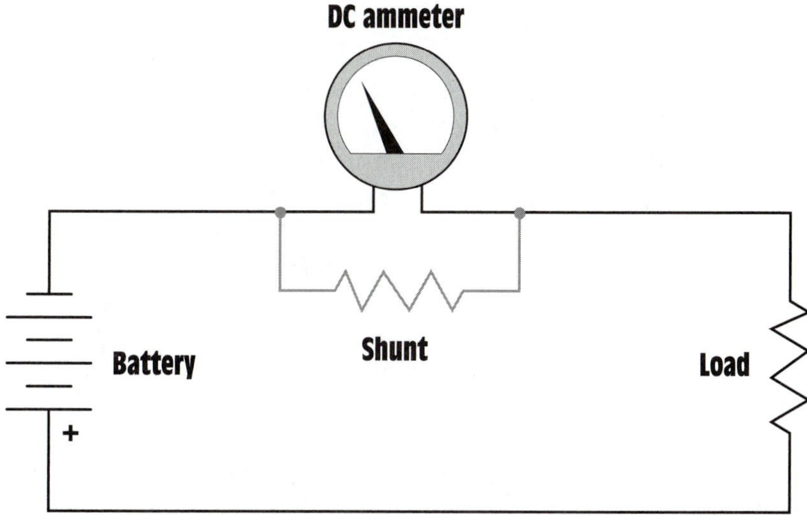

Figure 8-14 A shunt is used to set the value of the ammeter.

Figure 8-15 Ammeter shunt.

In the previous problem, no consideration was given to the electrical values of the meter movement. The reason is that the amount of current needed to operate the meter movement is so small compared to the 100 A circuit current, it could have no meaningful effect on the resistance value of the shunt. When computing the value for a low-current shunt, however, the meter values must be taken into consideration.

MULTIRANGE AMMETERS

multirange ammeters

When a multirange meter is used, care must be taken that the shunt is never disconnected from the meter.

Many ammeters, called multirange ammeters, are designed to operate on more than one range. This is done by connecting the meter movement to different shunts. **When a multirange meter is used, care must be taken that the shunt is never disconnected from the meter.** Disconnection would cause the meter movement to be inserted in series with the circuit, and full circuit current would flow through the meter. Two basic methods are used for connecting shunts to a meter movement. One method is to use a make-before-break switch. This type of switch is designed so that it will make contact with the next shunt before it breaks connection with the shunt to which it is connected, *Figure 8-16*. This method does, however, present a problem: contact resistance. Notice in Figure 8-16 that the rotary switch is in series with the shunt resistors. This causes the contact resistance to be added to the shunt resistance and can cause inaccuracy in the meter reading.

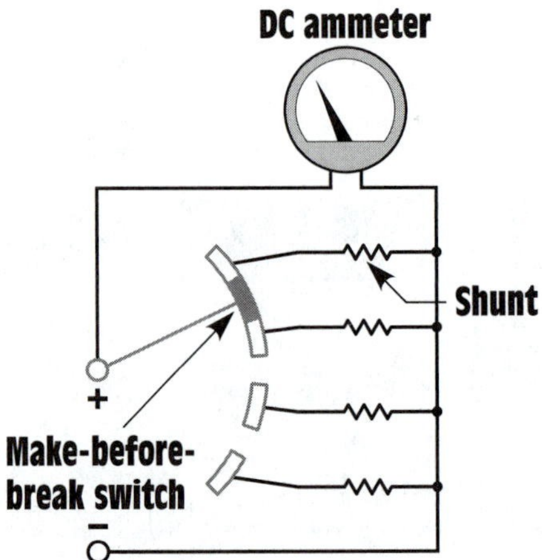

Figure 8-16 A make-before-break switch is used to change meter shunts.

THE AYRTON SHUNT

Ayrton shunt

The second method is to use an **Ayrton shunt,** *Figure 8-17.* In this type of circuit, connection is made to different parts of the shunt, and the meter movement is never disconnected from the shunt. Also, notice that the switch connections are made external to the shunt and meter. This prevents contact resistance from affecting the accuracy of the meter.

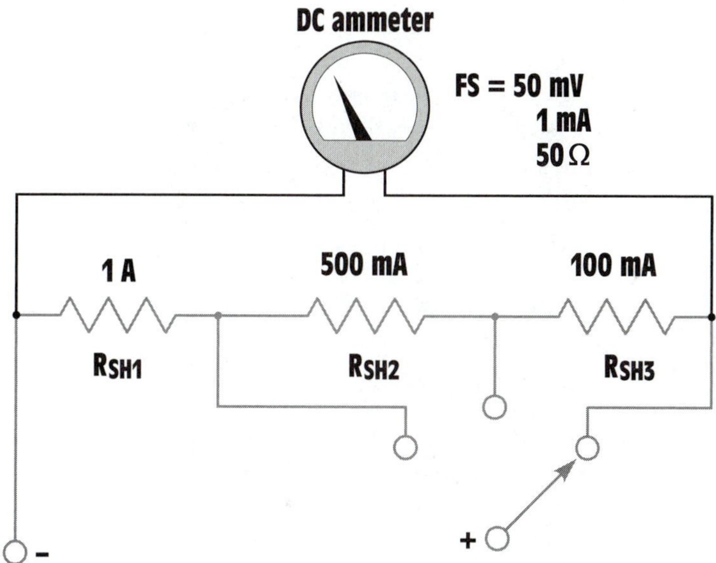

Figure 8-17 An Ayrton shunt.

ALTERNATING CURRENT (AC) AMMETERS

current transformer

Shunts can be used with AC ammeters to increase their range but cannot be used to decrease their range. Most AC ammeters use a **current transformer** instead of shunts to change scale values. This type of ammeter is shown in *Figure 8-18.* The primary of the transformer is connected

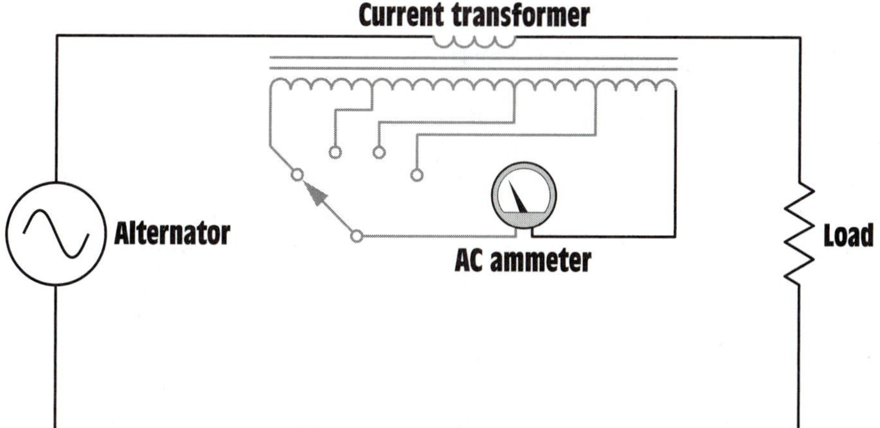

Figure 8-18 A current transformer is used to change the range of an AC ammeter.

Figure 8-19 Multirange AC ammeter.

in series with the load, and the ammeter is connected to the secondary of the transformer. Notice that the range of the meter is changed by selecting different taps on the secondary of the current transformer. The different taps on the transformer provide different turns ratios between the primary and secondary of the transformer. An ammeter of this type is shown in *Figure 8-19*.

CURRENT TRANSFORMERS (CTS)

When a large amount of AC current must be measured, a different type of current transformer is connected in the power line. These transformers have ratios that start at 200:5 and can have ratios of several thousand to five. These current transformers, generally referred to in industry as *CTs,* have a standard secondary current rating of 5-A AC. They are designed to be operated with a 5-A AC ammeter connected directly to their secondary winding, which produces a short circuit. CTs are designed to operate with the secondary winding shorted. **The secondary winding of a CT should never be opened when power is being applied to the primary. This will cause the transformer to produce a step-up in voltage that could be high enough to kill anyone who comes in contact with it.**

The secondary winding of a CT should never be opened when power is being applied to the primary. This will cause the transformer to produce a step-up in voltage that could be high enough to kill anyone who comes in contact with it.

Figure 8-20 Current transformer with a ratio of 600:5.

A current transformer is basically a toroid transformer. A toroid transformer is constructed with a hollow core similar to a donut, *Figure 8-20.* When current transformers are used, the main power line is inserted through the opening in the transformer, *Figure 8-21.* The power line acts as the primary of the transformer and is considered to be one turn.

The turns ratio of the transformer can be changed by looping the power wire through the opening in the transformer to produce a primary winding of more than one turn. For example, assume a current transformer has a ratio of 600:5. If the primary power wire is inserted through the opening, it will require a current of 600 A to deflect the meter full scale. If the primary power conductor is looped around and inserted through the window a second time, the primary now contains two turns of wire instead of one, *Figure 8-22.* It now requires 300 A of current flow in the primary to deflect the meter full scale. If the primary conductor is looped through the opening a third time, it would require only 200 A of current flow to deflect the meter full scale.

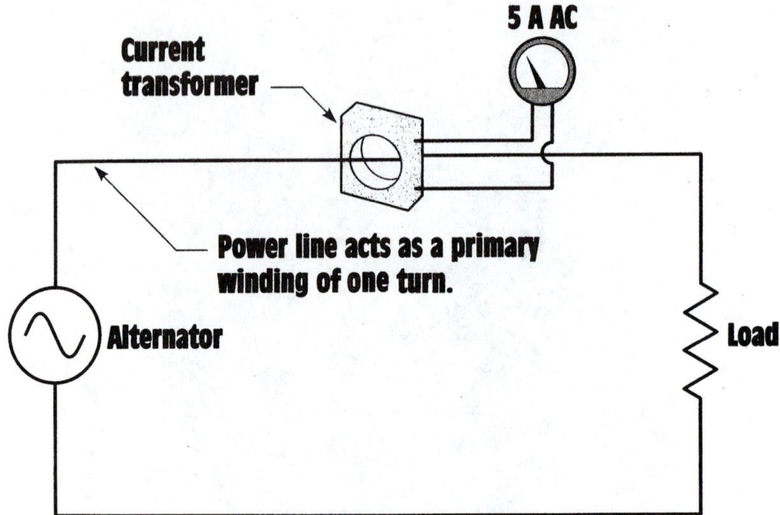

Figure 8-21 The main power line is run through the opening of the current transformer.

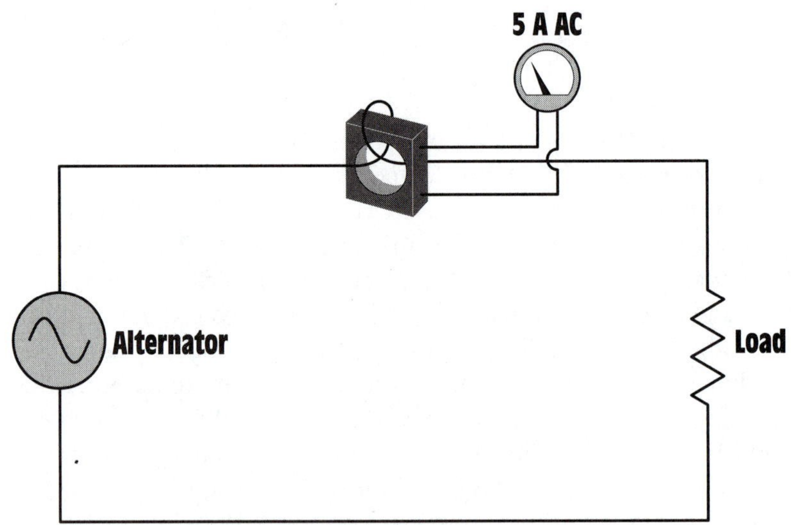

Figure 8-22 The primary conductor loops through the current transformer to produce a second turn, which changes the ratio.

CLAMP-ON AMMETERS

clamp-on ammeter

Many electricians use a type of AC ammeter called a **clamp-on ammeter,** *Figure 8-23.* To use this type of meter, the jaw of the meter is clamped around one of the conductors supplying power to the load, *Figure 8-24.* The meter is clamped around only one of the lines. If the meter is clamped around more than one line, the magnetic fields of the wires cancel each other and the meter indicates zero.

The clamp-on meter also uses a current transformer to operate. The jaw of the meter is part of the core material of the transformer. When the meter is connected around the current-carrying wire, the changing magnetic field produced by the AC current induces a voltage into the current transformer. The strength and frequency of the magnetic field determines the amount of voltage induced in the current transformer. Because 60 Hz is a standard frequency throughout the country, the amount of induced voltage is proportional to the strength of the magnetic field.

The clamp-on ammeter can be used with different range settings by changing the turns ratio of the secondary of the transformer, just as is done

Figure 8-23 Clamp-on AC ammeter. *(Courtesy of Amprobe Instrument®)*

Figure 8-24 The clamp-on ammeter connects around only one conductor.

on the in-line ammeter. The primary of the transformer is the conductor around which the movable jaw is connected. If the ammeter is connected around one wire, the primary has one turn of wire compared to the turns of the secondary. The turns ratio can be changed in the same manner as the ratio of the CT. If two turns of wire are wrapped around the jaw of the ammeter, *Figure 8-25,* the primary winding now contains two turns instead of one, and the turns ratio of the transformer is changed. The ammeter will now indicate double the amount of current in the circuit. The reading on the scale of the meter would have to be divided by two to get the correct reading. The ability to change the turns ratio of a clamp-on ammeter can be useful for measuring low currents. Changing the turns ratio is not limited to wrapping two turns of wire around the jaw of the ammeter. Any number of turns can be wrapped around the jaw, and the reading will be divided by that number.

A problem with many clamp-on type ammeters is that their lowest scale value is too high to accurately measure low current values. For example, an ammeter with a full scale range of 0–6 A would not be able to accurately measure a current flow of 0.2 A. The answer to this problem is to wrap many turns of wire around the jaw of the ammeter to change the secondary scale value. If ten turns of wire are wrapped around the jaw of a 0–6 A meter, the meter will now indicate a full scale value of 0.6 A. The meter could

Figure 8-25 Looping the conductor around the jaw of the ammeter changes the ratio.

now accurately measure a current of 0.2 A. In the field, however, it is not often that there is enough slack wire to make ten wraps around the jaw of an ammeter. A simple device can be constructed to overcome this problem. Assume that ten turns of wire are wound around a piece of non-conductive material, such as a piece of plastic pipe, *Figure 8-26*. Plastic tape is then used to prevent the wire from slipping off the plastic core. If alligator clips are attached to the ends of the wire, the device can be inserted in series with the load in a similar manner to using an in-line ammeter. The jaw of the ammeter can then be inserted through the opening in the plastic pipe, *Figure 8-27*. The current value is then read by moving the decimal one place to the left. A full scale value of 0–6 A becomes a full scale value of 0–0.6 A, and a 0–30 A scale becomes a 0–3 A scale.

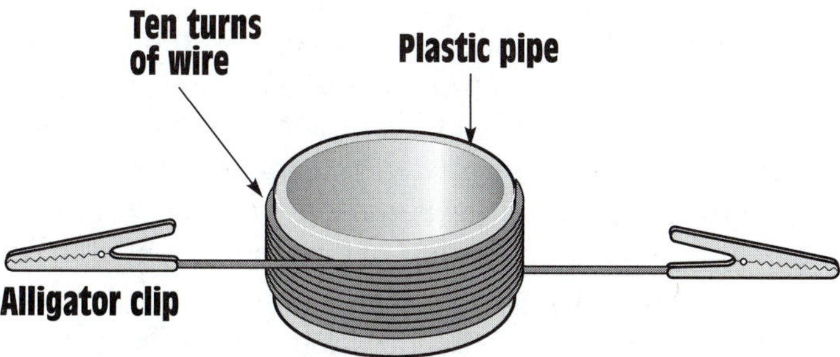

Figure 8-26 Ten turns of wire are wrapped around a plastic pipe and two alligator clips are attached.

Figure 8-27 A simple device for increasing the range of a clamp-on ammeter.

DC-AC CLAMP-ON AMMETERS

Most clamp-on ammeters that have the ability to measure both direct and alternating current do not operate on the principle of the current transformer. Current transformers depend on induction, which means that the current in the line must change direction periodically to provide a change of magnetic field polarity. It is the continuous change of field strength and direction that permits the current transformer to operate. The current in a DC circuit is unidirectional and does not change polarity, so it would not permit the current transformer to operate.

DC-AC clamp-on ammeters, *Figure 8-28,* use the *Hall effect* as the basic principle of operation. The Hall effect was discovered by Edward H. Hall at Johns Hopkins University in 1879. Hall originally used a piece of pure gold to produce the Hall effect, but today a semiconductor material is used because it has better operating characteristics and is less expensive. The device is often referred to as a *Hall generator. Figure 8-29* illustrates the operating principle of the Hall generator. A constant current generator is used to supply a continuous current to the semiconductor chip. The leads of a zero-center voltmeter are connected across the opposite sides of the chip. As long as the current flows through the center of the semiconductor chip, there is no potential difference or voltage developed across the chip.

Figure 8-28 DC-AC clamp-on ammeter. *(Courtesy of Amprobe Instrument)*
Digital clamp-on ammeter. *(Courtesy of Advanced Test Products)*

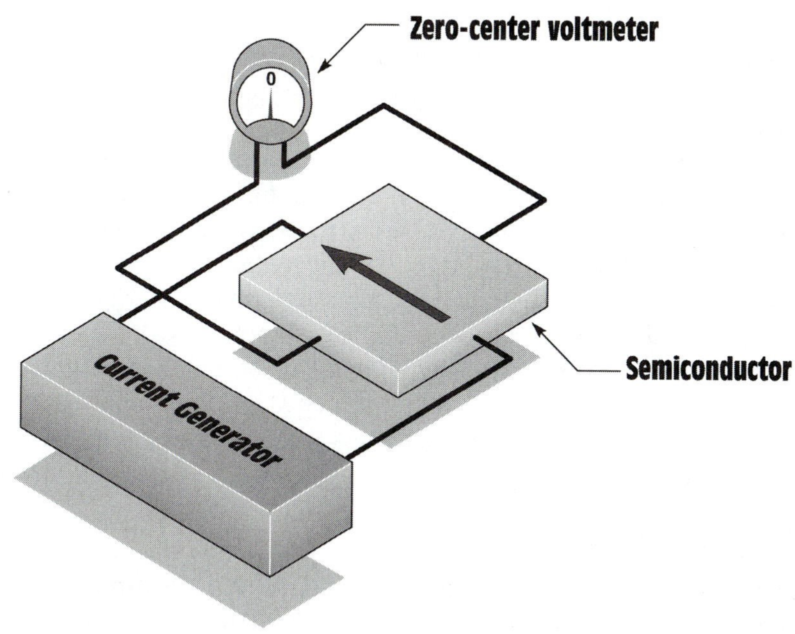

Figure 8-29 Basic Hall generator.

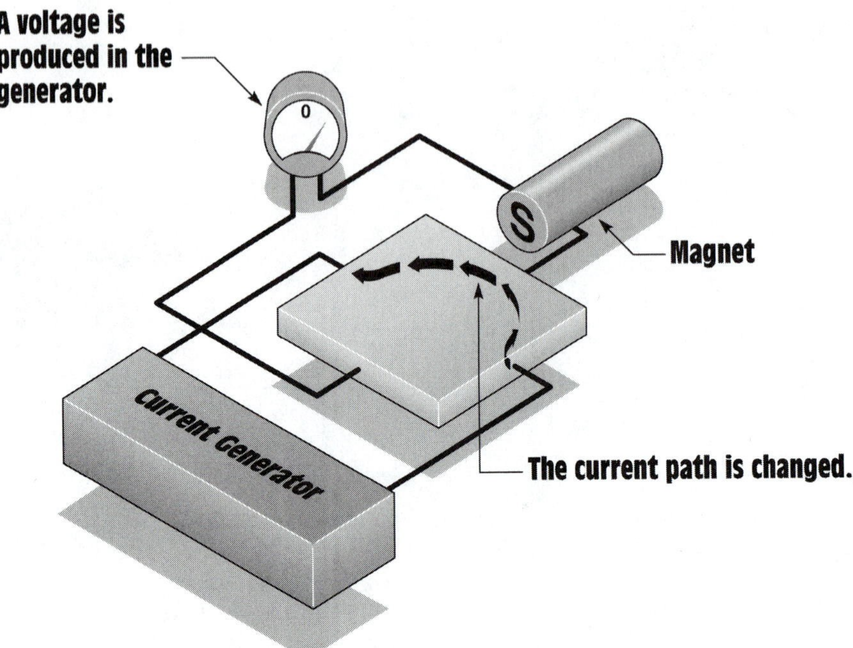

Figure 8-30 The presence of a magnetic field causes the Hall generator to produce a voltage.

If a magnetic field comes near the chip, *Figure 8-30,* the electron path is distorted and the current no longer flows through the center of the chip. A voltage across the sides of the chip is produced. The voltage is proportional to the amount of current flow and the amount of current distortion. Since the current remains constant and the amount of distortion is proportional to the strength of the magnetic field, the voltage produced across the chip is proportional to the strength of the magnetic field.

If the polarity of the magnetic field should be reversed, *Figure 8-31,* the current path would be distorted in the opposite direction, producing a voltage of the opposite polarity. Notice that the Hall generator produces a voltage in the presence of a magnetic field. It makes no difference whether the field is moving or stationary. The Hall effect can, therefore, be used to measure direct or alternating current.

THE OHMMETER

ohmmeter

The **ohmmeter** is used to measure resistance. The common VOM (volt-ohm-milliammeter) contains an ohmmeter. The ohmmeter has the only scale on a VOM that is nonlinear. The scale numbers increase in value as

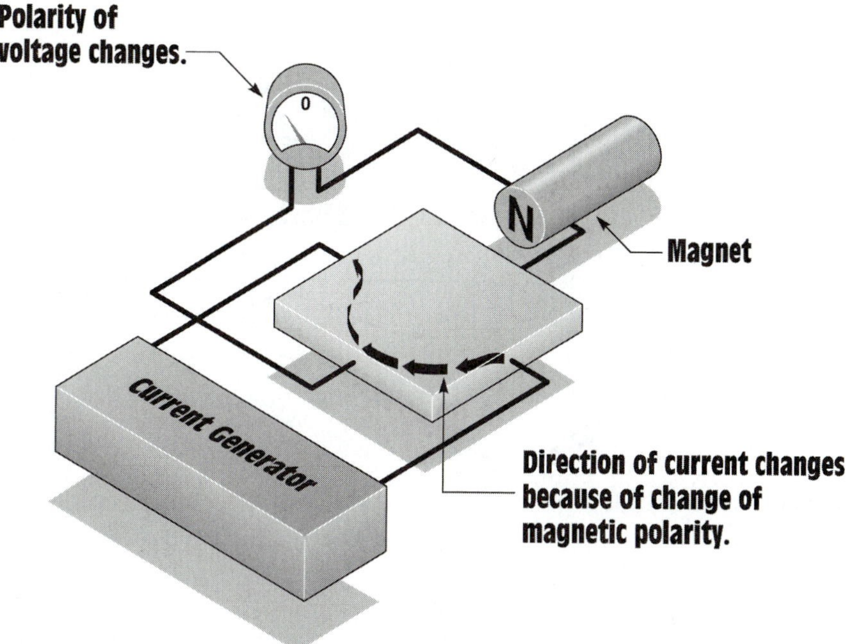

Figure 8-31 If the polarity of the magnetic field changes, the polarity of the voltage changes.

they progress from right to left. There are two basic types of analog ohmmeters, the series and the shunt. The series ohmmeter is used to measure high values of resistance, and the shunt type is used to measure low values of resistance. Regardless of the type used, the meter must provide its own power source to measure resistance. The power is provided by batteries located inside the instrument.

THE SERIES OHMMETER

A schematic for a basic series ohmmeter is shown in *Figure 8-32*. It is assumed that the meter movement has a resistance of 1,000 Ω and requires a current of 50 μA to deflect the meter full scale. The power source will be a 3-V battery. A fixed resistor with a value of 54 KΩ, R_1 is connected in series with the meter movement, and a variable resistor with a value of 10 KΩ, R_2, is connected in series with the meter and R_1. These resistance values were chosen to ensure there would be enough resistance in the circuit to limit the current flow through the meter movement to 50 μa. If Ohm's Law is used to compute the resistance needed (3 V/0.000050 A = 60,000 Ω), it will be seen that a value of 60K Ω is needed. This circuit contains a total of 65,000 Ω (1,000 [meter] + 54,000 + 10,000). The circuit resistance can be

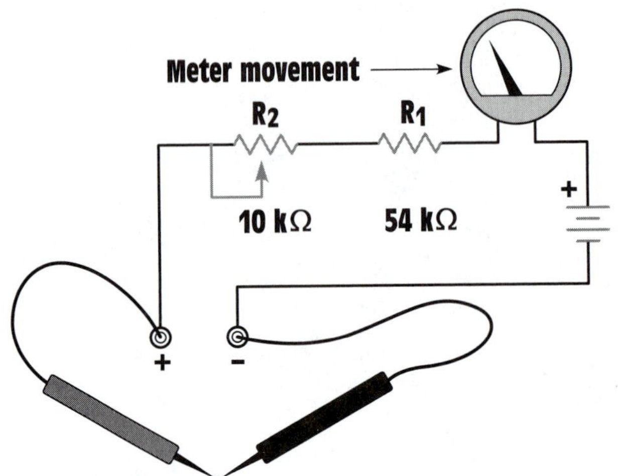

Figure 8-32 Basic series ohmmeter.

changed by adjusting the variable resistor to a value as low as 55,000 Ω, however, to compensate for the battery as it ages and becomes weaker.

When resistance is to be measured, the meter must first be zeroed. This is done with the ohms-adjust control, the variable resistor located on the front of the meter. To zero the meter, connect the leads together, Figure 8-32, and turn the ohms-adjust knob until the meter indicates zero at the far right end of the scale, *Figure 8-33*. When the leads are separated, the meter will again indicate infinity resistance at the left side of the meter scale. When the leads are connected across a resistance, the meter will again go up the scale. Since resistance has been added to the circuit, less than 50 μA of current will flow, and the meter will indicate some value other than zero.

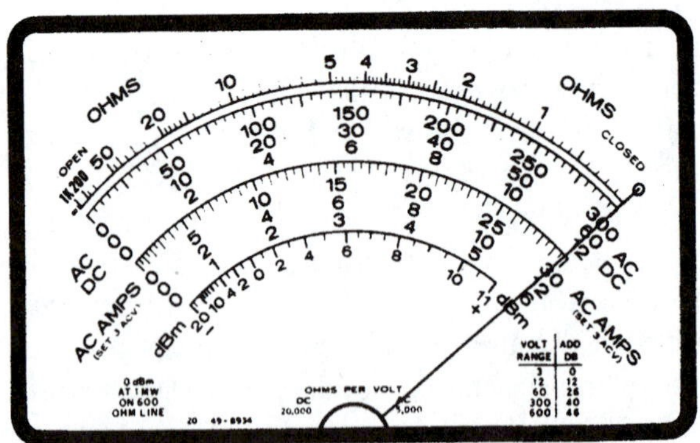

Figure 8-33 Adjusting the ohmmeter to zero.

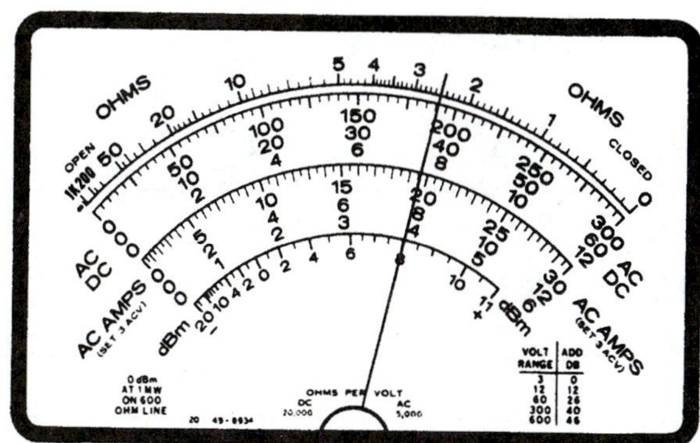

Figure 8-34 Reading the ohmmeter.

Figure 8-34 shows a meter indicating a resistance of 25 Ω, assuming the range setting is R_X1.

Ohmmeters can have different range settings, such as R_x1, R_x100, R_x1000, or $R_x10,000$. These different scales can be obtained by adding different values of resistance in the meter circuit and resetting the meter to zero. **An ohmmeter should always be readjusted to zero when the scale is changed.** On the R_x1 setting, the resistance is measured straight off the resistance scale located at the top of the meter. If the range is set for R_x1000, however, the reading must be multiplied by 1000. The ohmmeter reading shown in Figure 8-34 would be indicating a resistance of 25,000 Ω if the range had been set for R_x1000. Notice that the ohmmeter scale is read backward from the other scales: zero ohms is located on the far right side of the scale and maximum ohms is located at the far left side. It generally takes a little time and practice to read the ohmmeter properly.

An ohmmeter should always be readjusted to zero when the scale is changed.

SHUNT TYPE OHMMETERS

As stated previously, the shunt type ohmmeter is used for measuring low values of resistance. It operates on the same basic principle as an ammeter shunt. When using a shunt type ohmmeter, place the unknown value of resistance in parallel with the meter movement. This causes part of the circuit current to bypass the meter, *Figure 8-35*.

DIGITAL METERS

DIGITAL OHMMETERS

Digital ohmmeters display the resistance in figures instead of using a meter movement. When using a digital ohmmeter, care must be taken to notice the scale indication on the meter. For example, most digital meters

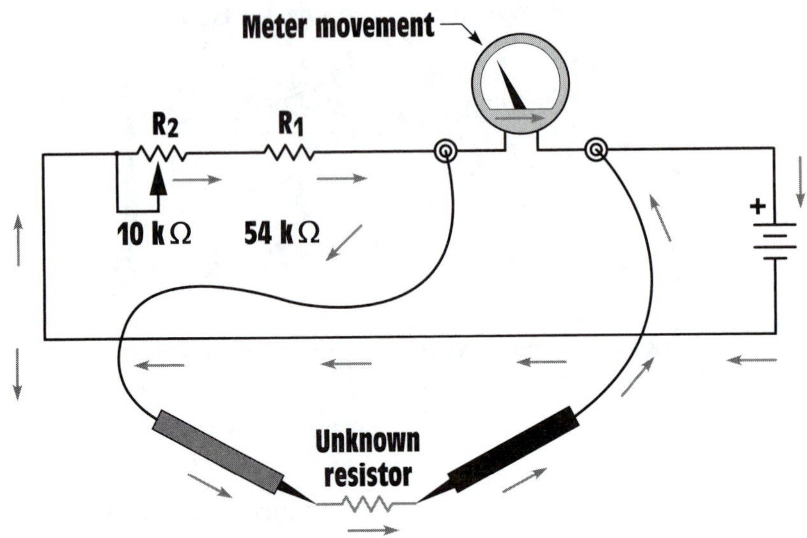

Figure 8-35 Shunt type ohmmeter.

will display a K on the scale to indicate kilohms or an M to indicate megohms (kilo means 1,000 and mega means 1,000,000) If the meter is showing a resistance of 0.200 K, it means 0.200 × 1000 or 200 Ω. If the meter indicates 1.65 M, it means 1.65 × 1,000,000 or 1,650,000 Ω.

Appearance is not the only difference between analog and digital ohmmeters. Their operating principle is different also. Analog meters operate by measuring the amount of current change in the circuit when an unknown value of resistance is added. Digital ohmmeters measure resistance by measuring the amount of voltage drop across an unknown resistance. In the circuit shown in *Figure 8-36,* a constant current generator is used to supply a known amount of current to a resistor, Rx. It will be assumed that the amount of current supplied is 1 mA. The voltage drop across the resistor is proportional to the resistance of the resistor and the amount of current flow. For example, assume the value of the unknown resistor is 4,700 Ω. The voltmeter would indicate a drop of 4.7 V when 1 mA of current flowed through the resistor. The scale factor of the ohmmeter can be changed by changing the amount of current flow through the resistor. Digital ohmmeters generally exhibit an accuracy of about 1%.

The ohmmeter, whether digital or analog, must never be connected to a circuit when the power is turned on.

The ohmmeter, whether digital or analog, must never be connected to a circuit when the power is turned on. Since the ohmmeter uses its own internal power supply, it has a very low operating voltage. Connecting a meter to power when it is set in the ohms position will probably damage or destroy the meter.

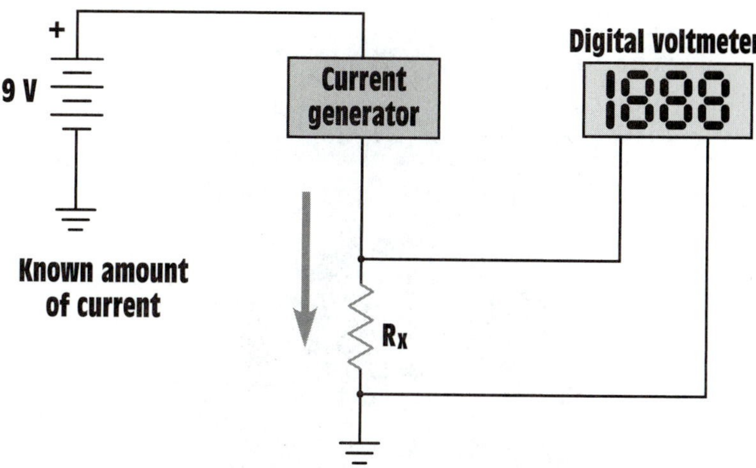

Figure 8-36 Digital ohmmeters operate by measuring the voltage drop across a resistor when a known amount of current flows through it.

DIGITAL MULTIMETERS

Digital multimeters have become increasingly popular in the past few years. The most apparent difference between digital meters and analog meters is that digital meters display their reading in discrete digits instead of with a pointer and scale. A digital multimeter is shown in *Figure 8-37*. Some digital meters have a range switch similar to the range switch used with analog meters. This switch sets the full range value of the meter. Many digital meters have voltage range settings from 200 mV to 2000 V. The lower ranges are used for accuracy. For example, assume it is necessary to measure a voltage of 16 V. The meter will be able to make a more accurate measurement when set on the 20-V range than when set on the 2000-V range.

Some digital meters do not contain a range-setting control. These meters are known as autoranging meters. They contain a function control switch that permits selection of the electrical quantity to be measured, such as AC volts, DC volts, ohms, and so on. When the meter probes are connected to the object to be tested, the meter automatically selects the proper range and displays the value.

Analog meters change scale value by inserting or removing resistance from the meter circuit, Figure 8-6. The typical resistance of an analog meter is 20,000 Ω per volt for DC and 5000 Ω per volt for AC. If the meter is set for a full scale value of 60 V, there will be 1.2 MΩ of resistance connected in series with the meter if it is being used to measure DC (60 × 20,000 = 1,200,000) and 300 kΩ if is it is being used to measure AC (60 × 5000 = 300,000). The impedance of the meter is of little concern if it is used

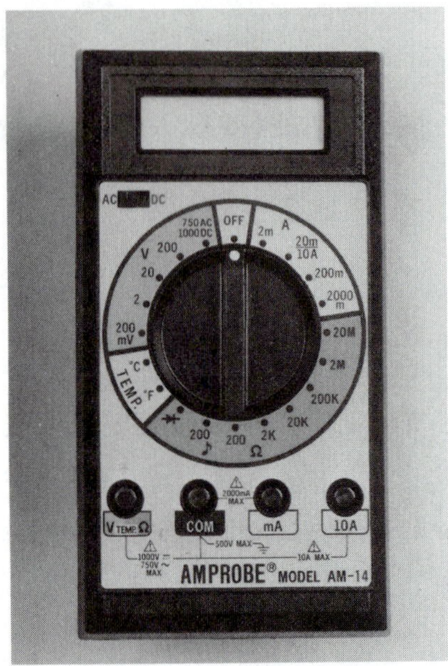

Figure 8-37 Digital multimeter. *(Courtesy: Amprobe Instrument)*

to measure circuits that are connected to a high current source. For example, assume the voltage of a 480-V panel is to be measured with a multimeter that has a resistance of 5000 Ω per volt. If the meter is set on the 600-V range, the resistance connected in series with the meter is 3 MΩ (600 × 5000 = 3,000,000). This will permit a current of 160 μA to flow in the meter circuit (480/3,000,000 = 0.000160). The 160 μA of current are not enough to affect the circuit being tested.

Now assume that this meter is to be used to test a 24-V circuit that has a current flow of 100 μA. If the 60-V range is used, the meter circuit contains a resistance of 300 kΩ (60 × 5000 = 300,000). This means that a current of 80 μA will flow when the meter is connected to the circuit (24/300,000 = 0.000080). The connection of the meter to the circuit has changed the entire circuit operation. This is known as the *loading effect.*

Digital meters do not have a loading effect. Most digital meters have an input impedance of about 10 MΩ on all ranges. This impedance is accomplished by using field effect transistors (FETs) and a voltage divider circuit. A simple schematic for this circuit is shown in *Figure 8-38.* Notice that the meter input is connected across 10 MΩ of resistance regardless of the range setting of the meter. If this meter is used to measure the voltage of the 24 V circuit, a current of 2.4 μA will flow through the meter. This is not enough current to upset the rest of the circuit, and voltage measurements can be made accurately.

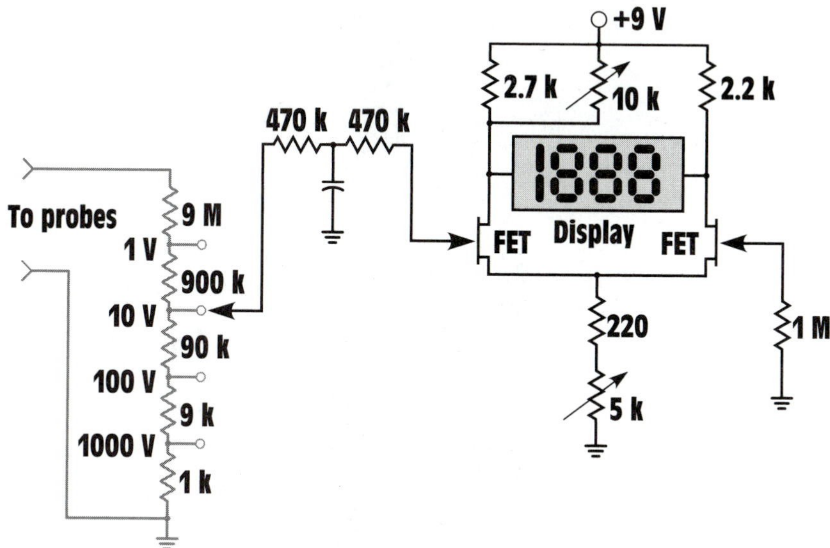

Figure 8-38 Digital voltmeter.

THE LOW-IMPEDANCE VOLTAGE TESTER

Another device used to test voltage is often referred to as a voltage tester. This device does measure voltage, but it does not contain a meter movement or digital display. It contains a coil and a plunger. The coil produces a magnetic field that is proportional to the voltage to which the tester is connected. The higher the voltage to which the tester is connected, the stronger the magnetic field becomes. The plunger must overcome the force of a spring as it is drawn into the coil, *Figure 8-39*. The plunger acts as a pointer to indicate the amount of voltage to which the tester is connected. The tester has an impedance of approximately 5000 Ω and can generally be used to measure voltages as high as 600 V. **The low-impedance voltage tester has a very large current draw when compared to other types of voltmeters and should never be used to test low-power circuits.**

The low-impedance voltage tester has a very large current draw compared with other types of voltmeters and should never be used to test low-power circuits.

The relatively high current draw of the voltage tester can be an advantage when testing certain types of circuits. This is true because it is not susceptible to giving the misleading voltage readings caused by high impedance ground paths or feedback voltages that affect other types of voltmeters. An example of this advantage is shown in *Figure 8-40*. A transformer is used to supply power to a load. Notice that neither the output side of the transformer nor the load are connected to ground. If a high impedance voltmeter is used to measure between one side of the transformer and a grounded point, it will most likely indicate some amount of voltage. This

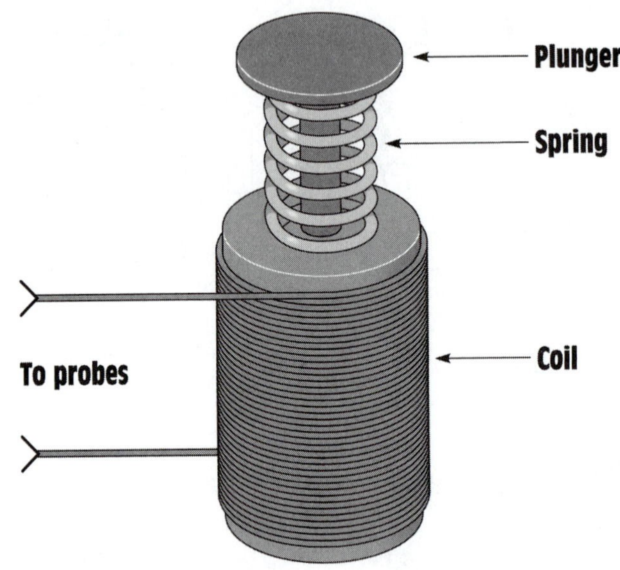

Figure 8-39 Low-impedance voltage tester.

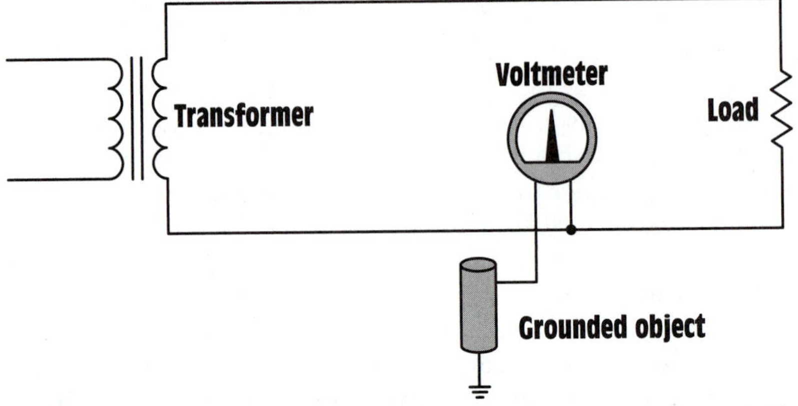

Figure 8-40 High-impedance ground paths can produce misleading voltage readings.

is because that ground can act as a large capacitor and permit a small amount of current to flow through the circuit created by the meter. This high impedance ground path can support only a few microamps of current flow, but it is enough to operate the meter movement. If a voltage tester is used to make the same measurement, it will not show a voltage because there cannot be enough current flow to attract the plunger. A voltage tester is shown in *Figure 8-41*.

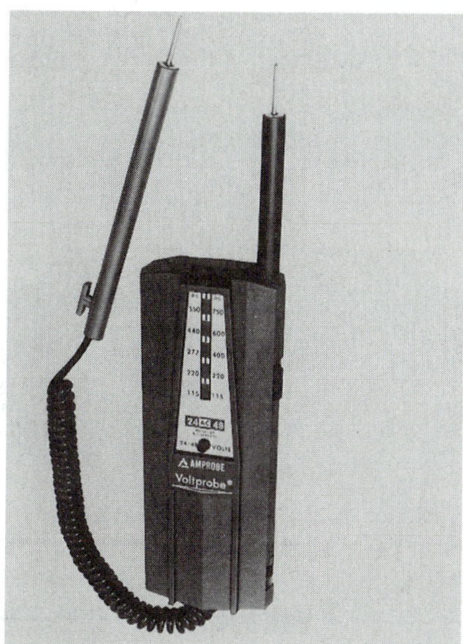

Figure 8-41 Voltage tester. *(Courtesy of Amprobe Instrument)*

THE WATTMETER

wattmeter

The **wattmeter** is used to measure the power in a circuit. This meter differs from the d'Arsonival type of meter because it does not contain a permanent magnet. This meter contains a set of electromagnets and a moving coil, *Figure 8-42*. The electromagnets are connected in series with the load in the same manner that an ammeter is connected. The moving coil has resistance connected in series with it, and is connected directly across the power source in the same manner as a voltmeter, *Figure 8-43*.

Since the electromagnet is connected in series with the load, the current flow through the load determines the magnetic field strength of the stationary magnet. The magnetic field strength of the moving coil is determined by the amount of line voltage. The turning force of the coil is proportional to the strength of these two magnetic fields. The deflection of the meter against the spring is proportional to the amount of current flow and voltage.

Since the wattmeter contains an electromagnet instead of a permanent magnet, the polarity of the magnetic field is determined by the direction of current flow. The same is true of the polarity of the moving coil connected across the source of voltage. If the wattmeter is connected into an AC circuit, the polarity of both coils will reverse at the same time, producing a continuous torque. For this reason, the wattmeter can be used to measure power in either a direct or alternating current circuit. It should be noted,

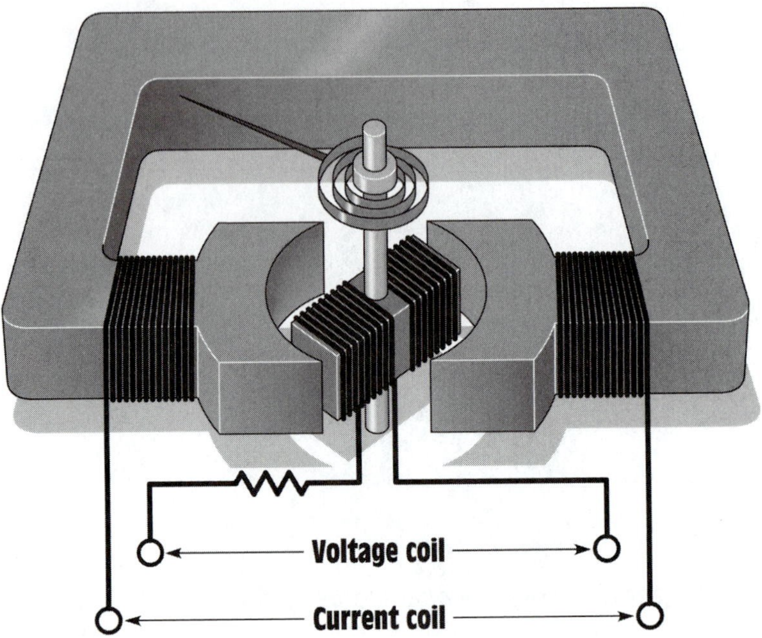

Figure 8-42 A dynamic wattmeter contains two coils—one for voltage and the other for current.

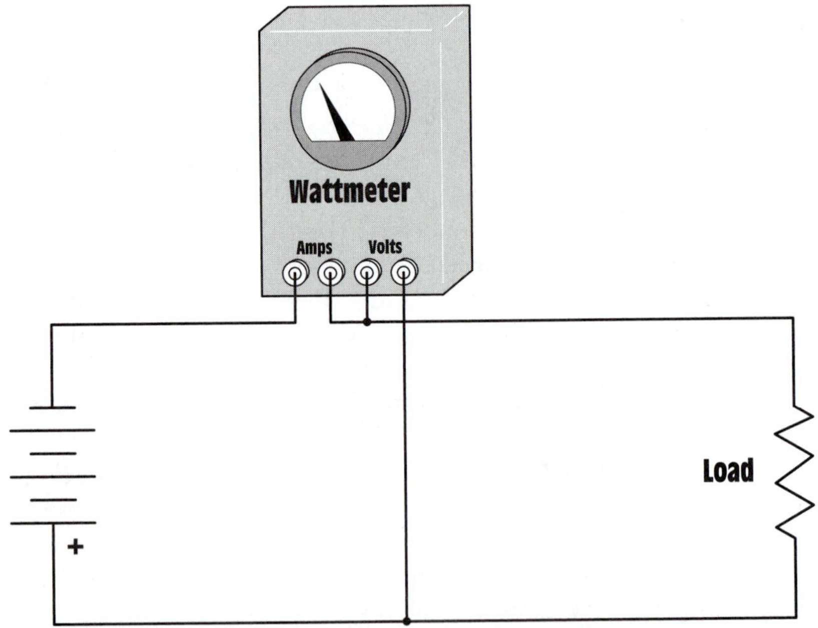

Figure 8-43 The current section of the wattmeter is connected in series with the load, and the voltage section is connected in parallel with the load.

however, that if the connection of the stationary coil or the moving coil should be reversed, the meter will attempt to read backward.

ELECTRONIC WATTMETERS

The wattmeter previously described is generally referred to as a *dynamic wattmeter.* Many manufacturers use an electronic circuit to determine watts because of the expense of producing dynamic wattmeters. This circuit is connected to a standard d'Arsonival meter movement that has been scaled to indicate watts. The advantage of an electronic wattmeter is that it costs much less. The disadvantage of an electronic wattmeter is that it cannot be connected into a DC circuit; it can be used in AC circuits only.

RECORDING METERS

There are occasions when it becomes necessary to make a recording of an electrical value over a long period of time. Recording meters produce a graph of metered values during a certain length of time. They are used to detect spike voltages, or currents of short duration, or sudden drops in voltage current, or power. Recording meters can show the amount of voltage or current, its duration, and the time of occurrence. Some meters have the ability to store information in memory over a period of several days. This information can be recalled later by the service technician. Two types of recording meters are shown in *Figure 8-44A* and *Figure 8-44B*.

BRIDGE CIRCUITS

One of the most common devices used to measure values of resistance, inductance, and capacitance accurately is a **bridge circuit.** A bridge is constructed by connecting four components to form a parallel-series circuit. All four components are of the same type, such as four resistors, four inductors, or four capacitors. The bridge used to measure resistance is called a **Wheatstone bridge.** The basic circuit for a Wheatstone bridge is shown in *Figure 8-45*. The bridge operates on the principle that the sum of the voltage drops in a series circuit must equal the applied voltage. A galvanometer is used to measure the voltage between points B and D. The galvanometer can be connected to different values of resistance or directly between points B and D. Values of resistance are used to determine the sensitivity of the meter circuit. When the meter is connected directly across the two points its sensitivity is maximum.

bridge circuit

Wheatstone bridge

In Figure 8-45, assume the battery has a voltage of 12 V, and that resistors R_1 and R_2 are precision resistors that have the same value of resistance. Since resistors R_1 and R_2 are connected in series and have the same value, each will have a voltage drop equal to one-half of the applied voltage or

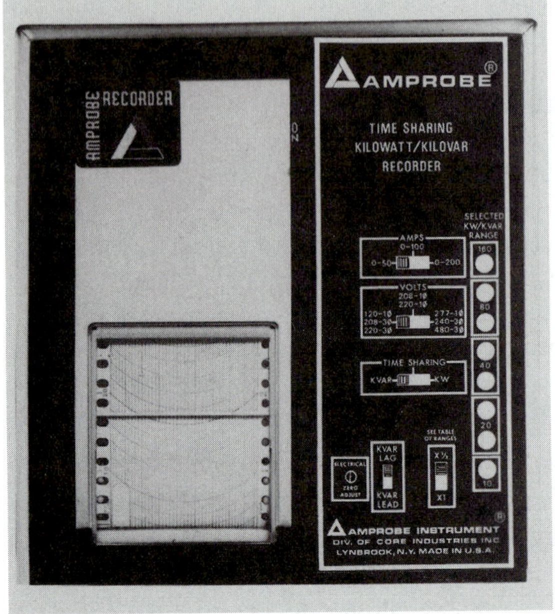

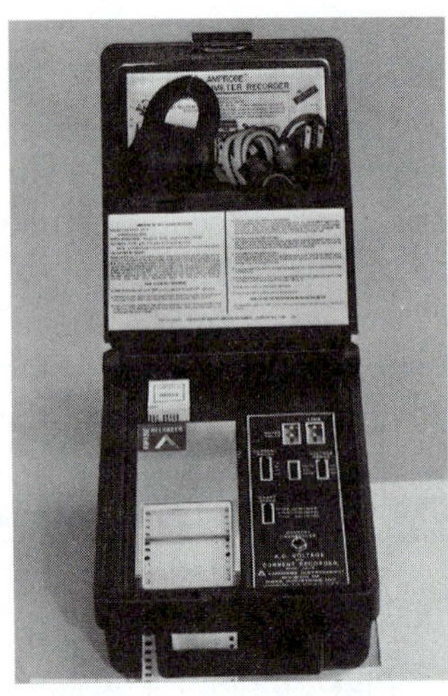

(A) (B)

Figure 8-44 Recording type meters *(Courtesy of Amprobe Instrument)*

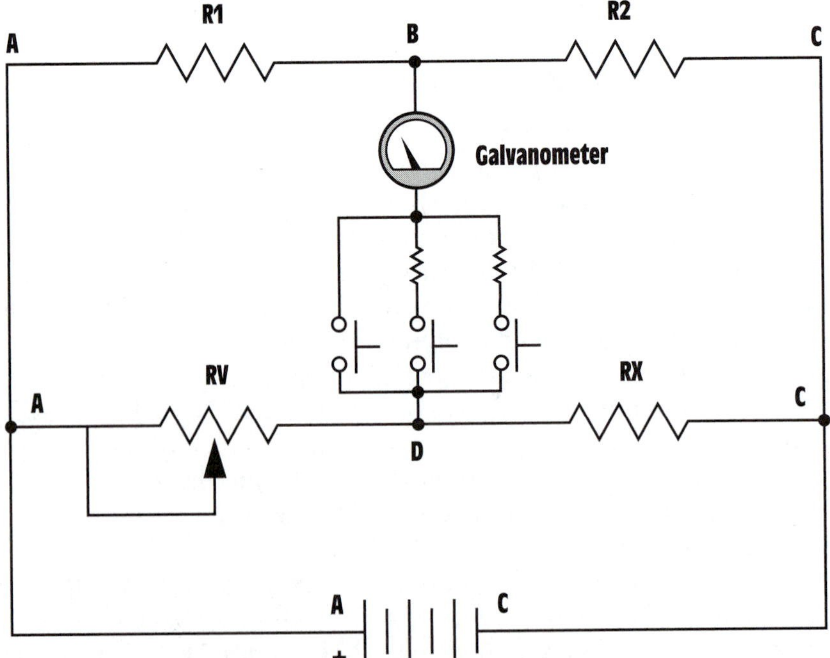

Figure 8-45 The Wheatstone bridge circuit is used to make accurate measurements of resistance and operates on the principle that the sum of the voltage drops in a series circuit must equal the applied voltage.

6 V. This means that point B is 6 V more negative than point A, and 6 V more positive than point C.

Resistors R_V (variable) and R_X (unknown) are connected in series. Resistor R_X represents the unknown value of resistance to be measured. Resistor R_V can be adjusted for different resistive values. If the value of R_V should be greater than the value of R_X, the voltage at point D will be more positive than the voltage at point B. This will cause the pointer of the zero center galvanometer to move in one direction. If the value of R_V is less than R_X, the voltage at point D will be more negative than the voltage at point B, causing the point to move in the opposite direction. When the value of R_V becomes equal to R_X, the voltage at point D will become equal to the voltage at point B. When this occurs, the galvanometer will indicate zero. A Wheatstone bridge is shown in *Figure 8-46*.

THE OSCILLOSCOPE

Many of the electronic control systems in today's industry produce voltage pulses that are meaningless to a VOM. In many instances, it is necessary to know not only the amount of voltage present at a particular point, but also the length or duration of the pulse and its frequency. Some pulses may be less than one volt and last for only a millisecond. A VOM would be

Figure 8-46 Wheatstone bridge.

useless for measuring such a pulse. It is therefore necessary to use an oscilloscope to learn what is actually happening in the circuit.

oscilloscope

The oscilloscope is a powerful tool in the hands of a trained technician. The first thing to understand is that an **oscilloscope** is a voltmeter. It does not measure current, resistance, or watts. The oscilloscope measures an amount of voltage during a period of time and produces a two-dimensional image.

Voltage Range Selection

The oscilloscope is divided into two main sections. One section is the voltage section, and the other is the time base. The display of the oscilloscope is divided by vertical and horizontal lines (*Figure 8-47*). Voltage is measured on the vertical, or *Y,* axis of the display, and time is measured on the horizontal, or *X,* axis. When using a VOM, a range selection switch is used to determine the full scale value of the meter. Ranges of 600 V, 300 V, 60 V, and 12 V are common. The ability to change ranges permits more-

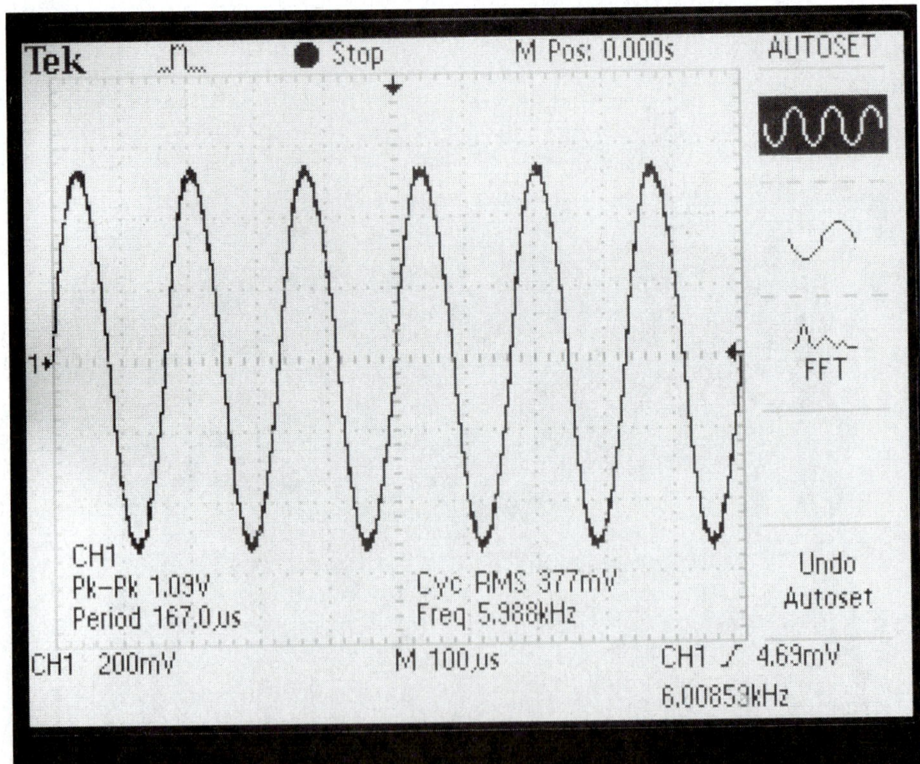

Figure 8-47 Oscilloscope display. *(Courtesy of Tektronix Inc.)*

accurate measurements to be made. The oscilloscope has a voltage range selection switch also (*Figure 8-48*). The voltage range selection switch on an oscilloscope selects volts per division instead of volts full scale. Assume that the voltage range switch shown in Figure 8-48 is set for 10mV at the 1X position. This means that each of the vertical lines on the *Y* axis of the display has a value of 10 mV. Assume the oscilloscope has been adjusted to permit 0 V to be shown on the center line of the display. If the oscilloscope probe were connected to a positive voltage of 30 mV, the trace would rise to the position shown in *Figure 8-49A*. If the probe were connected to a negative 30 mV, the trace will fall to the position shown in *Figure 8-49B*. Notice that the oscilloscope has the ability to display both a positive and a negative voltage. If the range switch were changed to 20 V per division, Figure 8-49A would be displaying 60 V positive.

The Time Base

The next section of the oscilloscope to become familiar with is the time base (*Figure 8-50*). The time base is calibrated in seconds per division and has range values from seconds to microseconds. The time base controls the value of the divisions of the lines in the horizontal direction (*X* axis). If the time base is set for 5 ms (milliseconds) per division, the trace will sweep

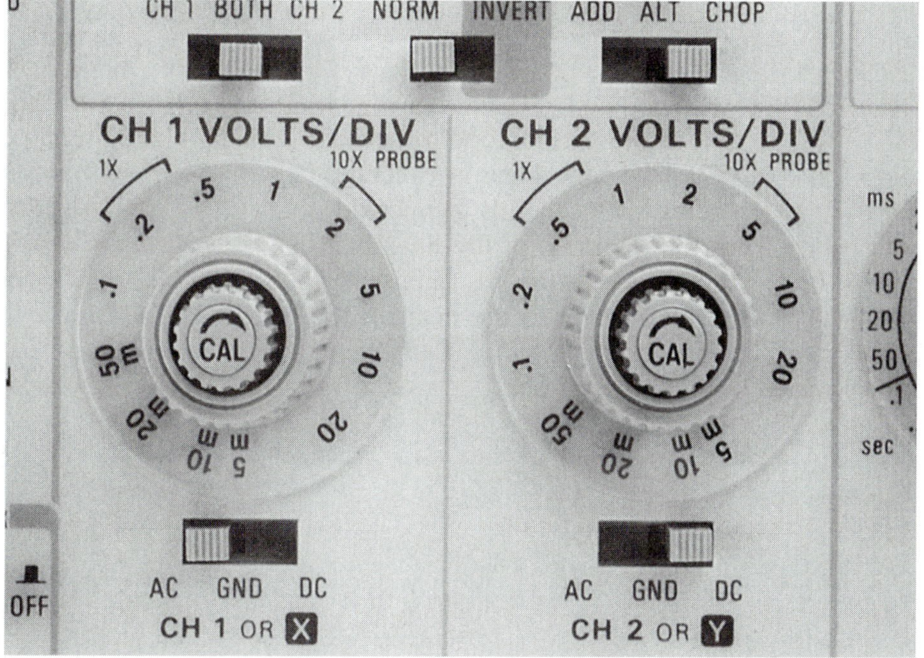

Figure 8-48 Voltage control. *(Courtesy of Tektronix Inc.)*

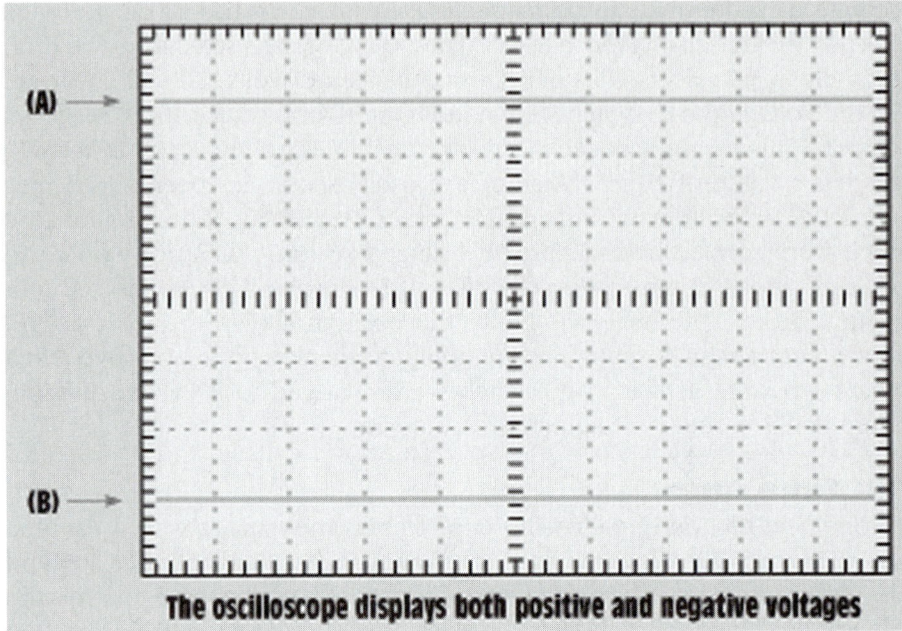

Figure 8-49 The oscilloscope displays both positive and negative voltages. *(Courtesy of Tektronix Inc.)*

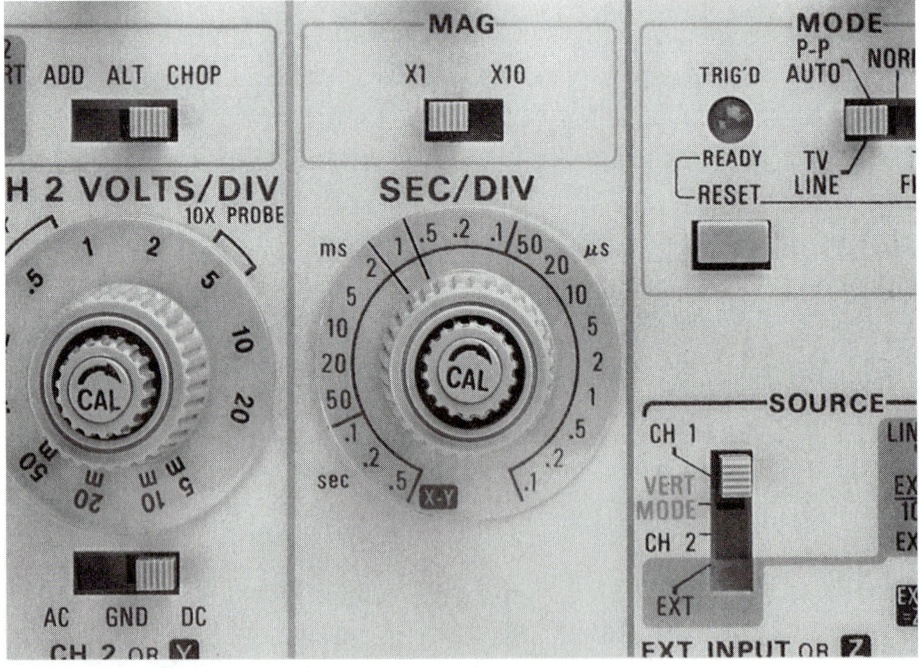

Figure 8-50 Time base. *(Courtesy of Tektronix Inc.)*

from one division to the next division in 5 ms. With the time base set in this position, it will take 50 ms to sweep from one side of the display screen to the other. If the time base is set for 2 μs (microseconds) per division, the trace will sweep the screen in 2 μs.

Measuring Frequency

Since the oscilloscope has the ability to measure the voltage with respect to time, it is possible to compute the frequency of the waveform. The frequency (F) of an AC wave form can be found by dividing 1 by the time (T) it takes to complete one cycle (F = 1/T). For example, assume the time base is set for 0.5 ms per division, and the voltage range is set for 20 V per division. If the oscilloscope has been set so that the center line of the display is 0 V, the AC wave form shown in *Figure 8-51* has a peak value of 55 V. The oscilloscope displays the peak or peak-to-peak value of voltage and not the root mean square (RMS), or effective, value. These terms will be discussed in greater detail in later units.

To measure the frequency, count the time it takes to complete one full cycle. A cycle is one complete waveform. Begin counting when the waveform starts to rise in the positive direction and stop when it again starts to rise in the positive direction. Since the time base is set for a value of 0.5 ms per division, the waveform shown in Figure 8–51 takes 4 ms to complete one full cycle. The frequency therefore is 250 Hz (1/0.004 = 250).

Attenuated Probes

Most oscilloscopes use a probe that acts as an attenuator. An attenuator is a device that divides or makes smaller the input signal (*Figure 8-52*). An attenuated probe is used to permit higher voltage readings than are normally possible. For example, most attenuated probes are 10 to 1. This means that if the voltage range switch is set for 5 V per division, the display would actually indicate 50 V per division. If the voltage range switch is set for 2 V per division, each division on the display actually has a value of 20 V per division.

Probe attenuators are made in different styles by different manufacturers. On some probes the attenuator is located in the probe head itself, while on others the attenuator is located at the scope input. Regardless of the type of attenuated probe used, it may have to be compensated or adjusted. In fact, probe compensation should be checked frequently. Different manufacturers use different methods for compensating their probes, so it is generally necessary to follow the procedures given in the operator's manual for the oscilloscope being used.

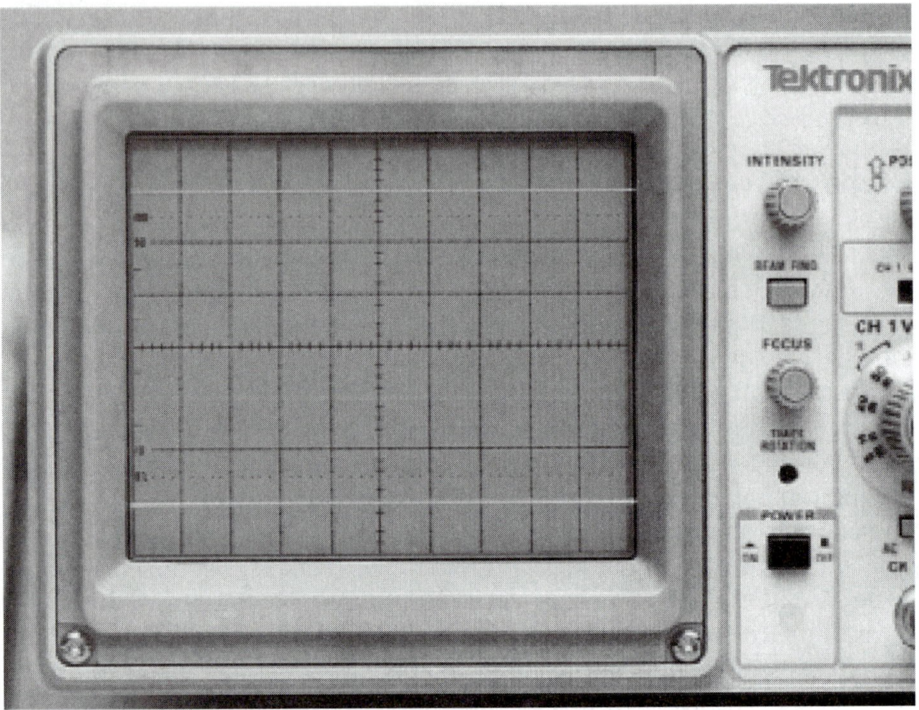

Figure 8-51 AC sine wave. *(Courtesy of Tektronix Inc.)*

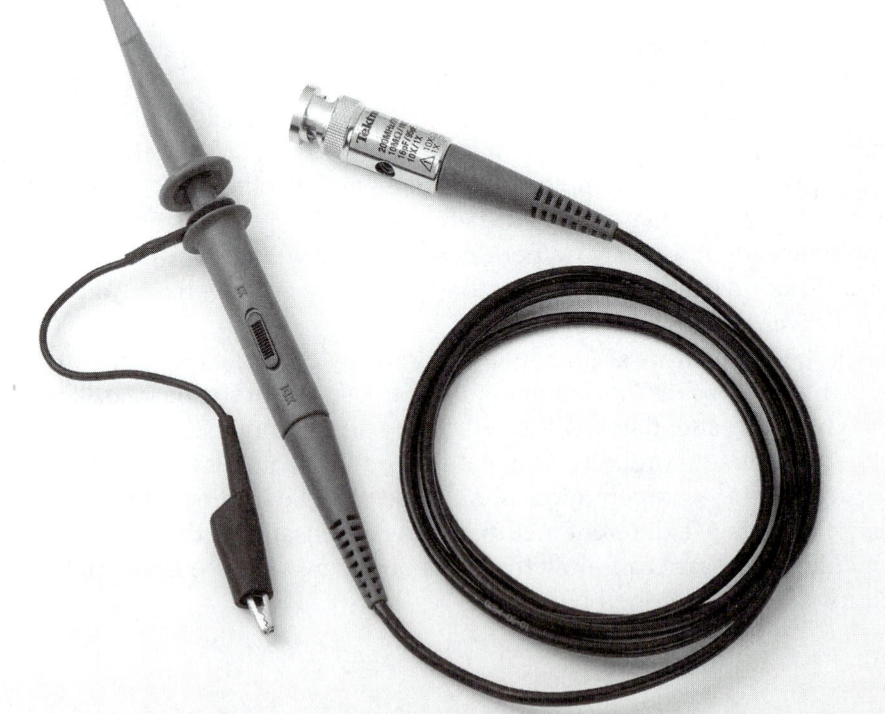

Figure 8-52 An oscilloscope probe and attenuator. *(Courtesy of Tektronix Inc.)*

OSCILLOSCOPE CONTROLS

The following is a list of common controls found on the oscilloscope. Refer to the oscilloscope shown in *Figure 8-53*.

1. **Power.** The power switch is used to turn the oscilloscope on or off.
2. **Beam finder.** This control is used to locate the position of the trace if it is off the display. The beam finder button will indicate the approximate location of the trace. The position controls are then used to move the trace back on the display.
3. **Probe adjust** (sometimes called calibrate). This is a reference voltage point used to compensate the probe. Most probe adjust points produce a square wave signal of about 0.5 V.
4. **Intensity and focus.** The intensity control adjusts the brightness of the trace. A bright spot should never be left on the display because it will burn a spot on the face of the CRT (cathode ray tube). This burned spot results in permanent damage to the CRT. The focus control sharpens the image of the trace.
5. **Vertical position.** This is used to adjust the trace up or down on the display. If a dual-trace oscilloscope is being used, there will be two vertical position controls. (A dual-trace oscilloscope contains two separate traces that can be used separately or together.)
6. **CH 1-both-CH 2.** This control determines which channel of a dual-trace oscilloscope is to be used, or if they are both to be used at the same time.
7. **ADD-ALT.-CHOP.** This control is active only when both traces are being displayed at the same time. The ADD adds the two waves together. ALT. stands for alternate. This alternates the sweep between channel 1 and channel 2. The CHOP mode alternates several times during one sweep. This generally makes the display appear more stable. The chop mode is generally used when displaying two traces at the same time.

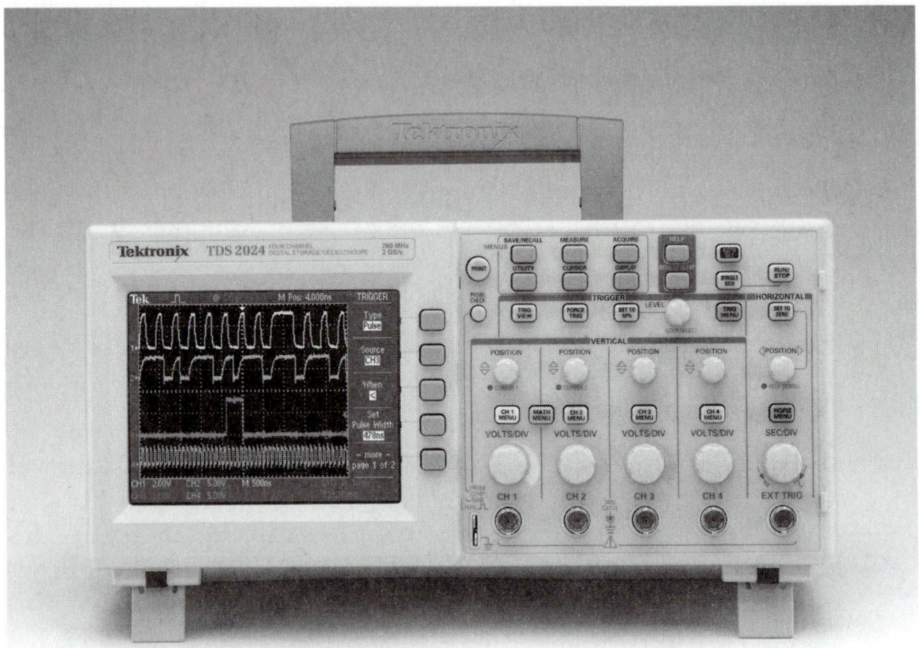

Figure 8-53 An oscilloscope. *(Courtesy of Tektronix Inc.)*

8. **AC-GRD-DC.** The AC is used to block any DC voltage when only the AC portion of the voltage is to be seen. For instance, assume an AC voltage of a few millivolts is riding on a DC voltage of several hundred volts. If the voltage range is set high enough so that 100 VDC can be seen on the display, the AC voltage cannot be seen. The AC section of this switch inserts a capacitor in series with the probe. The capacitor blocks the DC voltage and permits the AC voltage to pass. Since the 100 VDC has been blocked, the voltage range can be adjusted for millivolts per division, which will permit the AC signal to be seen.

   The GRD section of the switch stands for ground. This section grounds the input so the sweep can be adjusted for 0 V at any position on the display. The ground switch grounds at the scope and does not ground the probe. This permits the ground switch to be used when the probe is connected to a live circuit. The DC section permits the oscilloscope to display all of the voltage, both AC and DC, connected to the probe.

9. **Horizontal position.** This control adjusts the position of the trace from left to right.

10. **Auto-Normal.** This determines whether the time base will be triggered automatically or operated in a free-running mode. If this control is operated in the normal setting, the trigger signal is taken from the line to which the probe is connected. The scope is generally operated with the trigger set in the automatic position.
11. **Level.** The level control determines the amplitude the signal must be before the scope triggers.
12. **Slope.** The slope permits selection as to whether the trace is triggered by a negative or positive wave form.
13. **Int.-Line-Ext.** The *Int.* stands for internal. The scope is generally operated in this mode. In this setting, the trigger signal is provided by the scope. In the line mode, the trigger signal is provided from a sample of the line. The *Ext.*, or external, mode permits the trigger pulse to be applied from an external source.

These are not all the controls shown on the oscilloscope in *Figure 8-53,* but they are the major controls. Most oscilloscopes contain these controls.

Interpreting Waveforms

The ability to interpret the waveforms on the display of the oscilloscope takes time and practice. When using the oscilloscope, one must keep in mind that the display shows the voltage with respect to time.

In *Figure 8-54,* it is assumed that the voltage range has been set for 0.5 V per division, and the time base is set for 2 ms per division. It is also assumed that 0 V has been set on the center line of the display. The wave form shown is a square wave. The display shows that the voltage rises in the positive direction to a value of 1.4 V and remains there for 2 ms. The voltage then drops to 1.4 V negative and remains there for 2 ms before going back to positive. Since the voltage changes between positive and negative, it is an AC voltage. The length of one cycle is 4 ms. The frequency is therefore 250 Hz (1/0.004 = 250).

In *Figure 8-55,* the oscilloscope has been set for 50 mV per division and 20 μs (microseconds) per division. The display shows a voltage that is negative to the probe's ground lead and has a peak value of 150 mV. The wave form lasts for 20 μs and produces a frequency of 50 kHz (1/0.000020 = 50,000). The voltage is DC since it never crosses the zero reference and goes in the positive direction. This type of voltage is called *pulsating DC.*

In *Figure 8-56,* assume the scope has been set for a value of 50 V per division and 5 ms per division. The wave form shown rises from 0 to about 45 V in a period of about 1.5 ms. The voltage gradually increases to about 50 V in the space of 1 ms and then rises to a value of about 100 V in the

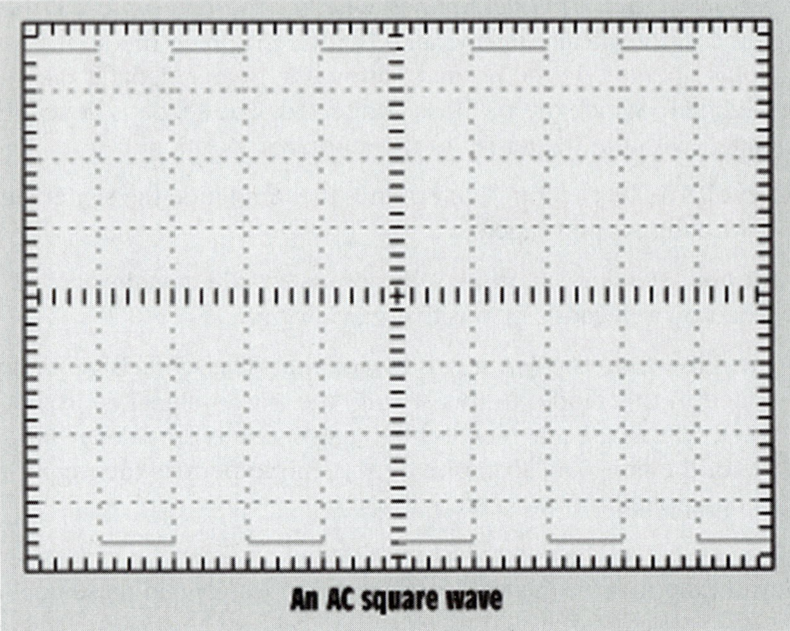

Figure 8-54 An AC square wave. *(Courtesy of Tektronix Inc.)*

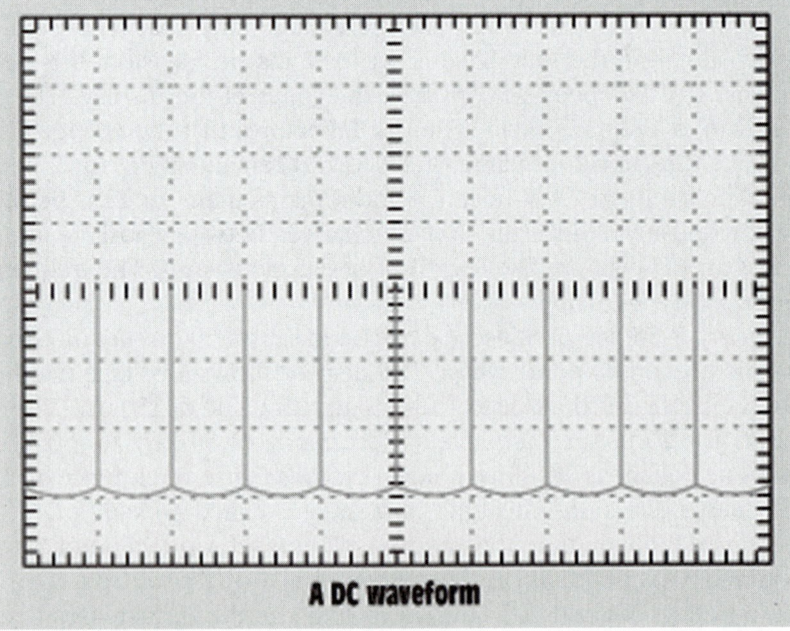

Figure 8-55 A DC waveform. *(Courtesy of Tektronix Inc.)*

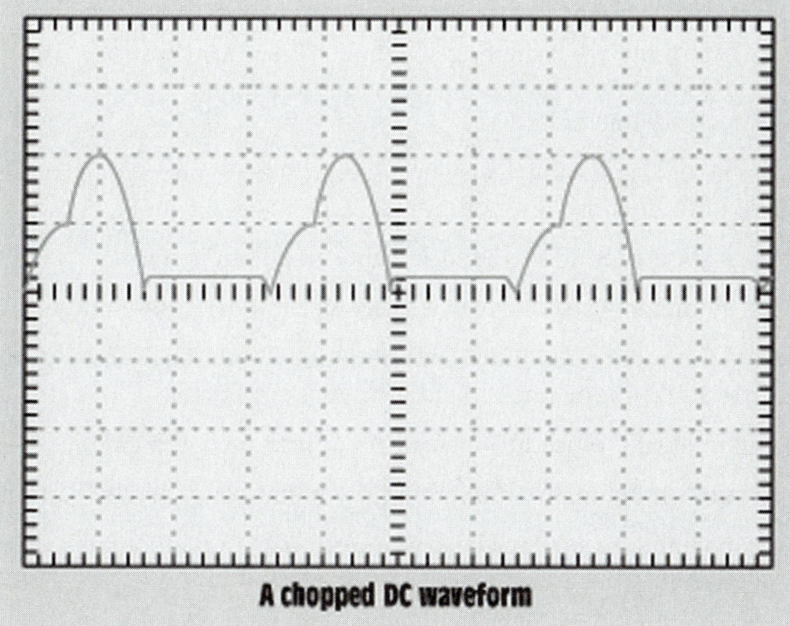

Figure 8-56 A chopped DC wave form. *(Courtesy of Tektronix Inc.)*

next 2 ms. The voltage then decreases to 0 in the next 4 ms. It then increases to a value of about 10 V in 0.5 ms and remains at that level for about 8 ms. This is one complete cycle for the waveform. The length of one cycle is about 16.6 ms, which is a frequency of 60.2 Hz (1/0.0166 = 60.2). The voltage is DC because it remains positive and never drops below the 0 line.

Learning to interpret the waveforms seen on the display of an oscilloscope will take time and practice, but it is well worth the effort. The oscilloscope is the only means by which many of the waveforms and voltages found in electronic circuits can be understood. Consequently, the oscilloscope is the single most valuable piece of equipment a technician can use.

SUMMARY

1. The d'Arsonival type of meter movement is based on the principle that like magnetic fields repel.
2. The d'Arsonival movement operates only on DC current.
3. Voltmeters have a high resistance and are designed to be connected directly across the power line.

4. The steps to reading a meter are:

   A. determine what quantity the meter is set to measure.
   B. determine the full range value of the meter.
   C. read the meter.

5. Ammeters have a low resistance and must be connected in series with a load to limit the flow of current.
6. Shunts are used to change the value of DC ammeters.
7. AC ammeters use a current transformer to change the range setting.
8. Clamp-on ammeters measure the flow of current by measuring the strength of the magnetic field around a conductor.
9. Ohmmeters are used to measure the resistance in a circuit.
10. Ohmmeters contain an internal power source, generally batteries.
11. Ohmmeters must never be connected to a circuit that has power applied to it.
12. Digital multimeters display their value in digits instead of using a meter movement.
13. Digital multimeters generally have an input impedance of 10 MΩ on all ranges.
14. The stationary coil of a wattmeter is connected in series with the load and the moving coil is connected to the line voltage.
15. The turning force of the wattmeter is proportional to the strength of the magnetic field of the stationary coil and the strength of the magnetic field of the moving coil.
16. Digital ohmmeters measure resistance by measuring the voltage drop across an unknown resistor when a known amount of current flows through it.
17. Low-impedance voltage testers are not susceptible to indicating a voltage caused by a high-impedance ground or a feedback.
18. A bridge circuit can be used to accurately measure values of resistance, inductance, and capacitance.
19. The oscilloscope measures the amplitude of voltage with respect to time.
20. The frequency of a waveform can be determined by dividing 1 by the time of one cycle ($F = 1/T$).

REVIEW QUESTIONS

1. To what is the turning force of a d'Arsonival meter movement proportional?
2. What type of voltage must be connected to a d'Arsonival meter movement?
3. A DC voltmeter has a resistance of 20,000 Ω per volt. What is the resistance of the meter if the range selection switch is set on the 250 V range?
4. What is the purpose of an ammeter shunt?
5. Name two methods used to make a DC multirange ammeter.
6. How is an ammeter connected into a circuit?
7. How is a voltmeter connected into a circuit?
8. An ammeter shunt has a voltage drop of 50 mV when 50 A of current flows through it. What is the resistance of the shunt?
9. What type of meter contains its own, separate power source?
10. What is the major difference between a wattmeter and a d'Arsonival meter?
11. What two factors determine the turning force of a wattmeter?
12. What electrical quantity is measured with an oscilloscope?
13. What is measured on the Y axis of an oscilloscope?
14. What is measured on the X axis of an oscilloscope?

9 Unit

Alternating Current

objectives

fter studying this unit, you should be able to:

- Discuss differences between direct and alternating current.
- Be able to compute instantaneous values of voltage and current for a sine wave.
- Be able to compute peak, RMS, and average values of voltage and current.
- Discuss the phase relationship of voltage and current in a pure resistive circuit.

Most of the electrical power produced in the world is alternating current. It is used to operate everything from home appliances such as television sets, computers, microwave ovens, and electric toasters, to the largest motors found in industry. Alternating current has several advantages over direct current that make it a better choice for the large-scale production of electrical power.

ADVANTAGES OF ALTERNATING CURRENT

Probably the single greatest advantage of alternating current is the fact that AC current can be transformed and DC current cannot. A transformer permits voltage to be stepped up or down. Voltage can be stepped up for the purpose of transmission and then stepped back down when it is to be used by some device. Transmission voltages of 69 kV, 138 kV, and 345 kV are common. The advantage of high voltage transmission is that less current is required to produce the same amount of power. The reduction of current permits smaller wires to be used, which results in a savings of material.

In the very early days of electric power generation, Thomas Edison, an American inventor, proposed powering the country with low-voltage direct current. He reasoned that low-voltage direct current was safer for people to use than the higher-voltage alternating current. A Serbian immigrant named Nikola Tesla, however, argued that direct current was impractical to use for large-scale applications. The disagreement was finally settled at the 1904 World's Fair held in St. Louis, Missouri. The 1904 World's Fair not only introduced the first ice cream cone and the first iced tea, it was also the first World's Fair to be lighted with "electric candles." At that time, the only two companies capable of providing electric lighting for the World's Fair were the Edison Company, headed by Thomas Edison, and the Westinghouse Company, headed by George Westinghouse, a close friend of Nikola Tesla. The Edison Company submitted a bid of over one dollar per lamp to light the fair with low-voltage direct current. The Westinghouse Company submitted a bid of less than 25 cents per lamp to light the fair using higher-voltage alternating current. This set the precedent for how electric power would be supplied throughout the world.

ALTERNATING CURRENT WAVEFORMS

SQUARE WAVES

Alternating current differs from direct current in that AC current reverses its direction of flow at periodic intervals, *Figure 9-1*. Alternating current waveforms can vary depending on how the current is produced. One waveform frequently encountered is the square wave, *Figure 9-2*. It is assumed that the oscilloscope, an instrument for measuring the polarity, amplitude, and duration of voltage in Figure 9-2 has been adjusted so that 0 V is represented by the center horizontal line. The waveform shows

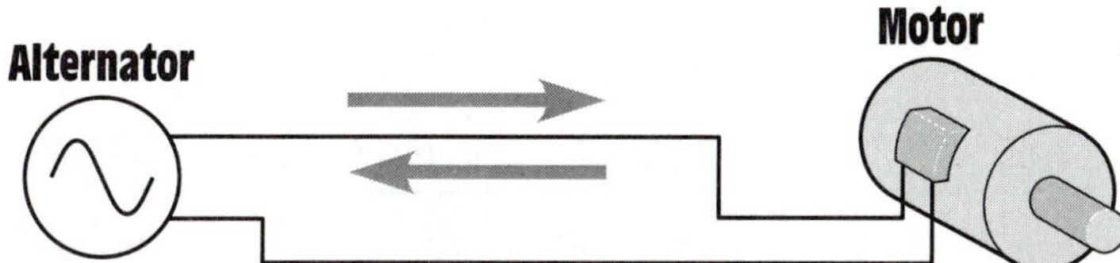

Figure 9-1 Alternating current flows first in one direction and then in the other.

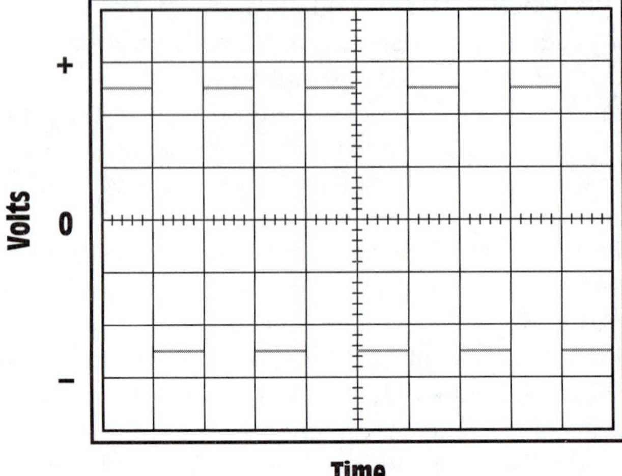

Figure 9-2 Square wave alternating current.

that the voltage is in the positive direction for some length of time and then changes polarity. The voltage remains negative for some length of time and then changes back to positive again. Each time the voltage reverses polarity, the current flow through the circuit changes direction. A square wave could be produced by a simple single-pole double-throw switch connected to two batteries, as shown in *Figure 9-3*. Each time the switch position is changed, current flows through the resistor in a different direction. Although this circuit will produce a square wave alternating

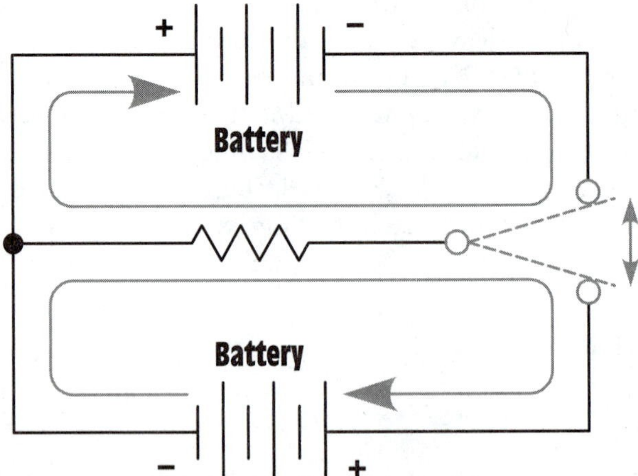

Figure 9-3 Square wave alternating current produced with a switch and two batteries.

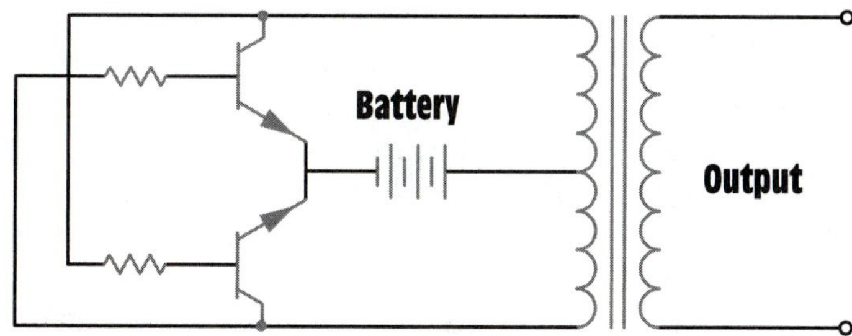

Figure 9-4 Square wave oscillator.

current, it is not practical. Square waves are generally produced by electronic devices called **oscillators.** The schematic diagram of a simple square wave oscillator is shown in *Figure 9-4*. In this circuit, two bipolar transistors are used as switches to reverse the direction of current flow through the windings of the transformer. This type of oscillator is often used to change the 12 V DC of an automobile battery into 120 V AC to operate electric hand tools such as drills and saws. This oscillator should not be used to power electronic devices such as television sets.

oscillators

TRIANGLE WAVES

triangle wave

linear wave

Another common AC waveform is the **triangle wave** shown in *Figure 9-5*. The triangle wave is a linear wave. A **linear wave** is one in which the voltage rises at a constant rate with respect to time. Linear waves form straight lines when plotted on a graph. For example, assume the waveform shown in Figure 9-5 reaches a maximum positive value of 100 V after 2 ms. The voltage will be 25 V after 0.5 ms, 50 V after 1 ms, and 75 V after 1.5 ms.

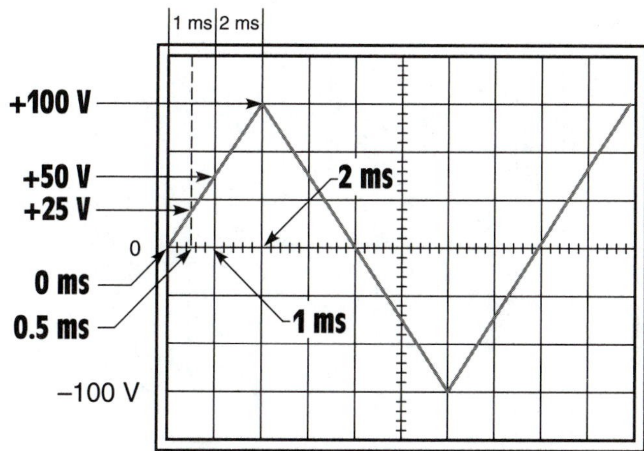

Figure 9-5 Triangle wave.

sine wave

cycle

frequency

hertz (Hz)

Sine waves are so named because the voltage at any point along the waveform is equal to the maximum, or peak, value times the sine of the angle of rotation.

SINE WAVES

The most common of all AC waveforms is the **sine wave,** *Figure 9-6.* They are produced by all rotating machines. The sine wave contains a total of 360 electrical degrees. It reaches its peak positive voltage at 90°, returns to a value of 0 volt at 180°, increases to its maximum negative voltage at 270°, and returns to 0 volt at 360°. Each complete waveform of 360° is called a **cycle.** The number of complete cycles that occur in one second is called the **frequency.** Frequency is measured in **hertz (Hz).** The most common frequency in the United States and Canada is 60 (Hz). This means that the voltage increases from zero to its maximum value in the positive direction, returns to zero, increases to its maximum value in the negative direction, and returns to zero 60 times each second.

Sine waves are so named because the voltage at any point along the waveform is equal to the maximum, or peak, value times the sine of the angle of rotation. *Figure 9-7* illustrates one-half of a loop of wire cutting through lines of magnetic flux. The flux lines are shown with equal spacing between each line, and the arrow denotes the arc of the loop as it cuts through the lines of flux. Notice the number of flux lines that are cut by the loop during the first 30° of rotation. Now notice the number of flux lines that are cut during the second and third 30° of rotation. Because the loop is cutting the flux lines at an angle, it must travel a greater distance between flux lines during the first degrees of rotation. This means that fewer flux lines are cut per second, which results in a lower induced voltage. *One V is induced in a conductor when it cuts lines of magnetic flux at a rate of 1 Wb/s.* One weber is equal to 100,000,000 lines of flux.

When the loop has rotated 90°, it is perpendicular to the flux lines and is cutting them at the maximum rate, which results in the highest, or peak,

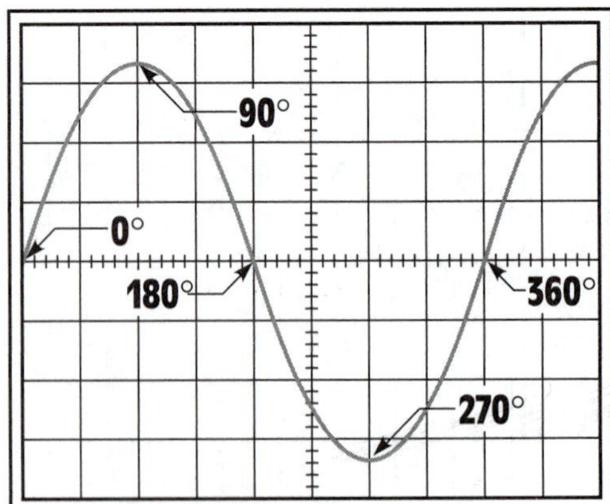

Figure 9-6 Sine wave.

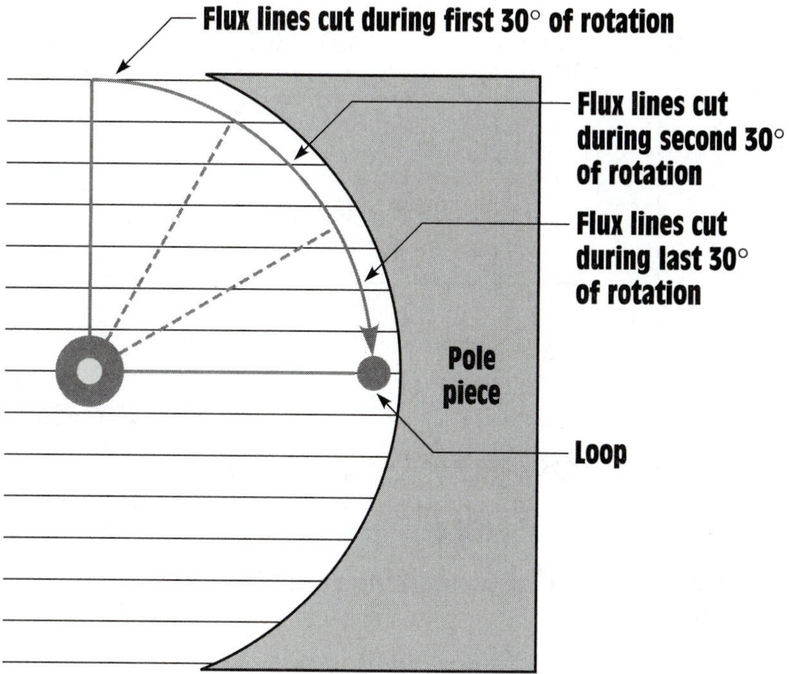

Figure 9-7 As the loop approaches 90° of rotation, the flux lines are cut at a faster rate.

voltage being induced in the loop. The voltage at any point during the rotation is equal to the maximum induced voltage times the sine of the angle of rotation. For example, if the induced voltage after 90° of rotation is 100 V, the voltage after 30° of rotation will be 50 V because the sine of a 30° angle is 0.5 ($100 \times 0.5 = 50$ V). The induced voltage after 45° of rotation is 70.7 V because the sine of a 45° angle is 0.707 ($100 \times 0.707 = 70.7$ V). A sine wave showing the instantaneous voltage values after different degrees of rotation is shown in *Figure 9-8*. The instantaneous voltage value is the value of voltage at any instant on the waveform.

The following formula can be used to determine the instantaneous value at any point along the sine wave:

$$E_{(INST)} = E_{(MAX)} \times SIN \angle$$

where

$E_{(INST)}$ = the voltage at any point on the waveform

$E_{(MAX)}$ = the maximum, or peak, voltage

$SIN \angle$ = the sine of the angle of rotation

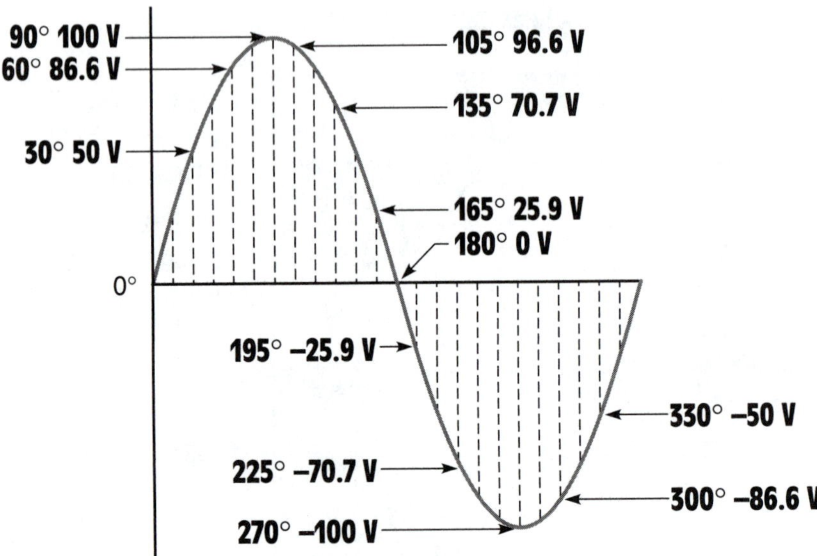

Figure 9-8 Instantaneous values of voltage along a sine wave.

Example 1

A sine wave has a maximum voltage of 138 V. What is the voltage after 78° of rotation?

Solution:

$$E_{(INST)} = E_{(MAX)} \times \text{SIN} \angle$$

$$E_{(INST)} = 138 \times 0.978 \text{ (SIN of 78°)}$$

$$E_{(INST)} = 134.96 \text{ V}$$

The formula can be changed to find the maximum value if the instantaneous value and the angle of rotation are known or to find the angle if the maximum and instantaneous values are known.

$$E_{(MAX)} = \frac{E_{(INST)}}{\text{SIN} \angle}$$

$$\text{SIN} \angle = \frac{E_{(INST)}}{E_{(MAX)}}$$

Example 2

A sine wave has an instantaneous voltage of 246 V after 53° of rotation. What is the maximum value the waveform will reach?

Solution:

$$E_{(MAX)} = \frac{E_{(INST)}}{SIN \angle}$$

$$E_{(MAX)} = \frac{246}{0.799}$$

$$E_{(MAX)} = 307.88 \text{ V}$$

Example 3

A sine wave has a maximum voltage of 350 V. At what angle of rotation will the voltage reach 53 V?

Solution:

$$SIN \angle = \frac{E_{(INST)}}{E_{(MAX)}}$$

$$SIN \angle = \frac{350}{53}$$

$$SIN \angle = 0.151$$

Note: 0.151 is the *sine* of the angle, not the angle. To find the angle that corresponds to a sine of 0.151 use the trigonometric function of a scientific calculator.

$$\angle = 8.71°$$

Sine Wave Values

Several measurements of voltage and current are associated with sine waves. These measurements are peak-to-peak, peak, RMS, and average. A sine wave showing these different measurements is shown in *Figure 9-9.*

Peak-to-Peak and Peak Values

The peak-to-peak value is measured from the maximum value in the positive direction to the maximum value in the negative direction. The peak-to-peak value is often the simplest measurement to make when using an oscilloscope.

peak

The **peak** value is measured from zero to the highest value obtained in either the positive or negative direction. The peak value is one-half of the peak-to-peak value.

RMS Values

In *Figure 9-10,* a 100-V battery is connected to a 100-Ω resistor. This connection will produce 1 A of current flow and the resistor will dissipate

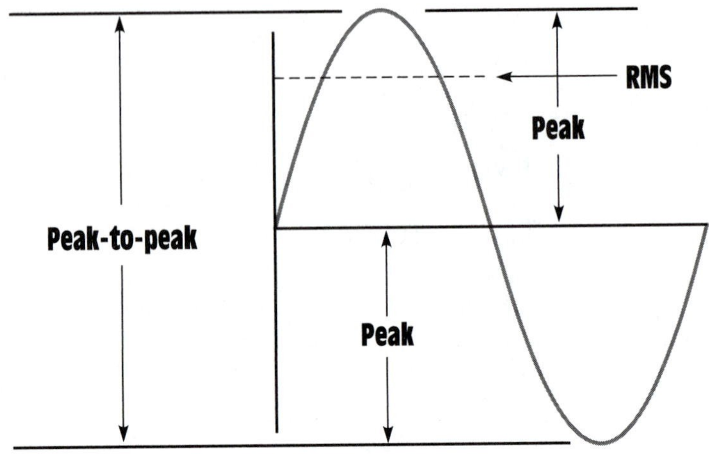

Figure 9-9 Sine wave values.

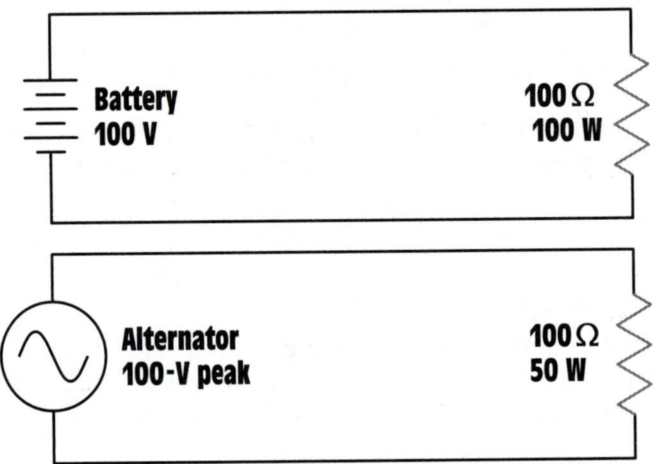

Figure 9-10 Direct current compared with a sine wave AC current.

100 W of power in the form of heat. An AC alternator that produces a peak voltage of 100 V is also shown connected to a 100-Ω resistor. A peak current of 1 A will flow in the circuit, but the resistor will dissipate only 50 W in the form of heat. The reason is that the voltage produced by a pure source of direct current, such as a battery, is one continuous value, *Figure 9-11*. The AC sine wave, however, begins at zero, increases to the maximum value, and decreases back to zero during the same period of time. Since the sine wave has a value of 100 V for only a short period of time and

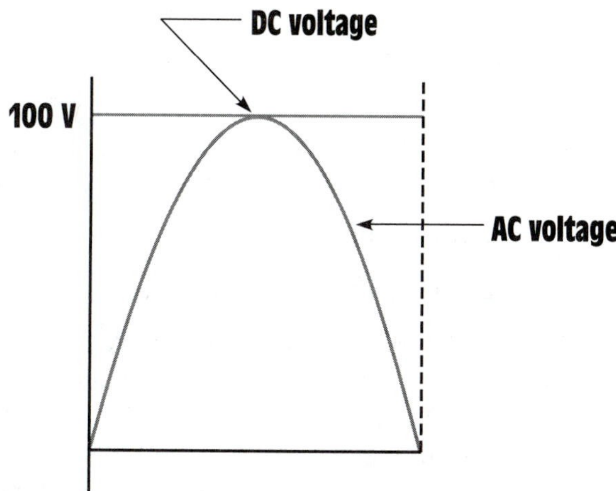

Figure 9-11 The DC voltage remains at a constant value during a half cycle of AC voltage.

is less than 100 V during the rest of the half cycle, it cannot produce as much power as 100 V of DC.

The solution to this problem is to use a value of AC voltage that will produce the same amount of power as a similar value of DC voltage. This AC value is called the **RMS,** or **effective,** value and is the value indicated by almost all AC voltmeters and ammeters. **RMS stands for root-mean-square, which is an abbreviation for the square root of the mean of the square of the instantaneous currents.** The RMS value can be found by dividing the peak value by the square root of (1.414), or by multiplying the peak value by 0.707 (the reciprocal of 1.414). The formulas for determining the RMS or peak value are:

RMS

effective

RMS stands for root-mean-square, which is an abbreviation for the square root of the mean of the square of the instantaneous currents.

$$\text{RMS} = \text{peak} \times 0.707$$

$$\text{peak} = \text{RMS} \times 1.414$$

Example 4

A sine wave has a peak value of 354 V. What is the RMS value?

Solution:

$$\text{RMS} = \text{peak} \times 0.707$$

$$\text{RMS} = 354 \times 0.707$$

$$\text{RMS} = 250.3\ \text{V}$$

Example 5

An AC voltage has a value of 120 V RMS. What is the peak value of voltage?

Solution:

$$\text{peak} = \text{RMS} \times 1.414$$

$$\text{peak} = 120 \times 1.414$$

$$\text{peak} = 169.7\ \text{V}$$

When the RMS value of voltage and current is used, it will produce the same amount of power as a similar value of DC voltage or current. If 100 volts RMS is applied to a 100-Ω resistor, the resistor will produce 100 watts of heat. AC voltmeters and ammeters indicate the RMS value, not the peak value. All values of AC voltage and current used in this text will be RMS values unless otherwise stated.

Average Values

average

Average values of voltage and current are actually direct current values. The average value must be found when a sine wave AC voltage is changed into DC with a rectifier, *Figure 9-12.* The rectifier shown is a bridge-type rectifier that produces full wave rectification. This means that both the positive and negative half of the AC waveform is changed into DC. The average value is the amount of voltage that would be indicated by a DC voltmeter if it were connected across the load resistor. The average voltage is proportional to the peak, or maximum, value of the waveform and to the length of time it is on as compared to the length of time it is off, *Figure*

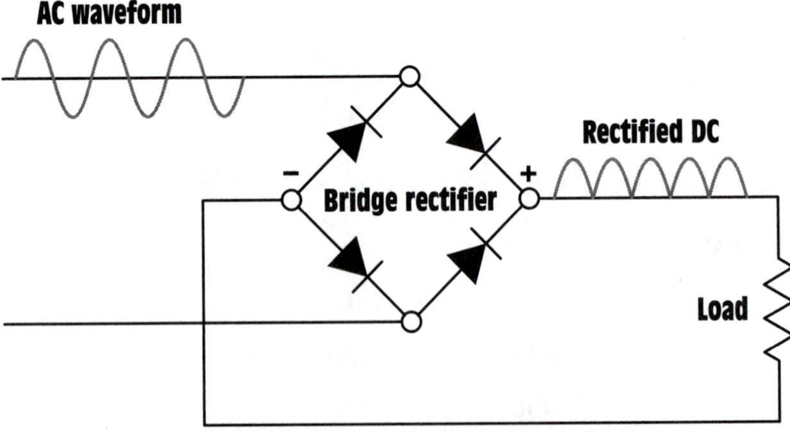

Figure 9-12 The bridge rectifier changes AC voltage into DC voltage.

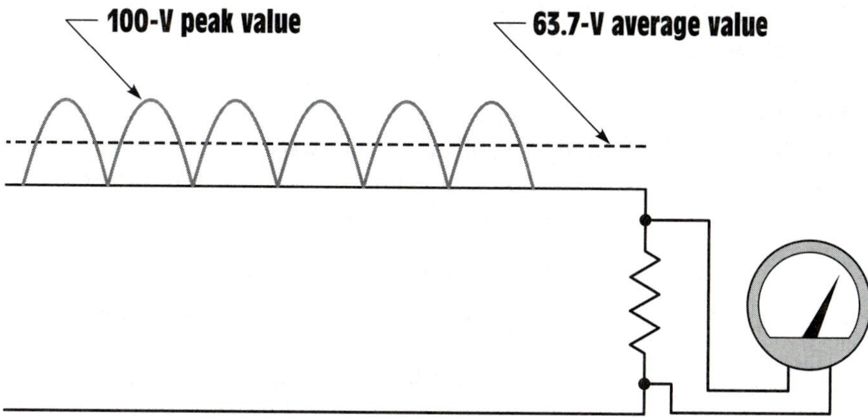

Figure 9-13 A DC voltmeter indicates the average value.

9-13. Notice in Figure 9-13 that the voltage waveform turns on and off, but it never changes polarity. The current, therefore, never reverses direction. This is called pulsating direct current. The pulses are often referred to as **ripple.** The average value of voltage will produce the same amount of power as a nonpulsating source of voltage such as a battery, *Figure 9-14.* For a sine wave, the average value of voltage is found by multiplying the peak value by 0.637, or by multiplying the RMS value by 0.9.

ripple

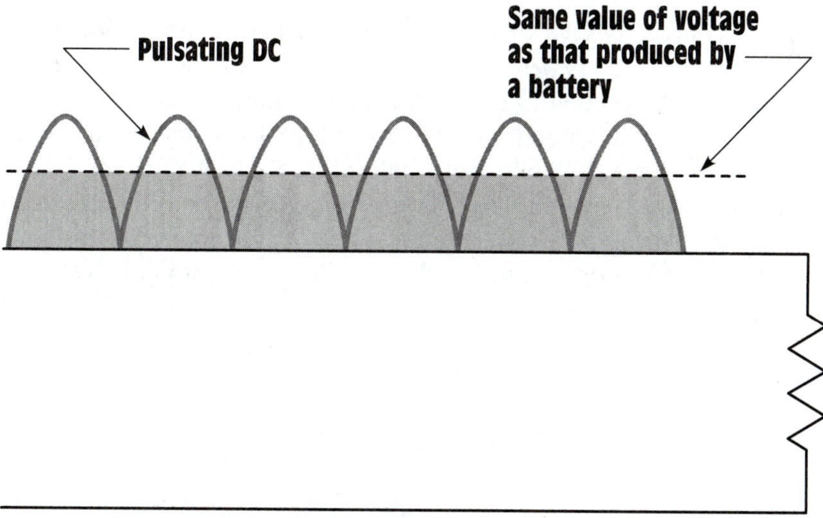

Figure 9-14 The average value produces the same amount of power as a nonpulsating source of voltage.

Example 6

An AC sine wave with an RMS value of 120 V is connected to a full wave rectifier. What is the average DC voltage?

Solution:

The problem can actually be solved in two ways. The RMS value can be changed into peak and then the peak value can be changed to the average value.

$$\text{peak} = \text{RMS} \times 1.414$$

$$\text{peak} = 120 \times 1.414$$

$$\text{peak} = 169.7\text{ V}$$

$$\text{average} = \text{peak} \times 0.637$$

$$\text{average} = 169.7 \times 0.637$$

$$\text{average} = 108\text{ V}$$

Example 7

The second method of determining the average value is to multiply the RMS value by 0.9.

$$\text{average} = \text{RMS} \times 0.9$$

$$\text{average} = 120 \times 0.9$$

$$\text{average} = 108\text{ V}$$

The conversion factors given are for full wave rectification. If a half-wave rectifier is used, *Figure 9-15,* only one-half of the AC waveform is converted into DC. To determine the average voltage for a half-wave rectifier, multiply the peak value by 0.637 or the RMS value by 0.9 and then divide the product by 2. Since only half of the AC waveform has been converted into direct current, the average voltage will be only half that of a full-wave rectifier, *Figure 9-16.*

A half-wave rectifier is connected to 277 V AC. What is the average DC voltage?

Solution:

$$\text{average} = \text{RMS} \times \frac{0.9}{2}$$

$$\text{average} = 277 \times \frac{0.9}{2}$$

$$\text{average} = 124.6\text{ V}$$

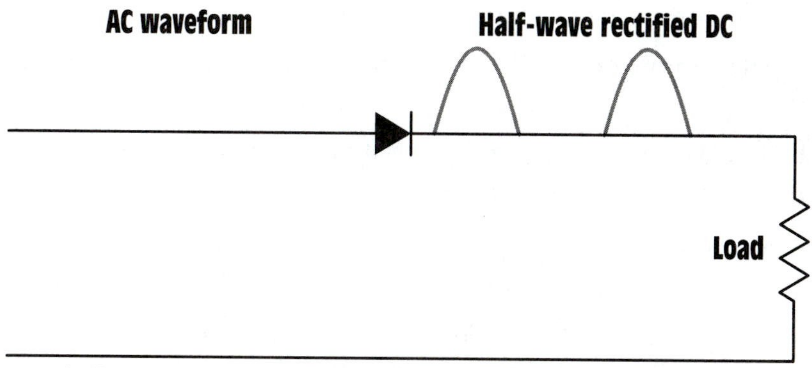

Figure 9-15 A half-wave rectifier converts only one-half of the AC waveform into DC.

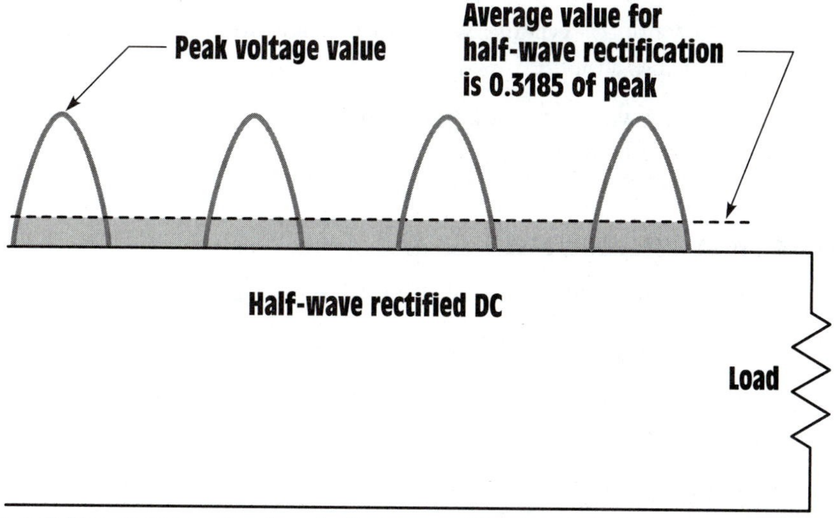

Figure 9-16 The average value for a half-wave rectifier is only half that of a full-wave rectifier.

SUMMARY

1. Most of the electrical power generated in the world is alternating current.
2. Alternating current can be transformed and direct current cannot.
3. Alternating current reverses its direction of flow at periodic intervals.
4. The most common AC waveform is the sine wave.
5. There are 360 degrees in one complete sine wave.

6. One complete waveform is called a cycle.
7. The number of complete cycles that occur in one second is called the frequency.
8. Sine waves are produced by rotating machines.
9. Frequency is measured in hertz (Hz).
10. The instantaneous voltage at any point on a sine wave is equal to the peak, or maximum, voltage times the sine of the angle of rotation.
11. The peak-to-peak voltage is the amount of voltage measured from the positive-most peak to the negative-most peak.
12. The peak value is the maximum amount of voltage attained by the waveform.
13. The RMS value of voltage will produce as much power as an equal amount of DC voltage.
14. The average value of voltage is used when an AC sine wave is changed into DC.

REVIEW QUESTIONS

1. What is the most common type of AC waveform?
2. How many degrees are there in one complete sine wave?
3. At what angle does the voltage reach its maximum negative value on a sine wave?
4. What is frequency?
5. A sine wave has a maximum value of 230 V. What is the voltage after 38° of rotation?
6. A sine wave has a voltage of 63 V after 22° of rotation. What is the maximum voltage reached by this waveform?
7. A sine wave has a maximum value of 560 V. At what angle of rotation will the voltage reach a value of 123 V?
8. A sine wave has a peak value of 433 V. What is the RMS value?
9. A sine wave has a peak-to-peak value of 88 V. What is the average value?
10. A DC voltage has an average value of 68 V. What is the RMS value?

Unit 10

Alternating Current Loads

objectives

After studying this unit, you should be able to:

- Discuss the relationships of voltage and current in a pure resistive load.
- Discuss the differences in true power, apparent power, and reactive power.
- Determine the phase angle of voltage and current in an AC circuit.
- Discuss the properties of inductance.
- Discuss the properties of capacitance.
- Compute values of inductive reactance.
- Compute values of capacitive reactance.
- Compute the value of impedance in an AC circuit.
- Discuss the differences between resistance and impedance.

watts

true power

In a direct current circuit, there is one basic type of load, which is resistive. In a DC circuit volts times amps always equals **watts** or **true power.** This is not true in an alternating current circuit. Alternating current circuits can have three different types of loads: resistive, inductive, and capacitive. Each of these loads produces a different circuit condition. In this unit it will be shown that volts times amps may not always equal watts in an AC circuit.

RESISTIVE LOADS

resistive loads

Resistive loads contain pure resistance; examples include electric heating equipment and incandescent lighting. Resistive loads are characterized by the fact that:

1. They produce heat.
2. The current and voltage are in phase with each other.
3. Only the resistive part of an AC circuit can produce true power or watts.

Any time that a circuit contains resistance, heat will be produced.

in phase

When an AC voltage is applied to a resistor, the current flow through the resistor will be a copy of the voltage, *Figure 10-1*. The current will rise and fall at the same rate as the voltage and will reverse the direction of flow when the voltage reverses polarity. In this condition, the current is said to be **in phase** with the voltage.

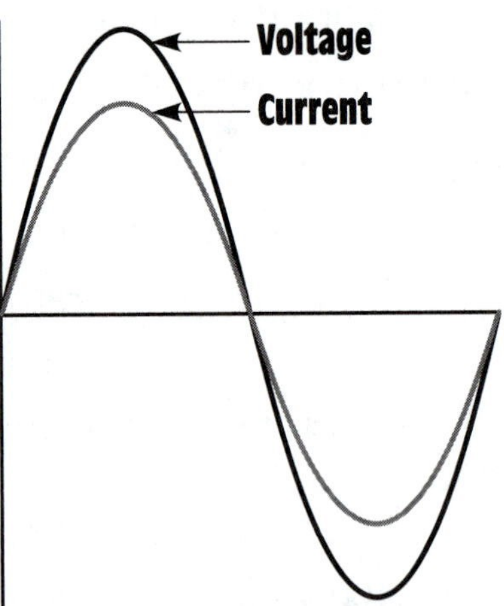

Figure 10-1 In a pure resistive circuit the voltage and current are in phase with each other.

POWER IN AN AC CIRCUIT

True power, or watts, can be produced only when both current and voltage are either positive or negative. When like signs are multiplied, the product is positive, (+ × + = + or − × − = +), and when unlike signs are multiplied the product is negative, (+ × − = −). Since current and voltage are either positive or negative at the same time, the product, watts, will always be positive, *Figure 10-2*.

True power, or watts, can be produced only when both current and voltage are either positive or negative.

INDUCTANCE

Inductance (L) is one of the primary types of loads in alternating current circuits. Some amount of inductance is present in all alternating current circuits because of the continually changing magnetic field, *Figure 10-3*. The amount of inductance of a single conductor is extremely small, and in most instances it is not considered in circuit calculations. Circuits are generally considered to contain inductance when any type of load that contains a coil is used. Loads such as motors, transformers, lighting ballast, and chokes all contain coils of wire.

In Unit 4, it was discussed that whenever current flows through a coil of wire, a magnetic field is created around the wire, *Figure 10-4*. If the

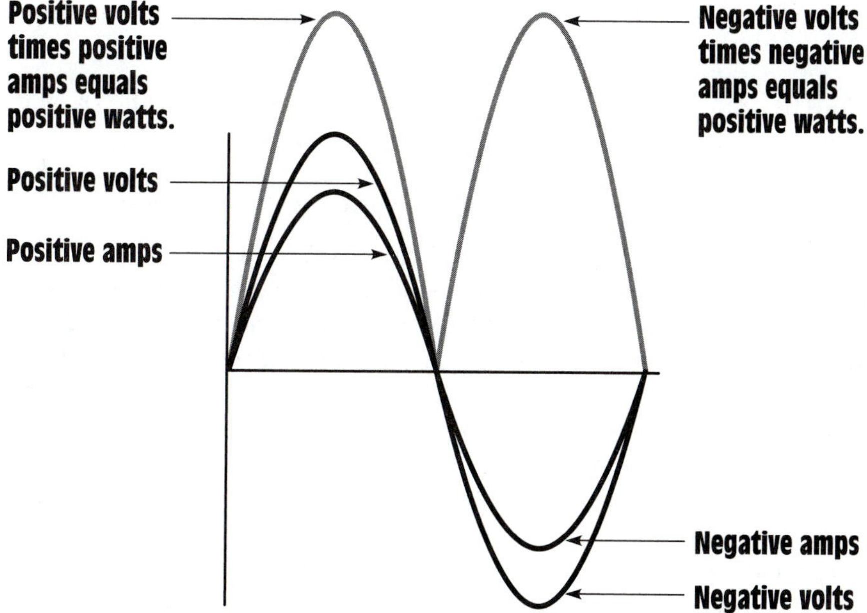

Figure 10-2 Power in a pure resistive AC circuit.

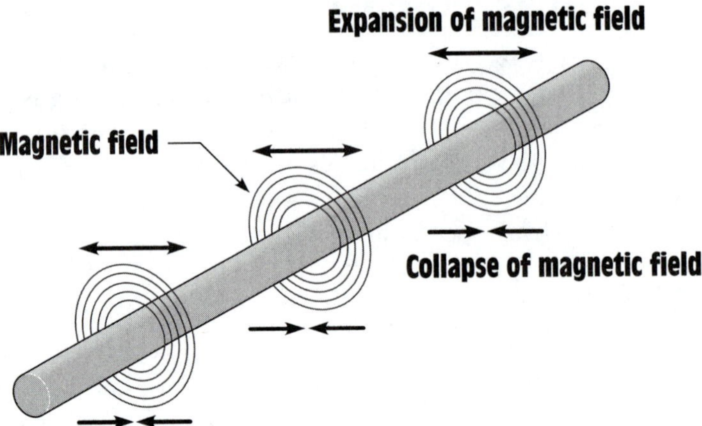

Figure 10-3 A continually changing magnetic field induces a voltage into any conductor.

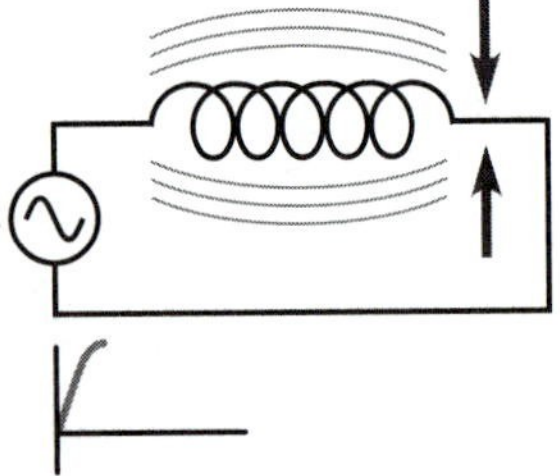

Figure 10-4 As current flows through a coil, a magnetic field is created around the coil.

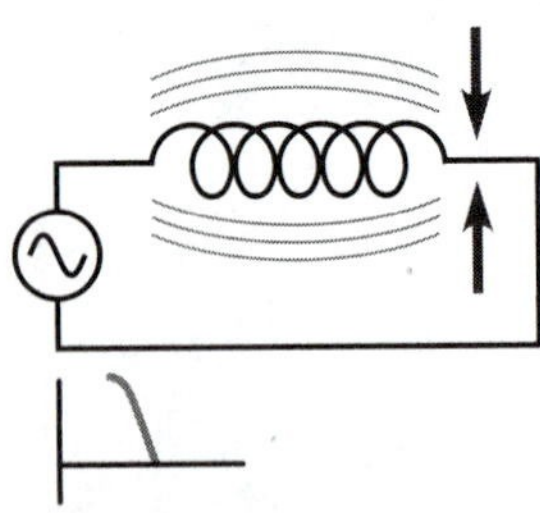

Figure 10-5 As current flow decreases, the magnetic field collapses.

amount of current decreases, the magnetic field will collapse, *Figure 10-5.* Several facts concerning inductance are:

1. When magnetic lines of flux cut through a coil, a voltage is induced in the coil.
2. An induced voltage is always opposite in polarity to the applied voltage.
3. The amount of induced voltage is proportional to the rate of change of current.
4. An inductor opposes a change of current.

The inductors in Figure 10-4 and Figure 10-5 are connected to an alternating voltage. This causes the magnetic field to continually increase, decrease, and reverse polarity. Since the magnetic field continually changes magnitude and direction, a voltage is continually being induced in the coil. This **induced voltage** is 180° out of phase with the applied voltage and is always in opposition to the applied voltage, *Figure 10-6.* Since the induced voltage is always in opposition to the applied voltage, the applied voltage must overcome the induced voltage before current can flow through the circuit. For example, assume an inductor is connected to a 120-V AC line. Now assume that the inductor has an induced voltage of 116 V. Since an equal amount of applied voltage must be used to overcome the induced voltage, there will be only 4 V to push current through the wire resistance of the coil $(120 - 116 = 4)$.

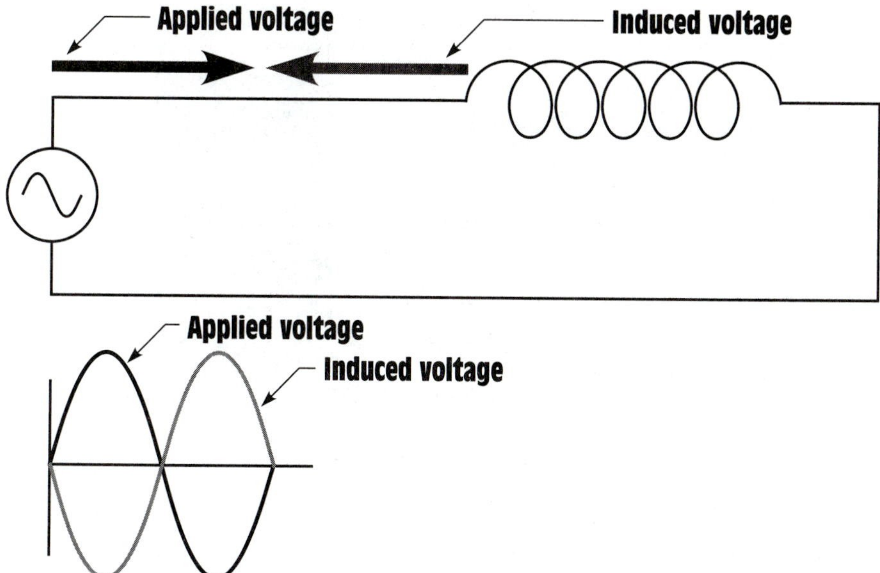

Figure 10-6 The applied voltage and induced voltage are 180° out of phase with each other.

COMPUTING THE INDUCED VOLTAGE

The amount of induced voltage in an inductor can be computed if the resistance of the wire in the coil and the amount of circuit current are known. For example, assume that an ohmmeter is used to measure the actual amount of resistance in a coil and the coil is found to contain 6 Ω of wire resistance, *Figure 10-7*. Now assume that the coil is connected to a 120-V AC circuit and an ammeter measures a current flow of 0.8 A,

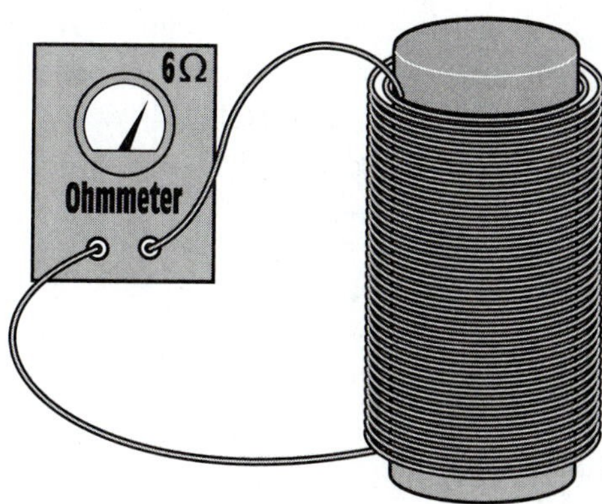

Figure 10-7 Measuring the resistance of a coil.

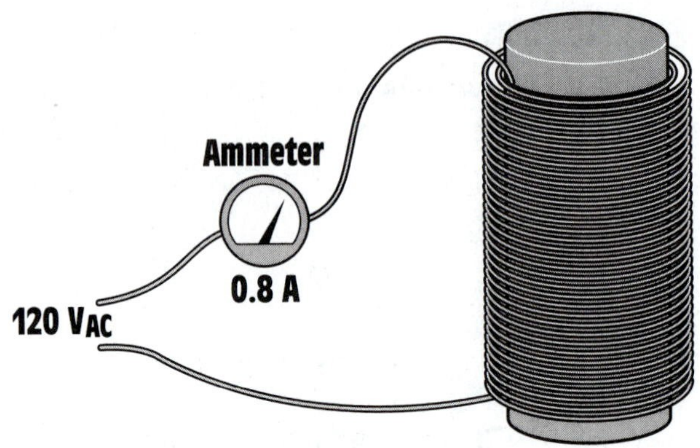

Figure 10-8 Measuring circuit current with an ammeter.

Figure 10-8. Ohm's Law can now be used to determine the amount of voltage necessary to push 0.8 A of current through 6 Ω of resistance.

$$E = I \times R$$

$$E = 0.8 \times 6$$

$$E = 4.8 \text{ V}$$

Since only 4.8 V is needed to push the current through the wire resistance of the inductor, the remainder of the 120 V is used to overcome the coil's induced voltage of 119.9 V ($\sqrt{120^2 - 4.8^2} = 119.9$).

INDUCTIVE REACTANCE

reactance

inductive reactance (X_L)

Notice that the induced voltage can limit the flow of current through the circuit in a manner similar to resistance. This induced voltage is *not* resistance, but it can limit the flow of current just as resistance does. This current-limiting property of.the inductor is called **reactance** and is symbolized by the letter *X*. Since this reactance is caused by inductance, it is called **inductive reactance** and is symbolized by $\mathbf{X_L}$, pronounced "X sub L." Inductive reactance is measured in ohms just as resistance is, and it can be computed when the values of inductance and frequency are known. The following formula can be used to find inductive reactance:

$$X_L = 2\pi FL$$

where

X_L = inductive reactance

2 = a constant

$\pi = 3.1416$

F = frequency in hertz (Hz)

L = inductance in henrys (H)

Inductive reactance is an inducted voltage and is, therefore, proportional to the three factors that determine induced voltage:

1. The *number* of turns of wire
2. The *strength* of the magnetic field
3. The *speed* of the cutting action (relative motion between the inductor and the magnetic lines of flux)

The number of turns of wire and the strength of the magnetic field are determined by the physical construction of the inductor. Factors such as the size of wire used, the number of turns, how close the turns are to each other, and the type of core material determine the amount of inductance (measured in henrys) of the coil, *Figure 10-9.* The speed of the cutting action is proportional to the frequency (Hz). An increase of frequency will cause the magnetic lines of flux to cut the conductors at a faster rate. This will produce a higher induced voltage or more inductive reactance.

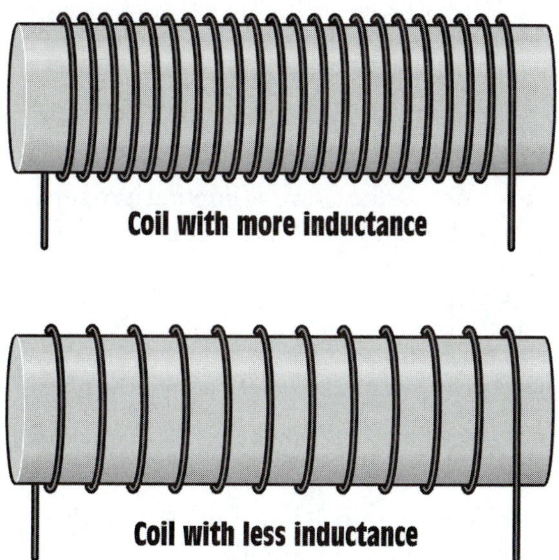

Figure 10-9 Coils with turns closer together produce more inductance than coils with turns far apart.

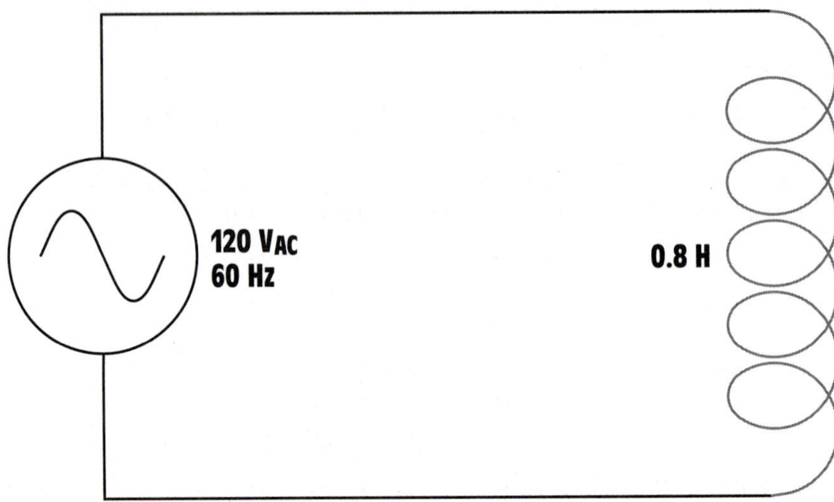

Figure 10-10 Circuit current is limited by inductive reactance.

Example 1

The inductor shown in *Figure 10-10* has an inductance of 0.8 H and is connected to a 120-V, 60-Hz line. How much current will flow in this circuit if the wire resistance of the inductor is negligible?

Solution:

The first step in solving this problem is to determine the amount of inductive reactance of the inductor.

$$X_L = 2\pi FL$$

$$X_L = 2 \times 3.1416 \times 60 \times 0.8$$

$$X_L = 301.6\ \Omega$$

Since inductive reactance is the current-limiting property of this circuit, it can be substituted for the value of R in an Ohm's Law formula.

$$I = \frac{E}{X_L}$$

$$I = \frac{120}{301.6}$$

$$I = 0.398\ A$$

If the amount of inductive reactance is known, the inductance of the coil can be determined using the formula:

$$L = \frac{X_L}{2\pi F}$$

Example 2

Assume an inductor with a negligible resistance is connected to a 36-V, 400-Hz line. If the circuit has a current flow of 0.2 A what is the inductance of the inductor?

Solution:

The first step is to determine the inductive reactance of the circuit.

$$X_L = \frac{E}{I}$$

$$X_L = \frac{36}{0.2}$$

$$X_L = 180\ \Omega$$

Now that the inductive reactance of the inductor is known, the inductance can be determined.

$$L = \frac{X_L}{2\pi F}$$

$$L = \frac{180}{2 \times 3.1416 \times 400}$$

$$L = 0.0716\ \text{H}$$

SCHEMATIC SYMBOLS

The schematic symbol used to represent an inductor depicts a coil of wire. Several symbols for inductors are shown in *Figure 10-11*. Although these symbols are different, they are similar. The symbols shown with the two parallel lines represent iron core inductors, and the symbols without the parallel lines represent air core inductors.

Figure 10-11 Schematic symbols for inductors.

VOLTAGE AND CURRENT RELATIONSHIPS IN AN INDUCTIVE CIRCUIT

When current flows through a pure resistive circuit, the current and voltage are in phase with each other. **In a pure inductive circuit the current lags the voltage by 90°.** At first this may seem to be an impossible condition until the relationship of applied voltage and induced voltage is considered. To understand how the current and applied voltage can become 90° out of phase with each other can best be explained by comparing the relationship of the current and induced voltage, *Figure 10-12*. Recall that the induced voltage is proportional to the rate of change of the current (speed of cutting action). At the beginning of the waveform, the current is shown at its maximum value in the negative direction. At this point, the

In a pure inductive circuit the current lags the voltage by 90°.

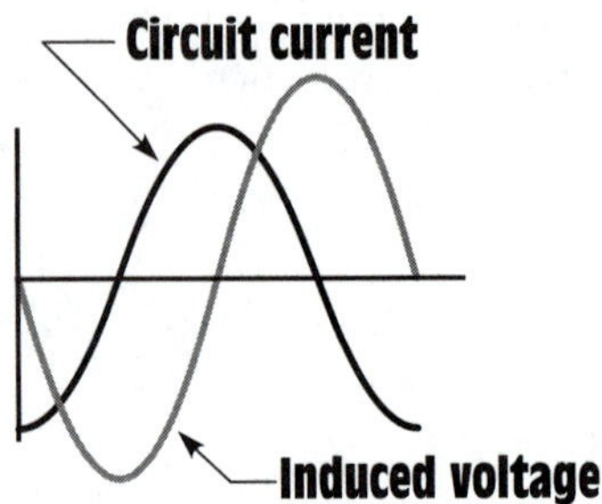

Figure 10-12 Induced voltage is proportional to the rate of change of current.

current is not changing, so induced voltage is zero. As the current begins to decrease in value, the magnetic field produced by the flow of current decreases or collapses and begins to induce a voltage into the coil as it cuts through the conductors, as shown in Figure 10-5.

The greatest rate of current change occurs when the current passes from negative through zero and begins to increase in the positive direction, *Figure 10-13*. Since the current is changing at the greatest rate, the induced voltage is maximum. As current approaches its peak value in the positive

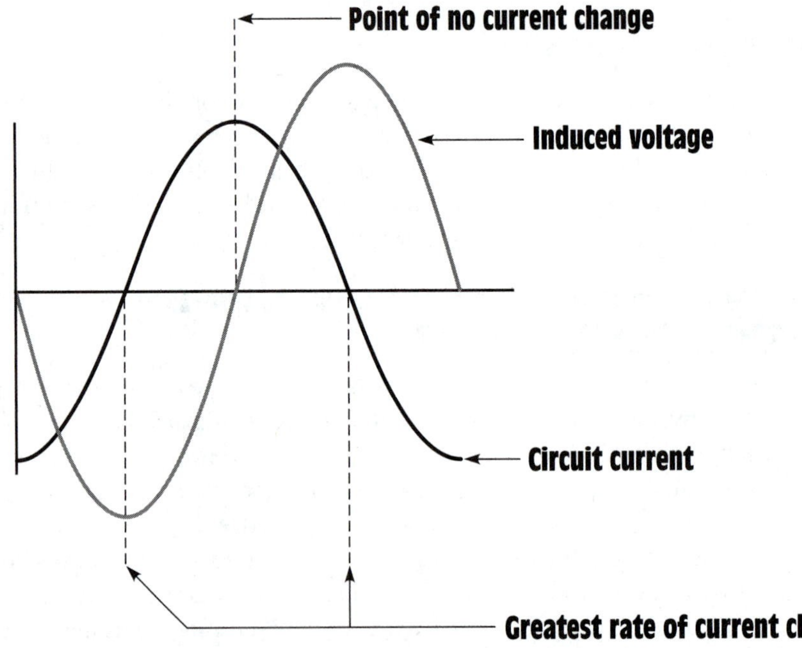

Figure 10-13 No voltage is induced when the current does not change.

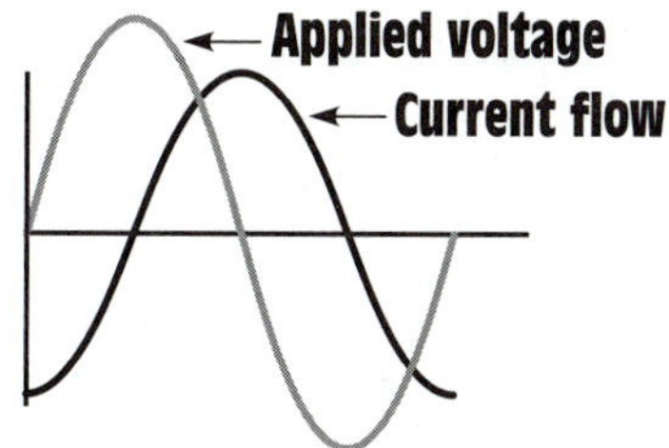

Figure 10-14 The current lags the applied voltage by 90°.

direction, the rate of change decreases, causing a decrease in the induced voltage. The induced voltage will again be zero when the current reaches its peak value and the magnetic field stops expanding.

It can be seen that the current flowing through the inductor is leading the induced voltage by 90°. Since the induced voltage is 180° out of phase with the applied voltage, the current will lag the applied voltage by 90°, *Figure 10-14*.

POWER IN AN INDUCTIVE CIRCUIT

In a pure resistive circuit, the true power, or watts, is equal to the product of the voltage and current. In a pure inductive circuit, however, no true power or watts is produced. Recall that voltage and current must both be either positive or negative before true power can be produced. Since the voltage and current are 90° out of phase with each other in a pure inductive circuit, the current and voltage will be at different polarities 50% of the time and at the same polarity 50% of the time. During the period of time when the current and voltage have the same polarity, power is being given to the circuit in the form of creating a magnetic field. When the current and voltage are opposite in polarity, power is being given back to the circuit as the magnetic field collapses and induces a voltage back into the circuit. Since power is stored in the form of a magnetic field and then given back, no power is used by the inductor. Any power used in an inductor is caused by losses such as the resistance of the wire used to construct the inductor; these are generally referred to as I^2R losses, eddy current losses, and hysteresis losses.

The current and voltage waveform in *Figure 10-15* has been divided into four sections; A, B, C, and D. During the first time period, indicated by A, the current is negative and the voltage is positive. During this period of time, energy is being given to the circuit as the magnetic field collapses. During the second time period, section B, both the voltage and current are positive. Power is being used to produce the magnetic field. In the third

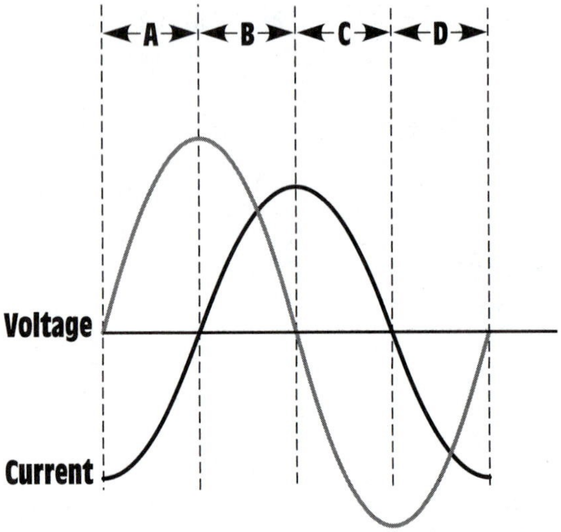

Figure 10-15 Voltage and current relationships during different parts of a cycle.

time period, indicated by C, the current is positive and the voltage is negative. Power is again being given back to the circuit as the field collapses. During the fourth time period, indicated by D, both the voltage and current are negative. Power is again being used to produce the magnetic field. If the amount of power used to produce the magnetic field is subtracted from the power given back, the result is zero.

REACTIVE POWER

reactive power (VARs)

Although essentially no true power is used in a pure inductive circuit (except by previously mentioned losses), an electrical measurement called **VARs** is used to measure the **reactive power** in a pure inductive circuit. VARs is an abbreviation for volt-amps-reactive. VARs can be computed in the same way as watts, except that inductive values are substituted for resistive values in the formulas. VARs is equal to the amount of current flowing through an inductive circuit times the voltage applied to the inductive part of the circuit. Several formulas for computing VARs are:

$$\text{VARs} = E_L \times I_L$$

$$\text{VARs} = \frac{E_L^2}{X_L}$$

$$\text{VARs} = I_L^2 \times X_L$$

where

E_L = voltage applied to an inductor

I_L = current flow through an inductor

X_L = inductive reactance

SUMMARY

1. The three basic types of alternating current loads are resistive, inductive, and capacitive.
2. In a pure resistive load, the current and voltage are in phase with each other.
3. In a pure resistive load, V × A = W.
4. Watts is a measure of true power.
5. True power, or watts, can be produced only during the periods of time that both the voltage and current have the same polarity.
6. An electric circuit produces true power when electrical energy is converted into some other form.
7. Induced voltage is proportional to the rate of change of current.
8. Induced voltage is always opposite in polarity to the applied voltage.
9. Inductive reactance is a counter-voltage that limits the flow of current like resistance.
10. Inductive reactance is measured in ohms (Ω).
11. Inductive reactance is proportional to the inductance of the coil and the frequency of the line.
12. Inductive reactance is symbolized by X_L.
13. Inductance is measured in henrys and is symbolized by the letter *L*.
14. The current lags the applied voltage by 90° in a pure inductive circuit.
15. All inductors contain some amount of resistance.
16. Pure inductive circuits contain no true power, or watts.
17. Reactive power is measured in VARs.
18. VARs is an abbreviation for volt-amps-reactive.

REVIEW QUESTIONS

1. How many degrees are the current and voltage out of phase with each other in a pure resistive circuit?
2. How many degrees are the current and voltage out of phase with each other in a pure inductive circuit?
3. To what is inductive reactance proportional?
4. An inductor is connected to a 240-V, 1000-Hz line. The circuit current is 0.6 A. What is the inductance of the inductor?
5. An inductor with an inductance of 3.6 H is connected to a 480-V, 60-Hz line. How much current will flow in this circuit?
6. If the frequency in question 5 is reduced to 50 Hz how much current will flow in the circuit?
7. An inductor has an inductive reactance of 250 Ω when connected to a 60-Hz line. What will be the inductive reactance if the inductor is connected to a 400-Hz line?
8. An electric heating element has a current draw of 12 A when connected to a 240-V AC line. What is the true power of this circuit?
9. Reactive power is measured in what units?
10. A pure inductive load has an inductive reactance of 8.4 Ω and is connected to a 480-V AC line. What is the true power, or W, produced in this circuit? How many VARs are produced in this circuit?

Practice Problems

INDUCTIVE CIRCUITS

Fill in all missing values in the chart. Refer to the following formulas:

$$X_L = 2\pi FL$$

$$L = \frac{X_L}{2\pi F}$$

$$F = \frac{X_L}{2\pi L}$$

Inductance (H)	Frequency (Hz)	Induct. Rct. (Ω)
1.2	60	
0.085		213.628
	1000	4712.389
0.65	600	
3.6		678.584
	25	411.459
0.5	60	
0.85		6408.849
	20	201.062
0.45	400	
4.8		2412.743
	1000	40.841

11 Unit

Capacitive Loads

objectives

*fter studying this unit, you should be able to:*

- Discuss differences between capacitive and inductive loads.
- Describe the construction of a capacitor.
- List the three factors that determine the capacitance of a capacitor.
- Discuss electrostatic charges.
- Compute values of capacitive reactance.
- Discuss the charge and discharge rates of a capacitor.
- Describe differences between AC and electrolytic capacitors.

Capacitive loads are the third type of loads that can be encountered in an alternating current circuit. Capacitive loads are the exact opposite of inductive loads. Inductors oppose a change of current and capacitors oppose a change of voltage. In a pure inductive circuit the current lags the applied voltage by 90°, and in a pure capacitive circuit the current leads the applied voltage by 90°. Inductors store electrical energy in the form of an electromagnetic field and then return it to the circuit. Capacitors store electrical energy in the form of an electrostatic field and then return it to the circuit. To better understand how a capacitor works it will first be necessary to discuss what a capacitor is.

CAPACITORS

Caution: It is the habit of some people to charge a capacitor to high voltage and then hand it to another person. While some people think this is comical, it is an extremely dangerous practice. Capacitors have the ability to supply an almost infinite amount of current. Under some conditions, a capacitor can have enough energy to cause a person's heart to go into fibrillation.

This statement is intended to make you realize the danger that capacitors can pose under certain conditions. Capacitors perform a variety of jobs, such as power factor correction, storing an electrical charge to produce a large current pulse, timing circuits, and electronic filters.

Capacitors are devices that oppose a change of voltage. The simplest type of capacitor is constructed by separating two metal **plates** with some type of insulating material called the **dielectric,** *Figure 11-1.* Three factors that determine the capacitance of a capacitor:

Capacitors are devices that oppose a change of voltage.

plates

dielectric

1. The area of the plates
2. The distance between the plates
3. The type of dielectric used

The greater the surface area of the plates, the more capacitance a capacitor will have. If a capacitor is charged by connecting it to a source of

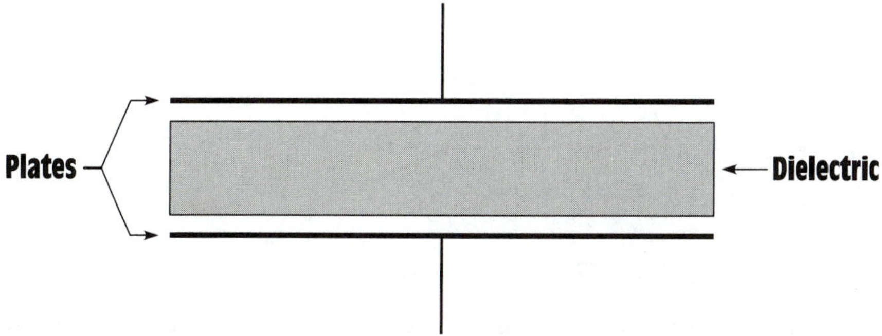

Figure 11-1 A capacitor is made by separating two metal plates with a dielectric.

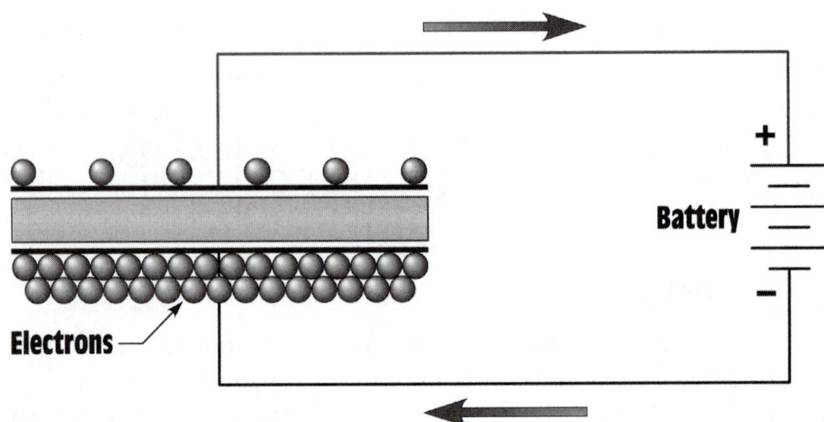

Figure 11-2 A capacitor can be charged by removing electrons from one plate and depositing electrons on the other plate.

Current can flow only during the period of time that a capacitor is either charging or discharging.

direct current, *Figure 11-2,* electrons are removed from the plate connected to the positive battery terminal and deposited on the plate connected to the negative terminal. This flow of current will continue until a voltage equal to the battery voltage is established across the plates of the capacitor, *Figure 11-3.* When these two voltages become equal, the flow of electrons will stop. The capacitor is now charged. If the battery is disconnected from the capacitor, the capacitor will remain charged as long as there is no path by which the electrons can move from one plate to the other, *Figure 11-4.* A good rule to remember concerning a capacitor and current flow is that **current can flow only during the period of time that a capacitor is either charging or discharging.**

In theory, it should be possible for a capacitor to remain in a charged condition forever. In actual practice, this is not the case. No dielectric is a perfect insulator, and electrons eventually move through the dielectric from the negative plate to the positive, causing the capacitor to discharge,

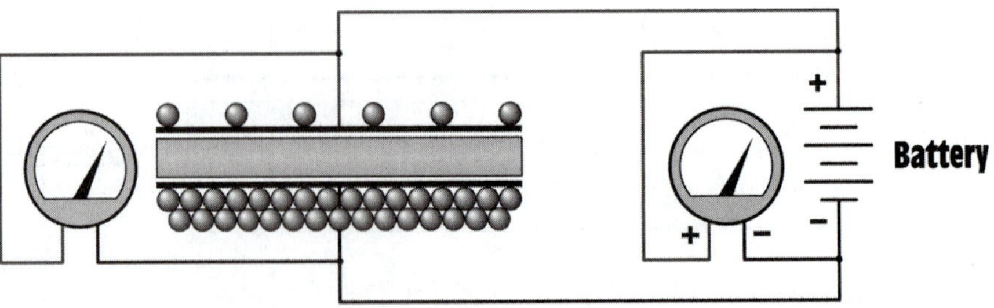

Figure 11-3 Current flows until the voltage across the capacitor is equal to the voltage of the battery.

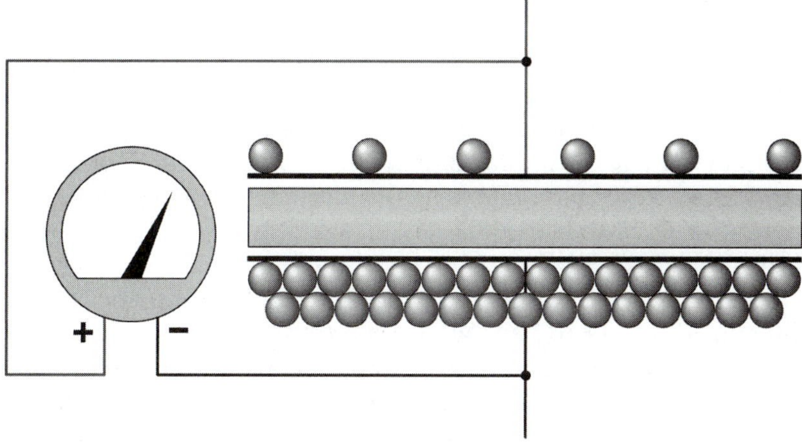

Figure 11-4 The capacitor remains charged after the battery is removed from the circuit.

Figure 11-5. This current flow through the dielectric is called **leakage current,** and is proportional to the resistance of the dielectric and the charge across the plates. If the dielectric of a capacitor becomes weak, it permits an excessive amount of leakage current to flow. A capacitor in this condition is often referred to as a leaky capacitor.

leakage current

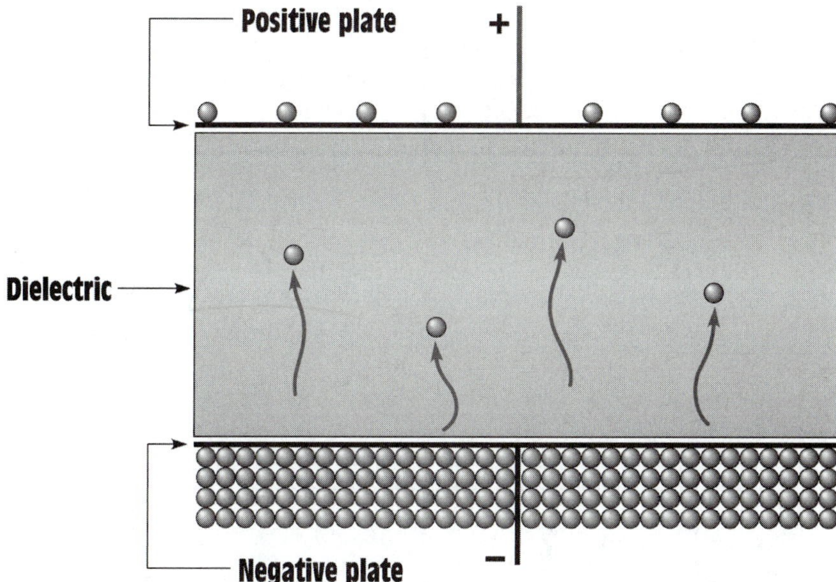

Figure 11-5 Electrons eventually leak through the dielectric. This flow of electrons is known as leakage current.

ELECTROSTATIC CHARGE

Two other factors that determine capacitance are the type of dielectric used and the distance between the plates. To understand these concepts, it is necessary to understand how a capacitor stores energy. In previous units it was discussed that an inductor stores energy in the form of an electromagnetic field. A capacitor stores energy in an electrostatic field.

DIELECTRIC STRESS

When a capacitor is not charged, the atoms of the dielectric are uniform as shown in *Figure 11-6*. The valence electrons orbit the nucleus in a circular pattern. When the capacitor becomes charged, however, a potential exists between the plates of the capacitor. The plate with the lack of electrons has a positive charge, and the plate with the excess of electrons has a negative charge. Since electrons are negative particles, they are repelled away from the negative plate and attracted to the positive plate. This attraction causes the electron orbit to become stretched as shown in *Figure 11-7*. This stretching of the atoms of the dielectric is called **dielectric stress.** Placing the atoms of the dielectric under stress has the same effect as drawing back a bow string and arrow, and holding it, *Figure 11-8*.

dielectric stress

The amount of dielectric stress is proportional to the voltage difference between the plates. The greater the voltage, the greater the dielectric stress.

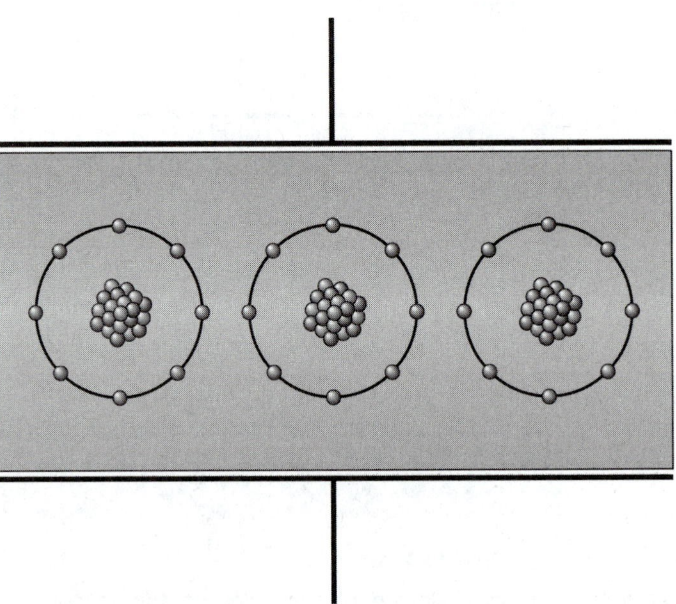

Figure 11-6 Atoms of the dielectric in an uncharged capacitor.

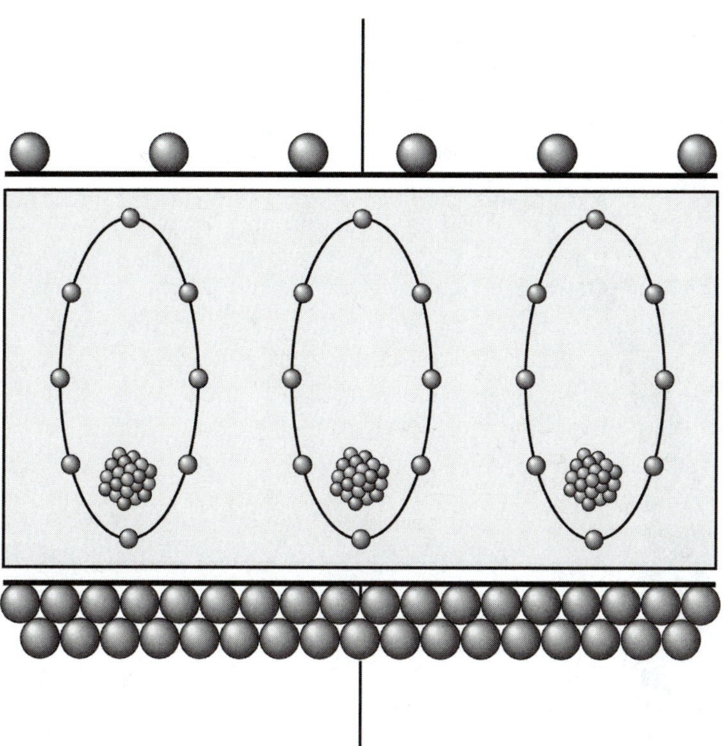

Figure 11-7 Atoms of the dielectric in a charged capacitor.

If the voltage becomes too great, the dielectric will break down and permit current to flow between the plates. At this point the capacitor becomes shorted. Capacitors have a voltage rating that should not be exceeded. The voltage rating indicates the maximum amount of voltage the dielectric is intended to withstand without breaking down. The amount of voltage applied to a capacitor is critical to its life span. Capacitors operated above their voltage rating will fail relatively quickly. Many years ago, the U.S. military made a study of the voltage rating of a capacitor as compared to its life span. The results showed that a capacitor operated at one-half its rated voltage will have a life span approximately eight times longer than a capacitor operated at the rated voltage.

electrostatic charge

The energy of the capacitor is stored in the dielectric in the form of an **electrostatic charge.** It is this electrostatic charge that permits the capacitor to produce extremely high currents under certain conditions. If the leads of a capacitor are shorted together, it has the effect of releasing the drawn-back bow string, Figure 11-8. When the bowstring is released, the arrow will be propelled forward at a high rate of speed. The same is true for the

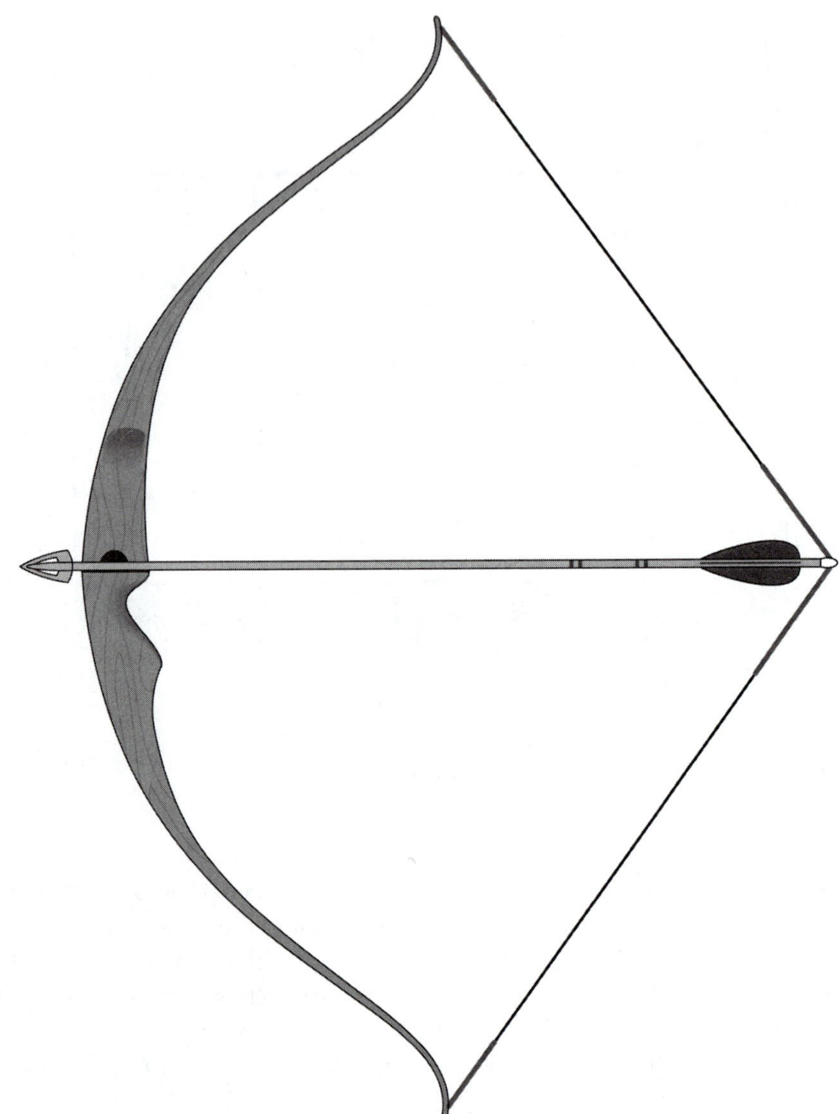

Figure 11-8 Dielectric stress is similar to drawing back a bowstring with an arrow and holding it.

capacitor. When the leads are shorted, the atoms of the dielectric snap back to their normal position. Shorting causes the electrons on the negative plate to be literally blown off and attracted to the positive plate. Capacitors can produce currents of thousands of amperes for short periods of time.

This principle is used to operate the electronic flash of many cameras. Electronic flash attachments contain a small glass tube filled with a gas

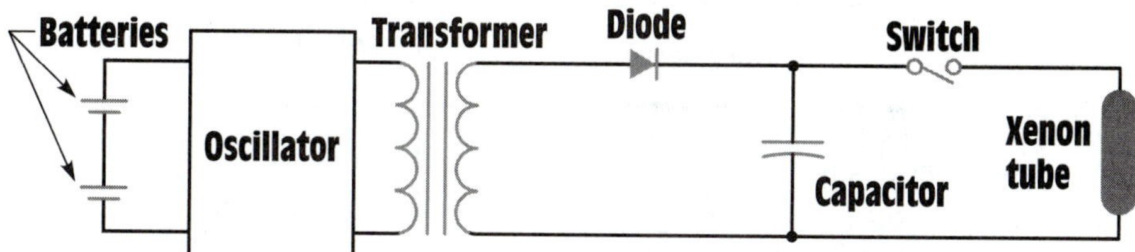

Figure 11-9 Energy is stored in a capacitor.

called xenon. Xenon produces a very bright white light similar to sunlight when the gas is ionized. A large amount of power is required, however, to produce a bright flash. A battery capable of directly ionizing the xenon would be very large, expensive, and have a potential of about 500 V. The simple circuit shown in *Figure 11-9* can be used to overcome the problem. In this circuit, two small 1.5-V batteries are connected to an oscillator. The oscillator changes the direct current of the batteries into square wave alternating current. The alternating current is then connected to a transformer and the voltage is increased to about 500 V peak. A diode changes the AC voltage back into DC and charges a capacitor. The capacitor charges to the peak value of the voltage waveform. When the switch is closed, the capacitor suddenly discharges through the xenon tube and supplies the power needed to ionize the gas. It may take several seconds to store enough energy in the capacitor to ionize the gas in the tube, but the capacitor can release the stored energy in a fraction of a second.

To understand how the capacitor can supply the energy needed, consider the amount of gunpowder contained in a 0.357 cartridge. If the powder were to be removed from the cartridge and burned in the open air, it would be found that the actual amount of energy contained in the powder is very small. This amount of energy would not even be able to raise the temperature by a noticeable amount in a small enclosed room. If this same amount of energy is converted into heat in a fraction of a second, however, enough force is developed to propel a heavy projectile with great force. This same principle is at work when a capacitor is charged over some period of time and then discharged in a fraction of a second.

DIELECTRIC CONSTANTS

Since the energy of the capacitor is stored in the dielectric, the type of dielectric used is extremely important in determining the amount of capacitance a capacitor will have. Different materials are assigned a number called the **dielectric constant.** Air is assigned the number 1 and is used as a reference for comparison. For example, assume a capacitor uses air as the

dielectric constant

Material	Dielectric constant
Air	1
Bakelite	4.0 –10.0
Castor oil	4.3 – 4.7
Cellulose acetate	7.0
Ceramic	1200
Dry paper	3.5
Hard rubber	2.8
Insulating oils	2.2 – 4.6
Lucite	2.4 – 3.0
Mica	6.4 – 7.0
Mycalex	8.0
Paraffin	1.9 – 2.2
Porcelain	5.5
Pure water	81
Pyrex glass	4.1 – 4.9
Rubber compounds	3.0 – 7.0
Teflon	2
Titanium dioxide compounds	90 – 170

Figure 11-10 Dielectric constant of different materials.

dielectric and its capacitance value is found to be 1 microfarad (μF). Now assume that some dielectric material is placed between the plates without changing the spacing and the capacitance value becomes 5 μF. This material has a dielectric constant of 5. A chart showing the dielectric constant of different materials is shown in *Figure 11-10.*

farad

A capacitor has a capacitance of one farad when a change of one volt across its plates results in a movement of one coulomb.

CAPACITOR RATINGS

The basic unit of capacitance is the **farad** and is symbolized by the letter *F.* It receives its name from a famous scientist named Michael Faraday. **A capacitor has a capacitance of one farad when a change of one volt across its plates results in a movement of one coulomb.**

$$Q = C \times V$$

where

Q = charge in coulombs

C = capacitance in farads

V = charging voltage

Although the farad is the basic unit of capacitance, it is seldom used because it is an extremely large amount of capacitance. The formula shown below can be used to determine the capacitance of a capacitor when the area of the plates, the dielectric constant, and the distance between the plates is known.

$$C = \frac{K \times A}{4.45\ D}$$

where

C = capacitance in pF (picofarads)

K = dielectric constant

A = area of one plate

D = distance between the plates

Example 1

What would be the plate area of a 1 F capacitor if air is used as the dielectric and the plates are separated by a distance of 1 in?

Solution:

The first step is to convert the above formula to solve for area.

$$A = \frac{C \times 4.45 \times D}{K}$$

$$A = \frac{1{,}000{,}000{,}000{,}000 \times 4.45 \times 1}{1}$$

$$A = 4{,}450{,}000{,}000{,}000 \text{ square inches}$$

$$A = 1108.5 \text{ square miles}$$

Since the basic unit of capacitance is so large, other units such as the microfarad (μF), nanofarad (nF), and picofarad (pF) are generally used.

$$\mu F = \frac{1}{1{,}000{,}000}\ (1 \times 10^{-6}) \text{ of a farad}$$

$$nF = \frac{1}{1{,}000{,}000{,}000}\ (1 \times 10^{-9}) \text{ of a farad}$$

$$pF = \frac{1}{1{,}000{,}000{,}000{,}000}\ (1 \times 10^{-12}) \text{ of a farad}$$

The picofarad is sometimes referred to as a micro-microfarad and is symbolized by μμF.

CAPACITORS CONNECTED IN PARALLEL

Connecting capacitors in parallel, *Figure 11-11,* has the same effect as increasing the plate area of one capacitor. In the example shown, three capacitors having a capacitance of 20 μF, 30 μF, and 60 μF are connected in parallel. The total capacitance of this connection is:

$$C_T = C_1 + C_2 + C_3$$

$$C_T = 20 + 30 + 60$$

$$C_T = 110\ \mu F$$

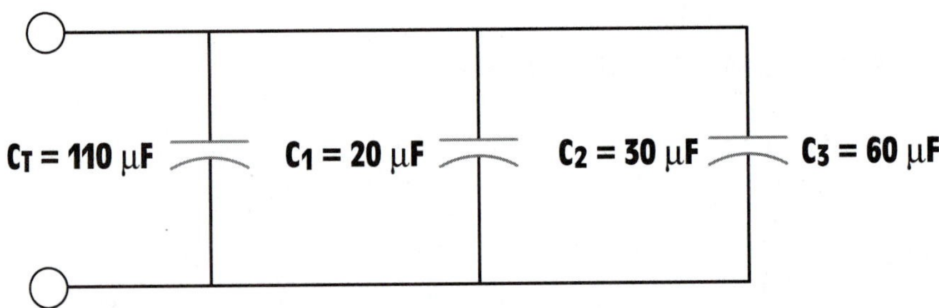

Figure 11-11 Capacitors connected in parallel.

CAPACITIVE CHARGE AND DISCHARGE RATES

exponential

Capacitors charge and discharge at an **exponential** rate. A charge curve for a capacitor is shown in *Figure 11-12.* The curve is divided into five time constants and each time constant is equal to 63.2% of the whole value. In Figure 11-12 it is assumed that a capacitor is to be charged to a total of 100 V. At the end of the first time constant the voltage has reached 63.2% of 100 or 63.2 V. At the end of the second time constant, the voltage reaches 63.2% of the remaining voltage, or 86.4 V. This continues until the capacitor has been charged to 100 V.

The capacitor discharges in the same manner, *Figure 11-13.* At the end of the first time constant the voltage will decrease to 63.2% of its charged value. In this example the voltage will decrease from 100 V to 36.8 V in the first time constant. At the end of the second time constant the voltage will drop to 13.6 V, and by the end of the third time constant the voltage has dropped to 5 V. The voltage will continue to drop at this rate until it reaches approximately 0 after five time constants.

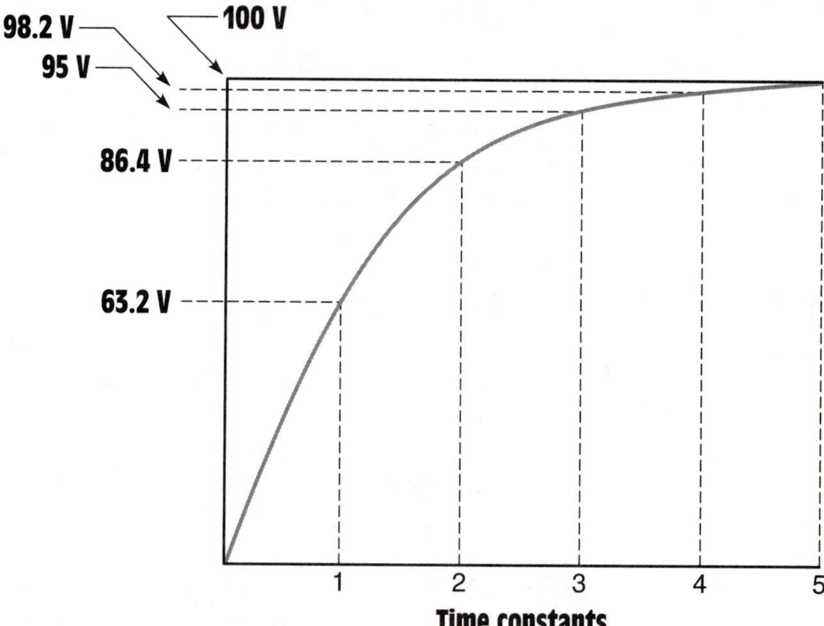

Figure 11-12 Capacitors charge at an exponential rate.

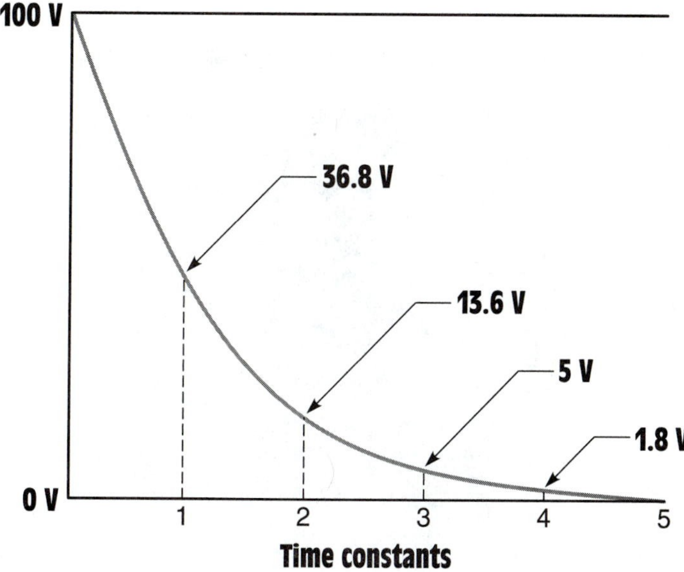

Figure 11-13 Capacitor discharge curve.

APPLICATIONS FOR CAPACITORS

Capacitors are among the most used of electrical components. They are used for power factor correction in industrial applications, in the start windings of many single phase AC motors, to produce phase shifts for SCR and triac circuits, to filter pulsating DC, and in RC timing circuits. Capacitors are used extensively in electronic circuits for control of frequency and pulse generation. The type of capacitor used is dictated by the circuit application.

NONPOLARIZED CAPACITORS

non-polarized capacitors

Capacitors can be divided into two basic groups, nonpolarized and polarized. **Nonpolarized capacitors** are often referred to as AC capacitors because they are not sensitive to polarity connection. Nonpolarized capacitors can be connected to either DC or AC circuits without harm to the capacitor. Most nonpolarized capacitors are constructed by separating metal plates by some type of dielectric. These capacitors can be obtained in many different styles and case types.

A common type of AC capacitor, called the paper capacitor or oil-filled capacitor, is often used in motor circuits, *Figure 11-14*. This capacitor derives its name from the type of dielectric used. This capacitor is constructed by separating plates made of metal foil with thin sheets of paper soaked in a dielectric oil, *Figure 11-15*. These capacitors are often used as the run or starting capacitor for single-phase motors. Many manufacturers of oil-filled

Figure 11-14 Oil-filled paper capacitor.

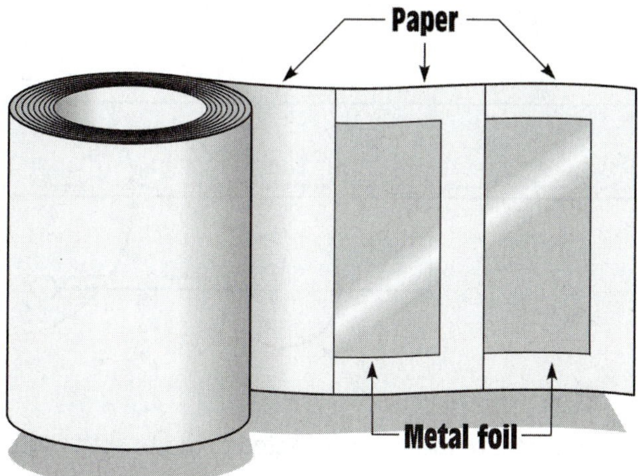

Figure 11-15 Oil-filled paper capacitor.

capacitors will identify one terminal with an arrow or painted dot or by stamping a dash in the capacitor can, *Figure 11-16*. This identified terminal marks the connection to the plate that is located nearer to the metal container or can. It has long been known that when a capacitor's dielectric breaks down and permits a short circuit to ground, it is most often the plate located nearer to the outside case that becomes grounded. For this reason,

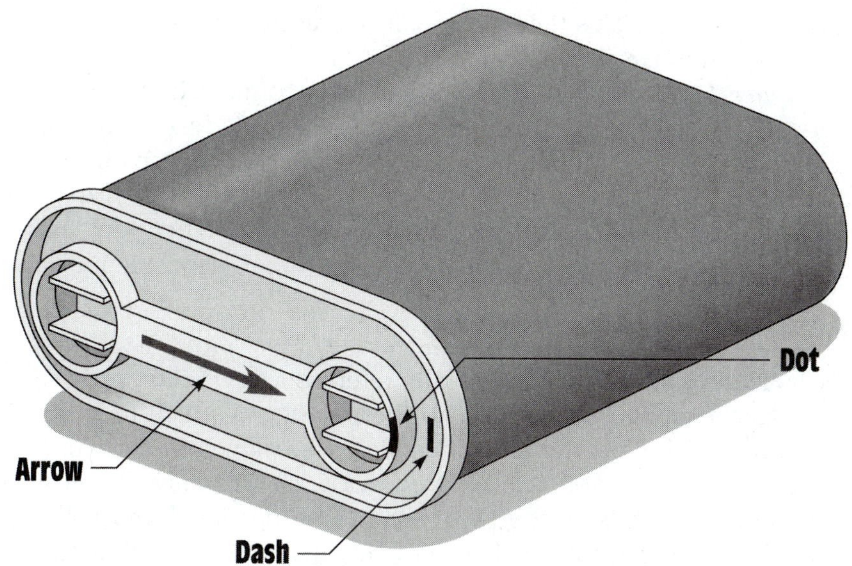

Figure 11-16 Marks indicate the plate nearest the capacitor case.

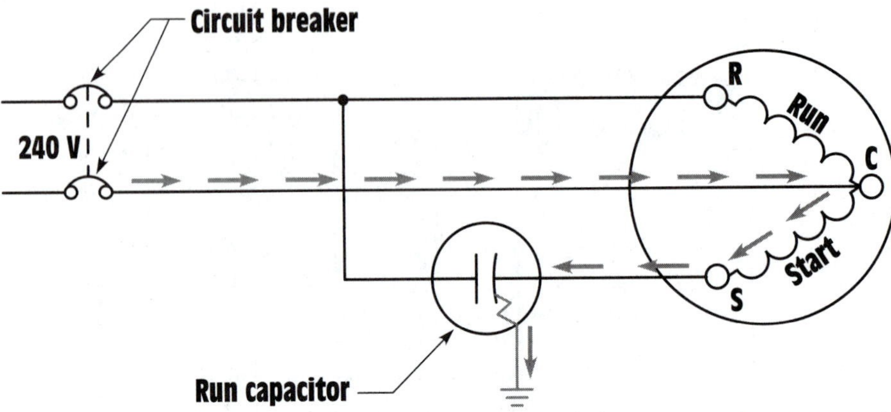

Figure 11-17 Identified capacitor terminal connected to motor start winding (incorrect connection).

it is generally desirable to connect the identified capacitor terminal to the line side instead of to the motor start winding.

In *Figure 11-17,* the run capacitor has been connected in the circuit in such a manner that the identified terminal is connected to the start winding of a single-phase motor. If the capacitor should become shorted to ground, a current path will exist through the motor start winding. The start winding is an inductive type load, and inductive reactance will limit the value of current flow to ground. Since the flow of current is limited, it will take the circuit breaker or fuse some time to open the circuit and disconnect the motor from the power line. This time delay can permit the start winding to overheat and become damaged.

In *Figure 11-18,* the run capacitor has been connected in the circuit in such a manner that the identified terminal is connected to the line side. If the capacitor should become shorted to ground, a current path exists directly to ground, bypassing the motor start winding. When the capacitor is connected in this manner, the start winding does not limit current flow and permits the fuse or circuit breaker to open the circuit almost immediately.

POLARIZED CAPACITORS

polarized capacitors

electrolytic

Polarized capacitors are generally referred to as **electrolytic** capacitors. These capacitors are sensitive to the polarity to which they are connected and will have one terminal identified as positive or negative, *Figure 11-19.* Polarized capacitors can be used in DC circuits only. If their polarity connection is reversed, the capacitor can be damaged and will sometimes explode. The advantage of electrolytic capacitors is that they can have very high capacitance in a small case.

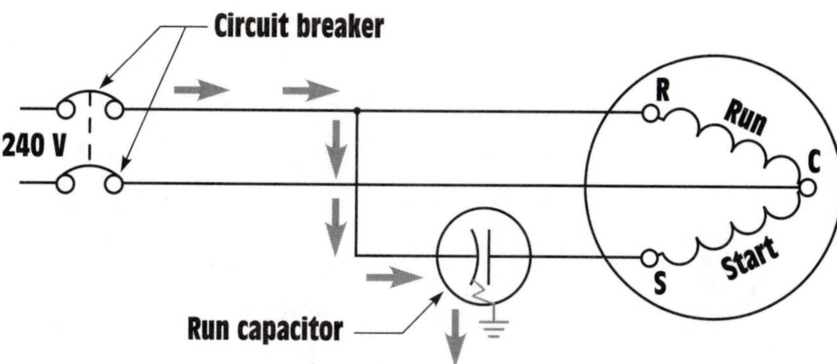

Figure 11-18 Identified terminal connected to the line (correct connections).

Figure 11-19 An electrolytic capacitor.

There are two basic types of electrolytic capacitors, the wet type and the dry type. The wet type electrolytic, *Figure 11-20,* has a positive plate made of aluminum foil. The negative plate is actually an electrolyte made from a borax solution. A second piece of aluminum foil is placed in contact

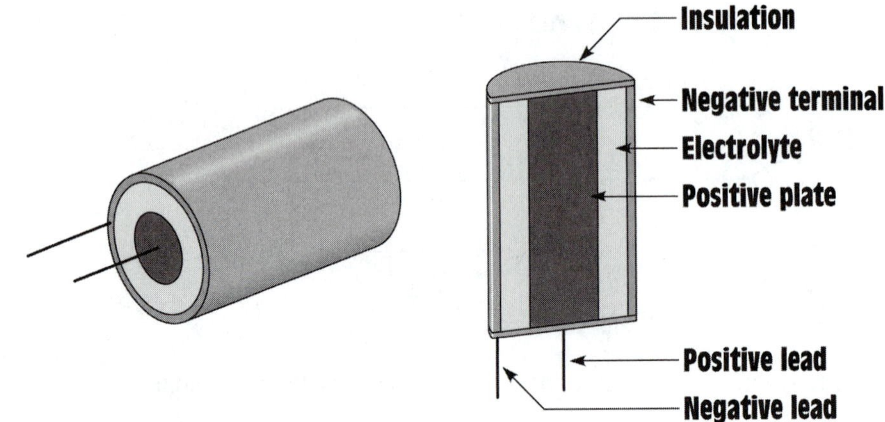

Figure 11-20 Wet type electrolytic capacitor.

with the electrolyte and becomes the negative terminal. When a source of direct current is connected to the capacitor, the borax solution forms an insulating oxide film on the positive plate. This film is only a few molecules thick and acts as the insulator to separate the plates. The capacitance is very high because the distance between the plates is so small.

If the polarity of the wet type electrolytic capacitor becomes reversed, the oxide insulating film will dissolve and the capacitor will become shorted. If the polarity connection is corrected, the film will reform and restore the capacitor.

AC ELECTROLYTIC CAPACITORS

This ability of the wet type electrolytic capacitor to be shorted and then reformed is the basis for a special type of nonpolarized electrolytic capacitor called the AC electrolytic capacitor. This capacitor is used as the starting capacitor for many small single-phase motors, as the run capacitor in many ceiling fan motors, and for low-power electronic circuits when a nonpolarized capacitor with a high capacitance is required. The AC electrolytic capacitor is made by connecting two wet type electrolytic capacitors inside the same case, *Figure 11-21*. In the example shown, the two wet type electrolytic capacitors have their negative terminals connected. When alternating current is applied to the leads, one capacitor will be connected to reverse polarity and become shorted. The other capacitor will be connected to the correct polarity and will form. During the next half cycle, the polarity changes: it forms the capacitor that was shorted and shorts the other capacitor. An AC electrolytic capacitor is shown in *Figure 11-22*.

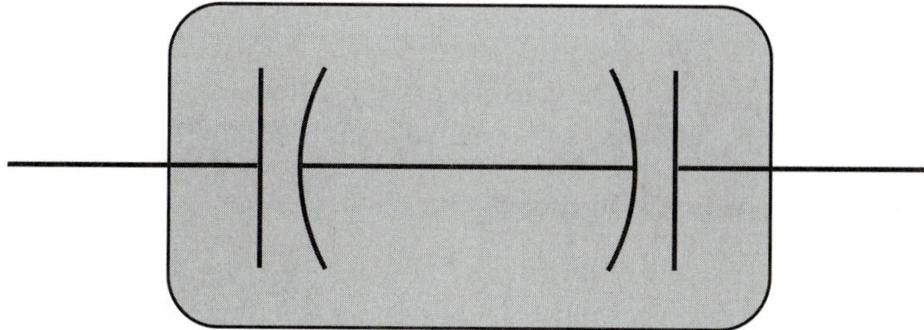

Figure 11-21 Two wet type electrolytic capacitors connect to form an AC electrolytic capacitor.

Figure 11-22 An AC electrolytic capacitor.

DRY TYPE ELECTROLYTIC CAPACITORS

The dry type electrolytic capacitor is very similar to the wet type except that gauze is used to hold the borax solution. This prevents the capacitor from leaking. Although the dry type electrolytic has the advantage of being relatively leak proof, it does have one disadvantage. If the polarity connection should become reversed and the oxide film is broken down, it will not reform when connected to the proper polarity. Reversing the polarity of a dry type electrolytic capacitor will permanently damage the capacitor.

VARIABLE CAPACITORS

Variable capacitors are constructed so that their capacitance value can be changed over a certain range. They generally contain a set of movable plates, which are connected to a shaft, and a set of stationary plates, *Figure 11-23*. The movable plates can be interleaved with the stationary plates to increase or decrease the capacitance value. Since air is used as the dielectric and the plate area is relatively small, variable capacitors are generally rated in picofarads. Another type of small variable capacitor is called the trimmer capacitor, also shown in Figure 11-23. This capacitor has one movable plate and one stationary plate. The capacitance value is changed by turning an adjustment screw that moves the movable plate closer to or farther away from the stationary plate. *Figure 11-24* shows schematic symbols used to represent variable capacitors.

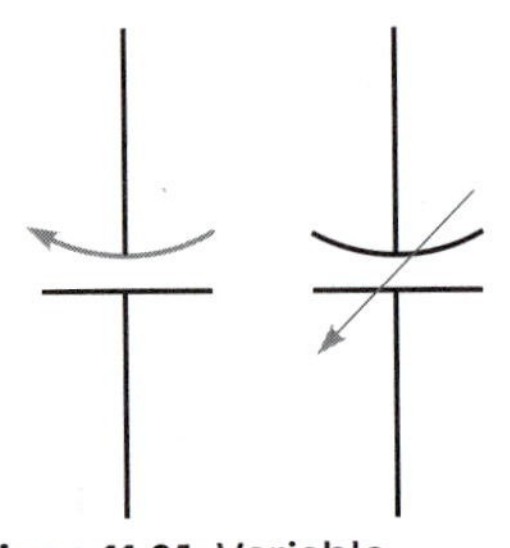

Figure 11-24 Variable capacitor symbols.

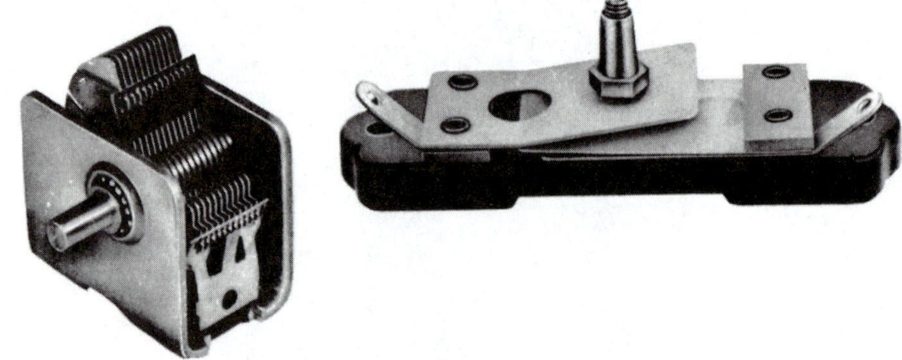

Figure 11-23 Variable capacitors.

TESTING CAPACITORS

Testing capacitors is difficult at best. Small electrolytic capacitors are generally tested for shorts with an ohmmeter. If the capacitor is not shorted, it should be tested for leakage using a variable DC power supply and a microammeter, *Figure 11-25*. When rated voltage is applied to the capacitor, the microammeter should indicate zero current flow.

Large AC oil-filled capacitors can be tested in a similar manner. To test the capacitor accurately, two measurements must be made. One is to measure the capacitance value of the capacitor to determine if it is the same or

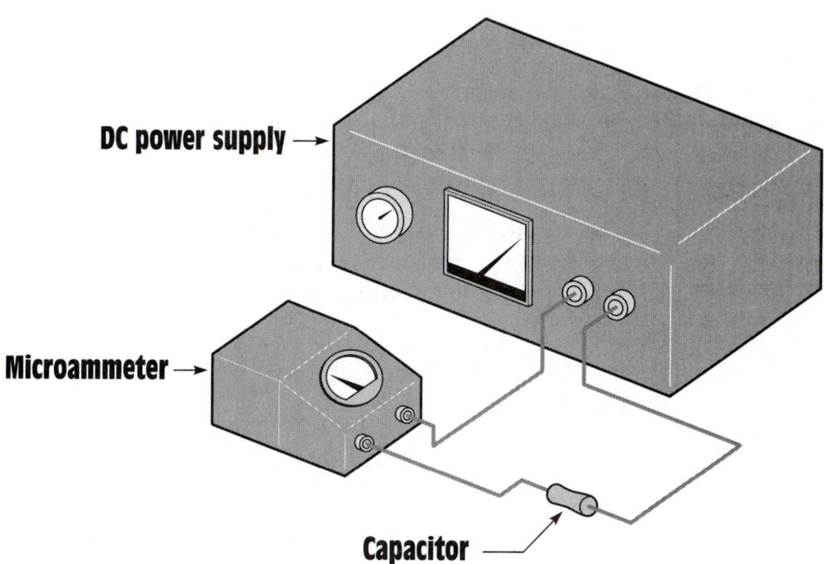

Figure 11-25 Testing a capacitor for leakage.

approximately the same as the rate value. The other is to test the strength of the dielectric.

The first test should be made with an ohmmeter. With the power disconnected, connect the terminals of an ohmmeter directly across the capacitor terminals, *Figure 11-26*. This test determines if the dielectric is shorted. When the ohmmeter is connected, the needle should swing up scale and return to infinity. The amount of needle swing is determined by the capacitance of the capacitor. Then reverse the ohmmeter connection,

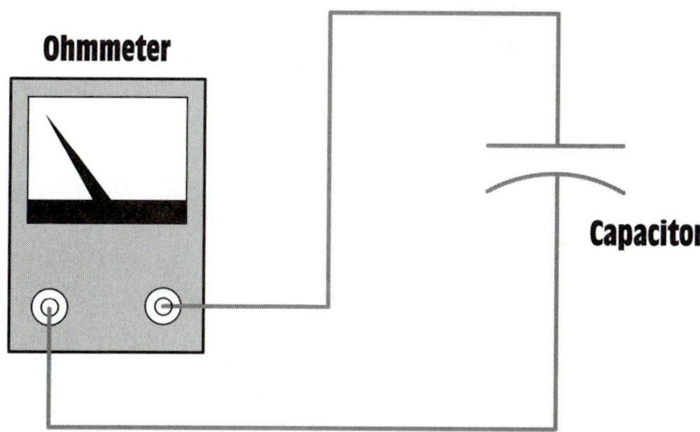

Figure 11-26 Testing the capacitor with an ohmmeter.

and the needle should move twice as far up scale and then return to the infinity setting.

HIPOT

If the ohmmeter test is successful, the dielectric must be tested at its rated voltage. This is called a dielectric strength test. To make this test, a dielectric test set must be used. This device is often referred to as a **HIPOT** because of its ability to produce a high voltage or high potential. The dielectric test set contains a variable voltage control, a voltmeter, and a microammeter. To use the HIPOT, connect its terminal leads to the capacitor terminals. Increase the output voltage until the rated voltage is applied to the capacitor. The microammeter indicates any current flow between the plates of the dielectric. If the capacitor is good, the microammeter should indicate zero current flow.

The capacitance value must be measured to determine if there are any open plates in the capacitor. To measure the capacitance value of the capacitor, connect some value of AC voltage across the plates of the capacitor, *Figure 11-27*. This voltage must not be greater than the rated capacitor voltage. Then measure the amount of current flow in the circuit. Now that the voltage and current flow are known, the capacitive reactance of the capacitor can be computed using the formula:

$$X_C = \frac{E}{I}$$

After the capacitive reactance has been determined, the capacitance can be computed using the formula:

$$C = \frac{1}{2\pi F X_C}$$

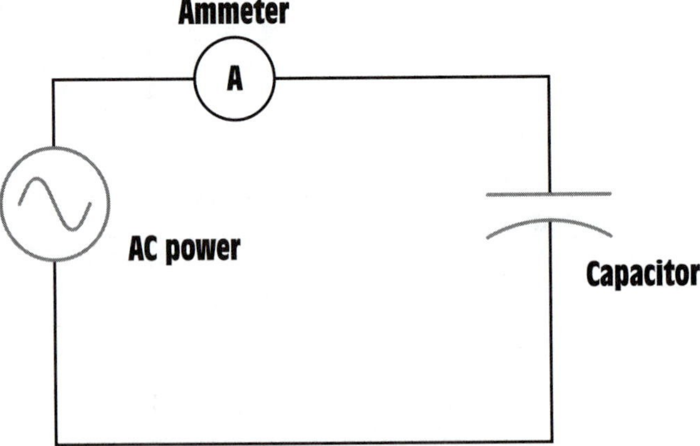

Figure 11-27 Determining the capacitance value.

CAPACITANCE IN ALTERNATING CURRENT CIRCUITS

When a capacitor is connected to an alternating current circuit, current will *appear to flow* through the capacitor. The reason is that in an AC circuit, the current is continually changing direction and polarity. To understand this concept, consider the hydraulic circuit shown in *Figure 11-28.* Two tanks are connected to a common pump. Assume tank A to be full and tank B to be empty. Now assume that the pump is used to pump water from tank A to tank B. When tank B becomes full, the pump reverses and pumps the water from tank B back into tank A. Each time a tank becomes filled, the pump reverses and pumps water back into the other tank. Notice that water is continually flowing in this circuit, but there is no direct connection between the two tanks.

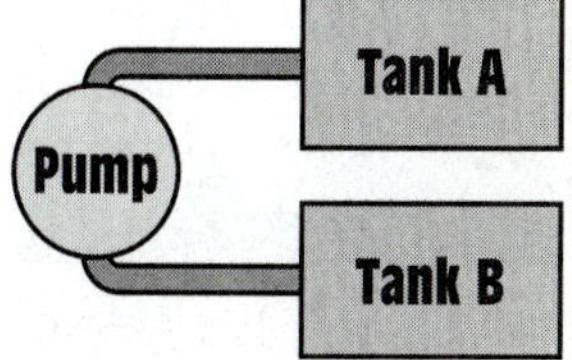

Figure 11-28 Water can flow continuously, but not between the two tanks.

A similar action takes place when a capacitor is connected to an alternating current circuit, *Figure 11-29.* In this circuit, the AC generator or alternator charges one plate of the capacitor positive and the other plate negative. During the next half cycle, the voltage will change polarity and the capacitor will discharge and recharge to the opposite polarity also. As long as the voltage continues to increase, decrease, and change polarity, current will flow from one plate of the capacitor to the other. If an ammeter were placed in the circuit, it would indicate a continuous flow of current, giving the appearance that current is flowing through the capacitor.

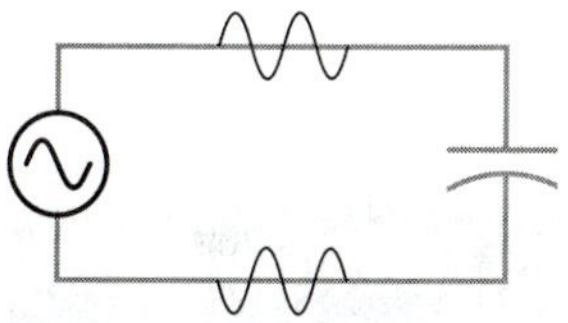

Figure 11-29 A capacitor connected to an AC circuit.

CAPACITIVE REACTANCE

As the capacitor is charged, an impressed voltage is developed across its plates as an electrostatic charge is built up, *Figure 11-30.* This impressed voltage opposes the applied voltage and limits the flow of current in the circuit. This counter-voltage is very similar to the counter-voltage produced by an inductor. The counter-voltage developed by the capacitor is called reactance also. Since this counter-voltage is caused by capacitance, it is called

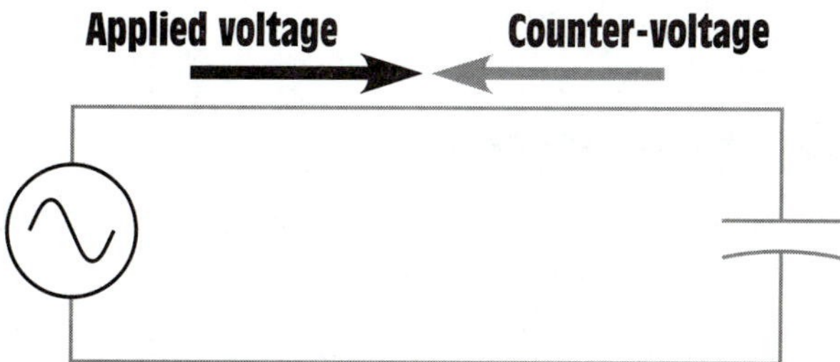

Figure 11-30 Counter-voltage limits the flow of current.

capacitive reactance (X_C)

capacitive reactance (X_C) and is measured in ohms. The formula for finding capacitive reactance is:

$$X_C = \frac{1}{2\pi FC}$$

where

X_C = capacitive reactance

π = 3.1416

F = frequency in hertz

C = capacitance in farads

Example 2

A 35-μF capacitor is connected to a 120-V, 60-Hz line. How much current will flow in this circuit?

Solution:

The first step is to compute the capacitive reactance. Recall that the value of C in the formula is given in farads. This must be changed to the capacitive units being used; in this case it will be microfarads.

$$X_C = \frac{1}{2 \times 3.1416 \times 60 \times (35 \times 10^{-6}}$$

$$X_C = 75.79\ \Omega$$

Now that the value of capacitive reactance is known, it can be used like resistance in an Ohm's Law formula. Since capacitive reactance is the current-limiting factor, it will replace the value of R.

$$I = \frac{E}{X_C}$$

$$I = \frac{120}{75.79}$$

$$I = 1.58\ A$$

COMPUTING CAPACITANCE

If the value of capacitive reactance is known, the capacitance of the capacitor can be found using the formula:

$$C = \frac{1}{2\pi FX_C}$$

Example 3

A capacitor is connected into a 480-V, 60-Hz circuit. An ammeter indicates a current flow of 2.6 A. What is the capacitance value of the capacitor?

Solution:

The first step is to compute the value of capacitive reactance. Since capacitive reactance, like resistance, limits current flow, it can be substituted for R in an Ohm's Law formula.

$$X_C = \frac{E}{I}$$

$$X_C = \frac{480}{2.6}$$

$$X_C = 184.61\ \Omega$$

Now that the capacitive reactance of the circuit is known, the value of capacitance can be found.

$$C = \frac{1}{2\pi F X_C}$$

$$C = \frac{1}{2 \times 3.1416 \times 60 \times 184.61}$$

$$C = \frac{1}{69{,}596.49}$$

$$C = 0.00001437\ F = 14.37\ \mu F$$

VOLTAGE AND CURRENT RELATIONSHIPS IN A PURE CAPACITIVE CIRCUIT

Earlier in this text it was shown that the current in a pure resistive circuit is in phase with the applied voltage, whereas current in a pure inductive circuit lags the applied voltage by 90°. In this unit it will be shown that in a pure capacitive circuit the current will *lead* the applied voltage by 90°.

When a capacitor is connected to an alternating current, the capacitor will charge and discharge at the same rate and time as the applied voltage. The charge in coulombs is equal to the capacitance of the capacitor times the applied voltage ($Q = C \times V$). When the applied voltage is zero, the charge in coulombs and impressed voltage will be zero also. When the applied voltage reaches its maximum value, positive or negative, the charge in coulombs and impressed voltage will reach maximum also, *Figure 11-31*. The impressed voltage will follow the same curve as the applied voltage.

In the waveform shown, voltage and charge are both shown to be zero at 0°. Since there is no charge on the capacitor, there is no opposition to

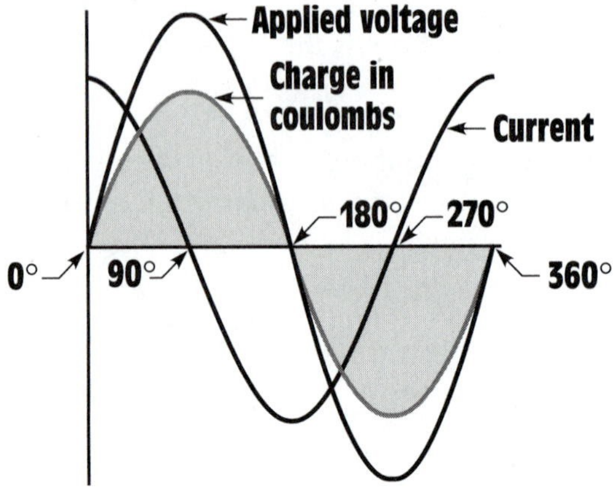

Figure 11-31 Capacitive current leads the applied voltage by 90°.

current flow, which is shown to be maximum. As the applied voltage increases from zero toward its positive peak at 90°, the capacitor begins to charge at the same time. The charge produces an impressed voltage across the plates of the capacitor that opposes the flow of current. The impressed voltage is 180° **out of phase** with the applied voltage, *Figure 11-32.* When the applied voltage reaches 90° in the positive direction, the charge reaches maximum, the impressed voltage reaches peak in the negative direction, and the current flow is zero.

out of phase

As the applied voltage begins to decrease, the capacitor begins to discharge, causing the current to flow in the opposite or negative direction. When the applied voltage and charge reach zero at 180°, the impressed voltage is zero also and the current flow is maximum in the negative direction. As the applied voltage and charge increase in the negative direction, the increase of the impressed voltage across the capacitor again causes the current to decrease. The applied voltage and charge reach maximum negative after 270° of rotation. The impressed voltage reaches maximum positive and the current has decreased to zero, *Figure 11-33.* As the applied voltage decreases from its maximum negative value, the capacitor again begins to discharge. This causes the current to flow in the positive direction. The current again reaches its maximum positive value when the applied voltage and charge reach zero after 360° of rotation.

POWER IN A PURE CAPACITIVE CIRCUIT

Since the current flow in a pure capacitive circuit is leading the applied voltage by 90°, the voltage and current will both have the same polarity for half the time during one cycle and have opposite polarities the other half

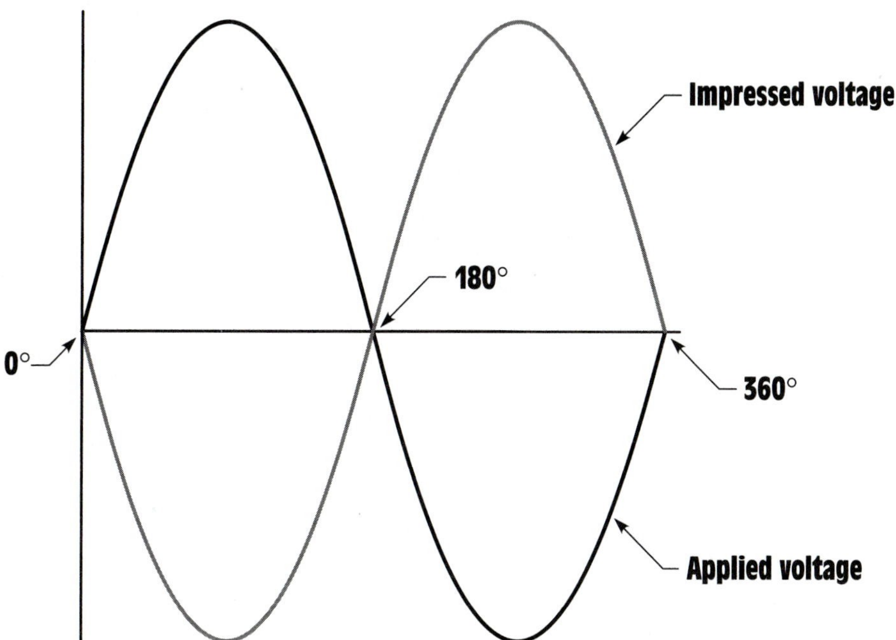

Figure 11-32 The impressed voltage is 180° out of phase with the applied voltage.

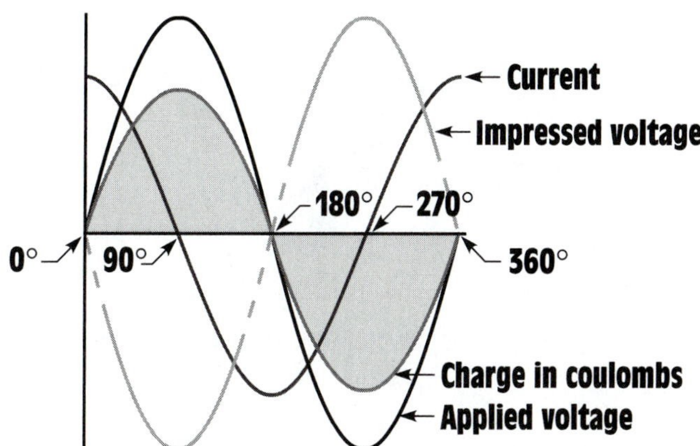

Figure 11-33 Voltage, current, and charge relationships for a capacitive circuit.

of the time, *Figure 11-34*. During the period of time when both the voltage and current have the same polarity, energy is being stored in the capacitor in the form of an electrostatic field. When the voltage and current have opposite polarities, the capacitor is discharging and the energy is returned to

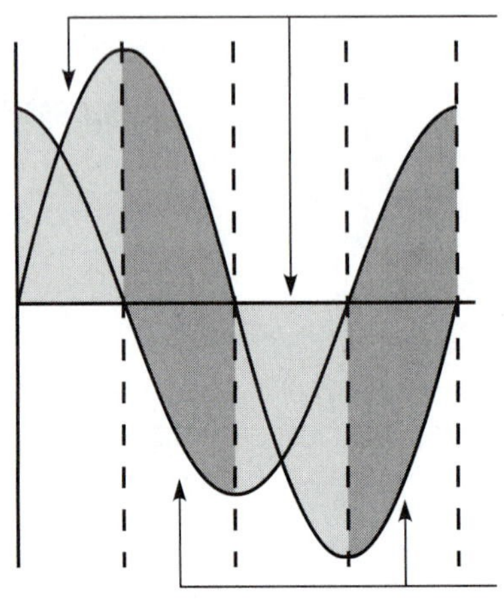

Figure 11-34 A pure capacitive circuit has no true power (watts). The power required to charge the capacitor is returned to the circuit when the capacitor discharges.

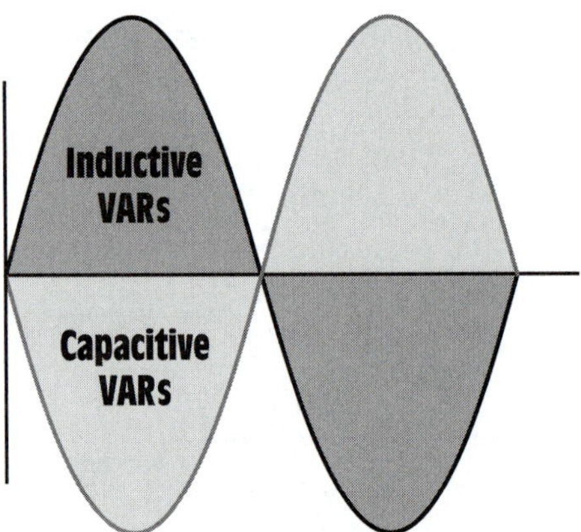

Figure 11-35 Inductive VARs and capacitive VARs are 180° out of phase with each other.

the circuit. When these values are added, the sum will equal zero just as it does with pure inductive circuits. Therefore, there is no true power, or watts, produced in a pure capacitive circuit.

The power value for a capacitor is reactive and is measured in VARs, just as it is for an inductor. Inductive VARs and capacitive VARs are 180° out of phase with each other, however, *Figure 11-35*.

SUMMARY

1. Capacitors are devices that oppose a change of voltage.
2. Three factors that determine the capacitance of a capacitor are:

   A. The surface area of the plates
   B. The distance between the plates
   C. The type of dielectric

3. A capacitor stores energy in an electrostatic field.
4. Current can flow only during the time a capacitor is charging or discharging.
5. Capacitors charge and discharge at an exponential rate.
6. The basic unit of capacitance is the farad.
7. Capacitors are generally rated in microfarads, nanofarads, or picofarads.
8. When capacitors are connected in parallel, their capacitance values are added together.
9. Five time constants are required to charge or discharge a capacitor.
10. Nonpolarized capacitors are often called AC capacitors.
11. Nonpolarized capacitors can be connected to direct or alternating current circuits.
12. Polarized capacitors are often referred to as electrolytic capacitors.
13. Polarized capacitors can be connected to direct current circuits only.
14. There are two basic types of electrolytic capacitors, the wet type and the dry type.
15. Wet type electrolytic capacitors can be reformed if reconnected to the correct polarity.
16. Dry type electrolytic capacitors will be permanently damaged if connected to the incorrect polarity.
17. To test a capacitor for leakage, a microammeter should be connected in series with the capacitor and rated voltage applied to the circuit.
18. When a capacitor is connected to an alternating current circuit, current will appear to flow through the capacitor.

19. Current appears to flow through a capacitor because of the continuous increase and decrease of voltage and also because of the continuous change of polarity in an AC circuit.
20. The current flow in a pure capacitive circuit is limited by capacitive reactance.
21. Capacitive reactance is proportional to the capacitance of the capacitor and the frequency of the AC line.
22. Capacitive reactance is measured in ohms.
23. In a pure capacitive circuit, the current leads the applied voltage by 90°.
24. There is no true power, or watts, in a pure capacitive circuit.
25. Capacitive power is reactive and is measured in VARs, as is inductance.
26. Capacitive and inductive VARs are 180° out of phase with each other.

REVIEW QUESTIONS

1. What is the dielectric?
2. List three factors that determine the capacitance of a capacitor.
3. A capacitor uses air as a dielectric and has a capacitance of 3 μF. A dielectric material is inserted between the plates without changing the spacing and the capacitance becomes 15 μF. What is the dielectric constant of this material?
4. In what form is the energy of a capacitor stored?
5. Four capacitors having values of 20 μF, 50 μF, 40 μF, and 60 μF are connected in parallel. What is the total capacitance of this circuit?
6. Can a nonpolarized capacitor be connected to a direct current circuit?
7. Explain how an AC electrolytic capacitor is constructed.
8. What type of electrolytic capacitor will be permanently damaged if connected to the incorrect polarity?
9. Can current flow through a capacitor?
10. What two factors determine the capacitive reactance of a capacitor?
11. How many degrees are the current and voltage out of phase in a pure capacitive circuit?

12. Does the current in a pure capacitive circuit lead or lag the applied voltage?
13. A 30 μF capacitor is connected into a 240 V 60 Hz circuit. What is the current flow in this circuit?
14. A capacitor is connected into a 1250 V 1000 Hz circuit. The current flow is 80 A. What is the capacitance of the capacitor?
15. On the average, how many times is the life expectancy of a capacitor increased if the capacitor is operated at half its voltage rating?
16. A capacitor is connected into a 277 V 400 Hz circuit. The circuit current is 12 A. What is the capacitance of the capacitor?

CAPACITIVE CIRCUITS

Practice Problems

Fill in all missing values in the chart using the following formulas.

$$X_C = \frac{1}{2\pi FC}$$

$$C = \frac{1}{2\pi F X_C}$$

$$F = \frac{1}{2\pi C X_C}$$

Capacitance	X_C	Frequency
38 μF		60 Hz
	78.8 Ω	400 Hz
250 pF	4.5 k Ω	
234 μF		10 kHz
	240 Ω	50 Hz
10 μF	36.8 Ω	
560 nF		2 MHz
	15 k Ω	60 Hz
75 μF	560 Ω	
470 pF		200 kHz
	6.8 k Ω	400 Hz
34 μF	450 Ω	

12 Unit

Three-Phase Circuits

objectives

fter studying this unit, you should be able to:

- Discuss the differences between three-phase and single-phase voltages.
- Discuss the characteristics of delta and wye connections.
- Compute voltage and current values for delta and wye circuits.
- Connect delta and wye circuits and make measurements with measuring instruments.

Most of the electrical power generated in the world today is three-phase. Three-phase power was first conceived by Nikola Tesla. In the early days of electric power generation, Tesla not only led the battle concerning whether the nation should be powered with low-voltage direct current or high-voltage alternating current, but he also proved that three-phase power was the most efficient way that electricity could be produced, transmitted, and consumed.

THREE-PHASE CIRCUITS

There are several reasons why three-phase power is superior to single-phase power.

1. The horsepower rating of three-phase motors and the KVA (kilo-volt-amp) rating of three-phase transformers is about 150% greater than for single-phase motors or transformers with a similar frame size.
2. The power delivered by a single-phase system pulsates, *Figure 12-1.* The power falls to zero three times during each cycle. The power delivered by a three-phase circuit pulsates also, but it never falls to zero, *Figure 12-2.* In a three-phase system, the power delivered to the load is the same at any instant. This produces superior operating characteristics for three-phase motors.
3. In a balanced three-phase system, the conductors need be only about 75% the size of conductors for a single-phase two-wire system of the same KVA rating. This helps offset the cost of supplying the third conductor required by three-phase systems.

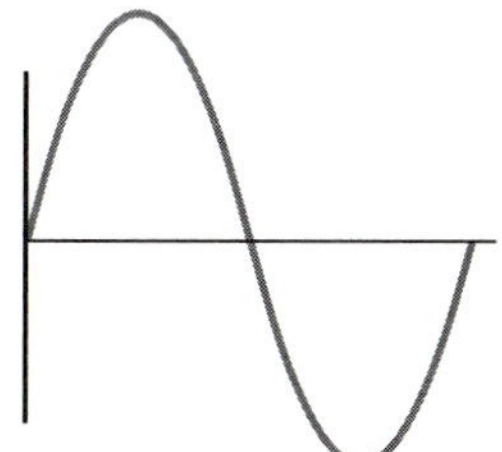

Figure 12-1 Single-phase power falls to zero three times each cycle.

A single-phase alternating voltage can be produced by rotating a magnetic field through the conductors of a stationary coil, as shown in *Figure 12-3.*

Since alternate polarities of the magnetic field cut through the conductors of the stationary coil, the induced voltage will change polarity at

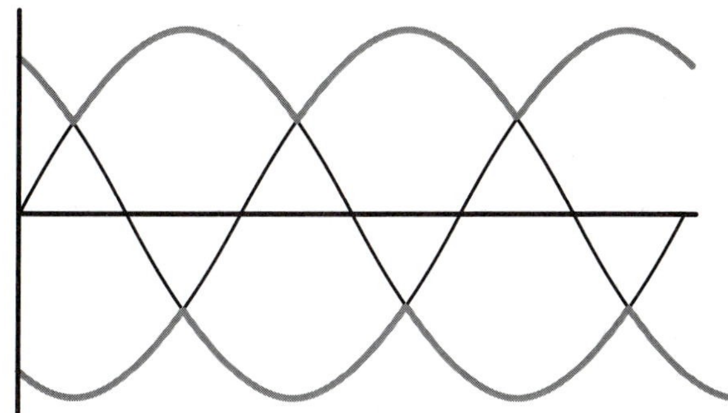

Figure 12-2 Three-phase power never falls to zero.

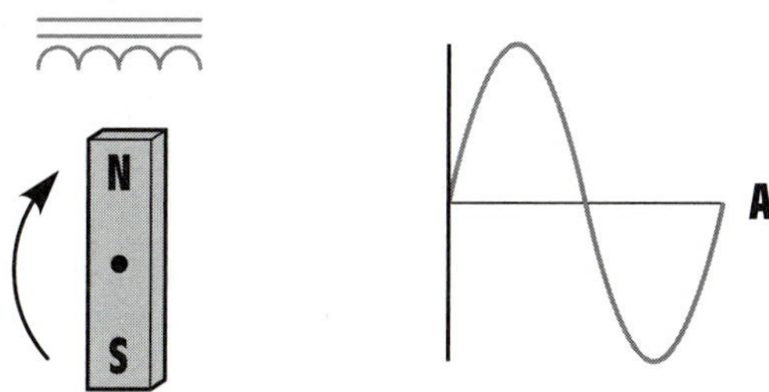

Figure 12-3 Producing a single-phase voltage.

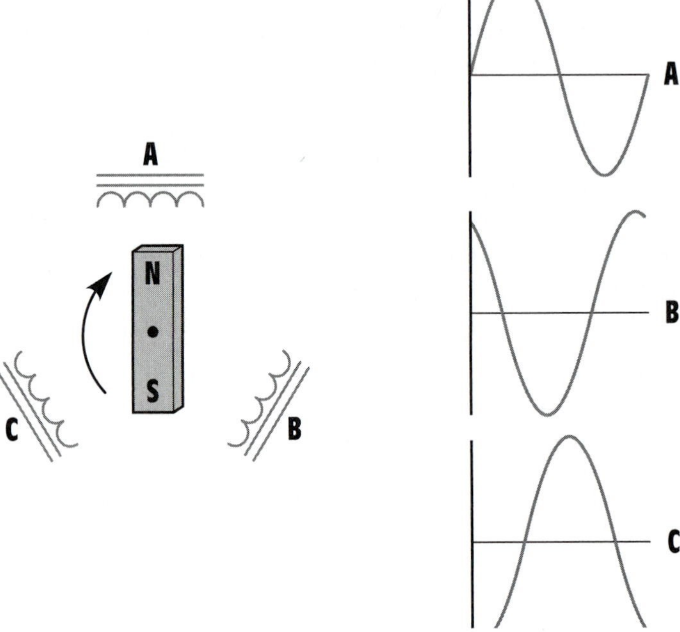

Figure 12-4 The voltages of a three-phase system are 120° out of phase with each other.

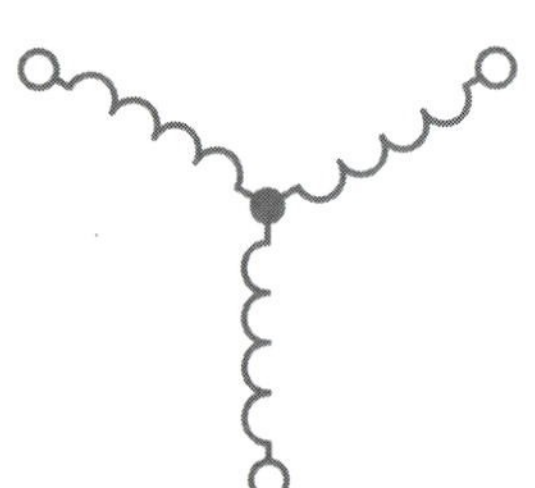

Figure 12-5 A wye connection is formed by joining one end of each of the windings together.

the same speed as the rotation of the magnetic field. The alternator shown in Figure 12-3 is single phase because it produces only one AC voltage.

If three separate coils are spaced 120° apart, as shown in *Figure 12-4,* three voltages 120° out of phase with each other will be produced when the magnetic field cuts through the coils. This is the manner in which a three-phase voltage is produced. There are two basic three-phase connections, the wye or star connection and the delta connection.

WYE CONNECTION

wye connection

star connection

phase voltage

line voltage

The **wye** or **star connection** is made by connecting one end of each of the three-phase windings together as shown in *Figure 12-5.* The voltage measured across a single winding or phase is known as the **phase voltage,** as shown in *Figure 12-6.* The voltage measured between the lines is known as the line-to-line voltage or simply as the **line voltage.**

In *Figure 12-7,* ammeters have been placed in the phase winding of a wye-connected load and in the line supplying power to the load. Voltmeters have been connected across the input to the load and across the phase. A

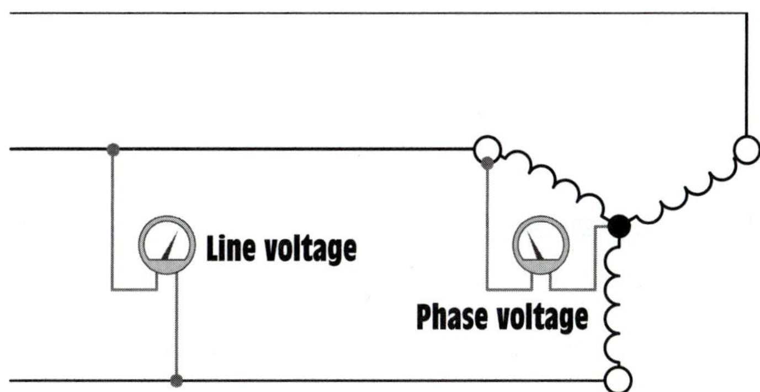

Figure 12-6 Line and phase voltages are different in a wye connection.

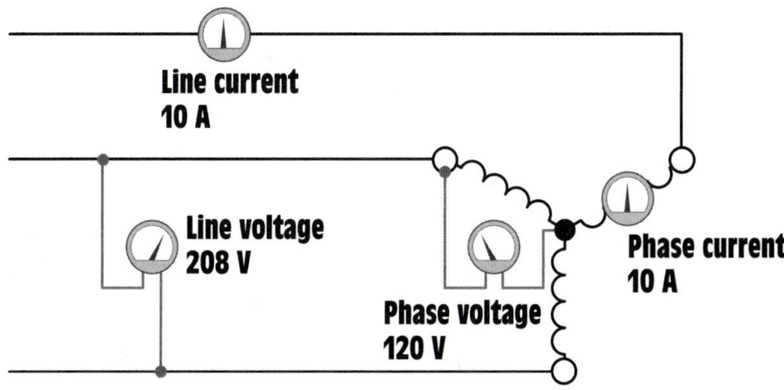

Figure 12-7 Line current and phase current are the same in a wye connection.

line voltage of 208 V has been applied to the load. Notice that the voltmeter connected across the lines indicates a value of 208 V, but the voltmeter connected across the phase indicates a value of 120 V.

In a wye connected system, the line voltage is higher than the phase voltage by a factor of the square root of 3 (1.732). Two formulas used to compute the voltage in a wye connected system are:

$$E_{Line} = E_{Phase} \times 1.732$$

and

$$E_{phase} = \frac{E_{Line}}{1.732}$$

In a wye-connected system, the line voltage is higher than the phase voltage by a factor of the square root of 3 (1.732).

In a wye-connected system, phase current and line current are the same.

Notice in Figure 12-7 that 10 A of current flow in both the phase and the line. **In a wye-connected system, phase current and line current are the same.**

$$I_{Line} = I_{Phase}$$

VOLTAGE RELATIONSHIPS IN A WYE CONNECTION

Many students of electricity have difficulty at first understanding why the line voltage of the wye connection used in this illustration is 208 V instead of 240 V. Since line voltage is measured across two phases that have a value of 120 V each, it would appear that the sum of the two voltages should be 240 V. One cause of this misconception is that many students are familiar with the 240/120 V connection supplied to most homes. If voltage is measured across the two incoming lines, a voltage of 240 V will be seen. If voltage is measured from either of the two lines to the neutral, a voltage of 120 V will be seen. The reason for this is that this is a single-phase connection derived from the center tap of a transformer, *Figure 12-8*. If the center tap is used as a common point, the two line voltages on either side of it will be 180° apart and opposite in polarity, *Figure 12-9*. The vector sum of these two voltages would be 240 V.

Three-phase voltages are 120° apart, not 180°. If the three voltages are drawn 120° apart, it will be seen that the vector sum of these voltages is 208 V, *Figure 12-10*. Another illustration of vector addition is shown in

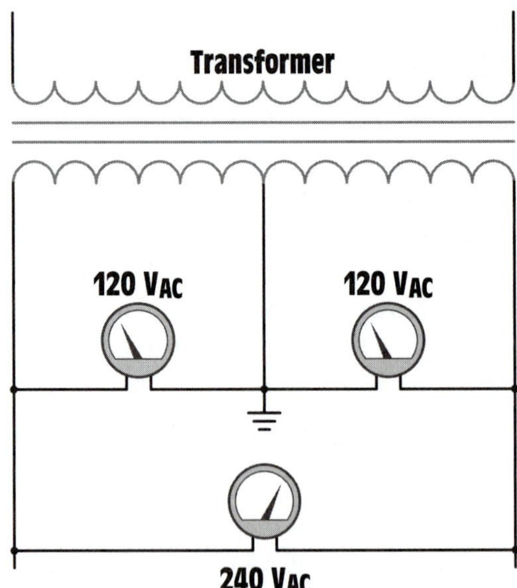

Figure 12-8 Single-phase transformer with grounded center tap.

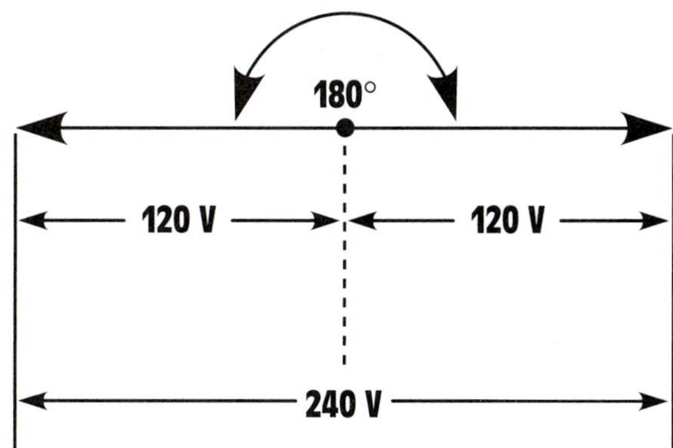

Figure 12-9 The voltages of a single-phase system are 180° out of phase with each other.

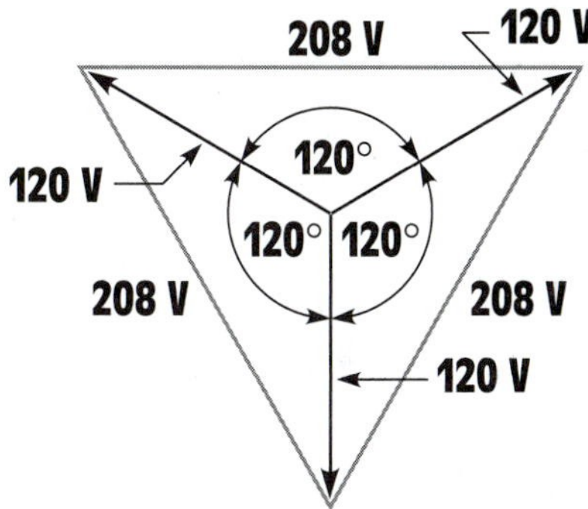

Figure 12-10 Vector sum of the voltages in a three-phase wye connection.

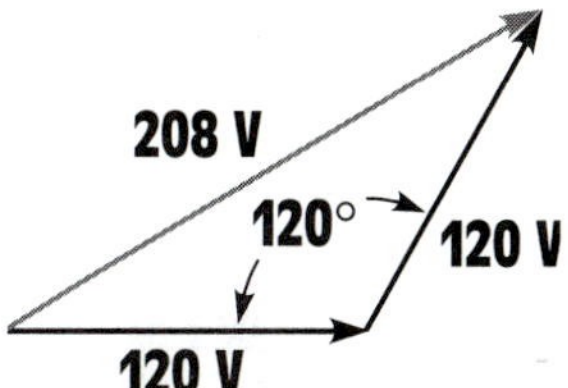

Figure 12-11 Adding voltage vectors of two-phase voltage values.

Figure 12-11. In this illustration two-phase voltage vectors are added and the resultant is drawn from the starting point of one vector to the end point of the other. The parallelogram method of vector addition for the voltages in a wye-connected three-phase system is shown in *Figure 12-12.*

DELTA CONNECTIONS

delta connection

In *Figure 12-13,* three separate inductive loads have been connected to form a **delta connection.** This connection receives its name from the fact that a schematic diagram of this connection resembles the Greek letter delta (Δ). In *Figure 12-14,* voltmeters have been connected across the lines and across the phase. Ammeters have been connected in the line and in the phase. **In the delta connection, line voltage and phase voltage are the same.** Notice that both voltmeters indicate a value of 480 V.

In a delta connection, line voltage and phase voltage are the same.

$$E_{Line} = E_{Phase}$$

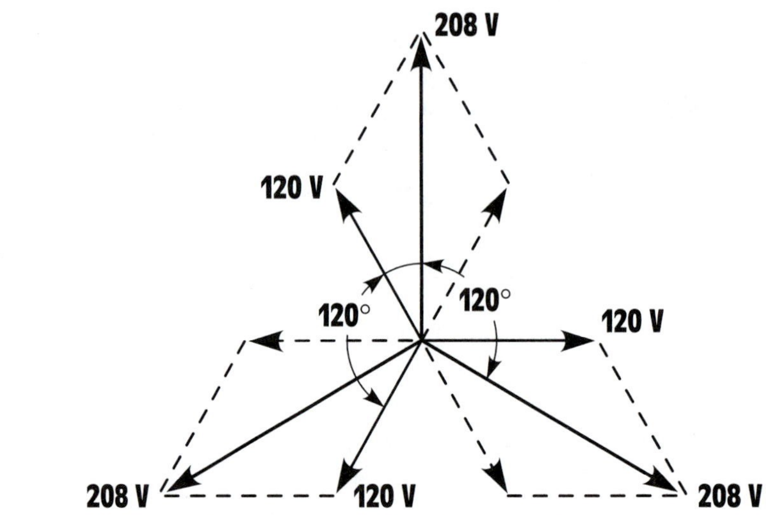

Figure 12-12 The parallelogram method of adding three-phase vectors.

Figure 12-13 Three-phase delta connection.

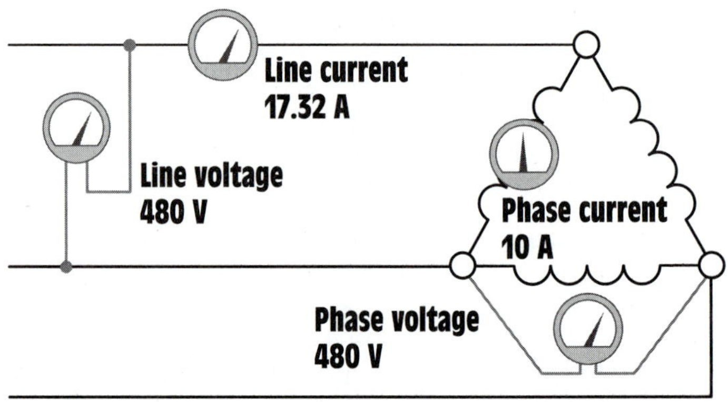

Figure 12-14 Voltage and current relationships in a delta connection.

The line current of a delta connection is higher than the phase current by a factor of the square root of 3 (1.732).

Notice that the line current and phase current are different, however. **The line current of a delta connection is higher than the phase current by a factor of the square root of 3 (1.732).** In the example shown, it is assumed that each of the phase windings has a current flow of 10 A. The current in each of the lines, however, is 17.32 A. The reason for this difference in current is that current flows through different windings at different times in a three-phase circuit. During some periods of time, current will flow between two lines only. At other times, current will flow from two lines to the third, *Figure 12-15*. The delta connection is similar to a parallel connection because there is always more than one path for current flow. Since these currents are 120° out of phase with each other, vector addition must be used when finding the sum of the currents, *Figure 12-16*. Formulas for determining the current in a delta connection are:

$$I_{Line} = I_{Phase} \times 1.732$$

and

$$I_{phase} = \frac{I_{Line}}{1.732}$$

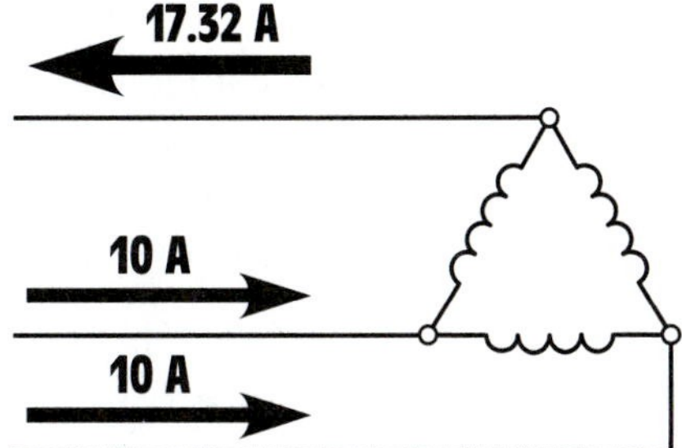

Figure 12-15 Division of currents in a delta connection.

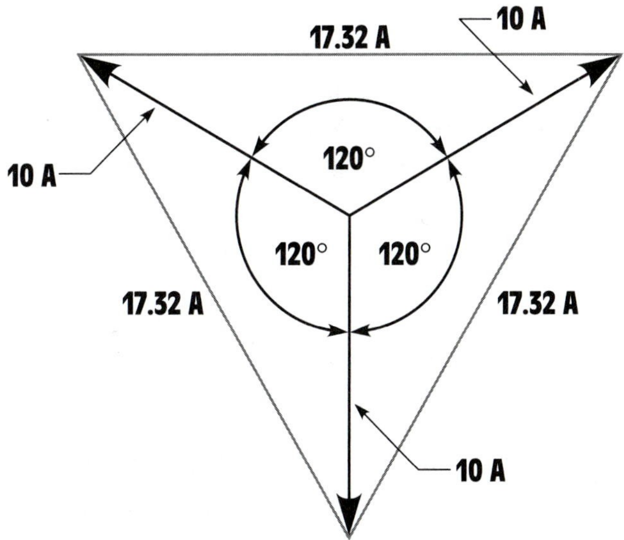

Figure 12-16 Vector addition is used to compute the sum of the currents in a delta connection.

THREE-PHASE POWER

Students sometimes become confused when computing power in three-phase circuits. One reason for this confusion is that there are actually two formulas that can be used. If *line* values of voltage and current are known, the power (watts) of a pure resistive load can be computed using the formula:

$$VA = \sqrt{3} \times E_{Line} \times I_{Line}$$

If the *phase* values of voltage and current are known, the apparent power can be computed using the formula:

$$VA = 3 \times E_{Phase} \times I_{Phase}$$

Notice that in the first formula, the line values of voltage and current are multiplied by the square root of 3. In the second formula, the phase values of voltage and current are multiplied by 3. The first formula is used more often because it is generally more convenient to obtain line values of voltage and current, which can be measured with a voltmeter and clamp-on ammeter.

THREE-PHASE CIRCUIT CALCULATIONS

In the following examples, values of line and phase voltage, line and phase current, and power will be computed for different types of three-phase connections.

Example 1

A wye-connected three-phase alternator supplies power to a delta-connected resistive load, *Figure 12-17*. The alternator has a line voltage of 480 V. Each resistor of the delta load has 8 Ω of resistance. Find the following values:

$E_{L(Load)}$ — line voltage of the load
$E_{P(Load)}$ — phase voltage of the load
$I_{P(Load)}$ — phase current of the load
$I_{L(Load)}$ — line current to the load
$I_{L(Alt)}$ — line current delivered by the alternator
$I_{P(Alt)}$ — phase current of the alternator
$E_{P(Alt)}$ — phase voltage of the alternator
P — true power

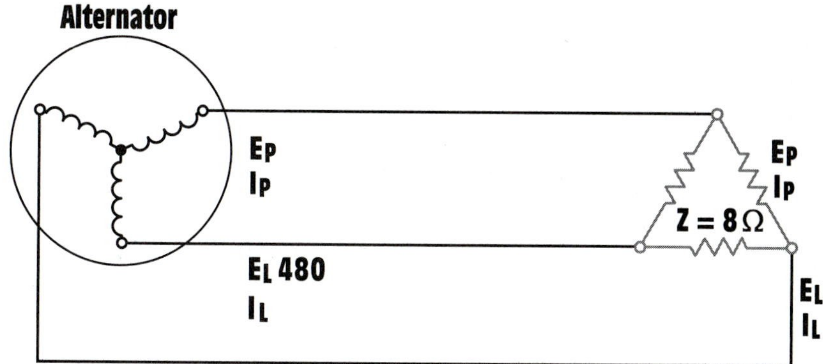

Figure 12-17 Computing three-phase values using a wye-connected power source and a delta-connected load (Example 1 circuit).

Solution:

The load is connected directly to the alternator. Therefore, the line voltage supplied by the alternator is the line voltage of the load.

$$E_{L(Load)} = 480 \text{ V}$$

The three resistors of the load are connected in a delta connection. In a delta connection, the phase voltage is the same as the line voltage.

$$E_{P(Load)} = E_{L(Load)}$$

$$E_{P(Load)} = 480 \text{ V}$$

Each of the three resistors in the load is one phase of the load. Now that the phase voltage is known (480 V), the amount of phase current can be computed using Ohm's Law.

$$I_{P(Load)} = \frac{E_{P(Load)}}{Z}$$

$$I_{P(Load)} = \frac{480}{8}$$

$$I_{P(Load)} = 60 \text{ A}$$

The three load resistors are connected as a delta with 60 A of current flow in each phase. The line current supplying a delta connection must be 1.732 times greater than the phase current.

$$I_{L(Load)} = I_{P(Load)} \times 1.732$$

$$I_{L(Load)} = 60 \times 1.732$$

$$I_{L(Load)} = 103.92 \text{ A}$$

The alternator must supply the line current to the load or loads to which it is connected. In this example, only one load is connected to the alternator. Therefore, the line current of the load will be the same as the line current of the alternator.

$$I_{L(Alt)} = 103.92 \text{ A}$$

The phase windings of the alternator are connected in a wye connection. In a wye connection, the phase current and line current are equal. The phase current of the alternator will, therefore, be the same as the alternator line current.

$$I_{P(Alt)} = 103.92 \text{ A}$$

The phase voltage of a wye connection is less than the line voltage by a factor of the square root of 3. The phase voltage of the alternator will be

$$E_{P(Alt)} = \frac{E_{L(Alt)}}{1.732}$$

$$E_{P(Alt)} = \frac{480}{1.732}$$

$$E_{P(Alt)} = 277.13 \text{ V}$$

In this circuit, the load is pure resistive. The voltage and current are in phase with each other, which produces a unity power factor of 1. The true power in this circuit will be computed using the formula:

$$P = 1.732 \times E_{L(Alt)} \times I_{L(Alt)} \times PF$$

$$P = 1.732 \times 480 \times 103.92 \times 1$$

$$P = 86{,}394.93 \text{ W}$$

Example 2

A delta-connected alternator is connected to a wye-connected resistive load, *Figure 12-18*. The alternator produces a line voltage of 240 V and the resistors have a value of 6 Ω each. The following values will be found:

$E_{L(Load)}$ — line voltage of the load

$E_{P(Load)}$ — phase voltage of the load

$I_{P(Load)}$ — phase current of the load

$I_{L(Load)}$ — line current to the load

$I_{L(Alt)}$ — line current delivered by the alternator

$I_{P(Alt)}$ — phase current of the alternator

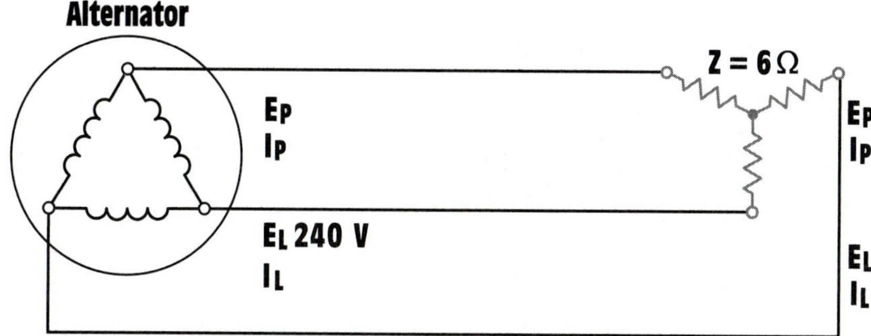

Figure 12-18 Computing three-phase values using a delta-connected source and a wye-connected load (Example 2 circuit).

$E_{P(Alt)}$ — phase voltage of the alternator

P — true power

Solution:

As was the case in Example 1, the load is connected directly to the output of the alternator. The line voltage of the load must, therefore, be the same as the line voltage of the alternator.

$$E_{L(Load)} = 240 \text{ V}$$

The phase voltage of a wye connection is less than the line voltage by a factor of 1.732.

$$E_{P(Load)} = \frac{240}{1.732}$$

$$E_{P(Load)} = 138.57 \text{ V}$$

Each of the three 6-Ω resistors is one phase of the wye-connected load. Since the phase voltage is 138.57 V, this voltage is applied to each of the three resistors. The amount of phase current can now be determined using Ohm's Law.

$$I_{P(Load)} = \frac{E_{P(Load)}}{Z}$$

$$I_{P(Load)} = \frac{138.57}{6}$$

$$I_{P(Load)} = 23.1 \text{ A}$$

The amount of line current needed to supply a wye-connected load is the same as the phase current of the load.

$$I_{L(Load)} = 23.1 \text{ A}$$

Only one load is connected to the alternator. The line current supplied to the load is the same as the line current of the alternator.

$$I_{L(Alt)} = 23.1 \text{ A}$$

The phase windings of the alternator are connected in delta. In a delta connection the phase current is less than the line current by a factor of 1.732.

$$I_{P(Alt)} = \frac{I_{L(Alt)}}{1.732}$$

$$I_{P(Alt)} = \frac{23.1}{1.732}$$

$$I_{P(Alt)} = 13.34 \text{ A}$$

The phase voltage of a delta is the same as the line voltage.

$$E_{P(Alt)} = 240 \text{ V}$$

Since the load in this example is pure resistive, the power factor has a value of unity, or 1. Power will be computed by using the line values of voltage and current.

$$P = 1.732 \times E_L \times I_L \times PF$$

$$P = 1.732 \times 240 \times 23.1 \times 1$$

$$P = 9{,}602.21 \text{ W}$$

Example 3

The phase windings of an alternator are connected in wye. The alternator produces a line voltage of 440 V, and supplies power to two resistive loads. One load contains resistors with a value of 4 Ω each, connected in wye. The second load contains resistors with a value of 6 Ω each, connected in delta, *Figure 12-19*. The following circuit values will be found.

$E_{L(Load\ 2)}$ — line voltage of load 2

$E_{P(Load\ 2)}$ — phase voltage of load 2

$I_{P(Load\ 2)}$ — phase current of load 2

$I_{L(Load\ 2)}$ — line current to load 2

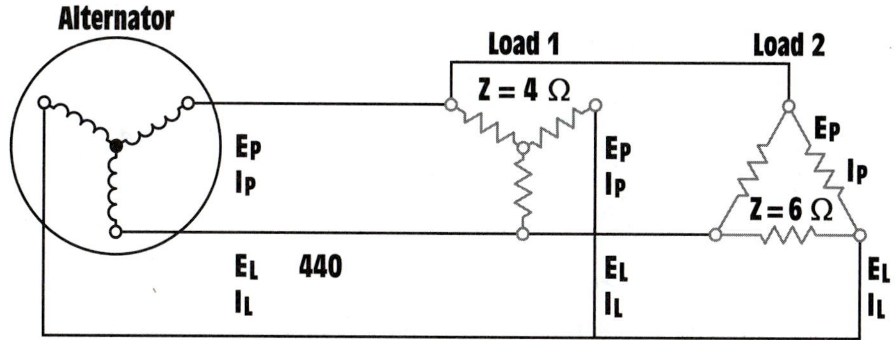

Figure 12-19 Computing three-phase values using a wye-connected source and two three-phase loads (Example 3 circuit).

$E_{P(Load\ 1)}$ — phase voltage of load 1
$I_{P(Load\ 1)}$ — phase current of load 1
$I_{L(Load\ 1)}$ — line current to load 1
$I_{L(Alt)}$ — line current delivered by the alternator
$I_{P(Alt)}$ — phase current of the alternator
$E_{P(Alt)}$ — phase voltage of the alternator
P — true power

Solution:

Both loads are connected directly to the output of the alternator. The line voltage for both loads 1 and 2 will be the same as the line voltage of the alternator.

$$E_{L(Load\ 2)} = 440\ V$$

$$E_{L(Load\ 1)} = 440\ V$$

Load 2 is connected as a delta. The phase voltage will be the same as the line voltage.

$$E_{P(Load\ 2)} = 440\ V$$

Each of the resistors that constitutes a phase of load 2 has a value of 6 Ω. The amount of phase current can be found using Ohm's Law.

$$I_{P(Load\ 2)} = \frac{E_{P(Load\ 2)}}{Z}$$

$$I_{P(Load\ 2)} = \frac{440}{6}$$

$$I_{P(Load\ 2)} = 73.33\ A$$

The line current supplying a delta-connected load is 1.732 times greater than the phase current. The amount of line current needed for load 2 can be computed by increasing the phase current value by 1.732.

$$I_{L(Load\ 2)} = I_{P(Load\ 2)} \times 1.732$$

$$I_{L(Load\ 2)} = 73.33 \times 1.732$$

$$I_{L(Load\ 2)} = 127.01\ A$$

The resistors of load 1 are connected to form a wye. The phase voltage of a wye connection is less than the line voltage by a factor of 1.732.

$$E_{P(Load\ 1)} = \frac{E_{L(Load\ 1)}}{1.732}$$

$$E_{P(Load\ 1)} = \frac{440}{1.732}$$

$$E_{P(Load\ 1)} = 254.04\ V$$

Now that the voltage applied to each of the 4 Ω resistors is known, the phase current can be computed using Ohm's Law.

$$I_{P(Load\ 1)} = \frac{E_{P(Load\ 1)}}{Z}$$

$$I_{P(Load\ 1)} = \frac{254.04}{4}$$

$$I_{P(Load\ 1)} = 63.51\ A$$

The line current supplying a wye-connected load is the same as the phase current. Therefore, the amount of line current needed to supply load 1 is:

$$I_{L(Load\ 1)} = 63.51\ A$$

The alternator must supply the line current needed to operate both loads. In this example, both loads are resistive. The total line current supplied by the alternator will be the sum of the line currents of the two loads.

$$I_{L(Alt)} = I_{L(Load\ 1)} + I_{L(Load\ 2)}$$

$$I_{L(Alt)} = 63.51 + 127.01$$

$$I_{L(Alt)} = 190.52\ A$$

Since the phase windings of the alternator in this example are connected in a wye, the phase current will be the same as the line current.

$$I_{P(Alt)} = 190.52\ A$$

The phase voltage of the alternator will be less than the line voltage by a factor of 1.732.

$$E_{P(Alt)} = \frac{440}{1.732}$$

$$E_{P(Alt)} = 254.04\ V$$

Both of the loads in this example are resistive and have a unity power factor of 1. The total power in this circuit can be found by using the line voltage and total line current supplied by the alternator.

$$P = 1.732 \times E_L \times I_L \times PF$$

$$P = 1.732 \times 440 \times 190.52 \times 1$$

$$P = 145{,}191.48\ W$$

SUMMARY

1. The voltages of a three-phase system are 120° out of phase with each other.
2. The two types of three-phase connections are wye and delta.
3. Wye connections are characterized by the fact that one terminal of each device is connected together.
4. In a wye connection, the phase voltage is less than the line voltage by a factor of 1.732. The phase current and line current are the same.
5. In a delta connection, the phase voltage is the same as the line voltage. The phase current is less than the line current by a factor of 1.732.

REVIEW QUESTIONS

1. How many degrees out of phase with each other are the voltages of a three-phase system?
2. What are the two main types of three-phase connections?
3. A wye-connected load has a voltage of 480 V applied to it. What is the voltage drop across each phase?
4. A wye-connected load has a phase current of 25 A. How much current is flowing through the lines supplying the load?
5. A delta connection has a voltage of 560 V connected to it. How much voltage is dropped across each phase?
6. A delta connection has 30 A of current flowing through each phase winding. How much current is flowing through each of the lines supplying power to the load?
7. A three-phase resistive load has a phase voltage of 240 V and a phase current of 18 A. What is the power of this load?
8. If the load in question 7 is connected in a wye, what would be the line voltage and line current supplying the load?
9. An alternator with a line voltage of 2400 V supplies a delta-connected load. The line current supplied to the load is 40 A. Assuming the load is a balanced three-phase load, what is the impedance of each phase?
10. If the load is pure resistive, what is the power of the circuit in question 9?

Practice Problems

THREE-PHASE CIRCUITS

1. Refer to the circuit shown in Figure 12-17 to answer the following questions. It is assumed the alternator has a line voltage of 240 V and the load has an impedance of 12 Ω per phase. Find all missing values.

$E_{P(A)}$ ______	$E_{P(L)}$ ______
$I_{P(A)}$ ______	$I_{P(L)}$ ______
$E_{L(A)}$ 240______	$E_{L(L)}$ ______
$I_{L(A)}$ ______	$I_{L(L)}$ ______
P ______	$Z_{(PHASE)}$ 12 Ω

2. Refer to the circuit shown in Figure 12-18 to answer the following questions. Assume the alternator has a line voltage of 4,160 V and the load has an impedance of 60 Ω per phase. Find all missing values.

$E_{P(A)}$ ________	$E_{P(L)}$ ________
$I_{P(A)}$ ________	$I_{P(L)}$ ________
$E_{L(A)}$ 4160 V	$E_{L(L)}$ ________
$I_{L(A)}$ ________	$I_{L(L)}$ ________
P ________	$Z_{(PHASE)}$ 60 Ω

3. Refer to the circuit shown in Figure 12-19 to answer the following questions. It is assumed that the alternator has a line voltage of 560 V. Load 1 has an impedance of 5 Ω per phase and load 2 has an impedance of 8 Ω per phase. Find all missing values.

$E_{P(A)}$ ________	$E_{P(L1)}$ ________	$E_{P(L2)}$ ________
$I_{P(A)}$ ________	$I_{P(L1)}$ ________	$I_{P(L2)}$ ________
$E_{L(A)}$ 560 V	$E_{L(L1)}$ ________	$E_{L(L2)}$ ________
$I_{L(A)}$ ________	$I_{L(L1)}$ ________	$I_{L(L2)}$ ________
P ________	$Z_{(PHASE)}$ 5 Ω	$Z_{(PHASE)}$ 8 Ω

13 **Unit**

Transformers

objectives

fter studying this unit, you should be able to:

- Discuss the different types of transformers.
- Calculate values of voltage, current, and turns for transformers using formulas.
- Calculate values of voltage, current, and turns for a transformer using the turns ratio.
- Connect a transformer and test the voltage output of different windings.

Transformers are one of the most common devices found in the electrical field. They range in size from less than one cubic inch to requiring rail cars to move them after they have been broken into sections. Their ratings can range from mVA (milli-volt-amps) to GVA (giga-volt-amps). It is imperative that anyone working in the electrical field have an understanding of transformer types and connections. This unit will present two main types of voltage transformers, isolation transformers and auto transformers.

SINGLE-PHASE TRANSFORMERS

transformer

A **transformer** is a magnetically operated machine that can change values of voltage, current, and impedance without a change of frequency. Transformers are the most efficient machines known. Their efficiencies commonly range from 90% to 99% at full load. Transformers can be divided into several classifications:

1. Isolation transformer
2. Auto transformer
3. Current transformer

All values of a transformer are proportional to its turns ratio.

A basic law concerning transformers is that **all values of a transformer are proportional to its turns ratio.** This does not mean that the exact number of turns of wire on each winding must be known to determine different values of voltage and current for a transformer. What must be known is the *ratio* of turns. For example, assume a transformer has two windings. One winding, the primary, has 1,000 turns of wire and the other, the secondary, has 250 turns of wire, *Figure 13-1*. The **turns ratio** of this transformer is 4 to 1 or 4:1 (1000/250 = 4). This indicates there are four turns of wire on the primary for every one turn of wire on the secondary.

turns ratio

TRANSFORMER FORMULAS

Different formulas can be used to find the values of voltage and current for a transformer. The following is a list of standard formulas:

where: N_P = number of turns in the primary

N_S = number of turns in the secondary

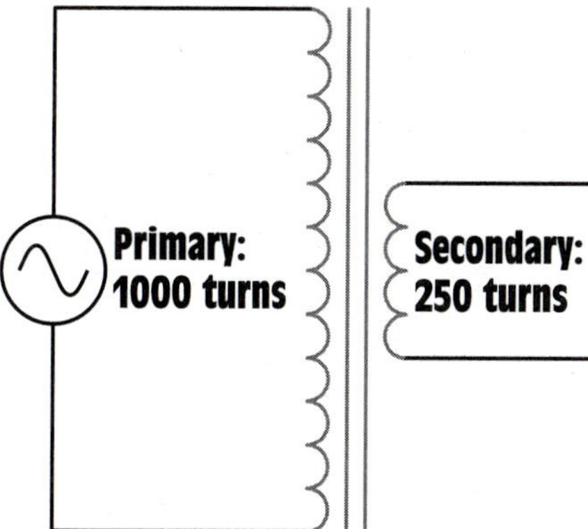

Figure 13-1 All values of a transformer are proportional to its turns ratio.

E_P = voltage of the primary

E_S = voltage of the secondary

I_P = current in the primary

I_S = current in the secondary

$$\frac{E_P}{E_S} = \frac{N_P}{N_S}$$

$$\frac{E_P}{E_S} = \frac{I_S}{I_P}$$

$$\frac{N_P}{N_S} = \frac{I_S}{I_P}$$

or

$$E_P \times N_S = E_S \times N_P$$

$$E_P \times I_P = E_S \times I_S$$

$$N_P \times I_P = N_S \times I_S$$

primary winding

The **primary winding** of a transformer is the power input winding. It is the winding that is connected to the incoming power supply. The **secondary winding** is the load winding or output winding. It is the side of the transformer that is connected to the driven load, *Figure 13-2*.

secondary winding

ISOLATION TRANSFORMERS

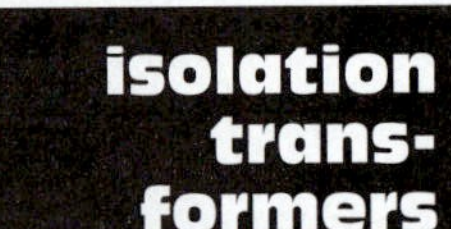

The transformers shown in Figure 13-1 and Figure 13-2 are **isolation transformers.** This means that the secondary winding is physically and electrically isolated from the primary winding. There is no electrical connection between the primary and secondary winding. This transformer is magnetically coupled, not electrically coupled. This line isolation is often a very desirable characteristic. Since there is no electrical connection between

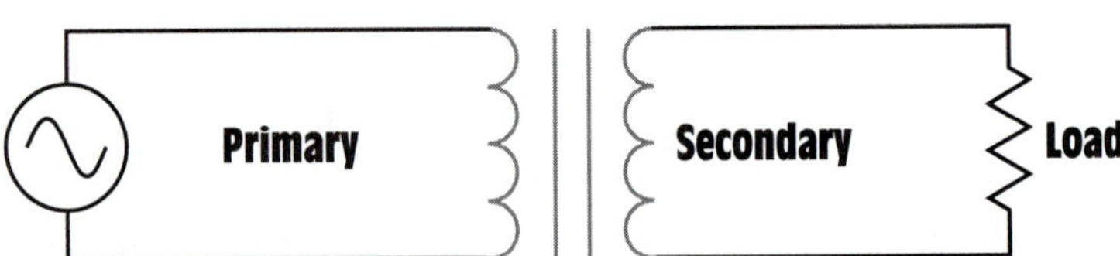

Figure 13-2 An isolation transformer has its primary and secondary winding electrically separated from each other.

the load and power supply, the transformer becomes a filter between the two. The isolation transformer will greatly reduce any voltage spikes that originate on the supply side before they are transferred to the load side. Some isolation transformers are built with a turns ratio of 1:1. A transformer of this type will have the same input and output voltage and is used for the purpose of isolation only.

The reason that the transformer can greatly reduce any voltage spikes before they reach the secondary is because of the rise time of current through an inductor. Current in an inductor rises at an exponential rate, *Figure 13-3*. As the current increases in value, the expanding magnetic field cuts through the conductors of the coil and induces a voltage that is opposed to the applied voltage. The amount of induced voltage is proportional to the rate of change of current. This simply means that the faster current attempts to increase, the greater the opposition to that increase will be. Spike voltages and currents are generally of very short duration, which means that they increase in value very rapidly, *Figure 13-4*. This rapid change of value causes the opposition to the change to increase just as rapidly. By the time the spike has been transferred to the secondary winding of the transformer it has been eliminated or greatly reduced, *Figure 13-5*.

The basic construction of an isolation transformer is shown in *Figure 13-6*. A metal core is used to provide good magnetic coupling between the two windings. The core is generally made of laminations stacked together. Laminating the core helps reduce power losses due to eddy current induction.

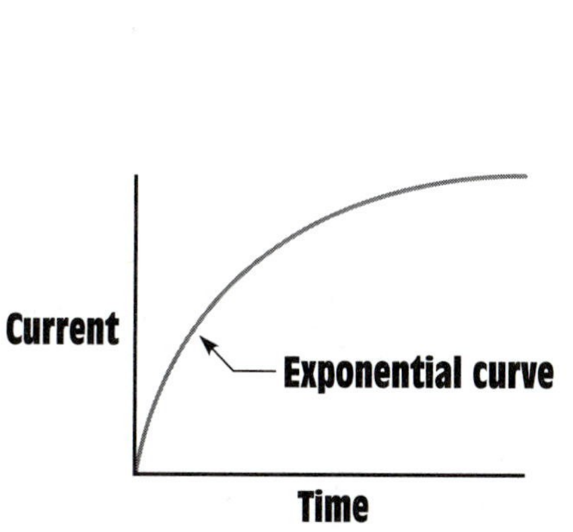

Figure 13-3 The current through an inductor rises at an exponential rate.

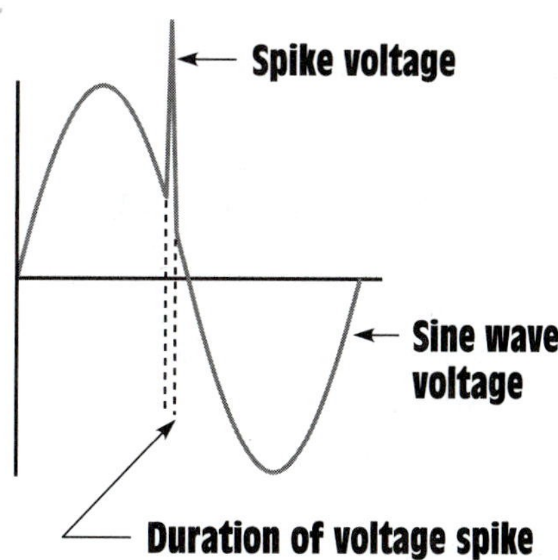

Figure 13-4 Voltage spikes are generally of very short duration.

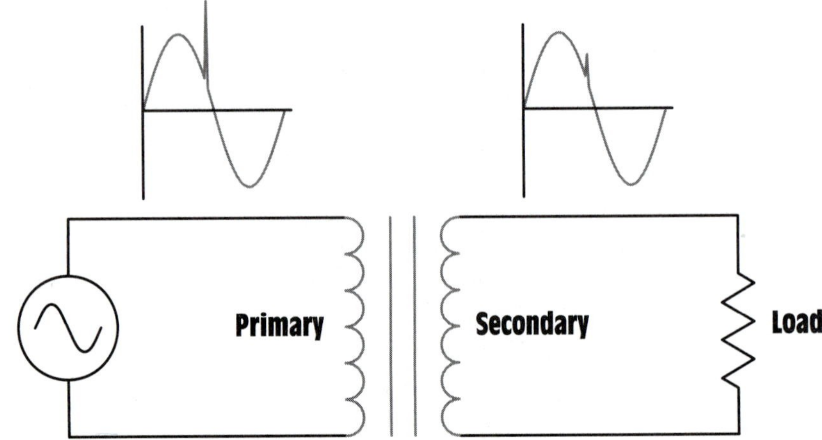

Figure 13-5 The isolation transformer greatly reduces the voltage spike.

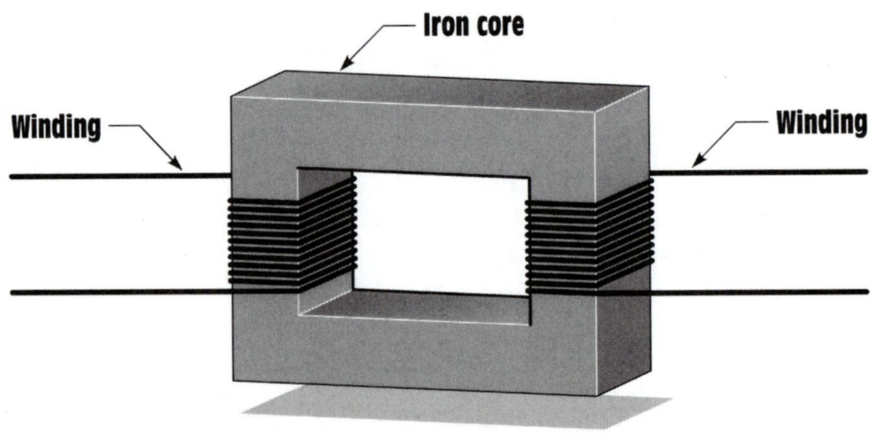

Figure 13-6 Basic construction of an isolation transformer.

BASIC OPERATING PRINCIPLES

In *Figure 13-7*, one winding of the transformer has been connected to an alternating current supply, and the other winding has been connected to a load. As current increases from zero to its peak positive point, a magnetic field expands outward around the coil. When the current decreases from its peak positive point toward zero, the magnetic field collapses. When the current increases toward its negative peak, the magnetic field again expands, but with an opposite polarity of that previously. The field again collapses when the current decreases from its negative peak toward zero. This continually expanding and collapsing magnetic field cuts the windings of the

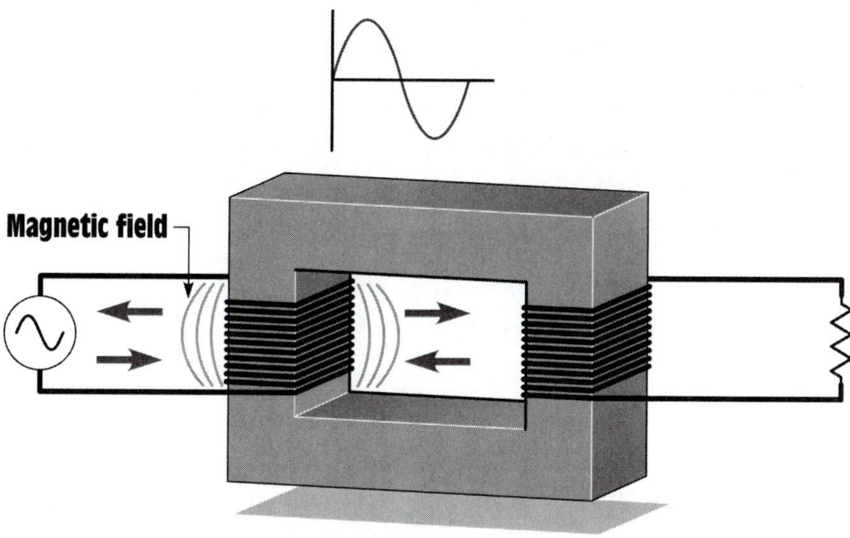

Figure 13-7 Magnetic field produced by alternating current.

primary and induces a voltage into it. This induced voltage opposes the applied voltage and limits the current flow of the primary. When a coil induces a voltage into itself, it is known as self-induction.

EXCITATION CURRENT

There will always be some amount of current flow in the primary of any voltage transformer regardless of type or size even if there is no load connected to the secondary. This is called the **excitation current** of the transformer. The excitation current is the amount of current required to magnetize the core of the transformer. The excitation current remains constant from no load to full load. As a general rule, the excitation current is such a small part of the full load current it is often omitted when making calculations.

excitation current

MUTUAL INDUCTION

Since the secondary windings are wound on the same core as the primary, the magnetic field produced by the primary winding cuts the windings of the secondary also, *Figure 13-8*. This continually changing magnetic field induces a voltage into the secondary winding. The ability of one coil to induce a voltage into another coil is called *mutual induction*. The amount of voltage induced in the secondary is determined by the number of turns of wire in the secondary as compared to the primary. For example, assume the primary has 240 turns of wire and is connected to 120 V AC.

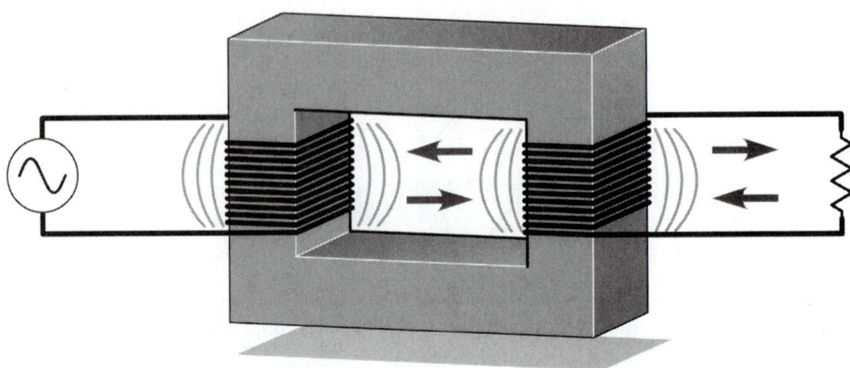

Figure 13-8 The magnetic field of the primary induces a voltage into the secondary.

volts-per-turn ratio

This gives the transformer a **volts-per-turn ratio** of 0.5 (120 V/240 turns = 0.5 volt per turn). Now assume the secondary winding contains 100 turns of wire. Since the transformer has a volt-per-turn ratio of 0.5, the secondary voltage will be 50 V (100 × 0.5 = 50).

TRANSFORMER CALCULATIONS

In the following examples, values of voltage, current, and turns for different transformers will be computed.

Example 1

Assume the isolation transformer shown in Figure 13-2 has 240 turns of wire on the primary and 60 turns of wire on the secondary. This is a ratio of 4:1 (240/60 = 4). Now assume that 120 V is connected to the primary winding. What is the voltage of the secondary winding?

$$\frac{E_P}{E_S} = \frac{N_P}{N_S}$$

$$\frac{120}{E_S} = \frac{240}{60}$$

$$240\ E_S = 7200$$

$$E_S = 30\ V$$

Solution:

step-down transformer

The transformer in this example is known as a **step-down transformer** because it has a lower secondary voltage than primary voltage.

Now assume that the load connected to the secondary winding has an impedance of 5 Ω. The next problem is to calculate the current flow in the secondary and primary windings. The current flow of the secondary can be computed using Ohm's Law since the voltage and impedance are known.

$$I = \frac{E}{Z}$$

$$I = \frac{30}{5}$$

$$I = 6\ A$$

Now that the amount of current flow in the secondary is known, the primary current can be computed using the formula:

$$\frac{E_P}{E_S} = \frac{I_S}{I_P}$$

$$\frac{120}{30} = \frac{6}{I_P}$$

$$120\ I_P = 180$$

$$I_P = 1.5\ A$$

Notice that the primary voltage is higher than the secondary voltage, but the primary current is much less than the secondary current. **A good rule for any type of transformer is that power in must equal power out.** If the primary voltage and current are multiplied together, the product should equal the product of the voltage and current of the secondary.

A good rule for any type of transformer is that power in must equal power out.

Primary	Secondary
$120 \times 1.5 = 180\ VA$	$30 \times 6 = 180\ VA$

Example 2

In the next example, assume that the primary winding contains 240 turns of wire and the secondary contains 1,200 turns of wire. This is a turns ratio of 1:5 (1200/240 = 5). Now assume that 120 V is connected to the primary winding. Compute the voltage output of the secondary winding.

$$\frac{E_P}{E_S} = \frac{N_P}{N_S}$$

$$\frac{120}{E_S} = \frac{240}{1200}$$

$$240\ E_S = 144{,}000$$

$$E_S = 600\ V$$

step-up transformer

Notice that the secondary voltage of this transformer is higher than the primary voltage. This type of transformer is known as a **step-up transformer**.

Now assume that the load connected to the secondary has an impedance of 2,400 Ω. Find the amount of current flow in the primary and secondary windings. The current flow in the secondary winding can be computed using Ohm's Law.

$$I = \frac{E}{Z}$$

$$I = \frac{600}{2400}$$

$$I = 0.25\ A$$

Now that the amount of current flow in the secondary is known, the primary current can be computed using the formula:

$$\frac{E_P}{E_S} = \frac{I_S}{I_P}$$

$$\frac{120}{600} = \frac{0.25}{I_P}$$

$$120\ I_P = 150$$

$$I_P = 1.25\ A$$

Notice that the amount of power input equals the amount of power output.

Primary	Secondary
$120 \times 1.25 = 150\ VA$	$600 \times 0.25 = 150\ VA$

CALCULATING TRANSFORMER VALUES USING THE TURNS RATIO

As illustrated in the previous examples, transformer values of voltage, current, and turns can be computed using formulas. It is also possible to compute these same values using the turns ratio. To make calculations using the turns ratio, a ratio is established that compares some number to 1 or 1 to some number. For example, assume a transformer has a primary rated at 240 V and a secondary rated at 96 V, *Figure 13-9*. The turns ratio can be computed by dividing the higher voltage by the lower voltage.

$$\text{Ratio} = \frac{240}{96}$$

$$\text{Ratio} = 2.5{:}1$$

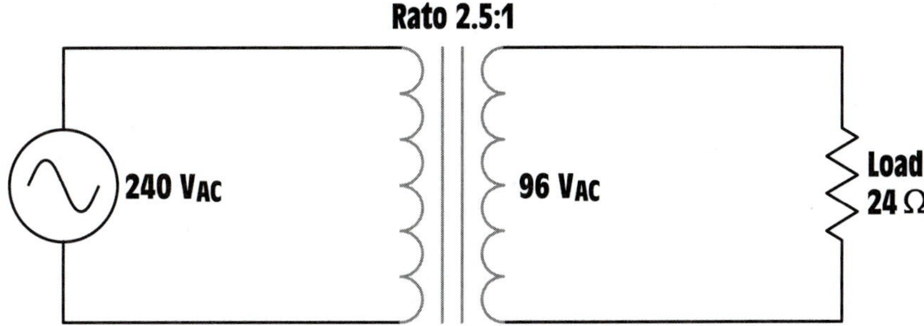

Figure 13-9 Computing transformer values using the turns ratio.

This ratio indicates that there are 2.5 turns of wire in the primary winding for every 1 turn of wire in the secondary. The side of the transformer with the lowest voltage will always have the lowest number (1) of the ratio.

Now assume that a resistance of 24 Ω is connected to the secondary winding. The amount of secondary current can be found using Ohm's Law.

$$I_S = \frac{96}{24}$$

$$I_S = 4 \text{ A}$$

The primary current can be found using the turns ratio. Recall that the volt-amps of the primary must equal the volt-amps of the secondary. Since the primary voltage is greater, the primary current will have to be less than the secondary current.

$$I_P = \frac{I_S}{\text{turns ratio}}$$

$$I_P = \frac{4}{2.5}$$

$$I = 1.6 \text{ A}$$

To check the answer, find the volt-amps of the primary and secondary.

Primary	Secondary
$240 \times 1.6 = 384$ VA	$96 \times 4 = 384$ VA

Now assume that the secondary winding contains 150 turns of wire. The primary turns can be found by using the turns ratio also. Since the primary voltage is higher than the secondary voltage, the primary must have more turns of wire.

$$N_P = N_S \times \text{turns ratio}$$

$$N_P = 150 \times 2.5$$

$$N_P = 375 \text{ turns}$$

In the next example, assume an isolation transformer has a primary voltage of 120 V and a secondary voltage of 500 V. The secondary has a load impedance of 1200 Ω. The secondary contains 800 turns of wire, *Figure 13-10*. The turns ratio can be found by dividing the higher voltage by the lower voltage.

$$\text{Ratio} = \frac{500}{120}$$

$$\text{Ratio} = 1{:}4.17$$

The secondary current can be found using Ohm's Law.

$$I_S = \frac{500}{1200}$$

$$I_S = 0.417 \text{ A}$$

In this example, the primary voltage is lower than the secondary voltage. Therefore, the primary current must be higher.

$$I_P = I_S \times \text{turns ratio}$$

$$I_P = 0.417 \times 4.17$$

$$I_P = 1.74 \text{ A}$$

To check this answer, compute the volt-amps of both windings.

Primary	Secondary
$120 \times 1.74 = 208.8$ VA	$500 \times 0.417 = 208.5$ VA

The slight difference in answers is caused by rounding off values.

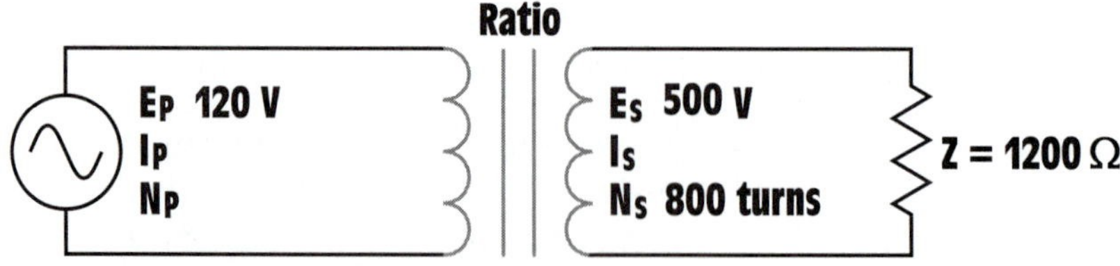

Figure 13-10 Calculating transformer values.

Since the primary voltage is less than the secondary voltage, the turns of wire in the primary will be less also.

$$N_P = \frac{N_S}{\text{turns ratio}}$$

$$N_P = \frac{800}{4.17}$$

$$N_P = 192 \text{ turns}$$

Figure 13-11 shows the transformer with all completed values.

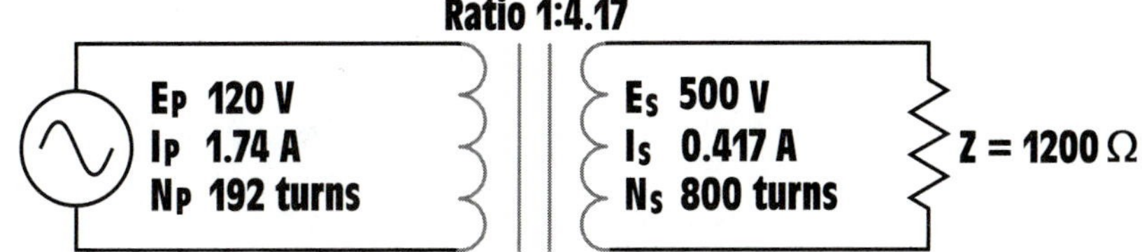

Figure 13-11 Transformer with completed values.

MULTIPLE-TAPPED WINDINGS

It is not uncommon for transformers to be designed with windings that have more than one set of lead wires connected to the primary or secondary. The transformer shown in *Figure 13-12* contains a secondary winding rated at 24 V. The primary winding contains several taps, however. One of the primary lead wires is labeled C and is the common for the other leads. The other leads are labeled 120, 208, and 240. This transformer is designed in such a manner that it can be connected to different primary voltages without changing the value of the secondary voltage. In this example, it is assumed that the secondary winding has a total of 120 turns of wire. To maintain the proper turns ratio, the primary would have 600 turns of wire between C and 120, 1040 turns between C and 208, and 1200 turns between C and 240.

A transformer with multiple primary taps is shown in *Figure 13-13*. This transformer has a 48-V center-tapped secondary winding. The primary winding is rated at 117 V and is marked as follows:

2-4	NOM	
1-4	−6%	OUTPUT
2-3	+12%	
1-3	+6%	

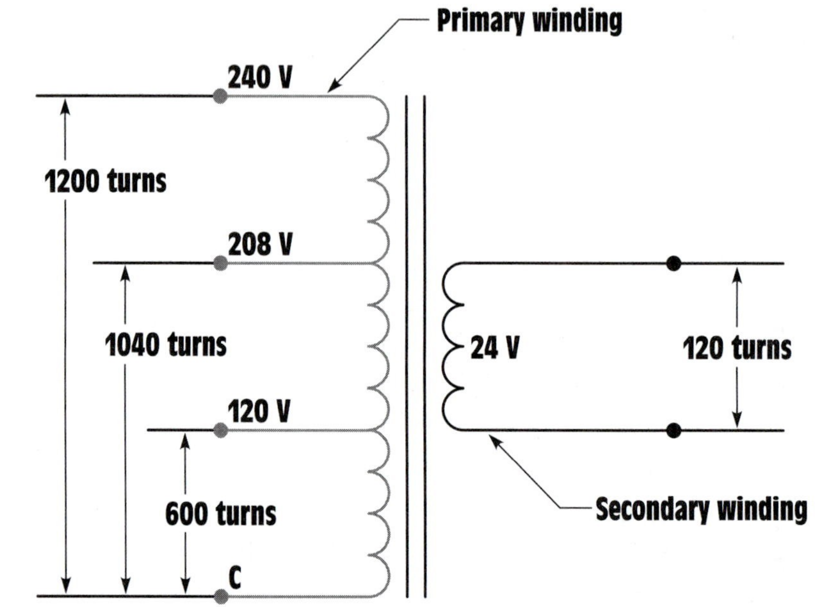

Figure 13-12 Transformer with multiple-tapped primary winding.

Figure 13-13 Transformer with multiple primary taps.

These markings indicate that if 117 V is connected to terminals 2 and 4, the secondary voltage will be the normal 48 V. If 117 V is connected to terminals 1 and 4 the output voltage will be 6% less than the rate voltage or 45.12 V. If 117 V is connected to terminals 2 and 3, the output voltage will be 12% higher than the rated voltage or 53.76 V. Transformers of this type are designed to permit some adjustment of the output voltage.

The transformer shown in *Figure 13-14* contains a single primary winding. The secondary winding, however, has been tapped at several points. One of the secondary lead wires is labeled C and is common to the other lead wires. When rated voltage is applied to the primary, voltages of 12, 24, and 48 V can be obtained at the secondary. It should also be noted that this arrangement of taps permits the transformer to be used as a center-tapped transformer for two of the voltages. If a load is placed across the lead wires labeled C and 24, the lead wire labeled 12 becomes a center tap. If a load is placed across the C and 48 lead wires, the 24 lead wire becomes a center tap.

In this example, it is assumed the primary winding has 300 turns of wire. In order to produce the proper turns ratio, it would require 30 turns of wire between C and 12, 60 turns of wire between C and 24, and 120 turns of wire between C and 48.

A transformer similar to the one just discussed is shown in *Figure 13-15*. This transformer, however, has taps labeled C, 24, 36, and 42.

The transformer shown in Figure *13-16* is similar to the transformer in Figure 13-14. The transformer in Figure 13-16, however, has multiple secondary windings instead of a single secondary winding with multiple taps. The advantage of the transformer in Figure 13-16 is that the secondary windings are electrically isolated from each other. These secondary windings can be either step-up or step-down depending on the application of the transformer.

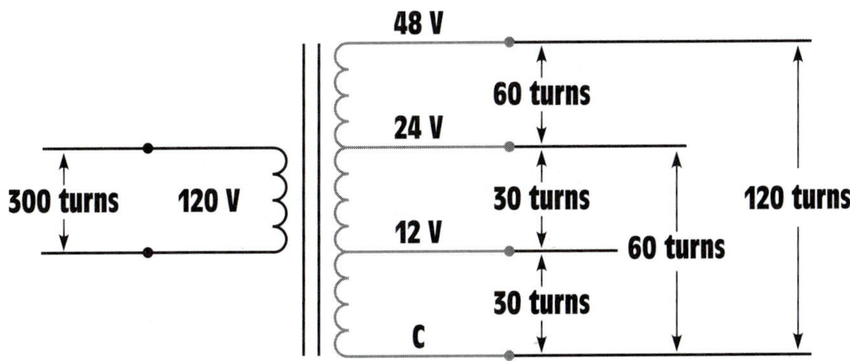

Figure 13-14 Transformer secondary with multiple taps.

Figure 13-15 Transformer with a multiple-tapped secondary winding.

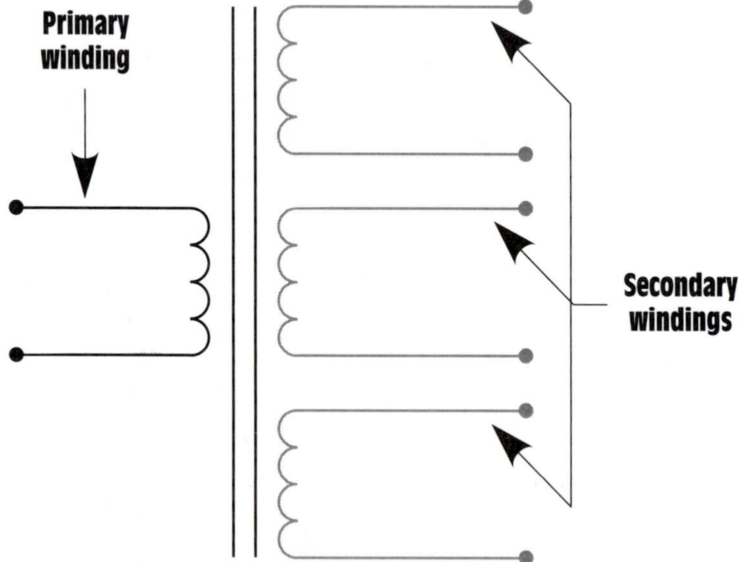

Figure 13-16 Transformer with multiple secondary windings.

COMPUTING VALUES FOR TRANSFORMERS WITH MULTIPLE SECONDARIES

When computing the values of a transformer with multiple secondary windings, each secondary must be treated as a different transformer. For example, the transformer in *Figure 13-17* contains one primary winding and three secondary windings. The primary is connected to 120 V AC and contains 300 turns of wire. One secondary has an output voltage of 560 V and a load impedance of 1000 Ω. The second secondary has an output voltage of 208 V and a load impedance of 400 Ω, and the third secondary has an output voltage of 24 V and a load impedance of 6 Ω. The current, turns of wire, and ratio for each secondary and the current of the primary will be found.

The first step will be to compute the turns ratio of the first secondary. The turns ratio can be found by dividing the smaller voltage into the larger.

$$\text{Ratio} = \frac{E_{S1}}{E_P}$$

$$\text{Ratio} = \frac{560}{120}$$

$$\text{Ratio} = 1{:}4.67$$

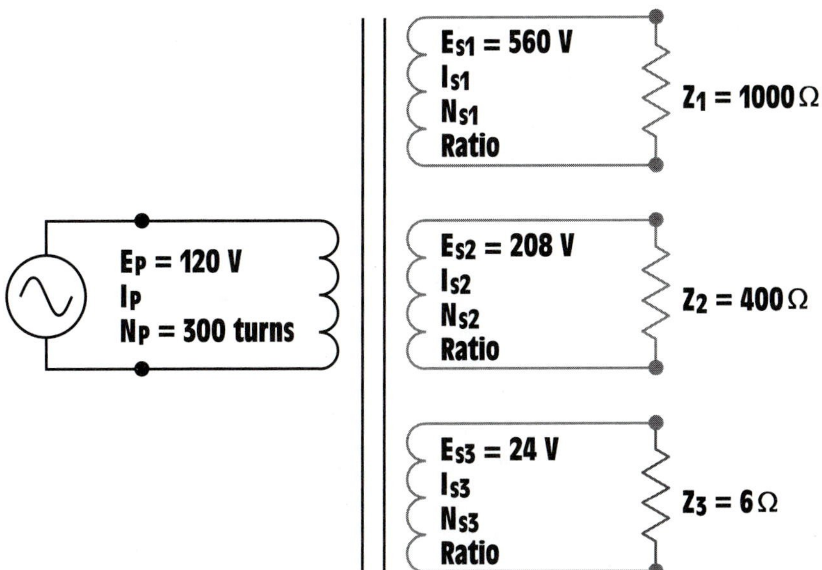

Figure 13-17 Computing values for a transformer with multiple secondary windings.

The current flow in the first secondary can be computed using Ohm's Law.

$$I_{S1} = \frac{560}{1000}$$

$$I_{S1} = 0.56 \text{ A}$$

The number of turns of wire in the first secondary winding will be found using the turns ratio. Since this secondary has a higher voltage than the primary, it must have more turns of wire.

$$N_{S1} = N_P \times \text{turns ratio}$$

$$N_{S1} = 300 \times 4.67$$

$$N_{S1} = 1401 \text{ turns}$$

The amount of primary current needed to supply this secondary winding can be found using the turns ratio also. Since the primary has less voltage, it will require more current.

$$I_{P(\text{FIRST SECONDARY})} = I_{S1} \times \text{turns ratio}$$

$$I_{P(\text{FIRST SECONDARY})} = 0.56 \times 4.67$$

$$I_{P(\text{FIRST SECONDARY})} = 2.61 \text{ A}$$

The turns ratio of the second secondary winding will be found by dividing the higher voltage by the lower.

$$\text{Ratio} = \frac{208}{120}$$

$$\text{Ratio} = 1{:}1.73$$

The amount of current flow in this secondary can be determined using Ohm's Law.

$$I_{S2} = \frac{208}{400}$$

$$I_{S2} = 0.52 \text{ A}$$

Since the voltage of this secondary is greater than the primary, it will have more turns of wire than the primary. The turns of this secondary will be found using the turns ratio.

$$N_{S2} = N_P \times \text{turns ratio}$$

$$N_{S2} = 300 \times 1.73$$

$$N_{S2} = 519 \text{ turns}$$

The voltage of the primary is less than this secondary. The primary will, therefore, require a greater amount of current. The amount of current required to operate this secondary will be computed using the turns ratio.

$$I_{P(SECOND\ SECONDARY)} = I_{S2} \times \text{turns ratio}$$

$$I_{P(SECOND\ SECONDARY)} = 0.52 \times 1.732$$

$$I_{P(SECOND\ SECONDARY)} = 0.9\ \text{A}$$

The turns ratio of the third secondary winding will be computed in the same way as the other two. The larger voltage will be divided by the smaller.

$$\text{Ratio} = \frac{120}{24}$$

$$\text{Ratio} = 5{:}1$$

The primary current will be found using Ohm's Law.

$$I_{S3} = \frac{24}{6}$$

$$I_{S3} = 4\ \text{A}$$

The output voltage of the third secondary is less than the primary. The number of turns of wire will, therefore, be less than the primary turns.

$$N_{S3} = \frac{N_P}{\text{turns ratio}}$$

$$N_{S3} = \frac{300}{5}$$

$$N_{S3} = 60\ \text{turns}$$

The primary has a higher voltage than this secondary. The primary current will, therefore, be less by the amount of the turns ratio.

$$I_{P(THIRD\ SECONDARY)} = \frac{I_{S3}}{\text{turns ratio}}$$

$$I_{P(THIRD\ SECONDARY)} = \frac{4}{5}$$

$$I_{P(THIRD\ SECONDARY)} = 0.8\ \text{A}$$

The primary must supply current to each of the three secondary windings. Therefore, the total amount of primary current will be the sum of the currents required to supply each secondary.

$$I_{P(TOTAL)} = I_{P1} + I_{P2} + I_{P3}$$

$$I_{P(TOTAL)} = 2.61 + 0.9 + 0.8$$

$$I_{P(TOTAL)} = 4.31\ \text{A}$$

The transformer with all computed values is shown in *Figure 13-18.*

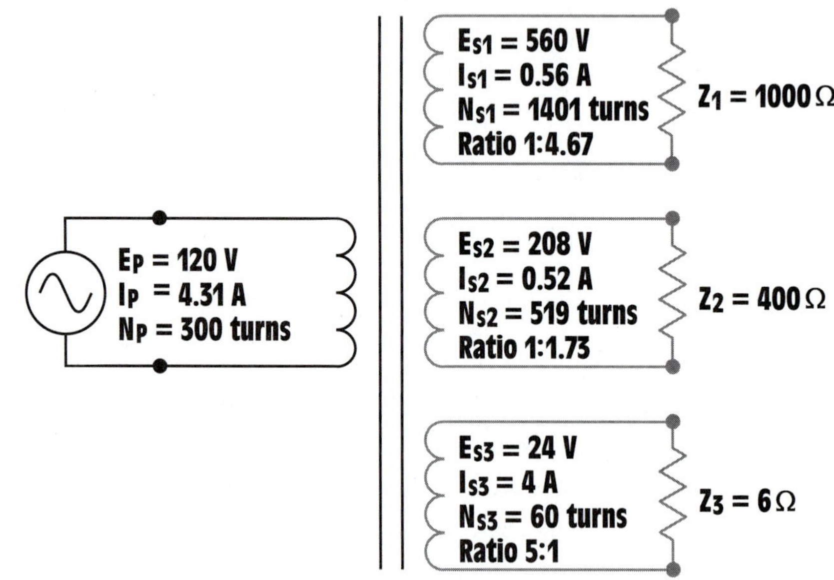

Figure 13-18 The transformer with all computed values.

DISTRIBUTION TRANSFORMERS

distribution transformer

neutral conductor

A very common type of isolation transformer is the **distribution transformer,** *Figure 13-19.* This transformer changes the high voltage of power company distribution lines to the common 240/120 V used to supply power to most homes and many businesses. In this example, it is assumed that the primary is connected to a 7200 V line. The secondary is 240 V with a center tap. The center tap is grounded and becomes the **neutral conductor.** If voltage is measured across the entire secondary, a voltage of 240 V will be seen. If voltage is measured from either line to the center tap, half of the secondary voltage, or 120 V, will be seen, *Figure 13-20.* The reason for this is that the voltages between the two secondary lines are in phase with each other. If a vector diagram is drawn to illustrate this condition, it will be seen that the grounded neutral conductor is connected to the axis point of the two voltage vectors, *Figure 13-21.* Loads that are intended to operate on

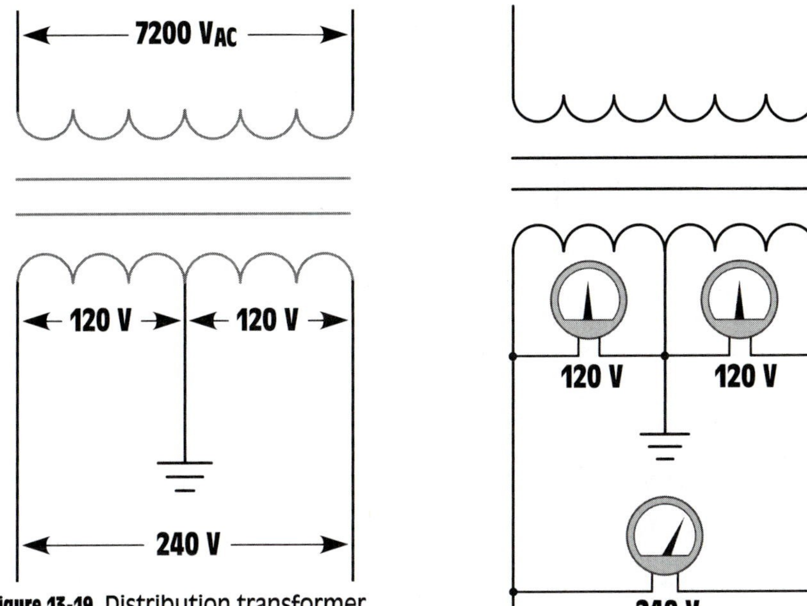

Figure 13-19 Distribution transformer.

Figure 13-20 The voltage from either line to neutral is 120 V. The voltage across the entire secondary winding is 240 V.

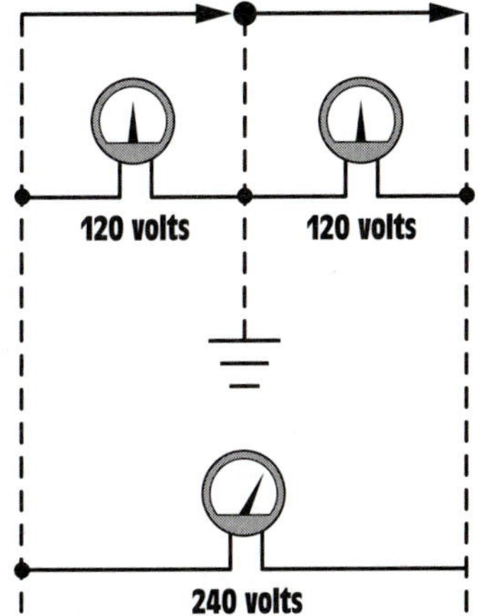

Figure 13-21 The voltages across the secondary are in phase with each other.

240 V, such as water heaters, electric resistance heating units, and central air conditioners are connected directly across the lines of the secondary, *Figure 13-22*.

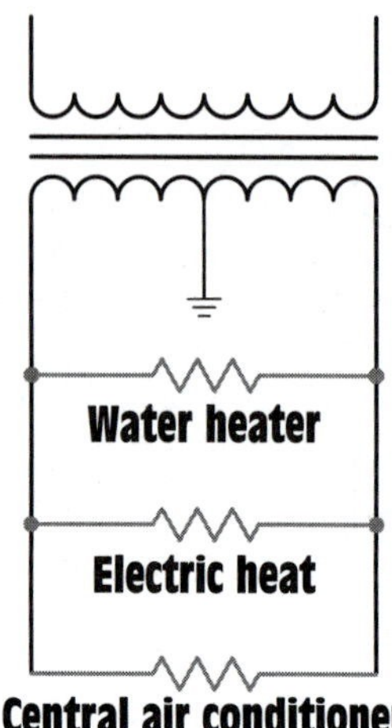

Figure 13-22 240-V loads connect directly across the secondary winding.

Loads that are intended to operate on 120 V connect from the center tap or neutral to one of the secondary lines. The function of the neutral is to carry the difference in current between the two secondary lines and maintain a balanced voltage. In the example shown, it is assumed that one of the secondary lines has a current flow of 30 A and the other has current flow of 24 A. The neutral will conduct the sum of the unbalanced load. In this example, the neutral current will be 6 A (30 − 24 = 6), *Figure 13-23*.

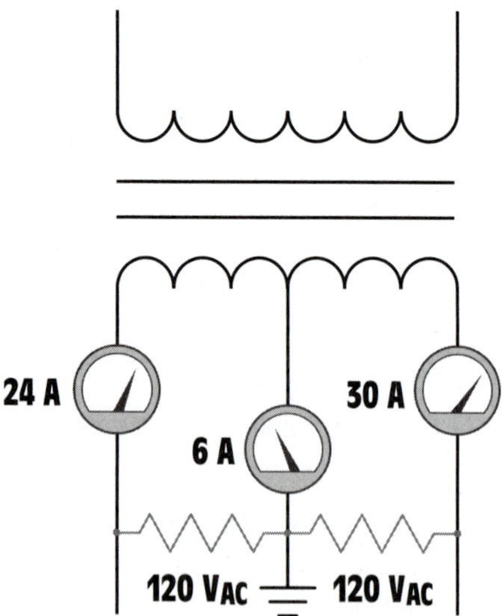

Figure 13-23 The neutral carries the sum of the unbalanced load.

AUTOTRANSFORMERS

autotrans-formers

Autotransformers are one-winding transformers. They use the same winding for both the primary and secondary, *Figure 13-24*. The primary winding is between points B and N, and has a voltage of 120 V applied to it. If the turns of wire are counted between points B and N, it can be seen there are 120 turns of wire. Now assume that the selector switch is set to point D. The load is now connected between points D and N. The secondary of this transformer contains 40 turns of wire. If the amount of voltage applied to the load is to be computed, the following formula can be used.

$$\frac{E_P}{E_S} = \frac{N_P}{N_S}$$

$$\frac{120}{E_S} = \frac{120}{40}$$

$$120\ E_S = 4800$$

$$E_S = 40\ V$$

Assume that the load connected to the secondary has an impedance of 10 Ω. The amount of current flow in the secondary circuit can be computed using the formula:

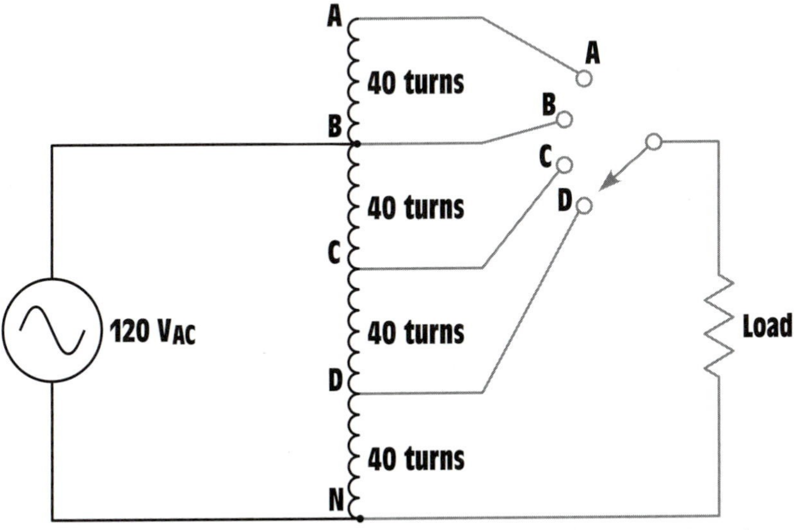

Figure 13-24 An autotransformer has only one winding for use by both the primary and secondary.

$$I = \frac{E}{Z}$$

$$I = \frac{40}{10}$$

$$I = 4\ \text{A}$$

The primary current can be computed by using the same formula that was used to compute primary current for an isolation type of transformer.

$$\frac{E_P}{E_S} = \frac{I_S}{I_P}$$

$$\frac{120}{40} = \frac{4}{I_P}$$

$$120\ I_P = 160$$

$$I_P = 1.333\ \text{A}$$

The amount of power input and output for the autotransformer must be the same, just as they are in an isolation transformer.

Primary	Secondary
$120 \times 1.333 = 160\ \text{VA}$	$40 \times 4 = 160\ \text{VA}$

Now assume that the rotary switch is connected to point A. The load is now connected to 160 turns of wire. The voltage applied to the load can be computed by:

$$\frac{E_P}{E_S} = \frac{N_P}{N_S}$$

$$\frac{120}{E_S} = \frac{120}{160}$$

$$120\ E_S = 19{,}200$$

$$E_S = 160\ \text{V}$$

Notice that the autotransformer, like the isolation transformer, can be either a step-up or step-down transformer.

If the rotary switch shown in Figure 13-24 were to be removed and replaced with a sliding tap that made contact directly to the transformer winding, the turns ratio could be adjusted continuously. This type of transformer is commonly referred to as a Variac or Powerstat, depending on the manufacturer.

The autotransformer does have one disadvantage. Since the load is connected to one side of the power line, there is no line isolation between the incoming power and the load. This can cause problems with certain types of equipment and must be a consideration when designing a power system.

TESTING THE TRANSFORMER

Several tests can be made to determine the condition of the transformer. A simple test for grounds, shorts, or opens can be made with an ohmmeter, *Figure 13-25*. Ohmmeter A is connected to one lead of the primary and one lead of the secondary. This test checks for shorted windings between the primary and secondary. The ohmmeter should indicate infinity. If there is more than one primary or secondary winding, all isolated windings should be tested for shorts. Ohmmeter B illustrates testing the windings for grounds. One lead of the ohmmeter is connected to the case of the transformer and the other is connected to the winding. All windings should be tested for grounds and the ohmmeter should indicate infinity for each winding. Ohmmeter C illustrates testing the windings for continuity. The wire resistance of the winding should be indicated by the ohmmeter.

If the transformer appears to be in good condition after the ohmmeter test, it should then be tested for shorts and grounds with a megohmmeter (megger). A megger will reveal problems of insulation breakdown

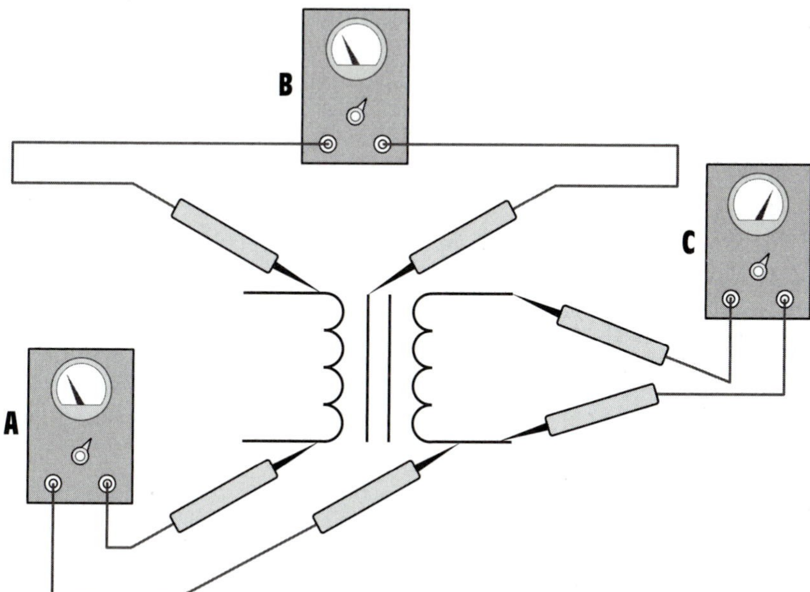

Figure 13-25 Testing a transformer with an ohmmeter.

that an ohmmeter will not. Large oil-filled transformers should have the condition of the dielectric oil tested at periodic intervals. This involves taking a sample of the oil and performing certain tests for dielectric strength and contamination.

TRANSFORMER RATINGS

Most transformers contain a nameplate that lists information concerning the transformer. The information listed is generally determined by the size, type, and manufacturer. Almost all nameplates will list the primary voltage, secondary voltage and KVA (kilo-volt-amps) rating. Transformers are rated in kilo-volt-amps and not kilowatts because the true power is determined by the power factor of the load. Other information that may be listed is frequency, temperature rise in C°, impedance, type of insulating oil, gallons of insulating oil, serial number, type number, model number, and whether the transformer is single phase or three phase.

DETERMINING MAXIMUM CURRENT

Notice that the nameplate does not list the current rating of the windings. Since power input must equal power output, the current rating for a winding can be determined by dividing the KVA rating by the winding voltage. For example, assume a transformer has a KVA rating of 0.5 KVA, a primary voltage of 480 V and a secondary voltage of 120 V. To determine the maximum current that can be supplied by the secondary, divide the KVA rating by the secondary voltage.

$$I_S = \frac{KVA}{E_S}$$

$$I_S = \frac{500}{120}$$

$$I_S = 4.16\text{ A}$$

The primary current can be computed in the same way.

$$I_P = \frac{KVA}{E_P}$$

$$I_P = \frac{500}{480}$$

$$I_P = 1.04\text{ A}$$

Transformers with multiple secondary windings will generally have the current rating listed with the voltage rating.

SUMMARY

1. All values of voltage, current, and impedance in a transformer are proportional to the turns ratio.
2. Transformers can change values of voltage, current, and impedance, but cannot change the frequency.
3. The primary winding of a transformer is connected to the power line.
4. The secondary winding is connected to the load.
5. A transformer that has a lower secondary voltage than primary voltage is a step-down transformer.
6. A transformer that has a higher secondary voltage than primary voltage is a step-up transformer.
7. An isolation transformer has its primary and secondary windings electrically and mechanically separated from each other.
8. When a coil induces a voltage into itself, it is known as self-induction.
9. When a coil induces a voltage into another coil it is known as mutual induction.
10. Autotransformers have only one winding, which is used as both the primary and secondary.
11. Autotransformers have a disadvantage in that they have no line isolation between the primary and secondary winding.
12. Isolation transformers help filter voltage and current spikes between the primary and secondary side.
13. The ampere rating of a primary or secondary winding can be determined by dividing the KVA rating by the rated voltage of the winding.

REVIEW QUESTIONS

1. What is a transformer?
2. What are common efficiencies for transformers?
3. What is an isolation transformer?
4. All values of a transformer are proportional to its ____________ ____________.
5. What is an autotransformer?
6. What is a disadvantage of an autotransformer?

7. Explain the difference between a step-up and a step-down transformer.
8. A transformer has a primary voltage of 240 V and a secondary voltage of 48 V. What is the turns ratio of this transformer?
9. A transformer has an output of 750 VA. The primary voltage is 120 V. What is the primary current?
10. A transformer has a turns ratio of 1:6. The primary current is 18 A. What is the secondary current?

Practice Problems

TRANSFORMERS

Refer to *Figure 13-26* to answer the following questions. Find all the missing values.

1.

E_P 120	E_S 24
I_P ______	I_S ______
N_P 300	N_S ______
Ratio ______	$Z = 3\ \Omega$

2.

E_P 240	E_S 320
I_P ______	I_S ______
N_P ______	N_S 280
Ratio ______	$Z = 500\ \Omega$

3.

E_P ______	E_S 160
I_P ______	I_S ______
N_P ______	N_S 80
Ratio 1:2.5	$Z = 12\ \Omega$

4.

E_P 48	E_S 240
I_P ______	I_S ______
N_P 220	N_S ______
Ratio ______	$Z = 360\ \Omega$

5.

E_P ______	E_S ______
I_P 16.5	I_S 3.25
N_P ______	N_S 450
Ratio ______	$Z = 56\ \Omega$

6.

E_P 480	E_S ______
I_P ______	I_S ______
N_P 275	N_S 525
Ratio ______	$Z = 1.2\ k\Omega$

Figure 13-26 Isolation transformer, Practice Problems 1–6.

Refer to *Figure 13-27* to answer the following questions. Find all the missing values.

7.

E_P 208	E_{S1} 320	E_{S2} 120	E_{S3} 24
I_P ______	I_{S1} ______	I_{S2} ______	I_{S3} ______
N_P 800	N_{S1} ______	N_{S2} ______	N_{S3} ______
	Ratio 1:	Ratio 2:	Ratio 3:
	R_1 12 kΩ	R_2 6 Ω	R_3 8 Ω

8.

E_P 277	E_{S1} 480	E_{S2} 208	E_{S3} 120
I_P ______	I_{S1} ______	I_{S2} ______	I_{S3} ______
N_P 350	N_{S1} ______	N_{S2} ______	N_{S3} ______
	Ratio 1:	Ratio 2:	Ratio 3:
	R_1 200 Ω	R_2 60 Ω	R_3 24 Ω

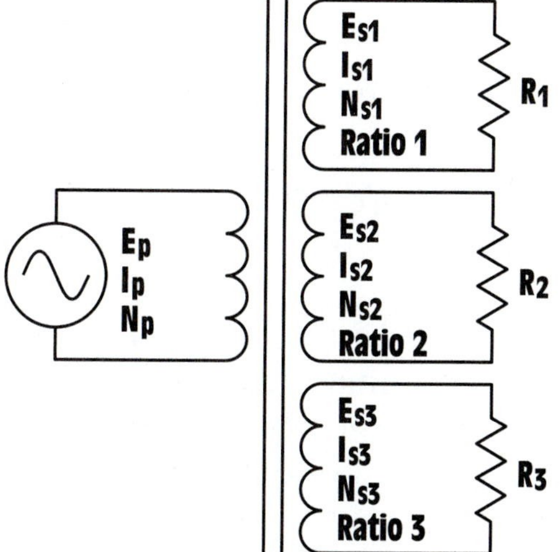

Figure 13-27 Single-phase transformer with multi-secondaries, Practice Problems 7–8.

14

Unit

Electrical Services

objectives

fter studying this unit, you should be able to:

- Describe the differences between three-phase and single-phase panels.
- Connect a three-phase branch circuit to a three-phase panel.
- Connect a 240-V single-phase circuit to a circuit breaker panel.
- Connect a 120-V single-phase circuit to a circuit breaker panel.
- Describe differences between ungrounded and neutral conductors.
- Select the proper conductor size for a particular circuit.

The heart of any electrical system is the service, or panel. All branch circuits originate from a fuse or circuit breaker panel. Industrial and large commercial services are generally three-phase and require the installation of a three-phase wattmeter and a three-phase panel. Residential and light commercial services are generally single phase.

THREE-PHASE SERVICES

Three-phase services generally fall into one of three categories:

1. Three-phase three-wire system.
2. Three-phase four-wire system with a grounded neutral conductor.
3. Three-phase four-wire system with a grounded neutral and a high leg.

The type of three-phase service is determined by the type of secondary transformer connection supplied by the power company. The three-phase three-wire system is made by connecting the secondary windings of a three-phase transformer bank in either wye or delta and connecting the three lines to a three-phase panel, *Figure 14-1*. This system does not contain a neutral or grounded conductor. The lines are generally designated as L_1, L_2, and L_3 or A, B, and C. Typical voltages for this type system are: 240, 440, 480, and 560.

A three-phase four-wire system is produced by connecting the secondary of a three-phase transformer bank in wye and connecting a fourth

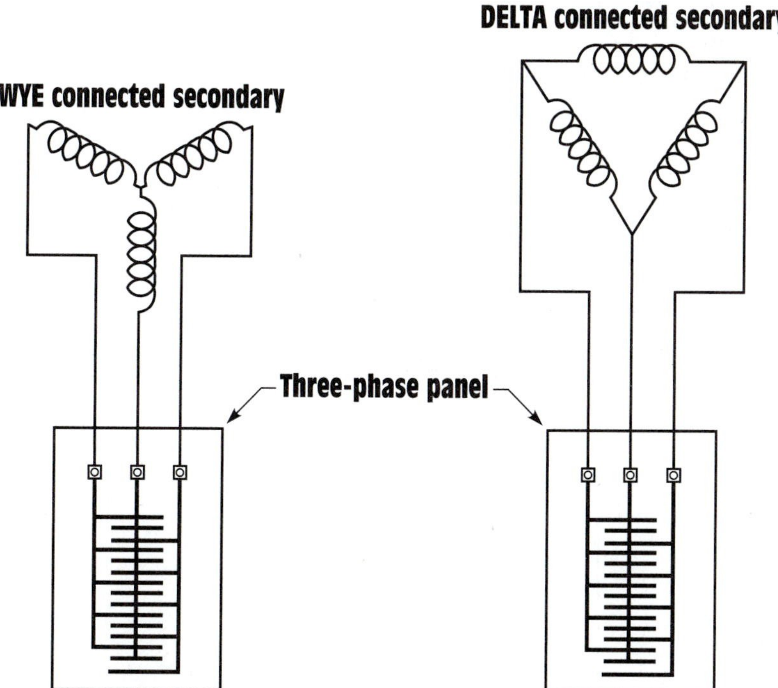

Figure 14-1 Three-phase three-wire system.

wire to the center tap of the wye, *Figure 14-2*. The fourth wire is grounded and becomes the neutral conductor. Phase voltage values can be obtained by connecting between any line and the neutral conductor. Line-to-line voltage values can be obtained by connecting between the lines. Typical voltages for this type of connection are 208/120 and 480/277. The 480/277-V connection is very popular for large commercial and office buildings because the 480-V three-phase capability is used to operate large air conditioning and heating loads, and the 277-V capability is used to operate fluorescent lighting.

The three-phase four-wire high leg connection is made by connecting the secondary windings of a three-phase transformer bank in delta or open delta and center tapping one of the transformer secondaries to provide a neutral conductor, *Figure 14-3*. The open delta connection is often used for light commercial and small industrial locations because it permits three-phase power to be provided with the use of only two transformers. The high leg connection is generally made when a major portion of the power needed is single phase, but there is some need for three-phase power. In this installation the transformer used to supply the single-phase power will be larger than the other transformer or transformers in the bank. The typical voltage for this system is 240 V between any of the three-phase lines. Since one of the transformers is center-tapped, single-phase voltages of 240/120 can be obtained by connecting between the two line conductors of the center-tapped transformer for 240 V, or between any of the line con-

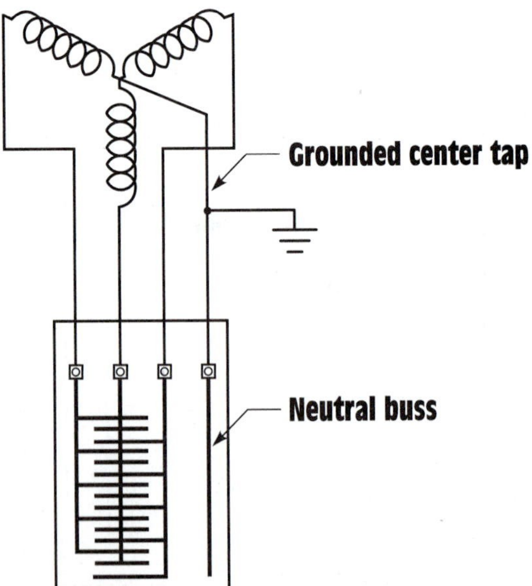

Figure 14-2 Three-phase four-wire system with grounded neutral conductor.

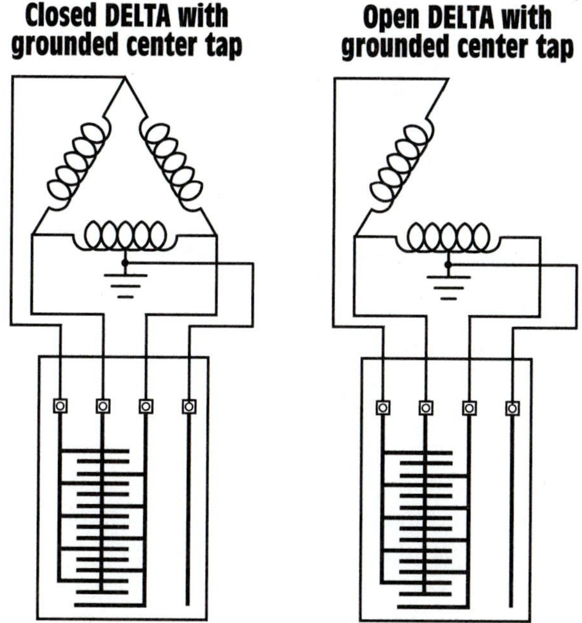

Figure 14-3 Three-phase four-wire system with high leg.

ductors and the center tap for 120 V, *Figure 14-4*. The high leg voltage is obtained by connecting between the grounded center tap and the third leg of the three-phase system. The high leg voltage for this system is 208 V.

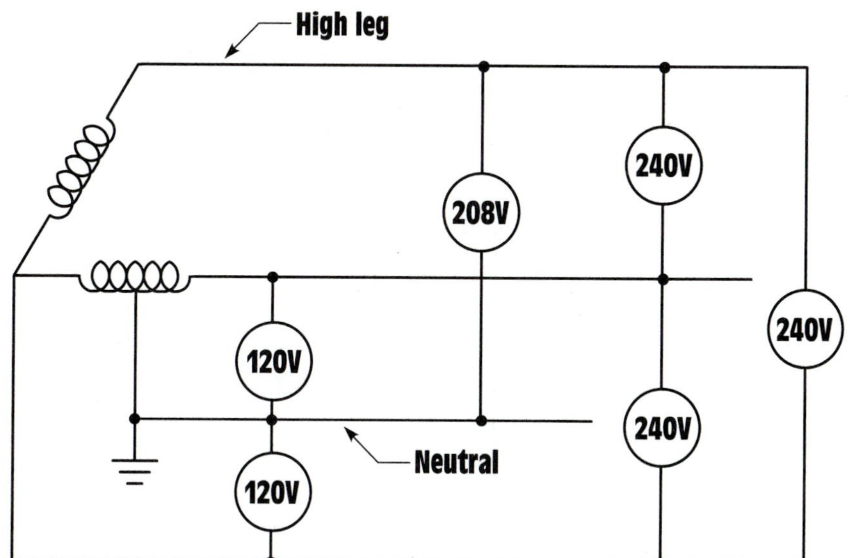

Figure 14-4 Voltages of a three-phase high leg system.

SINGLE-PHASE SERVICES

Single-phase services are derived from a single transformer with a center-tapped secondary winding, *Figure 14-5*. The center tap is grounded and becomes the neutral conductor. In this connection 240 V is available across the secondary winding of the transformer, and 120 V is available from the center tap to either end of the secondary winding.

FUSES AND CIRCUIT BREAKERS

As a general rule, service technicians are not required to install an electrical service, but are sometimes required to make connection to an existing panel. These branch circuits are actually connected to a fuse or circuit breaker that protects the circuit against short circuits and excessive amounts of current. Fuses are devices that contain fusible links, which are metal alloy strips that are designed to melt at a predetermined amount of current flow. Fuses are connected in series with the load, *Figure 14-6*. If an excessive amount of current should flow, the link will melt and open the circuit. Single-phase circuits contain one fuse connected in series with the load as shown in Figure 14-6. Three-phase circuits will contain a fuse in each of the three lines, *Figure 14-7*.

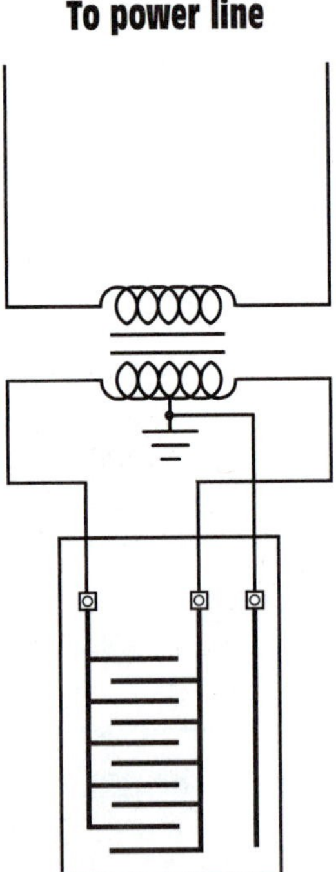

Figure 14-5 Single-phase service.

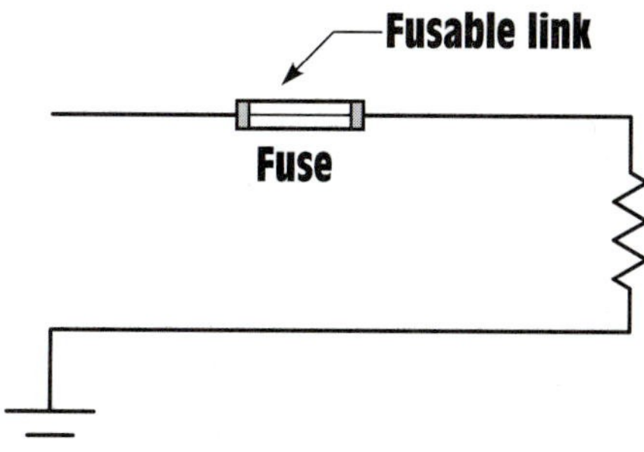

Figure 14-6 A single-phase circuit requires only one fuse in the ungrounded conductor.

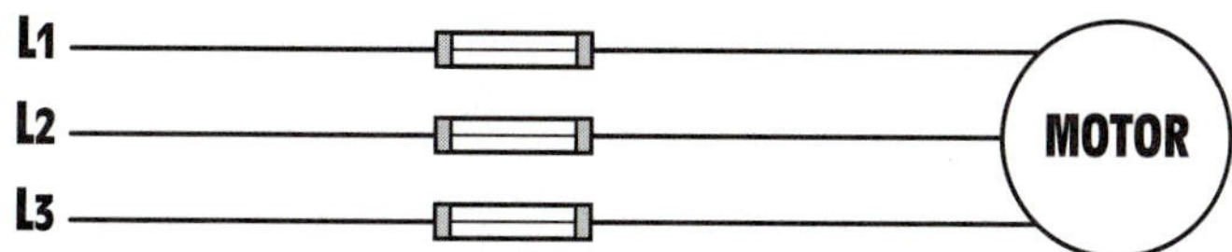

Figure 14-7 Three-phase circuits require a fuse in each of the three lines.

Fuses are rated by amperage and voltage. The more current and voltage a fuse must interrupt, the larger it will be. Two fuses are shown in *Figure 14-8*. The smaller fuse has a rating of 250 V and the larger a rating of 600 V. Higher voltages require a greater distance between points to interrupt an arc.

time delay fuses

slow blow fuses

Fuses are also rated other than by voltage and current. Some fuses are designed to delay for a period of time before they open. These fuses are often referred to as **time delay fuses** or **slow blow fuses.** They are generally used in motor circuits because they can give the motor an opportunity

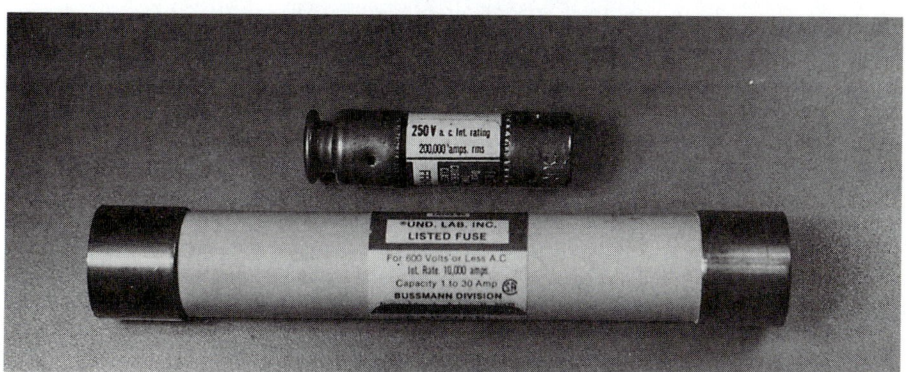

Figure 14-8 Fuse size is determined by voltage and current rating.

current limiters

to start before opening the circuit. Other types of fuses are designed to open immediately upon an overload. Some fuses have the ability to open so rapidly that they are referred to as **current limiters.** Current limiters actually open so fast that they limit the amount of current that can flow during a short circuit.

CIRCUIT BREAKERS

thermal circuit breakers

Another device used to limit the amount of current flow in a circuit is the circuit breaker. Circuit breakers are popular because they can be reset after tripping. There are two methods by which circuit breakers sense the amount of current flow, heat and magnetism. Circuit breakers that operate by heat are known as **thermal circuit breakers.** They contain a small heating element that is connected in series with the load. As current flows through the load, the heating element begins to heat. The heating element is located near a bimetal strip, *Figure 14-9.* If the current becomes high enough the bimetal strip will warp enough to open the contacts and disconnect the load from the line, *Figure 14-10.* Thermal circuit breakers require a certain amount of time to operate because the bimetal strip must heat to a certain temperature before it will open the contacts. Therefore, thermal circuit breakers are often referred to as time delay circuit breakers. The amount of time delay is proportional to the amount of overload. A large overload, such as a short circuit, will cause the circuit breaker to react very rapidly, while a small overload will permit the breaker to continue supply-

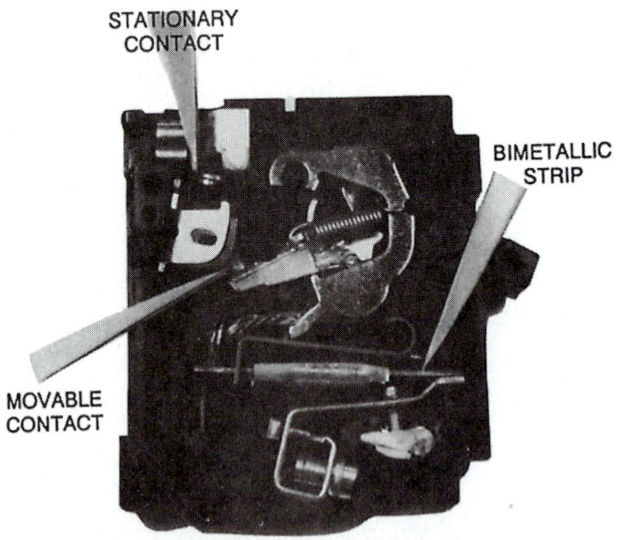

Figure 14-9 Single-pole thermal circuit breaker.

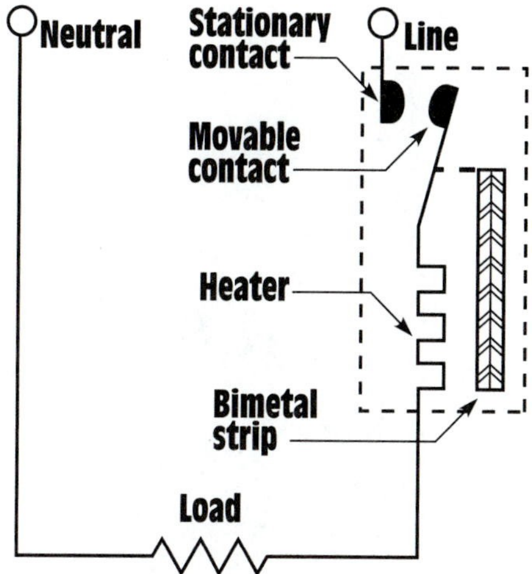

Figure 14-10 The thermal circuit breaker senses circuit current by inserting a heating element in series with the load.

ing current for a longer period of time. The schematic symbol for a thermal circuit breaker is shown in *Figure 14-11*.

Magnetic circuit breakers sense current by connecting a coil of large wire in series with the load, *Figure 14-12*. When current flows through the circuit, a magnetic field is developed around the wire. If the magnetic field

Figure 14-11 Schematic symbol used to represent a single-pole thermal circuit breaker.

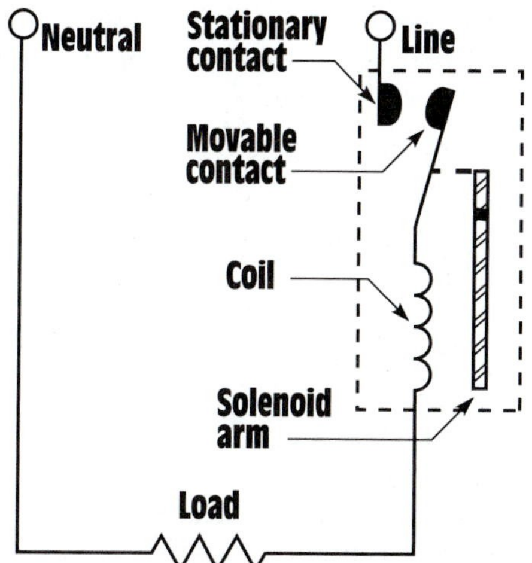

Figure 14-12 The magnetic circuit breaker senses circuit current by inserting a coil in series with the load.

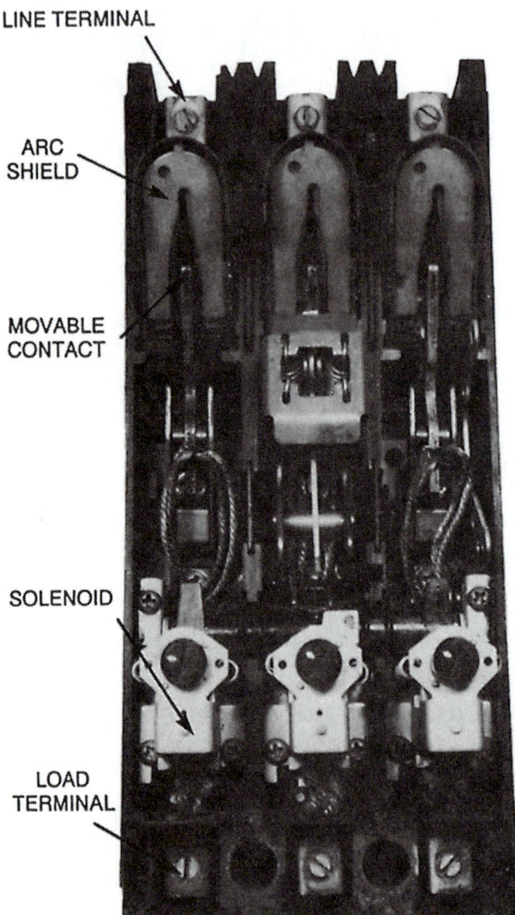

Figure 14-13 Internal construction of a three-pole magnetic circuit breaker.

Figure 14-14 Schematic symbol generally used to represent a magnetic circuit breaker.

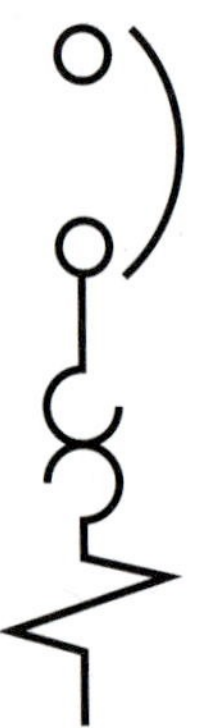

Figure 14-15 Schematic symbol generally used to represent a thermo-magnetic circuit breaker.

becomes strong enough, it will pull in a solenoid and open the contacts connected in series with the load, *Figure 14-13*. Magnetic circuit breakers are often referred to as instantaneous circuit breakers because there is almost no time delay upon opening when an overcurrent condition exists. The schematic symbol generally used to represent a magnetic circuit breaker is shown in *Figure 14-14*.

It should be noted that some circuit breakers are both thermal and magnetic. These circuit breakers contain both a heater element and bimetal strip, as well as a current-activated solenoid. The schematic symbol for this type of circuit breaker is shown in *Figure 14-15*.

CONNECTING BRANCH CIRCUITS TO THE PANEL

To connect a branch circuit to an electric panel several decisions must be made, such as the rating and type of circuit breaker to be installed and the size and type of wire to be used. The brand of circuit breaker to be purchased is generally determined by the type of panel that has been installed. Circuit breakers are not generic. Certain brands of circuit breakers are required to fit certain brands of panels.

The next decision is whether to purchase a three-pole circuit breaker, *Figure 14-16,* a double-pole circuit breaker, *Figure 14-17,* or a single-pole circuit breaker, *Figure 14-18.* Three-pole circuit breakers are used to connect three-phase circuits. When installing a three-phase three-wire branch circuit, one line of each phase is connected to a pole of the circuit breaker, *Figure 14-19.* If a three-phase four-wire branch circuit is to be installed, the three line conductors are connected to the poles of the circuit breaker and the fourth wire is connected to the neutral bus, *Figure 14-20.*

A 240 V single-phase connection can be made in two ways. If the circuit does not contain a neutral conductor, as in the installation of a water heater or central heating or central air conditioning, the insulated wires of a two conductor cable are connected to the poles of the circuit breaker, and the bare conductor is connected to the neutral bus, *Figure 14-21* and

(A)

(B)

Figure 14-16 Three-pole circuit breakers used to connect three-phase circuits.

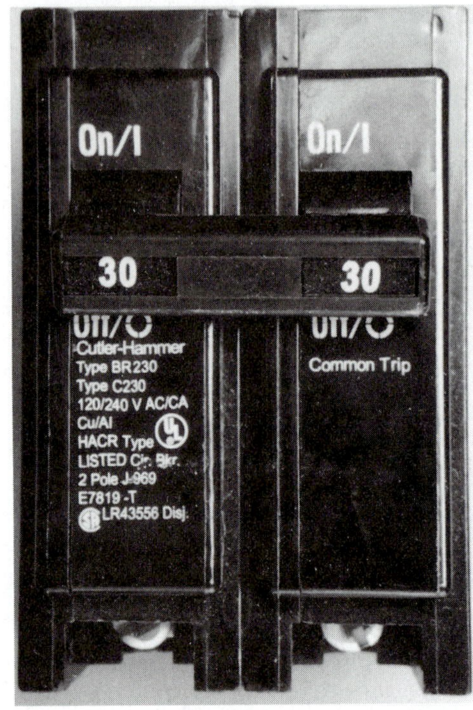

Figure 14-17 Double-pole circuit breaker used to connect 240-V single-phase circuits.

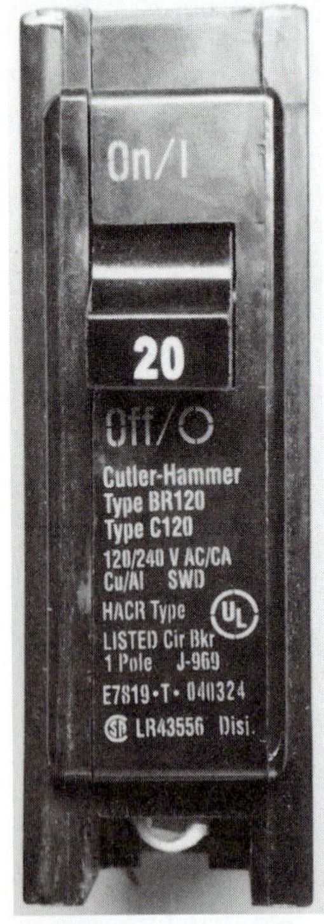

Figure 14-18 Single-phase circuit breaker used to connect 120-V single-phase circuits.

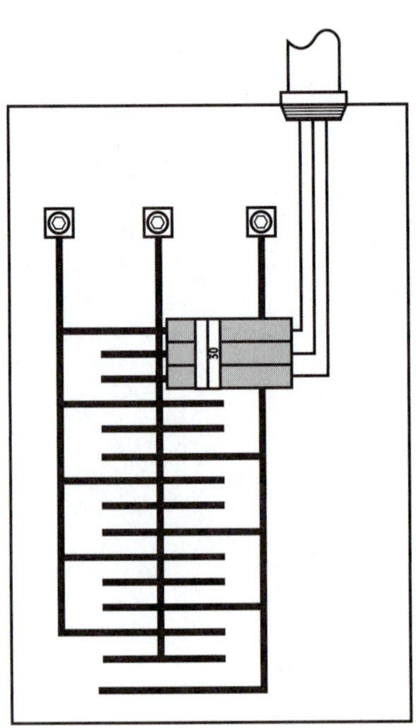

Figure 14-19 Three-phase three-wire connection.

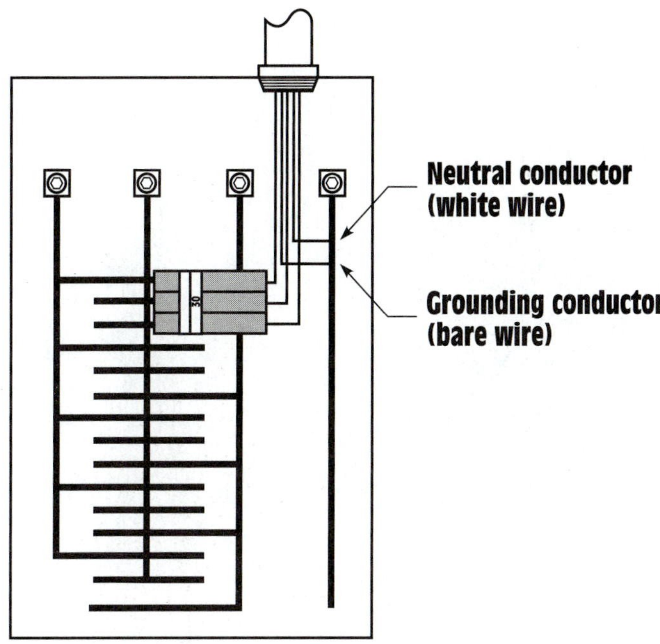

Figure 14-20 Three-phase four-wire connection.

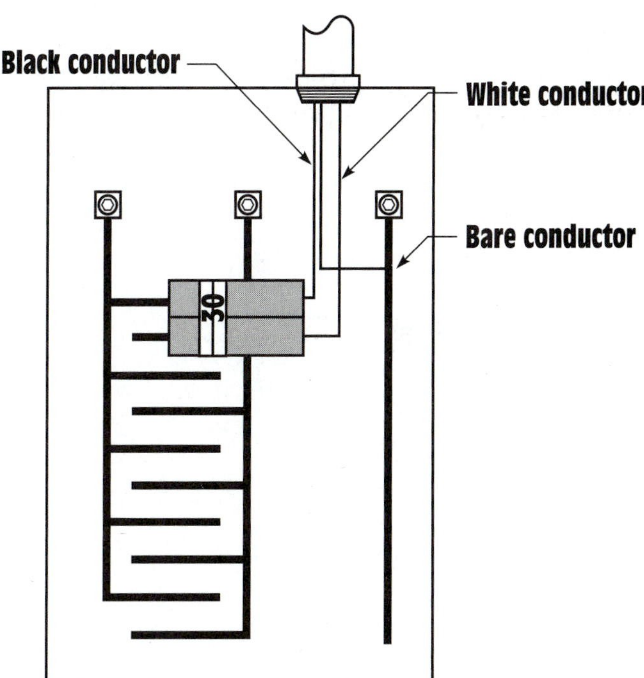

Figure 14-21 240-V single-phase circuit without neutral.

Figure 14-22 240-V single-phase connection with no neutral.

Figure 14-22. When this connection is made, the white conductor must be reidentified by marking it with a piece of colored tape to indicate that it is no longer a neutral wire. The tape should be a color other than white, green, or gray, and the conductor should be marked at both the panel and at the point where it attaches to the equipment.

A 240-V single-phase connection with neutral is made using a three-conductor cable with ground. The black and red conductors are connected to the poles of the circuit breaker, and the white and bare conductors are connected to the neutral bus, *Figure 14-23* and *Figure 14-24*. This type of branch circuit is used to operate equipment that operates on 240 V but also requires the use of a neutral conductor, as in clothes dryers and electric ranges.

A 120-V single-phase branch circuit is made by connecting the black wire of a two conductor cable to a single-pole circuit breaker, and connecting the white and bare wires to the neutral bus, *Figure 14-25* and *Figure 14-26*.

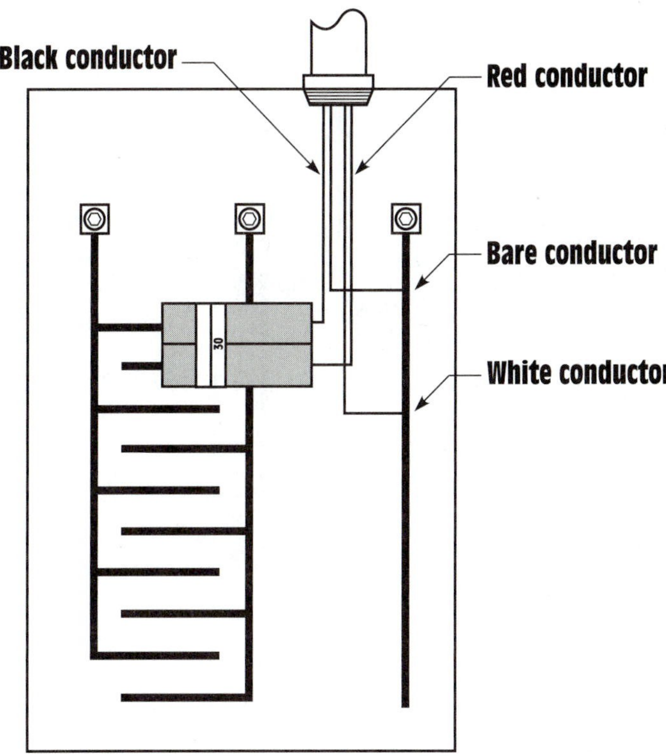

Figure 14-23 240-V single-phase circuit with neutral.

Figure 14-24 240-V single-phase connection with neutral.

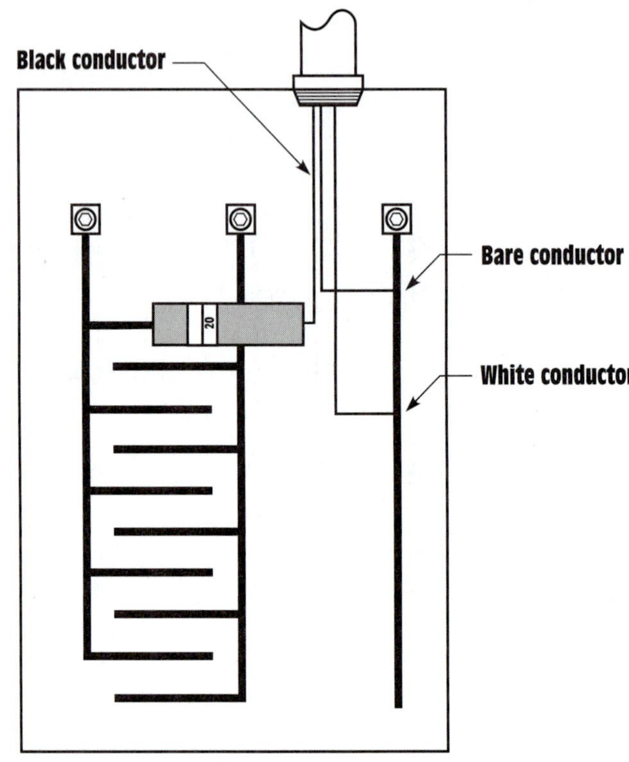

Figure 14-25 120-V single-phase circuit connection.

Figure 14-26 120-V single-phase connection.

DUPLEX AND HALF-SIZE CIRCUIT BREAKERS

The circuit breakers illustrated so far in this unit are called full size or standard size. They are approximately one inch in width. Many circuit breaker panels are designed to employ half-size circuit breakers, *Figure 14-27.* Half-size circuit breakers are the same length as a standard-size breaker but approximately half the width. Half-size breakers permit the panel box to be smaller but contain the same number of circuit breakers as a panel that uses standard-size breakers. Double-pole, half-size breakers are available and are approximately the same size as a standard single-pole breaker, *Figure 14-28.* Duplex circuit breakers are also available from some manufacturers. Duplex breakers are basically two half-size breakers in one case, *Figure 14-29.* The only difference between a duplex circuit breaker and a double-pole breaker is that the duplex breaker does not have a connection bar between the switch levers of the two circuit breakers. This permits them to be turned on, off, or tripped individually.

Circuit breaker panels that employ half-size and duplex circuit breakers are constructed with a different type of bus. Half-size and duplex breakers

Figure 14-27 Half size circuit breaker.

Figure 14-28 Double-pole half size circuit breaker.

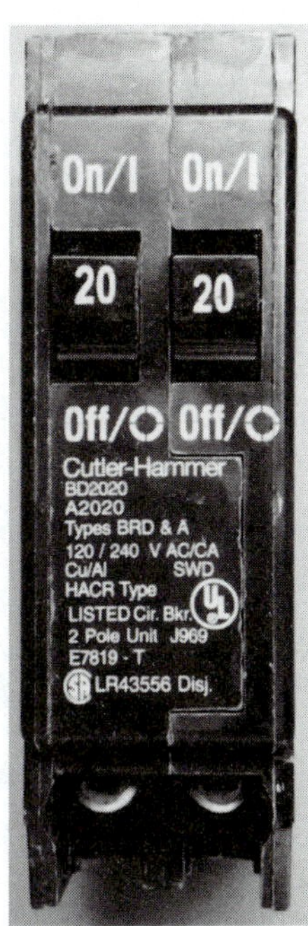

Figure 14-29 Duplex circuit breaker.

are so constructed that they will not plug into a standard breaker box. Using duplex breakers in a standard circuit breaker box could cause the panel to become overloaded. Some panels designed to use duplex circuit breakers will permit the use of full-size breakers also. The buses in panels designed to permit the use of duplex breakers are generally shaped differently than panels with a straight bus, *Figure 14-30.* A panel intended for use with half size circuit breakers is shown in *Figure 14-31.*

The material used to make the bus in a panel box can vary from one manufacturer to another or one model to another. Some contain a copper

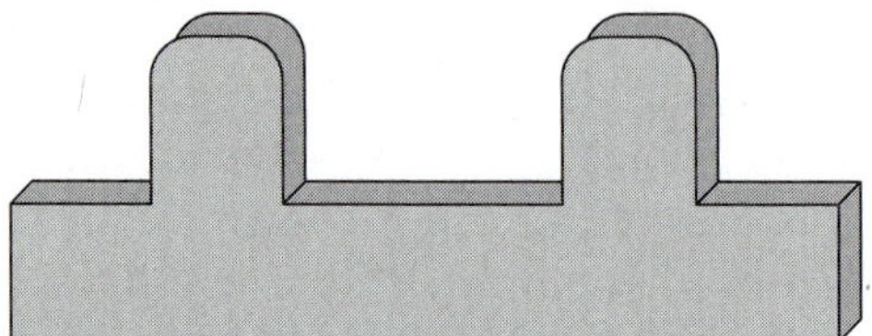

Figure 14-30 Panel bus intended to accept duplex circuit breaker.

Figure 14-31 Panel bus designed for half size circuit breakers.

bus and others may contain a bus made of cadmium-plated copper. A cadmium-plated copper bus look very similar to an aluminum bus, but they are more corrosion resistant. Some circuit breaker panels do contain an aluminum bus. When aluminum is used for the bus material, an antioxidant compound, *Figure 14-32,* should be used to prevent corrosion.

In the 1960s, the *National Electrical Code®* permitted the use of solid aluminum conductors for 15- and 20-ampere circuits in residential construction. The *National Electric Code®* was later modified to prevent the practice. The reason for the modification was the number of electrical fires that resulted from the use of aluminum conductors in #12 and #10 AWG sizes. The fires were not caused by the fact that aluminum is not as good of a conductor as copper, but because the aluminum conductor would cold flow away from a pressure connection. If an aluminum wire was placed under a screw connection, the pressure of the screw would cause the aluminum to cold flow and produce a poor connection. A round conductor would eventually appear to be flattened out as shown in *Figure 14-33*. The *National Electrical Code®* now permits aluminum conductors in sizes 8, 10, and 12 AWG to be used, provided they are made of type AA-8000 series

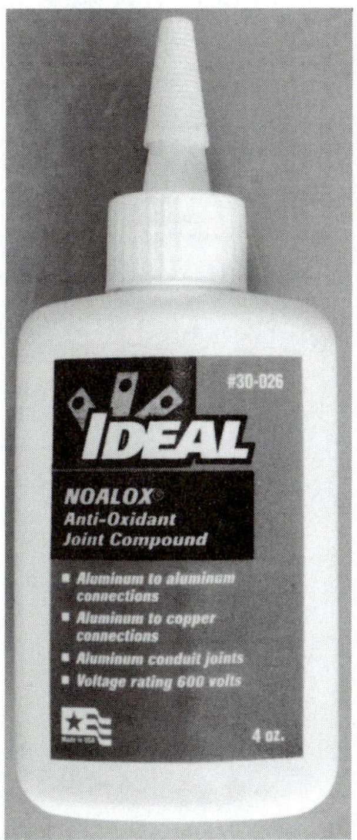

Figure 14-32 Anti-oxident compound should be used on aluminum bus.

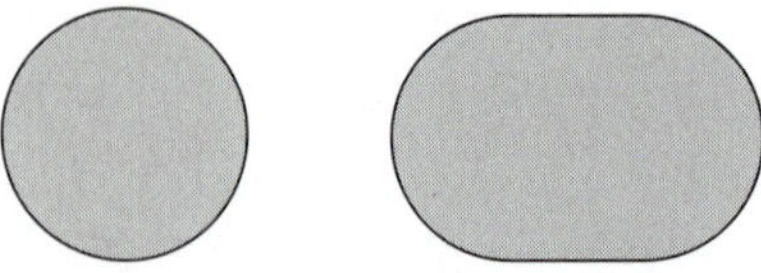

Figure 14-33 Round conductors eventually flatten out due to pressure.

electrical-grade aluminum alloy. The alloy is not as malleable as the older, pure aluminum conductors.

CONDUCTOR SIZE

When installing a branch circuit consider the size and type of wire to be used. The wire size is determined by the ampere rating of the circuit. The copper wire size generally used for most common branch circuits is shown in *Table 14-1*.

Although the wire sizes listed in Table 14-1 are generally used for residential and commercial branch circuits, it may be necessary to select conductors for branch circuits that have a larger ampere rating. This is especially true when working in an industrial environment. The ampere ratings and wire sizes listed in *Table 14-2* are a composite of values listed in the *National Electrical Code*® for not more than three conductors in a raceway or cable or for direct burial. Although some of the insulation types listed at the top of columns A and B permit the conductors to be operated at a higher temperature, and therefore carry more current, termination temperature limitations listed in the *National Electrical Code*® hold the ampere ratings to those listed in Table 14-2.

The table lists ampere ratings for both copper and aluminum wire. The table is divided into columns A and B. Insulation types are listed at the

Branch Circuit (Amperes)	American Wire Gauge Size
15	#14 AWG
20	#12 AWG
30	#10 AWG
40	#8 AWG
50	#6 AWG
60	#4 AWG

Table 14-1 Copper wire sizes for common ampere rating.

AWG or Kcmil	Copper Wire Amperes TW, UF (A)	Copper Wire Amperes FEPW, RH, RHW, THHW, THW, THWN, XHHW, USE, ZW, V, TA, TBS, SA, SIS, FEP, FEPB, RHH, THHN, THHW, XHHW (B)	Aluminum Wire Amperes TW, UF (A)	Aluminum Wire Amperes FEPW, RH, RHW, THHW, THW, THWN, XHHW, USE, ZW, V, TA, TBS, SA, SIS, FEP, FEPB, RHH, THHN, THHW, XHHW (B)
14 AWG	20	20	-----	-----
12 AWG	25	25	20	20
10 AWG	30	35	25	30
8 AWG	40	50	30	40
6 AWG	55	65	40	50
4 AWG	70	85	55	65
3 AWG	85	100	65	75
2 AWG	95	115	75	90
1 AWG	110	130	85	100
1/0 AWG	125	150	100	120
2/0 AWG	145	175	115	135
3/0 AWG	165	200	130	155
4/0 AWG	195	230	150	180
250 Kcmil	215	255	170	205
300 Kcmil	240	285	190	230
350 Kcmil	260	310	210	250
400 Kcmil	280	335	225	270
500 Kcmil	320	380	260	310
600 Kcmil	355	420	285	340
700 Kcmil	385	460	310	375
750 Kcmil	400	475	320	385
800 Kcmil	410	490	330	395
900 Kcmil	435	520	355	425
1000 Kcmil	455	545	375	445
1250 Kcmil	495	590	405	485
1500 Kcmil	520	625	435	520
1750 Kcmil	545	650	455	545
2000 Kcmil	560	665	470	560

Table 14-2 Ampere ratings for not more than three conductors in a raceway or cable or for direct burial. (Column A to be used for values less than 100 A and Column B to be used for current values of 100 A or greater.)

top of each column. Column A, for example, lists only types TW and UF. Conductor sizes should be chosen using the ampere ratings listed in column A for circuits that are rated less than 100 A, and from column B for circuits rated at 100 A or more. The only exception to this rule is if the insulation type is TW or UF, in which case column A will be used for all ampere values.

Example 1

Assume a branch circuit rated at 90 A is to be installed using copper wire with type THW insulation. What size conductor should be used for this installation?

Solution:

Since the ampere rating is less than 100 A, column A will be used. The nearest current value listed without going under 90 A is 95. A #2 AWG copper conductor will be used for this installation.

Example 2

A branch circuit rated at 150 A is to be installed using aluminum wire with type THHW insulation. What size conductor should be used for this installation?

Solution:

Column B of Aluminum Wire Amperes will be used because the circuit rating is greater than 100 A. The nearest current value listed without going under 150 A is 155. This corresponds to a #3/0 AWG conductor.

Example 3

A 200-A branch circuit is to be run underground using copper wire with type UF insulation. What size conductor should be used?

Solution:

Although the ampere rating is greater than 100 A, the insulation type is UF. Therefore, column A will be used to make the selection. The closest ampere rating without going under 200 is 215. A 250-Kcmil conductor will be used.

SUMMARY

1. The heart of any electrical system is the service or panel.
2. Large commercial and industrial locations are generally supplied with three-phase power.
3. Light commercial and residential locations are generally supplied with single-phase power.

4. Three basic types of three-phase service are:

   A. three-phase three-wire
   B. three-phase four-wire with grounded neutral (no high leg)
   C. three-phase four-wire with neutral and a high leg

5. Common voltage values for a three-phase three-wire system are 240, 440, 480, and 560.
6. Common voltage values for a three-phase four-wire wye connected system with a grounded neutral are 480/277 and 208/120.
7. The open delta connection is often used because it requires the use of only two transformers to supply three-phase power.
8. The three-phase four-wire high leg system generally provides 240 V between line conductors and 120 V between two of the line conductors and neutral.
9. Fuses and circuit breakers are used to protect the circuit against short circuits and excessive amounts of current.
10. Fuses protect the circuit by melting a metal alloy link and opening the circuit.
11. Fuses and circuit breakers are connected in series with the load.
12. The size of a fuse is determined by the ampere and voltage rating.
13. Two basic types of circuit breakers are thermal and magnetic.
14. Thermal circuit breakers sense circuit current by inserting a heating element in series with the load.
15. Magnetic circuit breakers sense circuit current by inserting a current-operated solenoid in series with the load.
16. Magnetic circuit breakers operate much faster than thermal circuit breakers.
17. The wire size used to supply power to a branch circuit is determined by the ampere rating of the circuit.

REVIEW QUESTIONS

1. An electrical service in a small commercial location is a three-phase four-wire system with a high leg. The voltage between the neutral conductor and two of the lines is 120 V. What will be the voltage between the neutral conductor and the high leg?

2. How does a fuse protect a circuit against short circuits?
3. What is used to sense current flow in a thermal circuit breaker?
4. What is used to sense current flow in a magnetic circuit breaker?
5. What is a current limiter?
6. In a thermal-type circuit breaker, what is used to actually cause the contacts to open and disconnect the load from the circuit?
7. When making a 240-V two-wire connection to a circuit breaker panel, what should be done to the white conductor?
8. Name two 240-V single-phase household appliances that require a neutral conductor.
9. What size copper conductor is generally used to supply power for a 240-V, 30-A circuit?
10. What size copper conductor is generally used to supply power for a 120-V, 30-A circuit?
11. An electric range is to be connected to a 50-A circuit breaker. What size copper conductor should be used to make this connection?
12. A 150-A branch circuit is to be connected to a service panel. The branch circuit conductor is aluminum with type THHW insulation. What size conductors should be used?
13. Assume that the circuit described in question 12 is to be connected with a copper conductor instead of aluminum. What size conductor should be used?
14. A 200-A branch circuit is to be connected to an electrical panel. The conductors are copper with type TW insulation. What size conductors should be used?
15. A service technician has a roll of #8 AWG UF cable. Can this cable be used to supply a 240-V, 30-A branch circuit?

Unit 15

General Wiring Practices

objectives

*fter studying this unit, you should be able to:*

- Discuss the types of cable used in most residential installations.
- Connect duplex receptacles.
- Connect single-pole switch circuits.
- List rules for connecting three-way switches.
- Make three-way switch connections.
- Make four-way switch connections.

This unit will present general wiring practices. Many times a service or maintenance technician needs to make general switch connections or connect a piece of equipment or an appliance to an electric service. It will be assumed that the connection is made with nonmetallic (NM) cable because nonmetallic cable is commonly used in residential applications. The same connection procedures would apply, however, if the conductors were run in conduit.

CABLES

In a residential occupancy, all circuits are connected with two- or three-conductor cable. A two-conductor cable actually contains three wires, a black, a white, and a bare, *Figure 15-1*. This cable is actually referred to as *two conductor with ground*. It is called *two-conductor* because the only two wires used as part of the circuit are the black and the white. The third wire is a grounding wire and is never intended to be used as part of the actual circuit. The bare wire is connected to the green screw of a receptacle or the case of an appliance at one end, and to the neutral bus in the main service panel at the other.

The three-conductor cable contains four wires, a black, a white, a red, and a bare, *Figure 15-2*. This cable is referred to as *three conductor with ground*. As with two-conductor cable, only three of the wires are to be used as circuit conductors: the black, the white, and the red. The bare wire is the grounding conductor.

RECEPTACLES

receptacles

duplex receptacles

Receptacles, or *outlets,* are used throughout a building for the connection of 120-V AC appliances. Most of these receptacles are **duplex receptacles,** which means that they contain two outlets on the same strap, *Figure 15-3*. Notice that each of the two outlets contains three openings, to permit the attachment of a male plug. Two of the openings are rectangular in shape and the third is similar to a semicircle. The smaller of the two rectangular openings is connected to the hot, or ungrounded, conductor. In a two-conductor circuit this opening will be connected to the black conductor via the gold screws located on the side of the outlet.

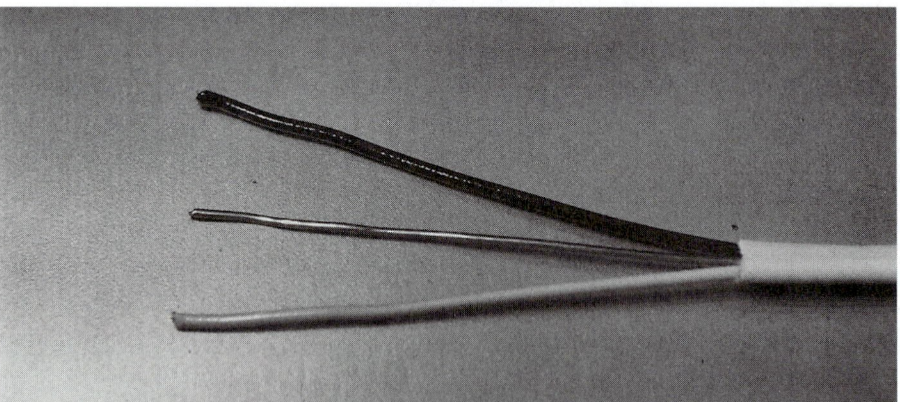

Figure 15-1 Two-conductor cable.

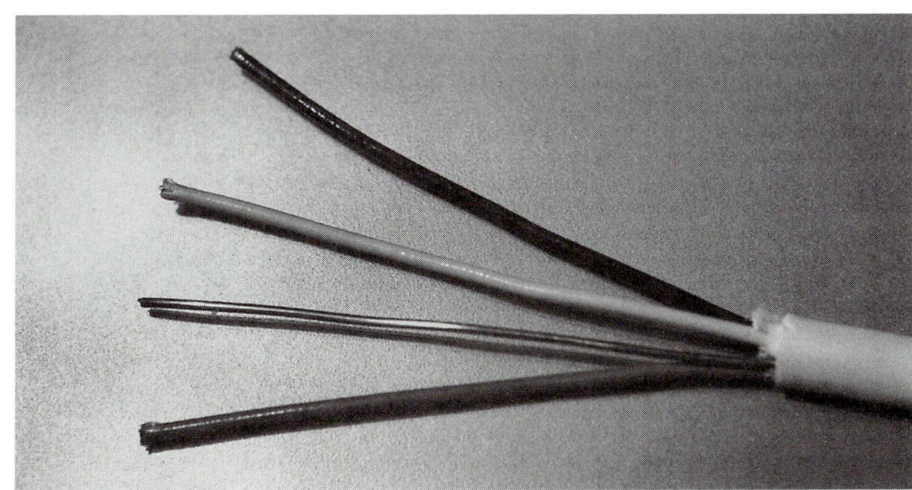

Figure 15-2 Three-conductor cable.

Figure 15-3 Duplex receptacle.

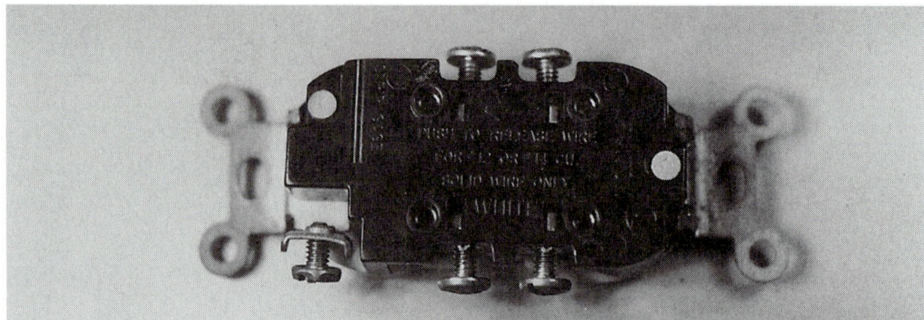

Figure 15-4 Stab-locks or quick-connects are located on the back of some receptacles.

The larger rectangular opening is connected to the neutral, or grounded, conductor. In a two-conductor circuit, this opening is connected to the white conductor via the silver screws. The semicircular opening is connected to the grounding, or bare, conductor via the green screw.

stab-locks

quick-connects

Many duplex receptacles contain a second method of making connection to the circuit conductors, *Figure 15-4*. Small, round holes located on the back of some receptacles can be used to make a connection to circuit conductors by removing about 3/4 inch of insulation from the conductor and inserting the bare wire into the hole. These connectors are generally referred to as **stab-locks** or **quick-connects,** and the advantage to using them is speed. Connection to the receptacle can be made by simply stripping about 3/4 inch of wire and inserting it into the proper hole. The disadvantage is contact resistance. Stab-locks make connection to the conductor with a piece of spring copper when the conductor is inserted into the hole. The spring copper is designed to bite into the copper wire and prevent it from slipping out, *Figure 15-5*. The only contact made with the conductor is the surface area of the spring copper holding the conductor against a small metal plate. Although the stab-lock connector provides the fastest method of making connection to the circuit conductors, it does not provide as good an electrical connection as bending a loop in the wire and tightening it under a screw. It should be noted that the primary cause of electrical fires is a poor connection.

CONNECTING RECEPTACLES

Receptacles or outlets are generally connected to a 15- or 20-A branch circuit. If a 15-A branch circuit is employed #14 AWG copper wire should be used. However, if a 20-A branch circuit is employed #12 AWG copper wire should be used. As a general rule, receptacle circuits are connected to a 20-A branch circuit and lighting circuits are connected to a 15-A branch circuit.

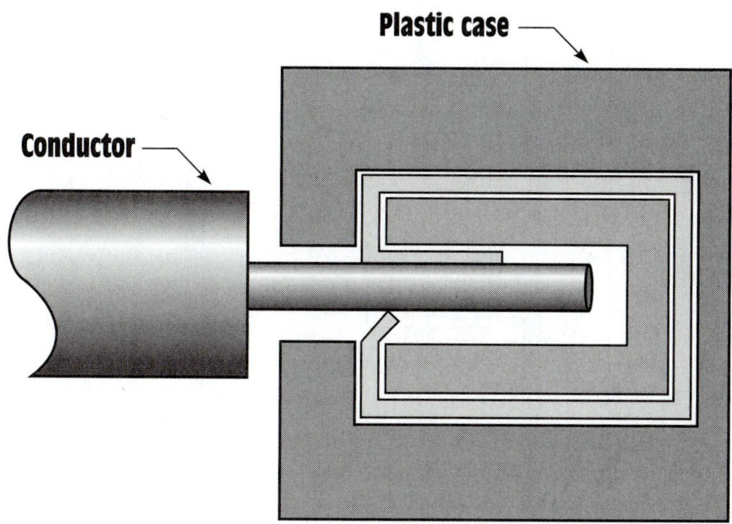

Figure 15-5 Stab-lock connectors make a relatively poor connection.

Receptacles are connected in parallel so that each is supplied with the same voltage, *Figure 15-6.* The ungrounded, or hot, conductor is connected to the gold screw; the grounded, or neutral, conductor is connected to the silver screw; and the grounding, or bare copper, conductor is connected to the green screw.

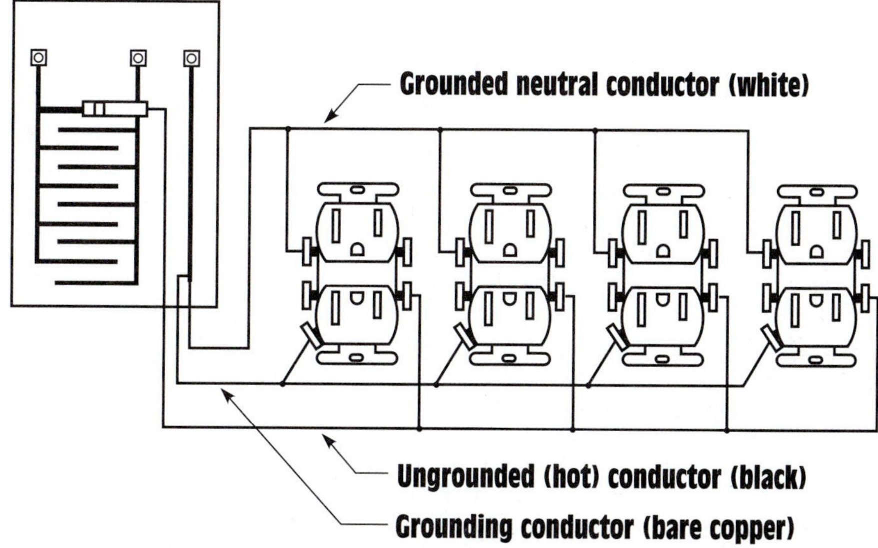

Figure 15-6 Electrical outlets are connected in parallel.

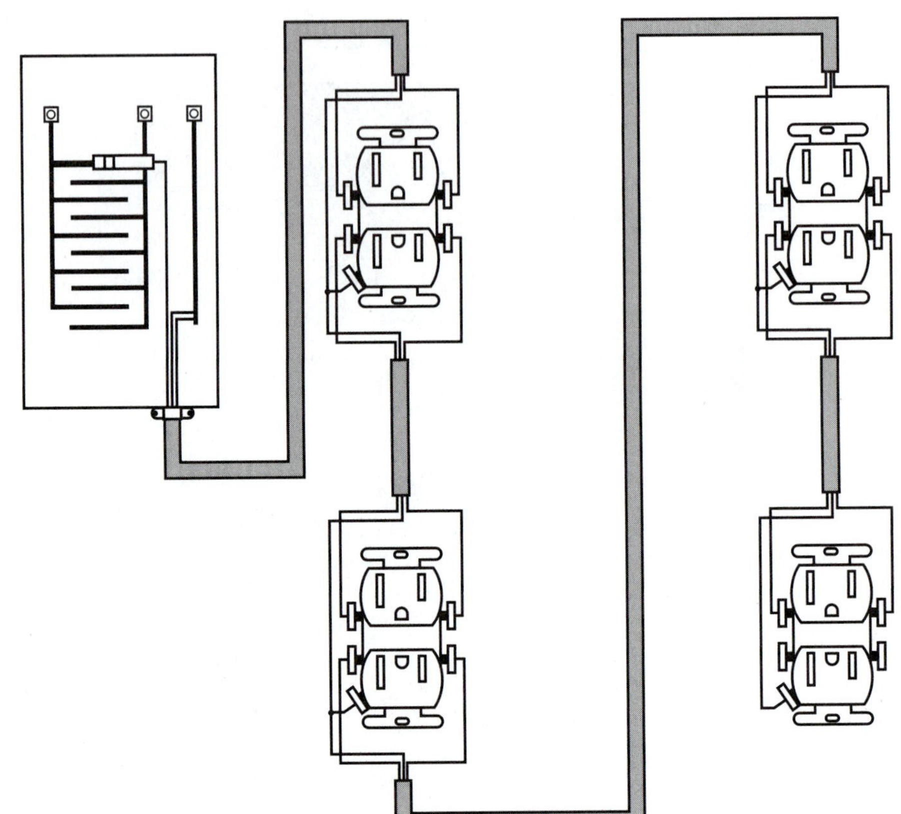

Figure 15-7 Receptacles are connected with two-conductor cable.

When outlets are installed, two-conductor cable is used to make connection from one outlet to another, *Figure 15-7.* The black wires are connected to the two gold screws, the white wires are connected to the silver screws, and the bare conductors are connected to the green screw(s). If the outlet contains only one green screw, grounding conductors must be connected together because it is not permissible to place more than one wire under a screw. When this is the case, a short piece of bare wire is connected with the other bare conductors. This short piece of wire is commonly called a **pig tail.** The pig tail is then connected to the green screw.

pig tail

SWITCH CONNECTIONS

Three basic types of switches are used in general wiring, the single-pole, the three-way, and the four-way. Each of these switches is used for a particular application. Single-pole switches are employed when it is necessary to control a light or other electrical device from one location. Single-pole switches can be identified because they have only two connection screws, although some switches may have a third, grounding screw, which will be green. Single-pole switches also have ON and OFF printed on the switch lever, *Figure 15-8.*

Switch connections are made in the ungrounded, or hot, conductor only, *Figure 15-9.* The neutral, or grounded, conductor is not to be broken. The only time the *National Electrical Code*® permits the neutral conductor to be broken is when both the ungrounded and grounded conductors are broken simultaneously.

Although the switch connection shown in Figure 15-9 is electrically correct, it is generally not possible to insert a switch directly in front of a light. The light is usually placed in the center of the ceiling, and the switch is

Figure 15-8 Single-pole switch.

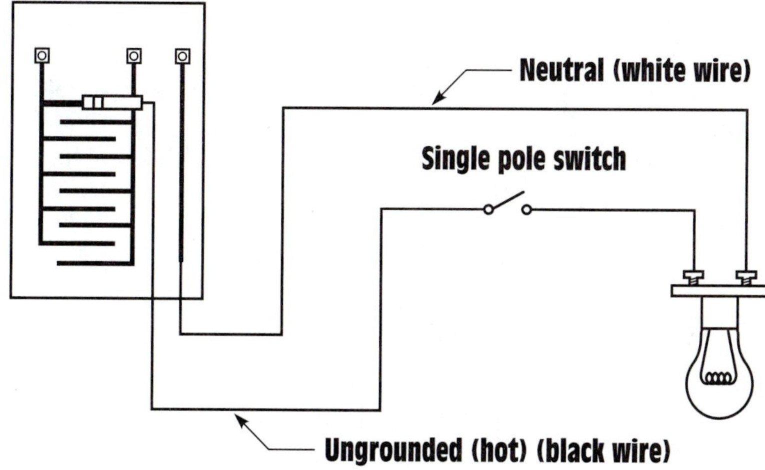

Figure 15-9 The switch connects in series with the ungrounded conductor.

placed beside a doorway. To accommodate the placement of the switch, a switch leg is used. A switch leg is actually an extension of the hot, or ungrounded, conductor, *Figure 15-10*.

As stated previously, all connections are made using two- or three-conductor cable, and all connections must be made inside a box, *Figure 15-11*.

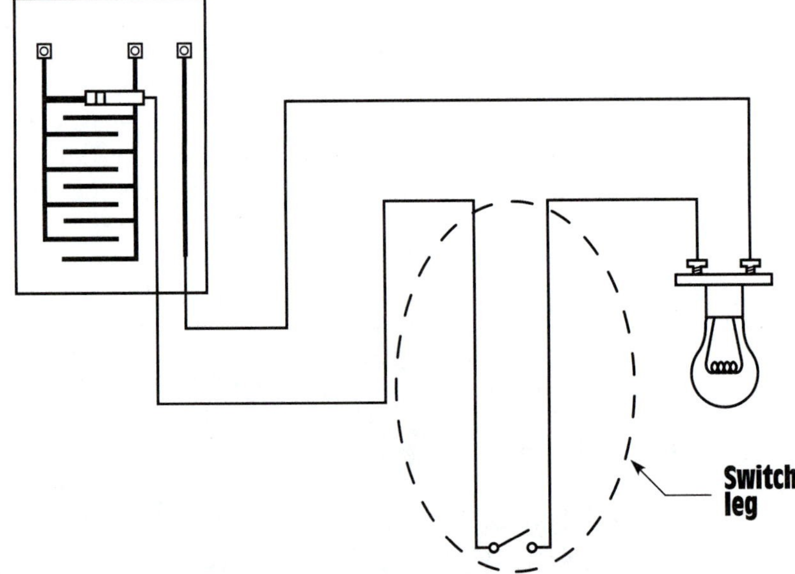

Figure 15-10 A switch leg is an extension of the ungrounded conductor.

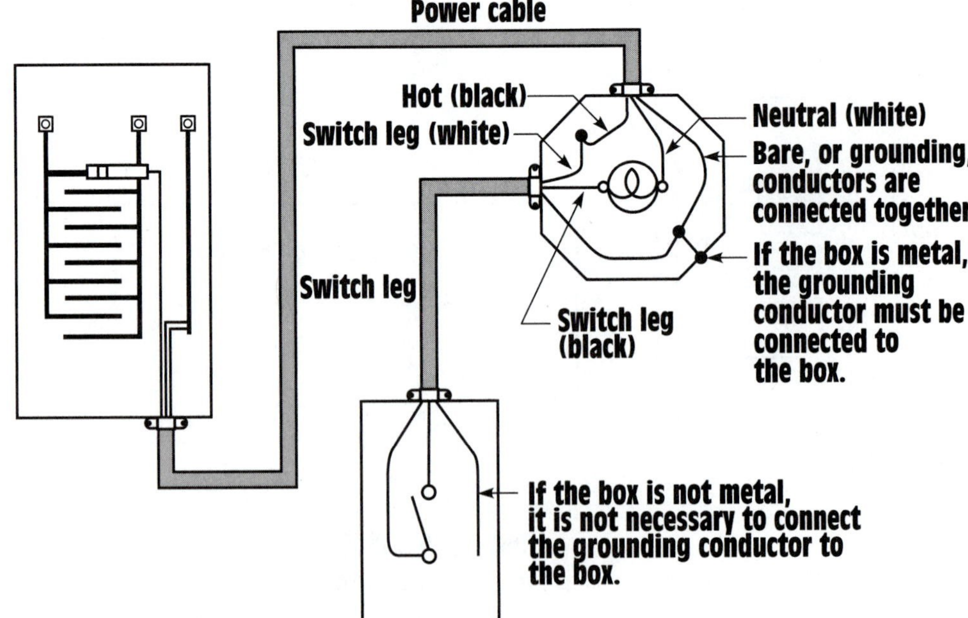

Figure 15-11 All connections are made inside a box.

SINGLE-POLE SWITCH CONNECTIONS

There are two basic methods of making a single-pole switch connection using cable. The first method is to supply power to the light and connect a switch leg between the light and the switch, as shown in Figure 15-11. When making this connection, several rules should be observed. One rule is that the neutral, or grounded, conductor cannot be broken. The neutral is, therefore, connected directly to the light. The second rule is that when making connection to a load, such as a light, the conductors must be identifiable. This means that it must be possible to identify which conductor is the neutral and which is the ungrounded in case it becomes necessary to work on the load at a later time. Therefore, when connection is made to the light, one wire must be white (neutral) and the other black or red (hot). To accomplish this, the white conductor of the switch leg is connected to the black conductor of the power cable. This is acceptable because the only time the *National Electrical Code®* permits a white wire to be connected to the ungrounded, or hot, conductor is in a switch leg. When connection is made to the switch, the black conductor of the switch leg is used to carry power back to the light. Notice that the light now has a white wire connected to one side and a black wire connected to the other.

The grounding or bare conductors must be electrically continuous throughout the entire building. When grounding conductors enter a box, they are all connected. If the box is metal, a grounding clip or some other method must be used to connect the grounding conductors to the box. If the box is made of a nonconductive material such as fiber or plastic, the grounding conductors are still tied together, but they do not have to be attached to the box. If the switch has a green grounding screw, it should be connected to the bare conductor.

The second method of making a single-pole switch connection is to supply power to the switch and then run a switch leg from the switch to the light, *Figure 15-12.* When this method is used, the white, or neutral, conductors are connected in the switch box to ensure they are electrically continuous. The black conductors are connected to the switch terminals. This permits the switch to break the connection of the hot conductor. At the light box, the black and white wires are connected to the light and the bare wire is connected to the box, if it is metal. If the box is not metal, it should be folded out of the way for future use.

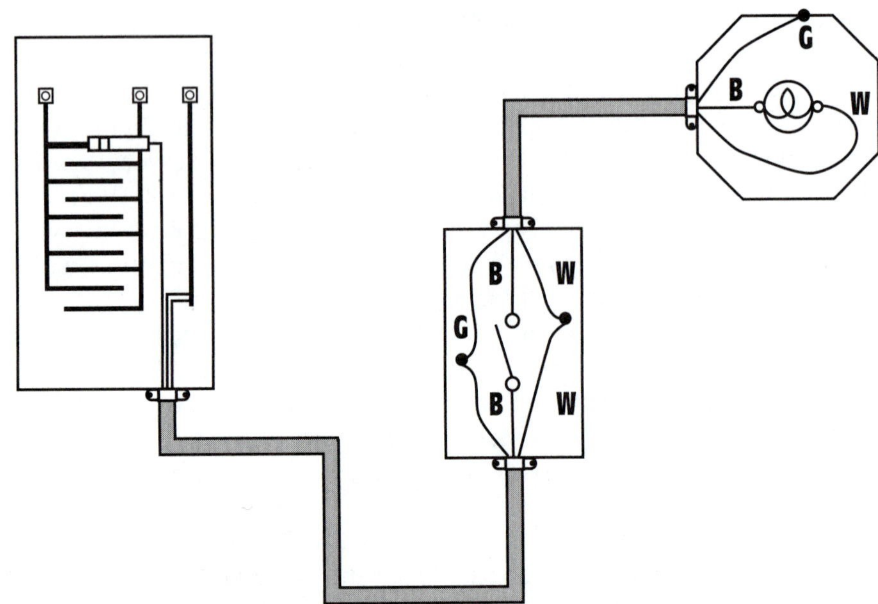

Figure 15-12 Power is brought to the switch.

THREE-WAY SWITCH CONNECTIONS

Three-way switches are used when it is desirable to control a light from two locations. The three-way switch is actually a single-pole double-throw (SPDT) switch, *Figure 15-13*. The switch contains two stationary contacts and one movable contact. The movable contact is referred to as the *pole* of the switch, and it is generally called the switch *common* because it can make connection between either of the two stationary contacts. The term *double-throw* means that it will make connection with a stationary contact when moved or thrown in either of two directions. The switch shown in Figure 15-13 makes connection between the movable, or common, contact and the bottom stationary contact. If the switch lever is flipped, or thrown, it will make connection between the common terminal and the top stationary contact, *Figure 15-14*.

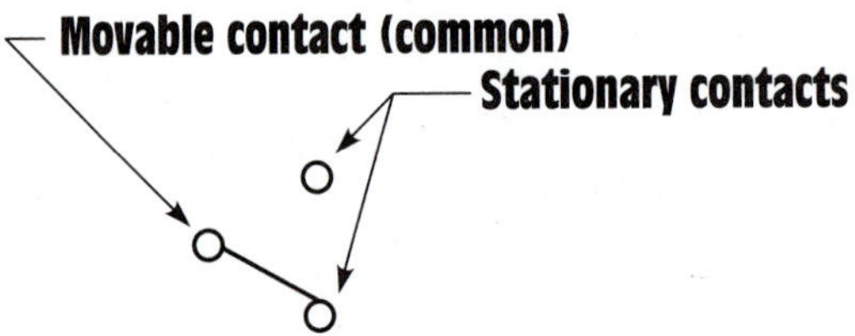

Figure 15-13 Schematic diagram of a three-way switch.

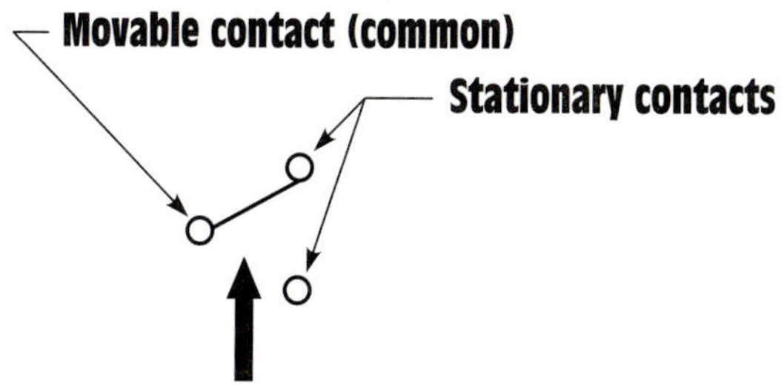

Figure 15-14 When the movable contact is changed, connection is made between the common terminal and the upper stationary contact.

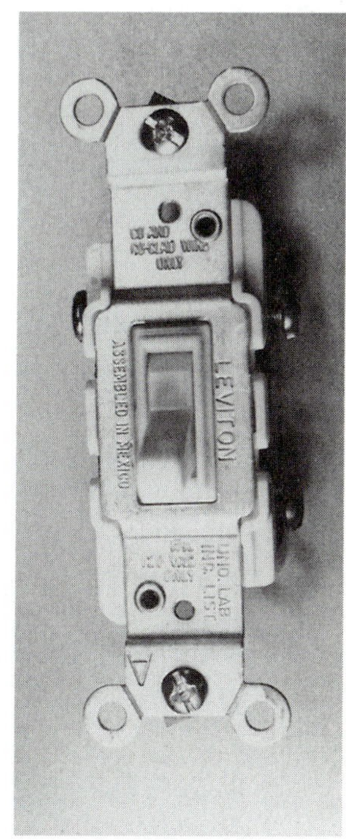

Figure 15-15 Three-way switch.

As shown in the schematic diagrams in Figure 15-13 and Figure 15-14, the three-way switch contains three terminals. One way of identifying a three-way switch is that it has three terminal connection screws, *Figure 15-15.* The common screw must be identified because it is necessary to know which is the common terminal when making a three-way switch connection. Some manufacturers identify the common terminal by making one screw a different color; other manufacturers mark the common terminal on the back side of the switch; and some manufacturers do both, *Figure 15-16.*

MAKING A THREE-WAY SWITCH CONNECTION

When making the connections for a three-way switch circuit, four rules can be used to simplify the procedure. Each of these rules will be presented with a schematic illustration.

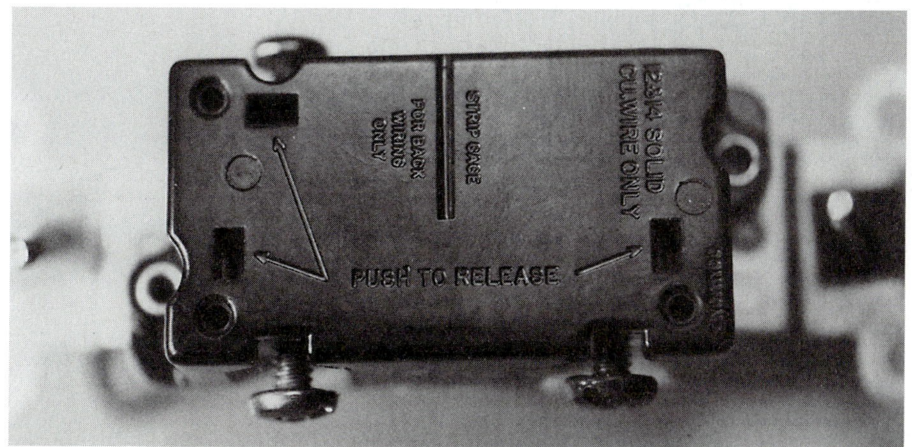

Figure 15-16 Identifying the common terminal.

The first rule is to connect the neutral to the light, *Figure 15-17*. The neutral conductor should never be broken and is, therefore, connected directly to the lighting fixture.

The second rule is to connect the hot conductor to the common terminal of one three-way switch, *Figure 15-18*. It makes no difference to which

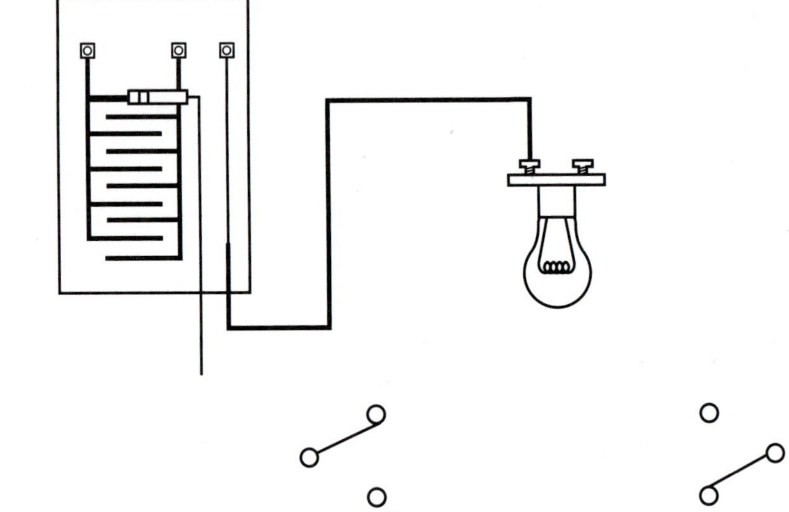

Figure 15-17 Rule 1: Connect the neutral to the light.

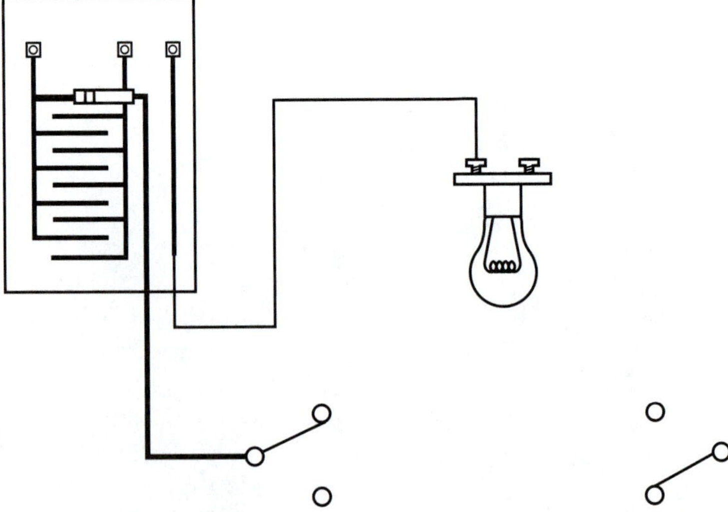

Figure 15-18 Rule 2: Connect the hot conductor to the common terminal of one three-way switch.

of the two switches the hot conductor is connected, but it must be connected to the common terminal of the switch.

The third rule is to connect the other side of the light to the common terminal of the second three-way switch, *Figure 15-19.* Notice that one side of the light has already been connected to the neutral conductor. The remaining side of the light is connected to the common terminal of the three-way switch that was not connected to the hot conductor.

The fourth rule is to connect the travelers, *Figure 15-20.* The traveler conductors run between the stationary contact terminals of the two three-way switches.

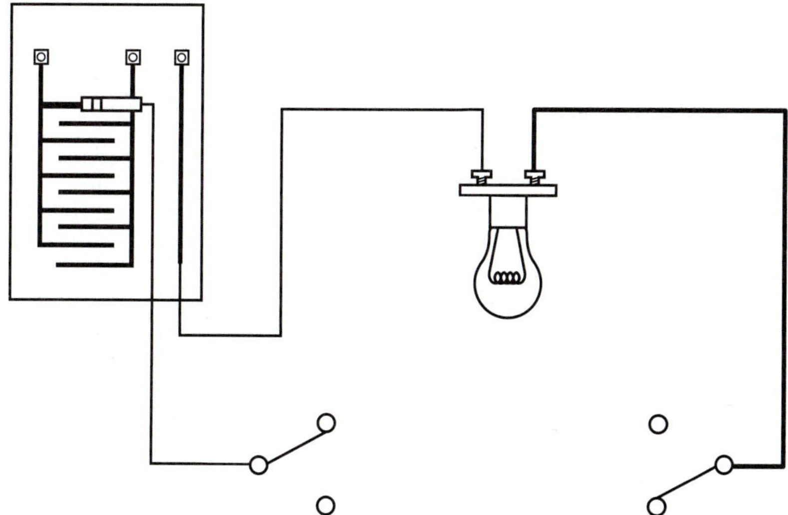

Figure 15-19 Rule 3: Connect the other side of the light to the common terminal of the second three-way switch.

TRACING A THREE-WAY SWITCH CIRCUIT

To understand how a three-way switch circuit operates, trace the current path starting at the circuit breaker. It will be assumed that current will flow from the circuit breaker and return to the neutral bus. In *Figure 15-21,* it can be seen that current cannot flow through this circuit because of the open switch. In this condition, the light is turned off.

If the movable contact of either switch should be changed, the circuit will toggle and a complete path will exist. In this example, it is assumed that the left switch position is changed, *Figure 15-22.* There is now a complete path through the circuit, and the light is turned on. The circuit will toggle between on and off regardless of which switch is changed. The light can now be controlled from either switch position.

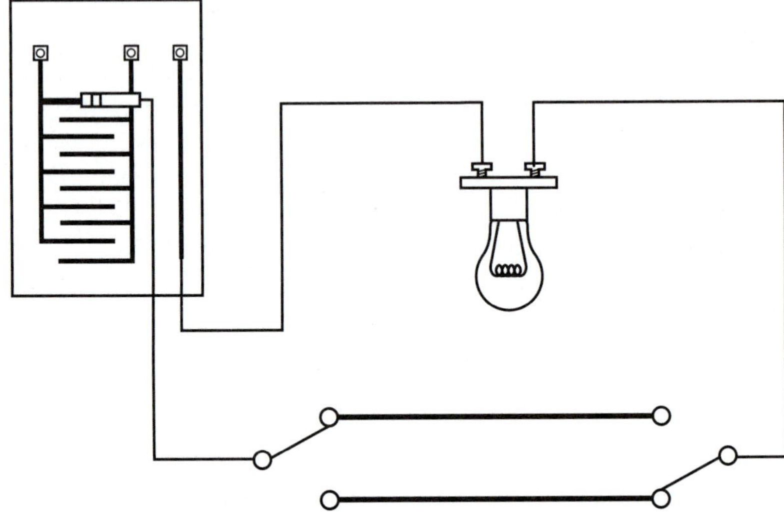

Figure 15-20 Rule 4: Connect the travelers.

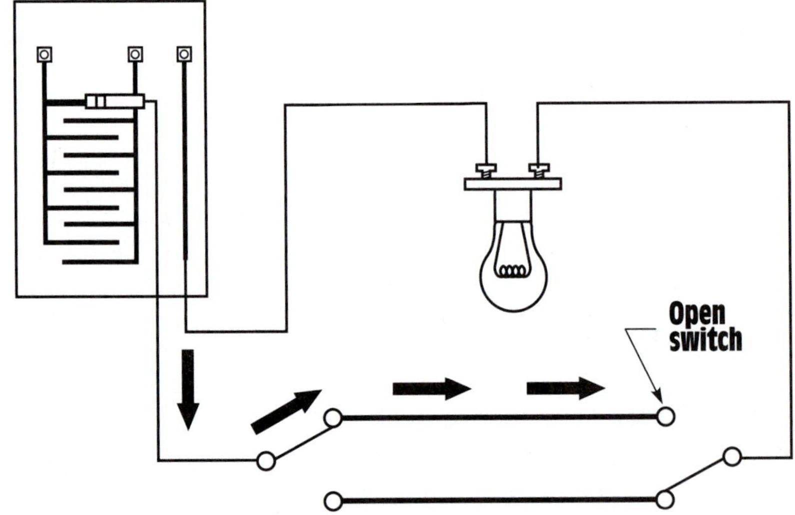

Figure 15-21 The circuit is open, and no current flows to the light.

MAKING THREE-WAY SWITCH CONNECTIONS WITH CABLE

The examples of switch connection and operation were done with electrical schematic diagrams to aid in understanding how the circuit is connected and how it works. In actual practice, however, all connections are made with two- and three-conductor cable. A good rule to remember when installing the wiring for a three-way switch connection is that a three-

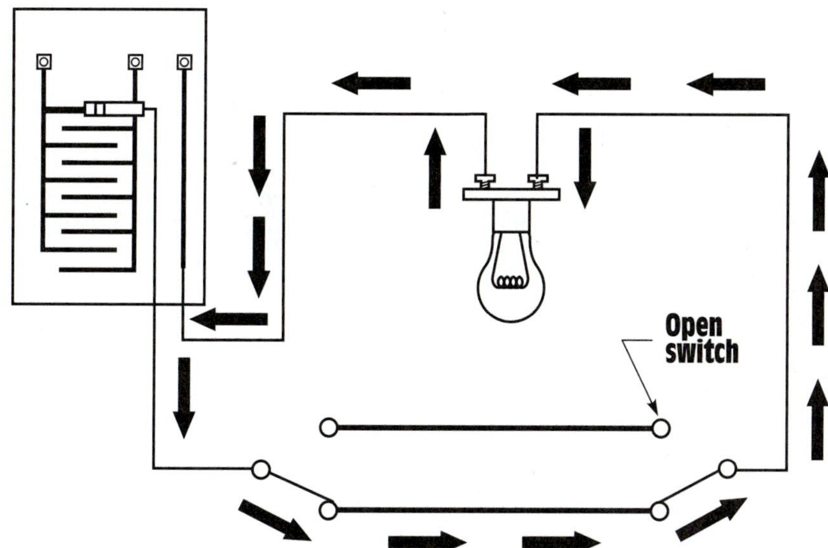

Figure 15-22 The circuit is closed, and current flows through the light.

conductor cable must be run between the switches. In the following examples of connecting three-way switches using cable, the grounding conductors have been omitted due to space limitations and to aid in clarity. When making these connections in the field, the grounding conductors are connected in the same manner as shown for single-pole switch connections. All grounding conductors must be connected to provide continuity throughout the entire system. If a metal box is used, a grounding conductor must be connected to the box.

When making three-way switch connections using two- and three-conductor cables, it is permissible to connect any of the conductors to the hot conductor, because it is a switch leg. A good rule to observe, however, is to always connect a black conductor to the common of a three-way switch. The reason for this is that a black conductor can always be used to connect to the common terminal, but it is not always possible to use a white or red conductor. If this is done there is no doubt as to which wire should be connected to the common terminal of the switch when the switches are physically connected in the circuit.

In the first example, a two-conductor power cable is supplied from the panel to the lighting fixture box and two three-conductor cables are run from the lighting outlet box to each switch box, *Figure 15-23*. The circuit will be connected using the four rules previously stated. The first step is to connect the neutral conductor to the light, *Figure 15-24*. The neutral conductor is the white wire of the power cable.

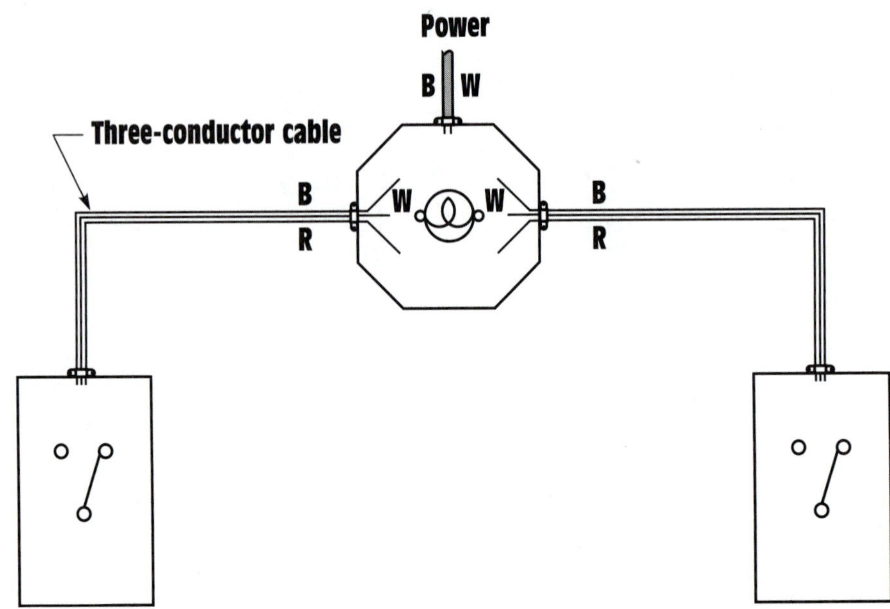

Figure 15-23 Example 1 for connecting three-way switches.

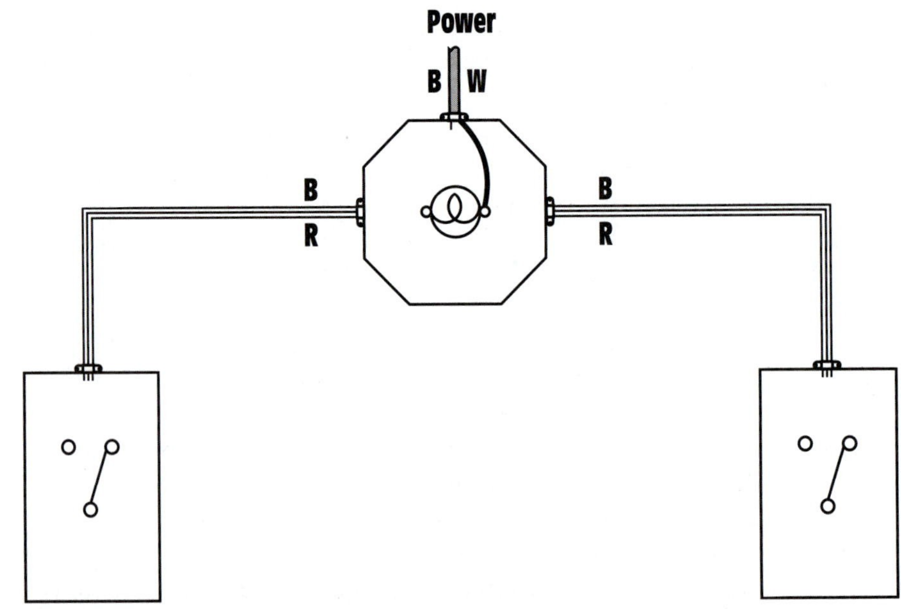

Figure 15-24 The neutral conductor is connected directly to the light.

The next step is to connect the ungrounded, or hot, conductor to the common terminal of one three-way switch. Since black conductors will be used to make connection to the common terminals of the three-way switches, the black wire of the power cable is connected to the black conductor of one of the three-wire cables, *Figure 15-25*. The other end of the black wire is then connected to the common terminal of a three-way switch.

The third step is to connect the other side of the light to the common terminal of the second three-way switch. The black wire of the other three-conductor cable will be used to make this connection, *Figure 15-26*.

The last step is to connect the traveler wires together. In this connection, the red and white wires of the three-conductor cables will be used as the travelers. The two white wires are connected and the two red wires are connected inside the lighting outlet box. These conductors are then connected to the traveler terminals of each switch, *Figure 15-27*.

In the second example of making a three-way switch connection using cable, the power cable enters the lighting fixture outlet box. A two-conductor cable is run between the lighting outlet box and one of the switch boxes, and a three-conductor cable is run between the two switch boxes, *Figure 15-28*. As in the previous example, the four rules will be followed to make this connection. The first step is to connect the grounded, or neutral, conductor to the light, *Figure 15-29*.

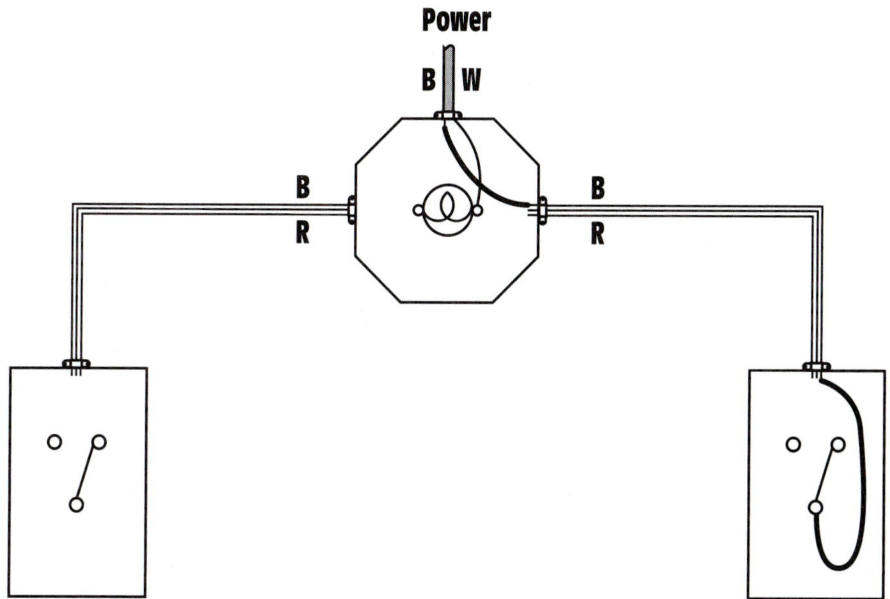

Figure 15-25 The ungrounded, or hot, conductor is connected to the common of one three-way switch.

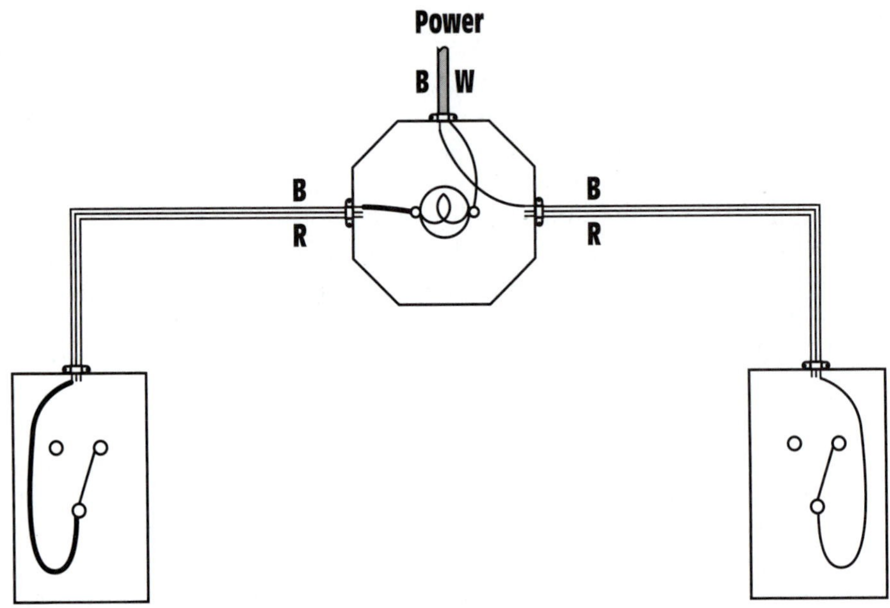

Figure 15-26 Connect the other side of the light to the common of the second three-way switch.

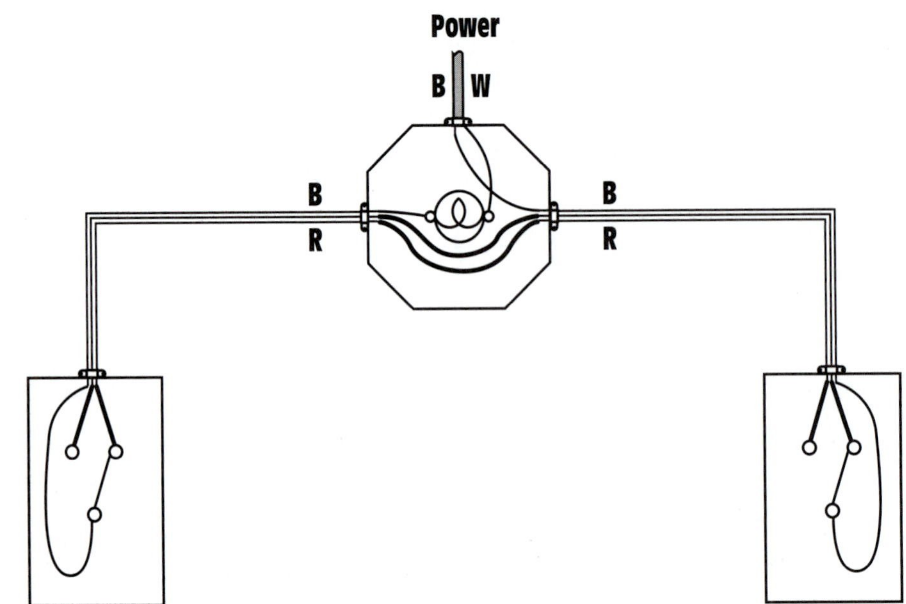

Figure 15-27 Connect the travelers.

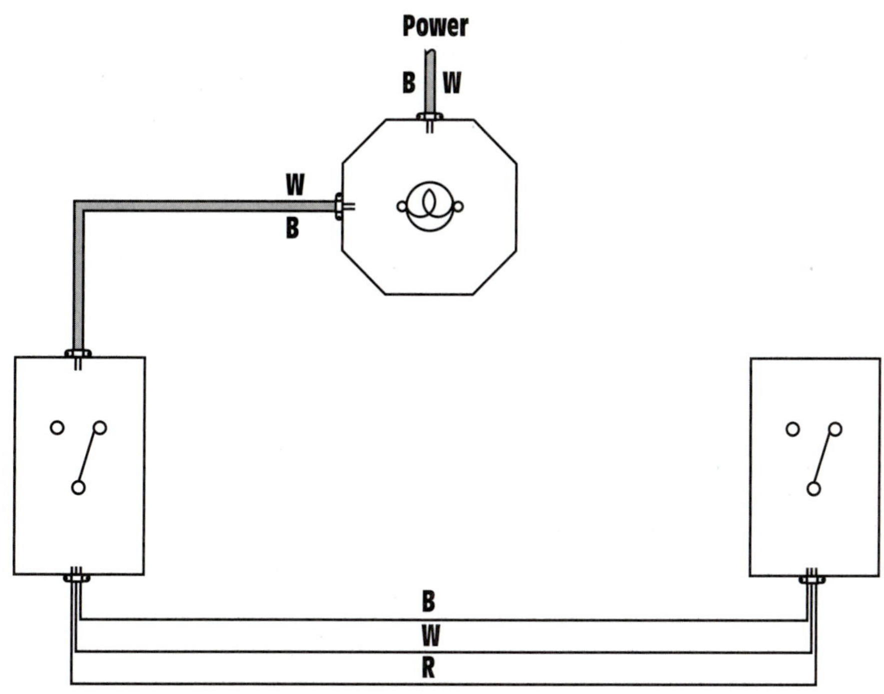

Figure 15-28 Example 2 for connecting three-way switches.

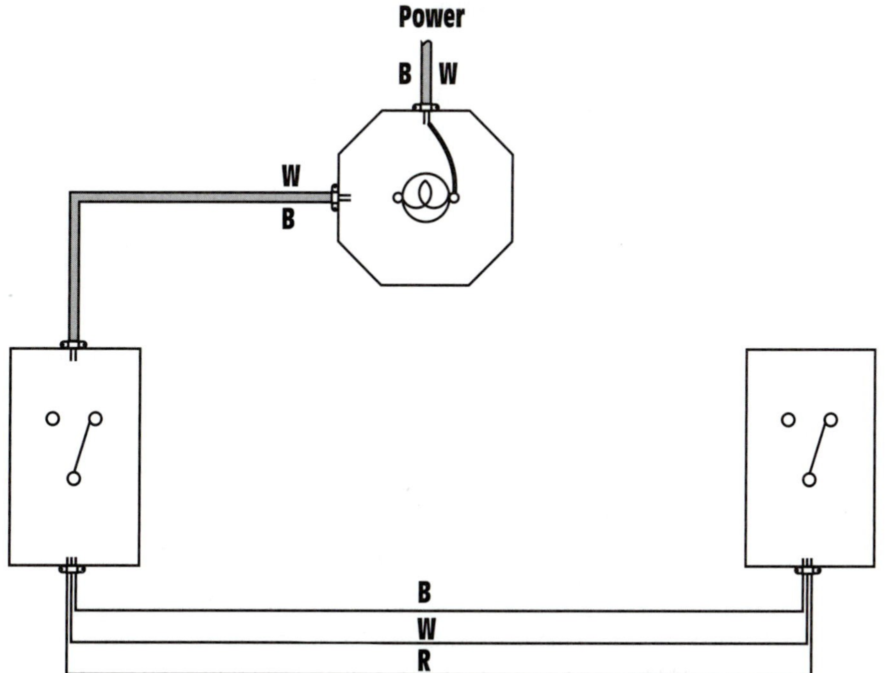

Figure 15-29 The neutral conductor is connected to the light.

The second step is to connect the hot conductor to the common terminal of one three-way switch, *Figure 15-30*. Since the conductors supplying power to the load must be identifiable, hot will be carried to the switch via the white wire of the two-conductor switch leg. For reasons stated previously, it is desirable that a black wire be used to make connection to the common terminal of a three-way switch. Therefore, the white wire of the two-conductor switch leg will be connected to the black wire of the three-conductor cable that runs between the two switches. The black wire of the three-conductor cable is then connected to the common terminal of the three-way switch located on the right side.

The third step is to connect the other side of the light to the common terminal of the second three-way switch. This connection is shown in *Figure 15-31*. The black wire of the two-conductor switch leg is used to make this connection. Notice that the light is connected to black and white conductors.

The last step is to connect the travelers. The red and white wires of the three-conductor cable connect the traveler screws of the two three-way switches, *Figure 15-32*.

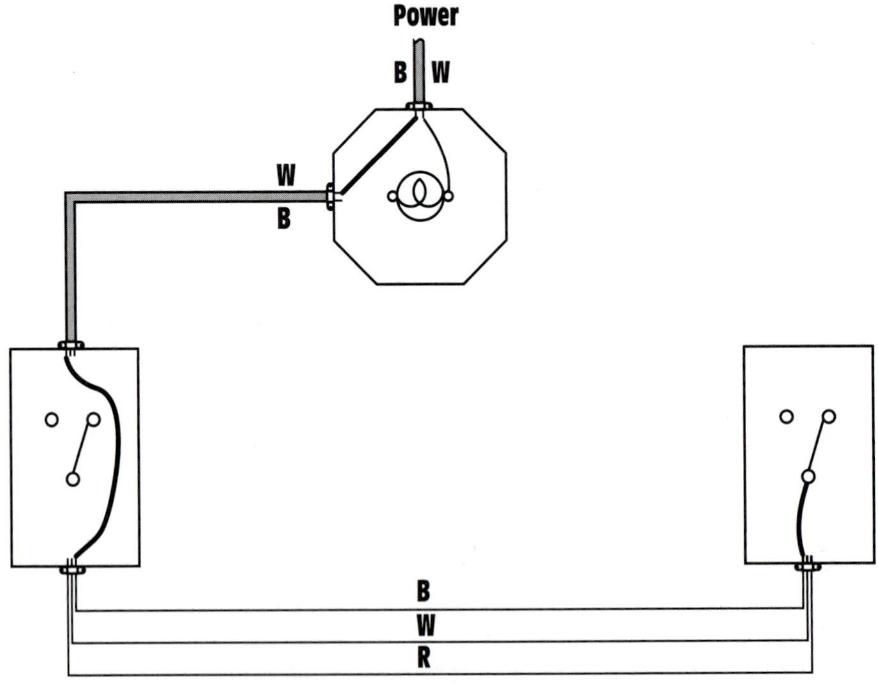

Figure 15-30 Connect the ungrounded conductor to the common terminal of one three-way switch.

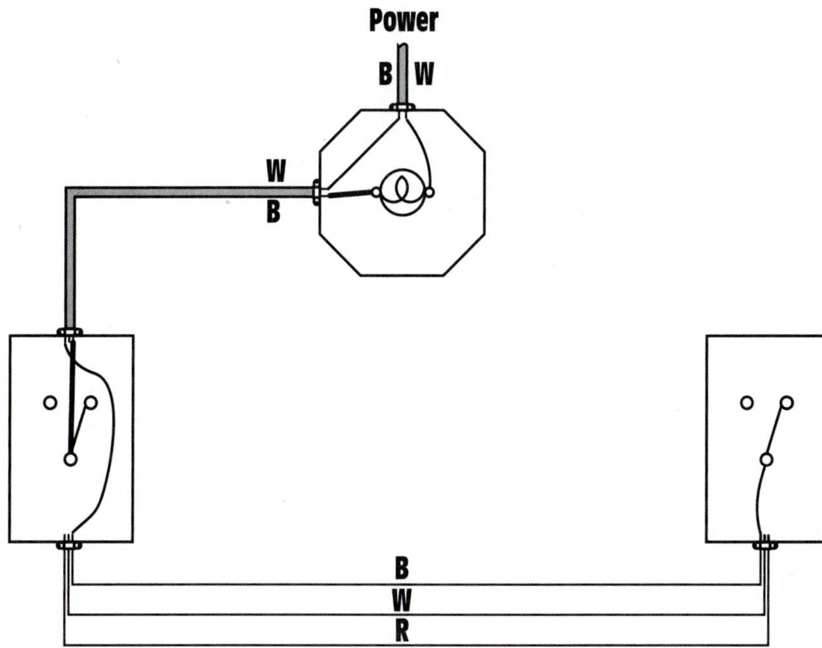

Figure 15-31 The other side of the light is connected to the common terminal of the second three-way switch.

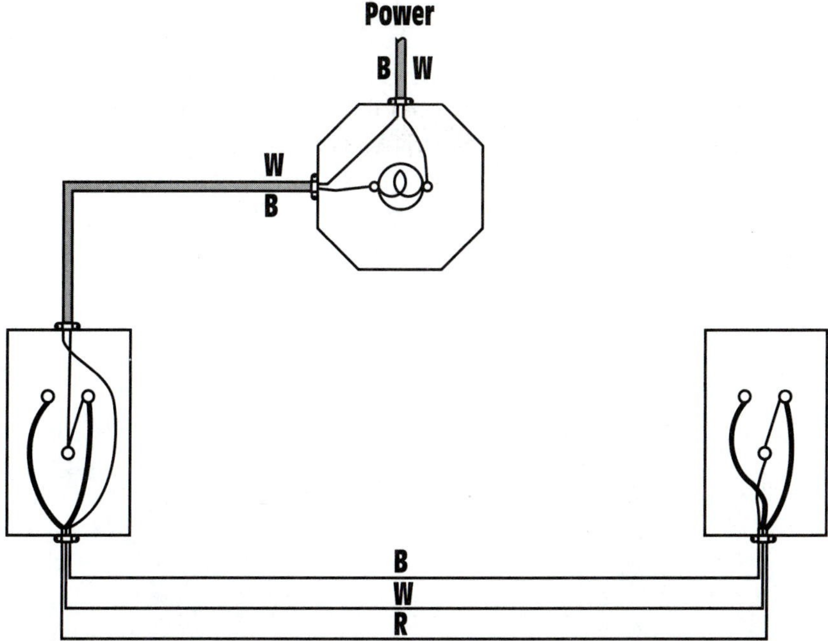

Figure 15-32 Connecting the traveler conductors between the two three-way switches.

FOUR-WAY SWITCHES

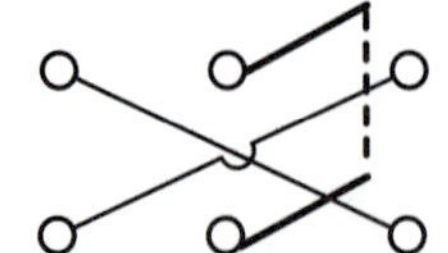

Figure 15-33 A four-way switch is a double-pole double-throw switch with an internal criss-cross connection.

Four-way switches are used to control a light from more than two locations. Four-way switches are actually double-pole double-throw (DPDT) switches with an internal criss-cross connection, *Figure 15-33*. Since the stationary contacts are connected inside the switch, only four screw terminals are required, *Figure 15-34*. When a four-way switch is thrown in one position, the terminal screws will be connected as shown in *Figure 15-35*. In this position, the switch terminals are connected straight across. If the switch is thrown in the other direction, connection will be made in a criss-cross fashion, as shown in *Figure 15-36*.

When making a four-way switch connection, the same four rules that apply to making three-way switch connections can be used. The only difference is that the four-way switch is connected in the traveler conductors, *Figure 15-37*. If a current path is traced in this circuit, it will be seen that an open switch prevents the flow of current to the light, *Figure 15-38*. The light is, therefore, turned off at this time.

Now assume that the position of the four-way switch is changed, as shown in *Figure 15-39*. A current path now exists through the light, and it is turned on. If the position of any one of the three switches should change, the light will be turned off. Changing the position of any of the three switches will alternately toggle the light from on to off and from off to on. Anytime a light is to be controlled from more than one location, two three-way switches must be used. Any number of switches above two, however, must be four-way. Virtually any number of four-way switches can be inserted between the travelers of the two three-way switches, *Figure 15-40*.

Like the other types of switch circuits previously discussed, four-way switch connections are actually made with two- and three-conductor cable. In this example, the power wire is brought from the panel and enters one of the three-way switch boxes. A three-conductor cable is connected between the first three-way switch and the four-way switch, and a second

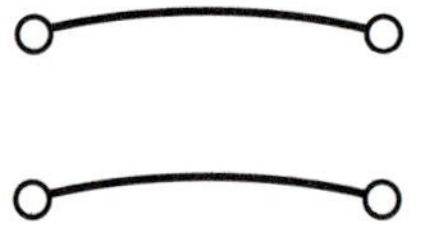

Figure 15-35 When thrown in one position, the four-way switch makes a connection straight across the terminals.

Figure 15-36 When thrown in the other position, the four-way switch makes a criss-cross connection.

Figure 15-34 Four-way switch.

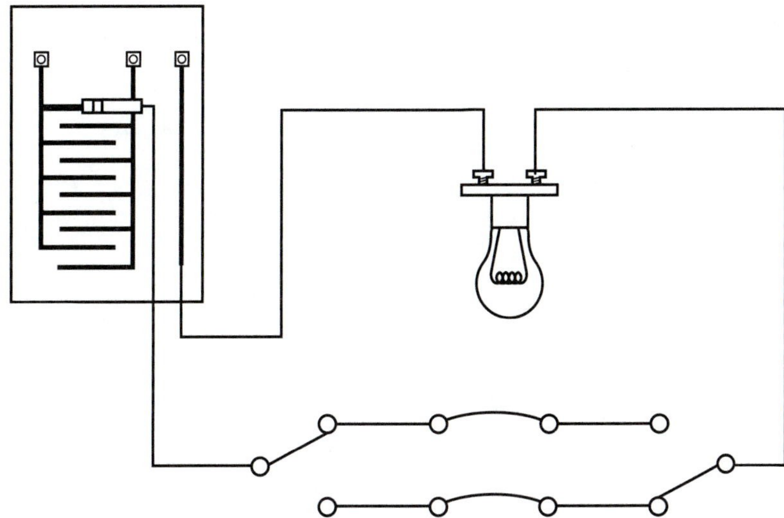

Figure 15-37 A four-way switch connects into the traveler conductors between the two three-way switches.

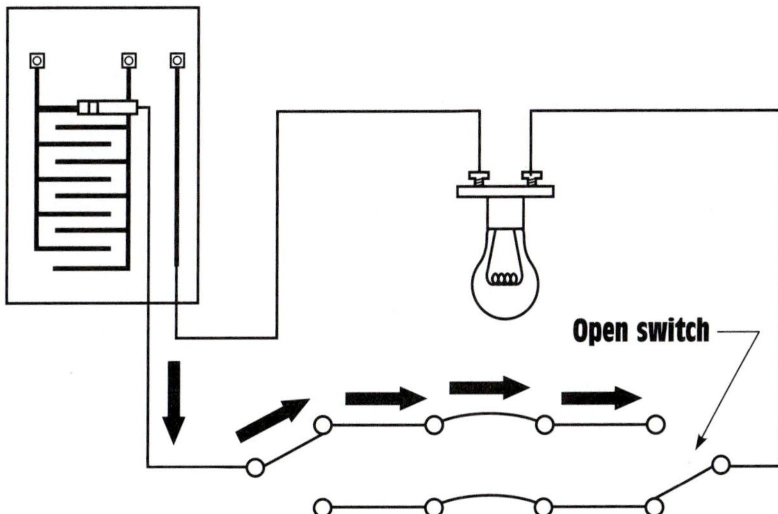

Figure 15-38 The current path is broken by an open switch.

three-conductor cable is connected between the four-way switch and the other three-way switch. A two-conductor cable is run from the second three-way switch to the lighting outlet box, *Figure 15-41*. The four rules for connecting three-way switches will be followed to make this connection.

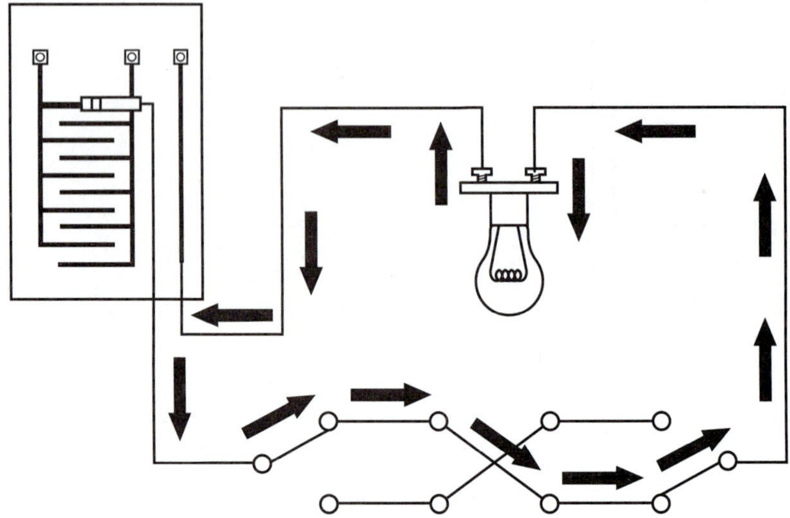

Figure 15-39 Changing the position of the four-way switch provides a current path for the circuit.

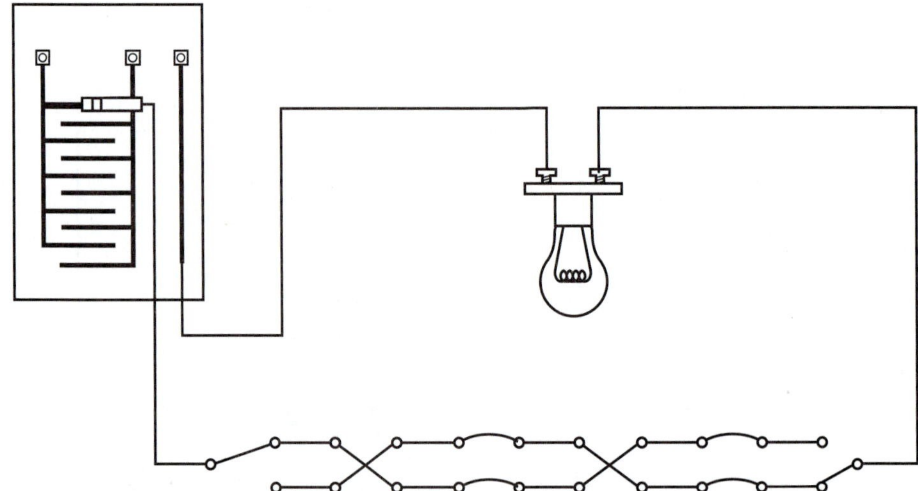

Figure 15-40 Any number of four-way switches can be connected between the travelers of two three-way switches.

The first step is to connect the neutral conductor to the light, *Figure 15-42*. The neutral conductor enters the three-way switch box on the right side. Since the *National Electrical Code*® requires that neutral, or grounded, conductors be white or gray, the white wire will be used to carry the neutral conductor through the circuit to the light.

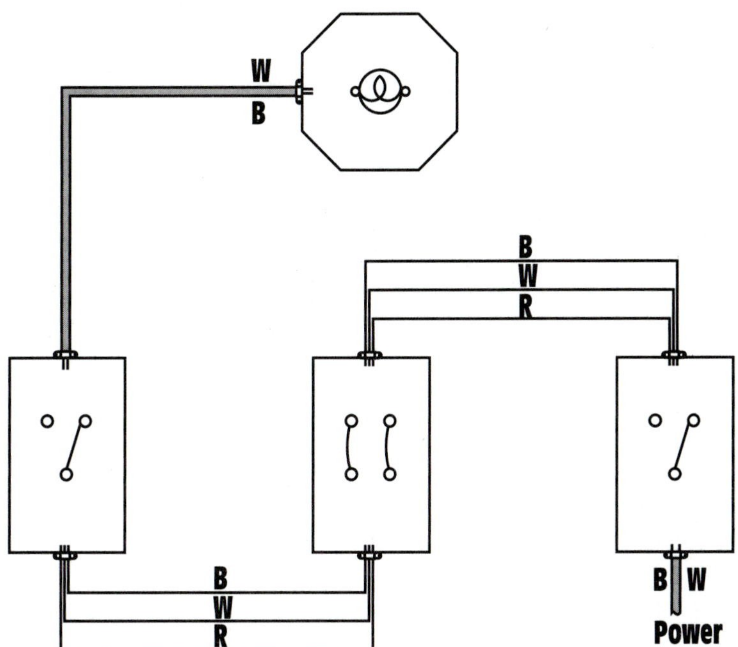

Figure 15-41 A four-way switch connection using two- and three-conductor cable.

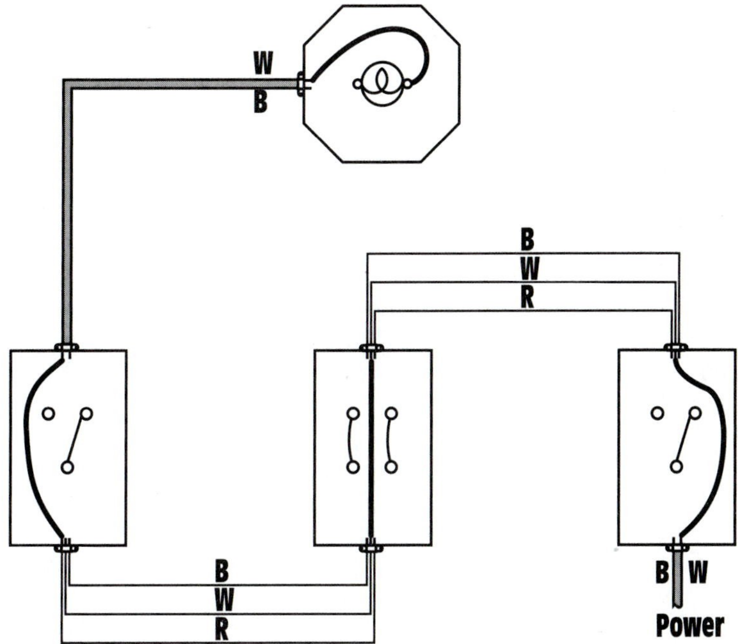

Figure 15-42 The neutral conductor is carried through the entire circuit and connected to the light.

The second step is to connect the ungrounded, or hot, conductor to the common terminal of one three-way switch. The black wire of the power cable is connected to the common terminal of the three-way switch located on the right side, *Figure 15-43*.

The third step is to connect the other side of the light to the common terminal of the second three-way switch. The black wire of the two-conductor runs between the three-way switch located on the left side and the lighting outlet box will be used to make this connection, *Figure 15-44*.

The final step is to connect the travelers. In this circuit, the four-way switch will be connected between the traveler conductors, *Figure 15-45*. The red and black wires of the three conductor cables will be used for this connection. When connecting the four-way switch into the circuit, it is important to connect the traveler conductors to the proper screw terminals. The traveler conductor from one of the three conductor cables is generally connected to the two top screws, and the traveler conductor from the other three conductor cables is connected to the two bottom screws. Some switches have the connection labeled on the switch, as shown in *Figure 15-46,* but some do not.

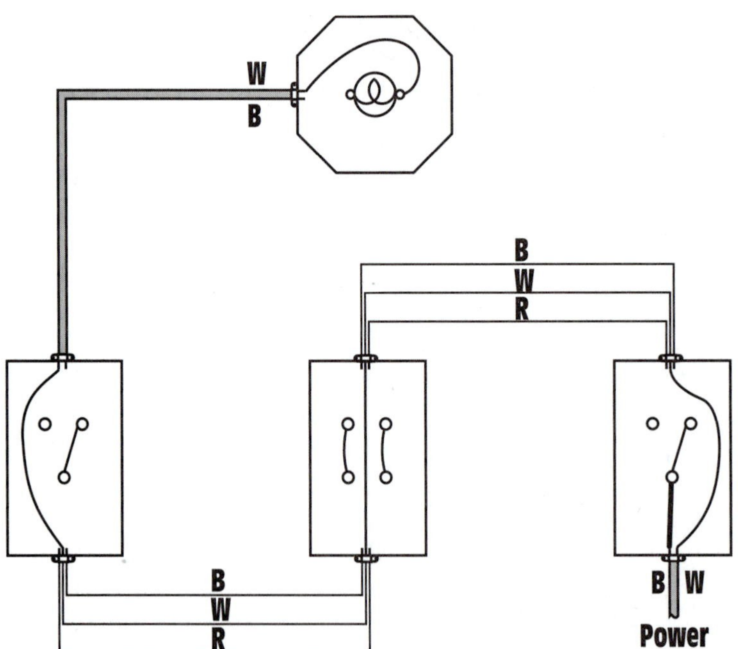

Figure 15-43 The ungrounded, or hot, conductor is connected to the common terminal of one three-way switch.

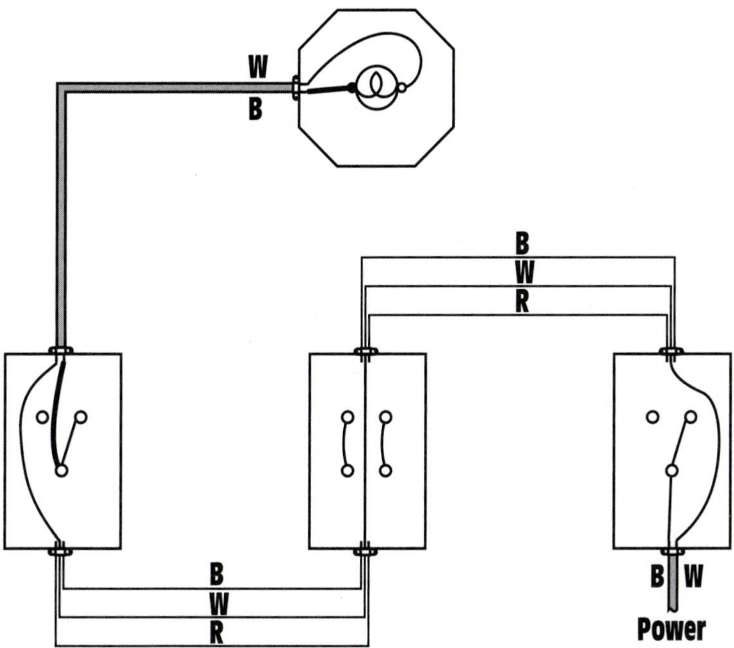

Figure 15-44 The other side of the light is connected to the common terminal of the second three-way switch.

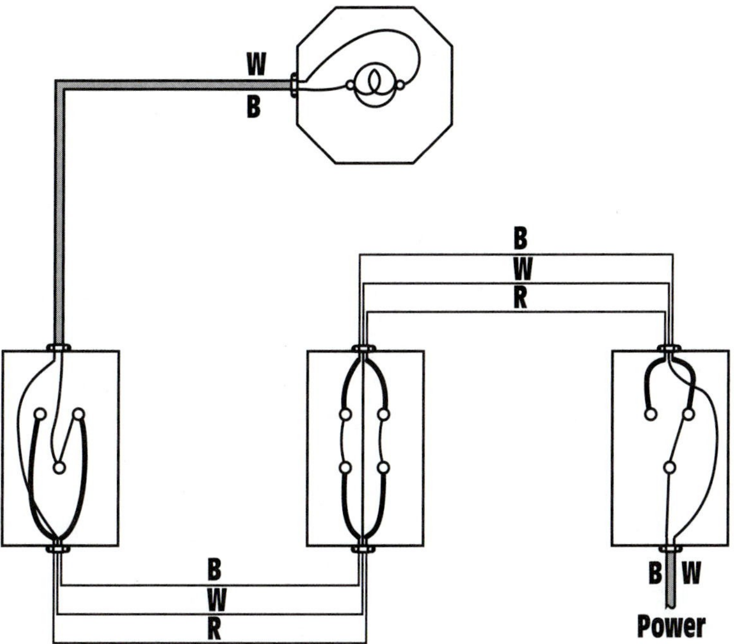

Figure 15-45 The four-way switch is connected in the travelers between the two three-way switches.

Figure 15-46 Some switches have the screws labeled.

GROUND FAULT CIRCUIT INTERRUPTERS (GFCIS)

There are two basic types of ground fault circuit interrupters (GFCIs). One type is designed to protect industrial equipment. This type is generally used on three-phase systems and is designed to disconnect the circuit if a ground should occur. It operates by passing all three-phase conductors through a toroid transformer, *Figure 15-47*. The circuit works on the principle that when current flows through a conductor, a magnetic field is developed around the conductor. As long as the current is the same in each of the three phases, the magnetic fields cancel each other. As a result there is no voltage induced in the toroid transformer. If one of the phase conductors becomes grounded, a large amount of current will flow in it, causing a more intense magnetic field in that conductor. The increased magnetic field induces a voltage into the transformer. The transformer then causes the control circuit to de-energize the power supplying the load. Ground fault

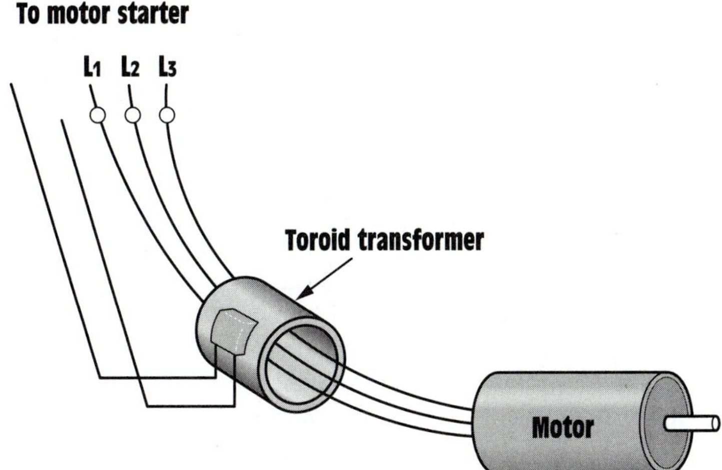

Figure 15-47 Industrial ground fault detectors protect equipment.

detectors of this type generally require from several to 1,000 or more amperes to open the circuit. These detectors are intended to protect equipment, not people.

PERSONNEL GFCI DETECTORS

The articles concerning ground fault protection in the *National Electrical Code®* deal mainly with personnel protection. A ground fault interrupter can generally be purchased as a circuit breaker or as a receptacle. When the circuit breaker type of protector is used, all devices connected to that circuit are ground fault protected, *Figure 15-48*. GFCI circuit breakers are

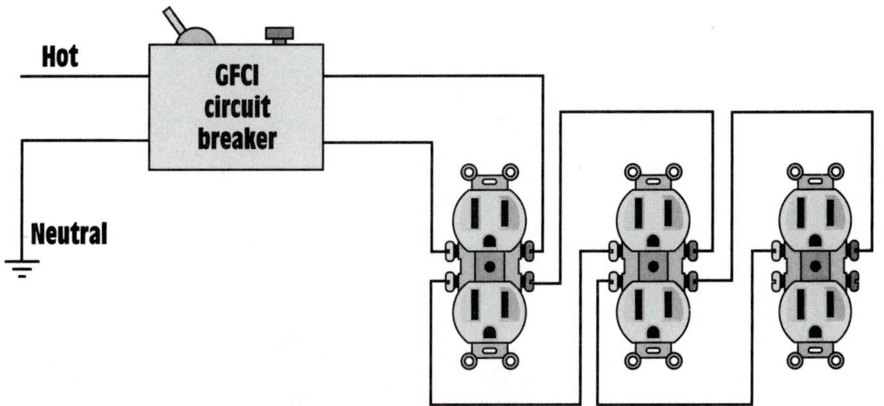

Figure 15-48 The GFCI circuit breaker protects all devices connected to it.

constructed differently than standard breakers. Standard single-pole circuit breakers contain only one connection screw for termination of the branch circuit ungrounded, or hot, conductor. Single-pole GFCI circuit breakers contain two connection screws, *Figure 15-49.* The hot, or ungrounded, branch circuit conductor connects to the brass-colored screw, and the neutral conductor connects to the silver screw. The white pig tail attached to the circuit breaker connects to the neutral bus bar in the service panel. The reason the hot and neutral conductors must connect to the circuit breaker is because both conductors must pass through a toroid transformer, *Figure 15-50.* The ground fault circuit interrupter operates by sensing the amount of current flow in each conductor. Assume that current flows from the circuit breaker to the load and back to neutral, *Figure 15-51.* In a closed circuit, the amount of current leaving the circuit breaker should be the same as the amount of current returning. Therefore, the current should be the same in both the hot and neutral conductors but will flow in opposite directions. Since the two currents flow in opposite directions, the magnetic fields produced by the current in each conductor cancel each other. This results in no induced voltage in the toroid transformer.

A ground fault is an unintentional path to ground. Ground faults can be caused by many different conditions, such as bad insulation on an appliance cord where it enters the case of the appliance or the winding of a mo-

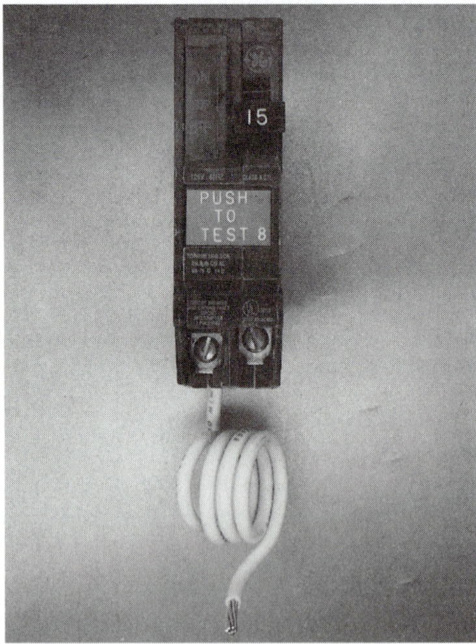

Figure 15-49 GFCI circuit breaker.

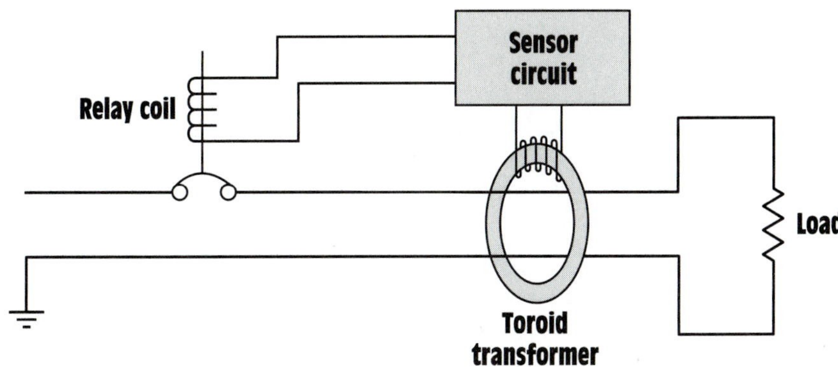

Figure 15-50 Both the hot and neutral conductors pass through the transformer.

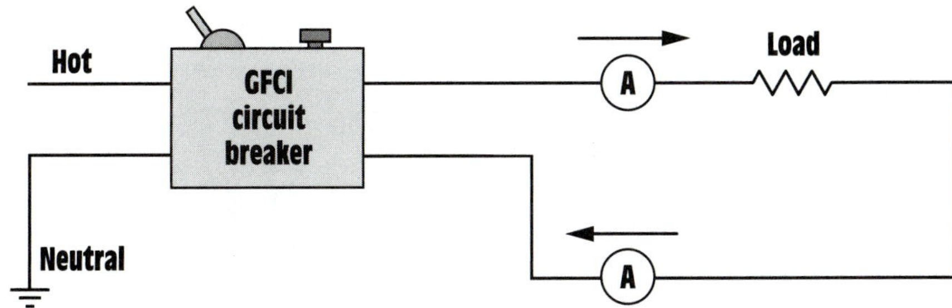

Figure 15-51 The current leaving the circuit breaker should be the same as the current returning.

tor coming in contact with the motor case. These conditions can be hazardous, as when a person touches the case of the appliance and a grounded water line. At that moment a current path exists through the person to ground, *Figure 15-52*. Since there is now a separate current path to ground (other than the neutral conductor), more current flows in the hot, or ungrounded, conductor than in the neutral conductor. This produces a stronger magnetic field in the hot conductor. The extra field strength of the hot conductor induces a voltage into the toroid transformer. The toroid transformer signals the sensor circuit that there is a ground fault. The sensor circuit activates a small relay coil and opens the circuit breaker. The sensor circuit for a ground fault circuit breaker is shown in *Figure 15-53*. Ground fault circuit interrupters for personnel protection will open the circuit with a ground fault of approximately 5 ma (0.005 amps). GFCI devices contain a test button that is used to check the device for proper operation. If the test button is pressed, the circuit breaker should trip open. The button should be tested periodically to ensure that the ground fault circuit operates properly.

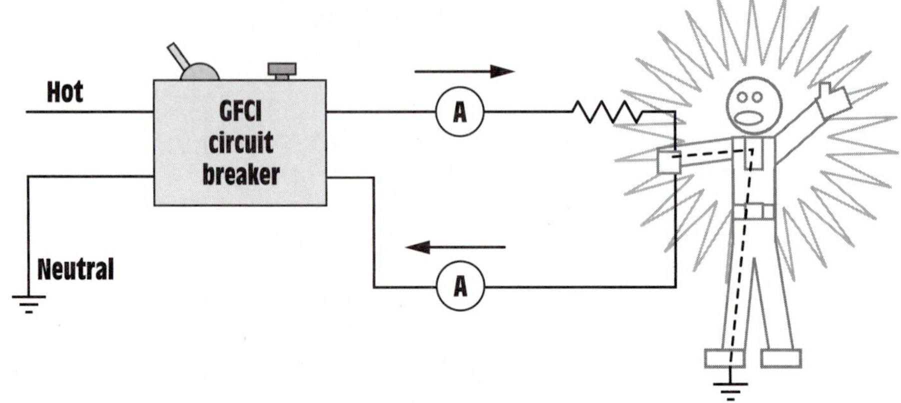

Figure 15-52 A ground fault is an unintended path to ground.

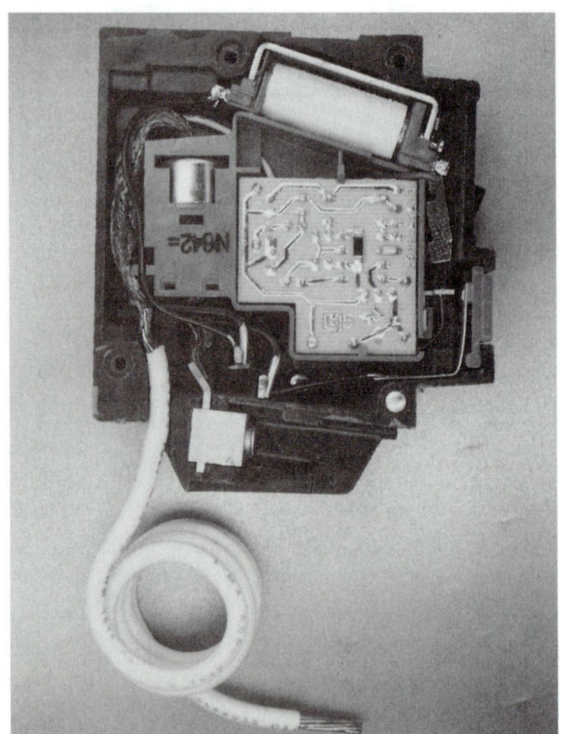

Figure 15-53 GFCI circuit breaker sensor circuit.

GROUND FAULT RECEPTACLES

Another device used for ground fault protection of personnel is the GFCI receptacle, *Figure 15-54*. The receptacle has an advantage over the circuit breaker in that ground fault protection starts at the place of usage in-

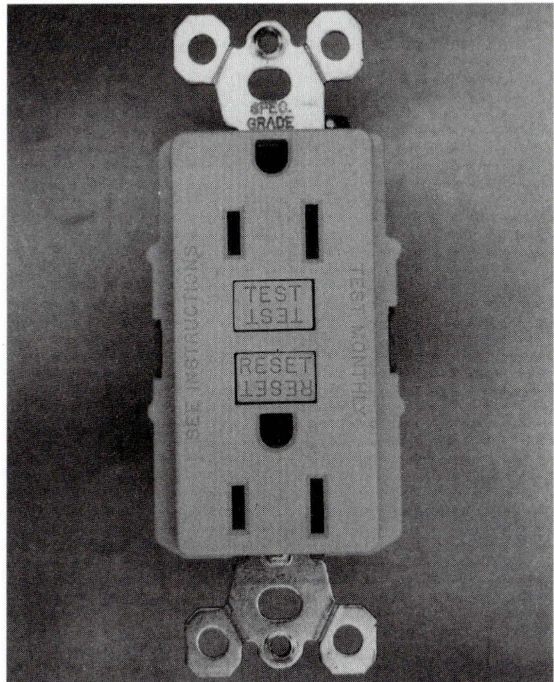

Figure 15-54 GFCI receptacle.

stead of in the service panel. Ground fault circuit breakers sometimes have a problem with nuisance tripping, caused mainly by long wire runs. Long wire runs can exhibit a fairly large amount of stray capacitance that causes leakage current to ground. The problem is generally more noticeable in damp weather than in dry and in areas close to the coast, because of salt content in the air.

Ground fault receptacles contain the same basic circuitry as the circuit breaker. They have a relay coil and a toroid transformer, *Figure 15-55,* and an electronic sensing circuit, *Figure 15-56.* When connecting GFCI receptacles, particular attention must be paid concerning how they are connected into the circuit. There are two important factors that must be considered.

1. The incoming power must be connected to the LINE side of the receptacle.
2. The hot and neutral conductors *must* be connected to the correct screws.

The screw terminals on a ground fault receptacle are labeled LINE and LOAD, *Figure 15-57.* The incoming power must be connected to the LINE terminals and the hot and neutral conductors must be connected to the proper terminals, otherwise the device will not work. Any devices connected

Figure 15-55 GFCI receptacles contain a relay coil and toroid transformer.

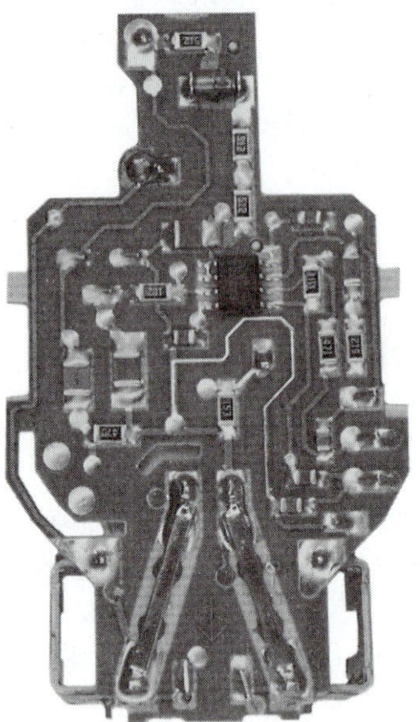

Figure 15-56 Electronic sensing circuit.

to the LOAD side of the receptacle are ground fault protected by the receptacle, *Figure 15-58*. Devices not connected to the load side are not protected.

The *National Electrical Code®* requires the use of ground fault interrupters on all outside receptacles and on receptacles in bathrooms, open garages, and boat docks; around swimming pools and fountains; within six

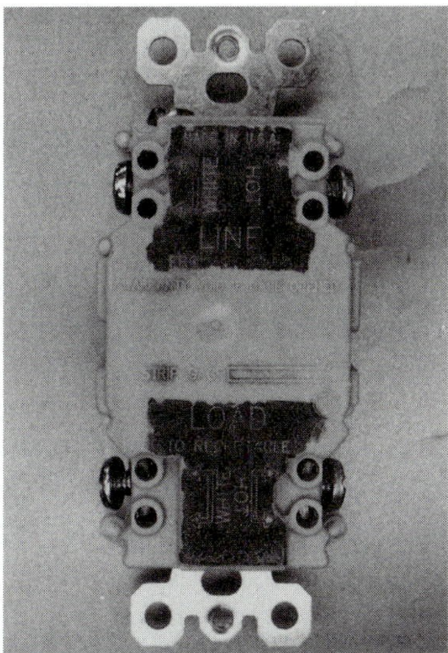

Figure 15-57 The screw terminals are labeled LINE and LOAD.

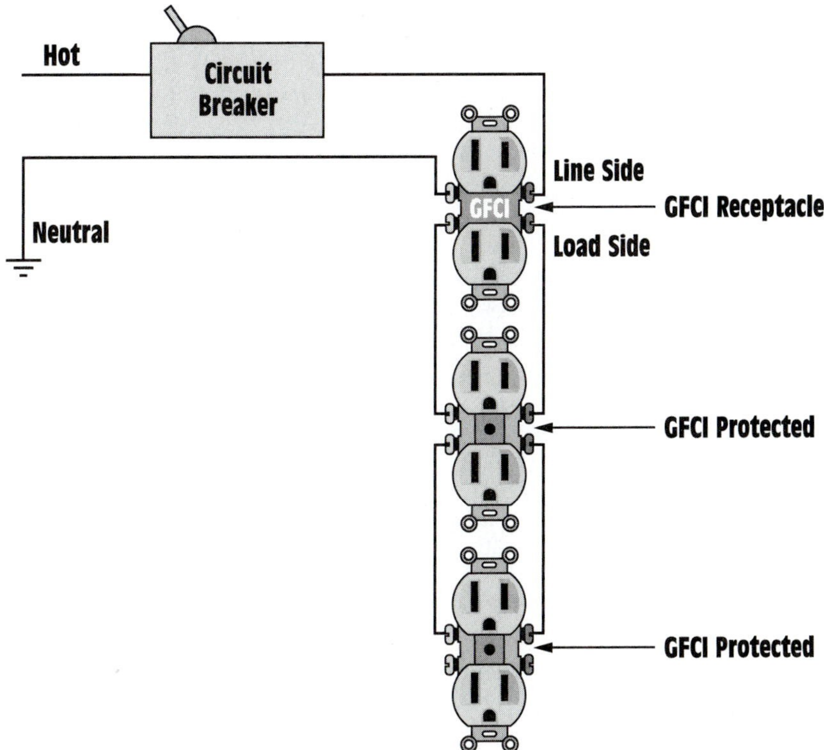

Figure 15-58 Devices connected to the LOAD side are GFCI protected.

feet of a kitchen sink; on snow-melting equipment; in crawl spaces and unfinished basements; and on recreational vehicles.

ARC-FAULT CIRCUIT INTERRUPTERS

Arc-fault circuit interrupters (AFCIs) are similar to ground fault circuit interrupters in that they are designed to protect people from a particular hazard. The group fault interrupter is designed to protect against electrocution, whereas the arc-fault interrupter is intended to protect against fire. Studies have shown that one-third of electrical-related fires are caused by an arc-fault condition. The *National Electrical Code*® requires that arc-fault circuit interrupters be used to protect circuits that supply 120-volt outlets in bedrooms. An arc-fault is a plasma flame that can develop temperatures in excess of 6,000° C or 10,832° F. Arc-faults occur when an intermittent gap between two conductors or a conductor and ground permits current to "jump" between the two conductive surfaces. There are two basic types of arc-faults, the parallel and the series.

PARALLEL ARC-FAULTS

Parallel arc-faults are caused by two conductors becoming shorted together, *Figure 15-59.* A prime example of this is when the insulation of a lamp cord or extension cord has become damaged and permits the two conductors to short together. The current in this type of fault is limited by the resistance of the conductors in the circuit, and is generally much higher

Figure 15-59 A parallel arc fault is caused by two conductors touching.

than the rated current of a typical thermo/magnetic circuit breaker. A continuous short will usually cause the circuit breaker to trip almost immediately because it will activate the magnetic part of the circuit breaker, but an intermittent short may take some time to heat the thermal part of the circuit breaker enough to cause it to trip open. Thermal/magnetic-type circuit breakers are generally effective in protecting against this type of arc-fault, but cords with small-size conductors, such as lamp and small extension cords, can add enough resistance to the circuit to permit the condition to exist long enough to produce sufficient heat to start a fire.

Parallel arc-faults are generally more hazardous than series arc-faults because they generate a greater amount of heat. Arc-faults of this type often cause hot metal to be ejected into combustible material. Parallel arc-faults, however, generally produce peak currents that are well above the normal current rating of a circuit breaker. This permits the electronic circuits in the arc-fault circuit interrupter to detect them very quickly and trip the breaker in a fraction of a second.

SERIES ARC-FAULTS

Series arc-faults are generally caused by loose connections. A loose screw on an outlet terminal or an improperly made wire nut connection are prime examples of this type of problem. They are called series arc-faults because the circuit contains some type of current-limiting resistance connected in series with the arc, *Figure 15-60*. Although the amount of electrical energy converted into heat is less than that of a parallel arc-fault, series arc-faults can be move dangerous. The fact that the current is limited by some type of load keeps the current below the thermal and magnetic trip rating of a common thermo/magnetic circuit breaker. Since the peak arc

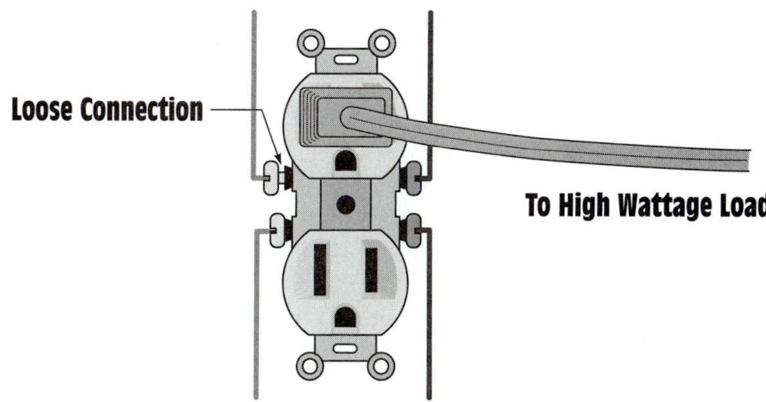

Figure 15-60 Loose connections are the cause of most series arc faults.

current is never greater than the normal steady current flow, series arcing is more difficult to detect than parallel arcing.

When the current of an arc remains below the normal range of a common thermo/magnetic circuit breaker it cannot provide protection. If a hair dryer, for example, normally has a current draw of 12 amperes, but the wall outlet has a loose screw at one terminal so that the circuit makes connection only one half of the time, the average circuit current is 6 amperes. This is well below the trip rating of a common circuit breaker. A 6-ampere arc, however, can produce a tremendous amount of heat in a small area.

ARC-FAULT DETECTION

There are conditions where arcing in an electrical circuit is normal, such as:

- Turning a light switch on or off.
- Switching of a motor relay.
- Plugging in an appliance that is already turned on.
- Changing a light bulb with the power turned on.
- The arcs caused by motors that contain a commutator and brushes.

The arc-fault circuit interrupter is designed to be able to distinguish between normally occurring arcs and an arc-fault. An arc caused by a toggle switch being used to turn a light on or off will produce a current spike of short duration as shown in *Figure 15-61*. An arc-fault, however, is an intermittent connection and will generally produce current spikes of various magnitudes and lengths of time, *Figure 15-62*. In order for an arc-fault circuit interrupter to determine the difference between a normally occurring arc and an arc-fault, a microprocessor and other related electronic components are employed to detect these differences. The AFCI contains current and temperature sensors as well as a microprocessor and nonvolatile (retains its information when power is switched off) memory. The current and temperature sensors permit the AFCI to operate as a normal circuit breaker in the event of a circuit overload or short circuit. The microprocessor continuously monitors the current and compares the waveform to information stored in the memory. The microprocessor monitors the current for the magnitude, duration, and length of time between pulses, not for a particular waveform. For this reason, there are some appliances that can produce waveforms similar to that of an arc-fault and may cause the AFCI to trip. Appliances containing motors that employ the use of brushes and a commutator, such as vacuum cleaners and hand drills, will produce a similar waveform.

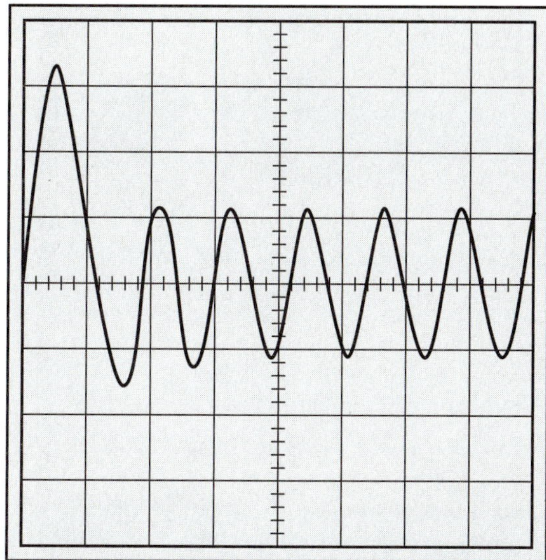

Figure 15-61 Typical current spike caused by turning a light switch on or off.

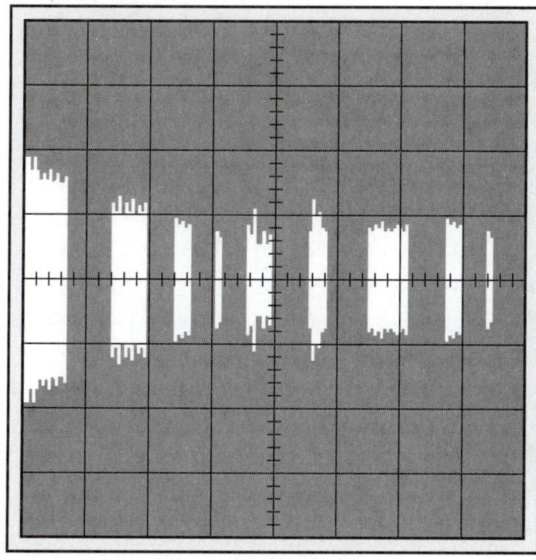

Figure 15-62 Typical waveform produced by an arc fault.

CONNECTING AN ARC-FAULT CIRCUIT INTERRUPTER

The AFCI is connected in the same manner as a ground fault circuit breaker. The AFCI contains a white pigtail, *Figure 15-63,* that is connected to the neutral bus bar in the panel box. Both the neutral and hot or

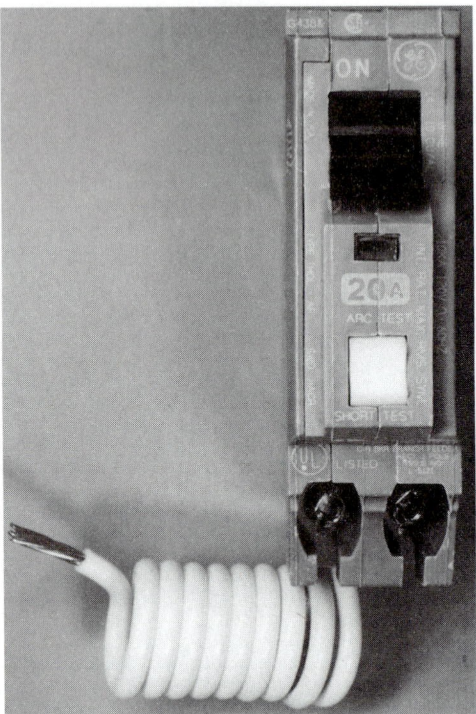

Figure 15-63 Arc fault circuit interrupter.

ungrounded conductors of the branch circuit are connected to the arc-fault circuit breaker. The circuit breaker contains a silver-colored and a brass-colored screw. The neutral or white wire of the branch circuit is inserted under the silver screw and the black wire is inserted under the brass screw. A rocker switch located on the front of the AFCI permits the breaker to be tested for both short and arc condition. In addition to the manual test switch, the microprocessor performs a self-test about once every 10 minutes.

LIGHT DIMMERS

Another common device used in general wiring is the light dimmer, *Figure 15-64*. Dimmers can be obtained that employ single-pole or three-way switches. Single-pole dimmers generally have one green and two black pig-tail leads. The two black leads are connected to the switch terminals, and the green is for the grounding. Three-way switch dimmers commonly use one black and two red pig tails plus a green for grounding. The two red pig tails are the traveler connections, and the black is the common. Most dimmers are for use on incandescent lamps only and should not be used to control fluorescent lights or ceiling fans. There are some

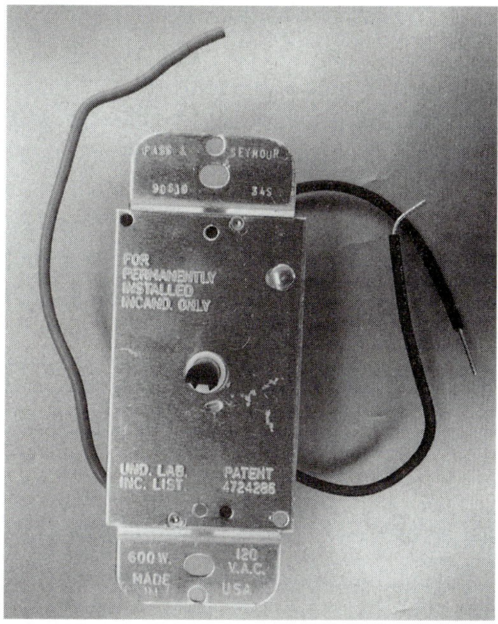

Figure 15-64 Light dimmer.

fluorescent lights that can be dimmed, but they require a special ballast and dimmer control.

Light dimmers operate by chopping the AC waveform applied to the lamp, *Figure 15-65*. If the waveform is on for 50% of the time and off for 50% of the time, the voltage applied to the load will be one-half the applied voltage. If the waveform is permitted to remain on for 75% of the time and off for 25% of the time, the voltage applied to the load will be 75% of the line value, or 90 volts if the circuit voltage is 120 volts. A solid-state component called a triac is used to control the AC waveform. The triac is a bi-directional device in that it will permit current to flow through

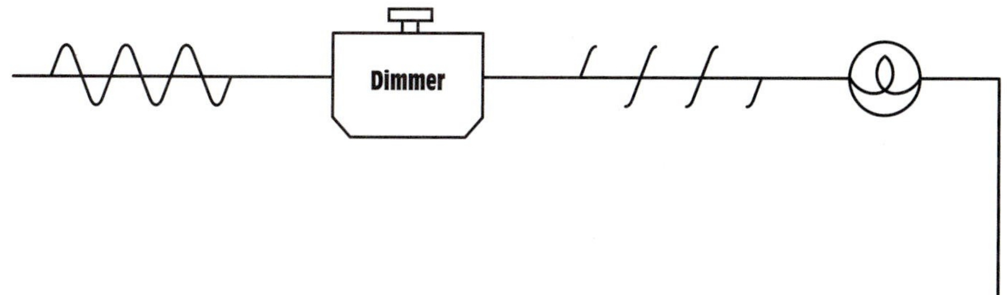

Figure 15-65 The dimmer operates by chopping the AC waveform.

it in either direction. The triac is a member of the thyristor family of devices. Thyristors are solid-state devices that exhibit only two states of operation, on or off. To turn the triac on, current must be applied to the gate circuit. A simple triac control circuit is shown in *Figure 15-66*. The control knob on the dimmer switch is connected to the variable resistor, or pot. A second resistor is used to limit current to the base of the triac in the event that the control knob should be turned to zero resistance. A second solid-state component, called a diac, is used to control when gate current is applied to the triac. The diac is a bi-directional, voltage-activated diode. The diac will not turn on until the voltage applied to it reaches a certain value. In this circuit, that value will be assumed to be 15 volts. When the diac turns on, it will remain on until the voltage drops to a value less than that needed to turn it on. That value will be assumed to be 5 volts. The circuit operates as follows:

1. When the switch is closed, the triac is turned off but current flows through the pot (variable resistor) and current-limiting fixed resistor to the capacitor. The resistance value of the pot and current-limiting resistor determines the charge time of the capacitor.
2. When the capacitor charges to a value of 15 volts, the diac turns on and discharges the capacitor through the gate of the triac, causing the triac to turn on and supply power to the lamp. When the capacitor discharges to a value of 5 volts, the diac turns off and the capacitor begins charging again.
3. At the end of each half-cycle, the triac will turn off and remain off until it receives another pulse of gate current.

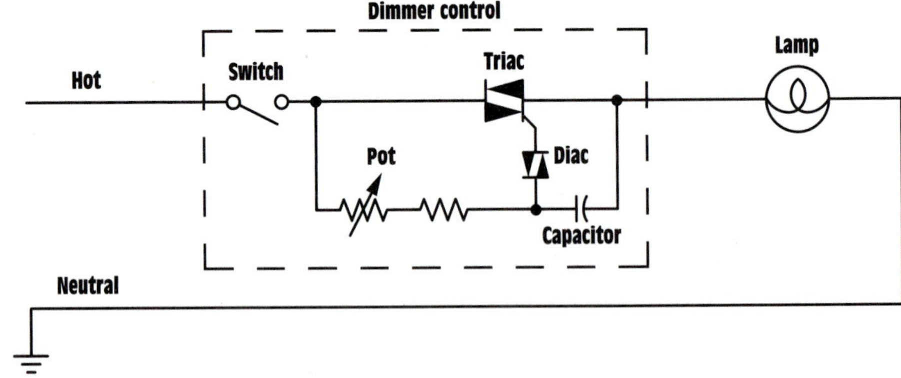

Figure 15-66 Basic dimmer switch control circuit.

Figure 15-67 Typical dimmer control circuit.

4. Since the value of the pot controls the charge time of the capacitor, the triac can be gated on at different times during the cycle of AC voltage.

The circuit board for a typical dimmer control is shown in *Figure 15-67.*

SUMMARY

1. A residential occupancy is wired with two- and three-conductor cable.
2. Receptacles are connected in parallel.
3. When connecting a receptacle, the black conductor is connected to the brass or gold screw, the white conductor is connected to the silver screw, and the bare conductor is connected to the green screw.
4. The neutral conductor is referred to as the grounded conductor.
5. The bare conductor is referred to as the grounding conductor.
6. The hot conductor is referred to as the ungrounded conductor.
7. A duplex receptacle contains two outlets on the same strap.

8. Receptacles and switches can be connected quickly with stab-locks or quick-connects, but these do not make as good a connection as screws.
9. The leading cause of electrical fires is poor connection.
10. Single-pole switches use two terminal screws for connection and have ON and OFF printed on the switch lever.
11. Single-pole switches are used to control a light from one location.
12. Three-way switches use three terminal screws for connection.
13. When connecting three-way switches, three-conductor cable must be run between the switches.
14. Four rules for connecting three- and four-way switches are:
    A. Connect the neutral to the light.
    B. Connect the hot to the common terminal of one three-way switch.
    C. Connect the other side of the light to the common terminal of the second three-way switch.
    D. Connect the travelers.
15. Four-way switches use four terminal screws for connection.
16. Four-way switches are used when it is necessary to control a light from more than two locations.
17. Four-way switches connect in the traveler conductors between two three-way switches.
18. A ground fault is an unintended path to ground.
19. Ground fault detectors for personnel protection will open the circuit with a fault current of approximately 5 ma.
20. Ground fault detectors for personnel protection can be obtained as circuit breakers or receptacles.
21. GFCI circuit breakers have two screws, instead of one, for connection to branch circuit conductors.
22. When installing a GFCI circuit breaker, both the hot and neutral branch circuit conductors must be terminated at the circuit breaker.
23. All devices connected to a GFCI circuit breaker are ground fault protected.
24. When installing a GFCI receptacle, the incoming power must be connected to the LINE side of the receptacle.

25. Devices connected to the LOAD side of a GFCI receptacle are ground fault protected by the receptacle.
26. The *National Electrical Code®* requires ground fault circuit interrupters on outside receptacles and on receptacles around swimming pools and fountains; in unfinished basements, garages, bathrooms, and kitchens; on snow-melting equipment; and in recreational vehicles (RVs).
27. Light dimmers control the voltage applied to the lamp by chopping the AC waveform.
28. Light dimmers use a triac as the main control device.
29. The *National Electrical Code®* requires that 120-volt outlets in bedrooms be protected by arc-fault circuit interrupters.
30. Arc-fault circuit interrupters contain a microprocessor that examines the current waveform.
31. Arc-fault circuit interrupters are intended to protect people against fire.
32. Thirty-three percent of electrical fires are caused by arc-faults.

REVIEW QUESTIONS

1. A two-conductor cable actually contains three wires. Why is it called a two-conductor cable if it has three wires?
2. What are the colors of the wires in a three-conductor cable?
3. When installing a duplex receptacle, which wire should be connected to the gold or brass screw?
4. What is the advantage of stab-lock connectors?
5. What is the disadvantage of stab-lock connectors?
6. What is the leading cause of electrical fires?
7. List four rules for the connection of three-way switches.
8. What are traveler conductors?
9. What is a switch leg?
10. How is the common screw terminal of a three-way switch generally identified?

11. A light is to have switches from five locations. How many single-pole, three-way, and four-way switches are required?
12. List two ways of identifying a single-pole switch.
13. A circuit is to supply power to a number of duplex receptacles. The circuit is to be protected by a 20 A circuit breaker. What size copper wire should be employed?
14. When does the *National Electrical Code*® permit a white wire to be connected to an ungrounded conductor?
15. When conductors are connected to a light, the *National Electrical Code*® requires that the wires be identified. Define the term *identified.*
16. What part of the circuit is used to detect the presence of a fault current in a ground fault circuit interrupter device?
17. Ground fault circuit interrupters intended for personnel protection open the circuit with ____ amperes of fault current.
18. When installing a GFCI receptacle, should the incoming power be connected to the LINE or LOAD side of the receptacle?
19. When installing a GFCI receptacle, how is it possible to use the receptacle to provide ground fault protection to other receptacles on the same circuit?
20. What type of solid state device does a light dimmer use to control the voltage to the light?
21. Both ground fault circuit interrupters and arc-fault circuit interrupters are intended to protect people from certain hazards. What hazard is each intended to protect people against?
22. Is it possible for a ground fault circuit interrupter to operate if the grounding conductor (bare or green wire) has not been connected? Explain your answer.
23. Arc-faults can develop temperatures in excess of ____°C.
24. The *National Electrical Code*® requires that arc-fault circuit interrupters be used in what type of circuits?
25. Between the parallel and series arc-faults, which is generally more dangerous and why?

Unit 16

Three-Phase Motors

objectives

fter studying this unit, you should be able to:

- Discuss the basic operating principles of three-phase motors.
- List factors that produce a rotating magnetic field.
- List different types of three-phase motors.
- Discuss the operating principles of squirrel-cage motors.
- Connect dual voltage motors for proper operation on the desired voltage.
- Discuss the operation of consequent pole motors.
- Discuss the operation of wound-rotor motors.
- Discuss the operation of synchronous motors.

Three-phase motors are used as the prime mover for industry throughout the United States and Canada. These motors convert the three-phase alternating current into mechanical energy to operate all types of machinery. Three-phase motors are smaller in size and lighter in weight, and they have higher efficiencies per horsepower than single-phase motors. They are extremely rugged and require very little maintenance. Many of these motors operate twenty-four hours a day, seven days a week for many years without a problem.

THREE-PHASE MOTORS

There are three basic types of three-phase motors:

1. the squirrel-cage induction motor.
2. the wound-rotor induction motor.
3. the synchronous motor.

All three motors operate on the same principle, and they all use the same basic design for the stator windings. The difference between these motors is the type of rotor used. Two of the three motors are induction motors and operate on the principle of electromagnetic induction in a manner similar to transformers. In fact, AC induction motors were patented as **rotating transformers** by Nikola Tesla. The stator winding of a motor is often referred to as the motor primary, and the rotor is referred to as the motor secondary.

rotating transformers

THE ROTATING MAGNETIC FIELD

The principle of operation for all three-phase motors is the **rotating magnetic field.** There are three factors that cause the magnetic field to rotate:

rotating magnetic field

1. voltages in a three-phase system that are 120° out of phase with each other.
2. the three voltages changing polarity at regular intervals.
3. the arrangement of the stator windings around the inside of the motor.

SYNCHRONOUS SPEED

The speed at which the magnetic field rotates is called the **synchronous speed.** Two factors that determine the speed of the rotating magnetic field are:

synchronous speed

1. the number of stator poles (per phase).
2. the frequency of the applied voltage.

The following chart shows the synchronous speed at 60 Hz for different numbers of stator poles.

RPM	Stator Poles
3600	2
1800	4
1200	6
900	8

The stator winding of a three-phase motor is shown in *Figure 16-1*. The synchronous speed can be calculated using the following formula:

$$S = \frac{120\ F}{P}$$

where

S = speed in RPM

F = frequency in Hz

P = number of stator poles (per phase)

Example 1

What is the synchronous speed of a four-pole motor connected to 50 Hz?

Figure 16-1 Stator of a three-phase motor.

Solution:

$$S = \frac{120 \times 50}{4}$$

$$S = 1500 \text{ RPM}$$

Example 2

What frequency should be applied to a six-pole motor to produce a synchronous speed of 400 RPM?

Solution:

First change the base formula to find frequency. Once that is done, known values can be substituted in the formula.

$$F = \frac{PS}{120}$$

$$F = \frac{6 \times 400}{120}$$

$$F = 20 \text{ Hz}$$

DIRECTION OF ROTATION FOR THREE-PHASE MOTORS

The direction of rotation of any three-phase motor can be changed by reversing two of its stator leads.

On many types of machinery, the direction of rotation of the motor is critical. **The direction of rotation of any three-phase motor can be changed by reversing two of its stator leads.** This causes the direction of the rotating magnetic field to reverse. When a motor is connected to a machine that will not be damaged when its direction of rotation is reversed, power can be momentarily applied to the motor to observe its direction of rotation. If the rotation is incorrect, any two line leads can be interchanged to reverse the motor's rotation.

When a motor is to be connected to a machine that can be damaged by incorrect rotation, however, the direction of rotation must be determined before the motor is connected to its load. This can be accomplished in two basic ways. One way is to make electrical connection to the motor before it is mechanically connected to the load. The **direction of rotation** can then be tested by momentarily applying power to the motor before it is coupled to the load.

direction of rotation

There may be occasions when this is not practical or is very inconvenient. It is possible to determine the direction of rotation of a motor before power is connected to it with the use of a **phase rotation meter.** The phase rotation meter is used to compare the phase rotation of two different three-phase connections. The meter contains six terminal leads. Three of the leads are connected to one side of the meter and labeled MOTOR.

phase rotation meter

These three motor leads are labeled A, B, or C. The LINE leads are located on the other side of the meter, and are labeled A, B, or C.

To determine the direction of rotation of the motor, first zero the meter by following the instructions provided by the manufacturer. Then set the meter selector switch to MOTOR, and connect the three MOTOR leads of the meter to the T leads of the motor as shown in *Figure 16-2*. The phase rotation meter contains a zero center voltmeter. One side of the voltmeter is labeled INCORRECT, and the other side is labeled CORRECT. While observing the zero center voltmeter, manually turn the motor shaft in the direction of desired rotation. The zero center voltmeter will immediately swing in the CORRECT or INCORRECT direction. When the motor shaft stops turning, the needle may swing in the opposite direction. It is the first indication of the voltmeter that is to be used.

If the voltmeter needle indicated CORRECT, label the motor T leads A, B, or C to correspond with the MOTOR leads from the phase rotation meter. If the voltmeter needle indicated INCORRECT, change any two of the MOTOR leads from the phase rotation meter and again turn the motor shaft. The voltmeter needle should now indicate CORRECT. The motor T leads can now be labeled to correspond with the MOTOR leads from the phase rotation meter.

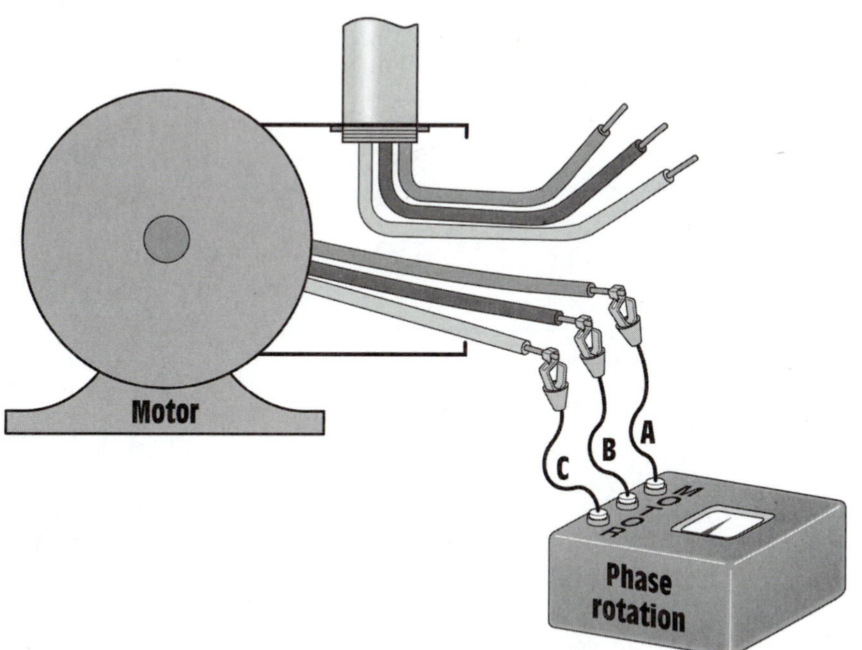

Figure 16-2 Connecting the phase rotation meter to the motor.

After the motor T leads have been labeled A, B, or C to correspond with the leads of the phase rotation meter, the rotation of the line supplying power to the motor must be determined. Set the selector switch on the phase rotation meter to the LINE position. After making certain the power has been turned off, connect the three LINE leads of the phase rotation meter to the incoming power line, *Figure 16-3*. Turn on the power and observe the zero center voltmeter. If the meter is pointing in the CORRECT direction, turn off the power and label the line leads A, B, or C to correspond with the LINE leads of the phase rotation meter.

If the voltmeter is pointing in the INCORRECT direction, turn off the power and change any two of the leads from the phase rotation meter. When the power is turned on, the voltmeter should point in the CORRECT direction. Turn off the power and label the line leads A, B, or C to correspond with the leads from the phase rotation meter.

Now that the motor T leads and the incoming power leads have been labeled, connect the line lead labeled A to the T lead labeled A, the line lead labeled B to the T lead labeled B, and the line lead labeled C to the T lead labeled C. When power is connected to the motor, it will operate in the proper direction.

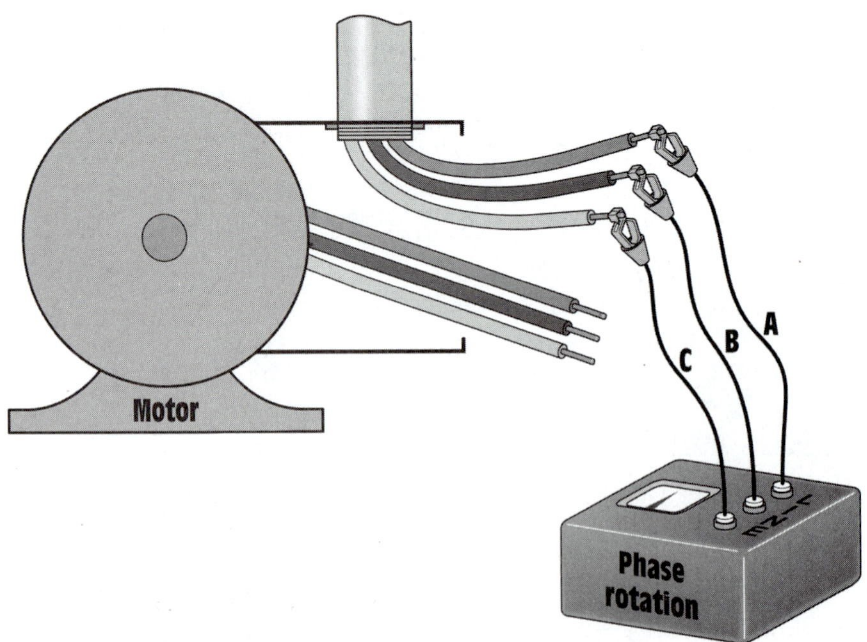

Figure 16-3 Connecting the phase rotation meter to the line.

CONNECTING DUAL-VOLTAGE THREE-PHASE MOTORS

Many of the three-phase motors used in industry are designed to be operated on two voltages, such as 240 or 480 V. Motors of this type contain two sets of windings per phase. Most dual voltage motors bring out nine T leads at the terminal box. There is a standard method used to number these leads, as shown in *Figure 16-4*. Starting with terminal 1, the leads are numbered in a decreasing spiral as shown. Another method of determining the proper lead numbers is to add three to each terminal. For example, starting with lead 1, add three to one. Three plus one equals four. The phase winding that begins with 1 ends with 4. Now add three to four. Three plus four equals seven. The beginning of the second winding for phase 1 is seven. This method will work for the windings of all phases. If in doubt, draw a diagram of the phase windings and number them in a spiral.

HIGH-VOLTAGE CONNECTIONS

Three-phase motors can be constructed to operate in either wye or delta. If a motor is to be connected to high voltage, the phase windings will be connected in series. In *Figure 16-5*, a schematic diagram and terminal connection chart for high voltage are shown for a wye-connected motor. In *Figure 16-6*, a schematic diagram and terminal connection chart for high voltage are shown for a delta-connected motor. Notice that in both cases the windings are connected in series.

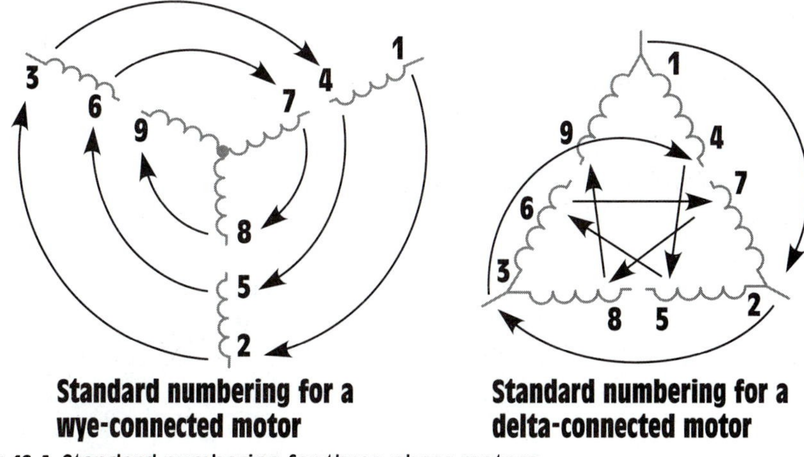

Figure 16-4 Standard numbering for three-phase motors.

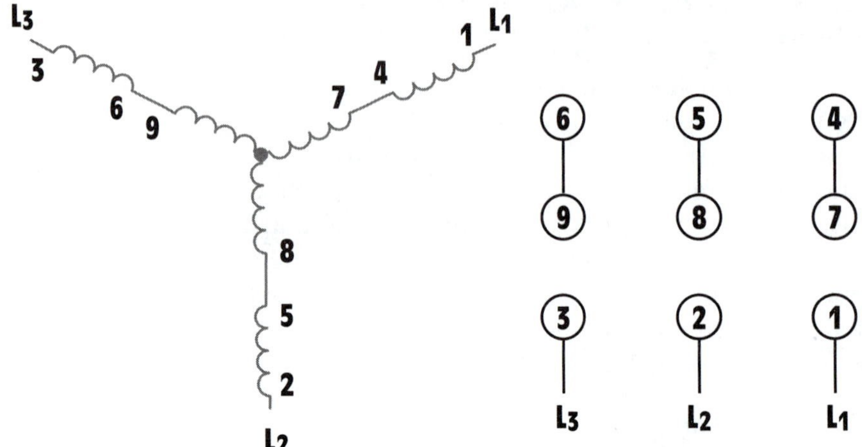

Figure 16-5 High-voltage wye connection.

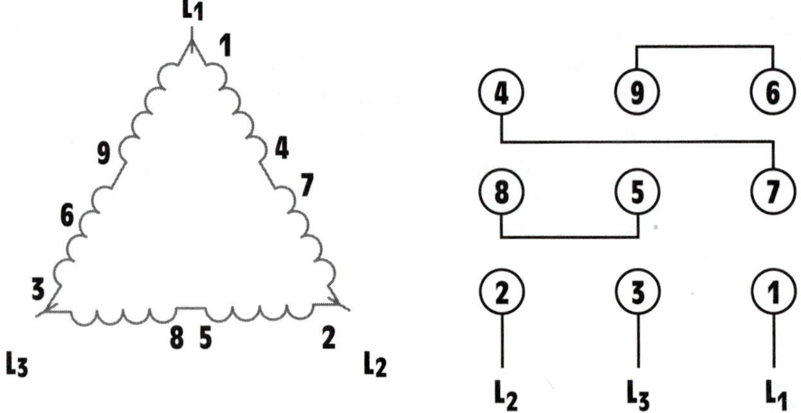

Figure 16-6 High-voltage delta connection.

LOW-VOLTAGE CONNECTIONS

When a motor is to be connected for low-voltage operation, the phase windings must connect in parallel. *Figure 16-7* shows the basic schematic diagram for a wye-connected motor with parallel phase windings. In actual practice, however, it is not possible to make this exact connection with a nine lead motor. The schematic shows that terminal 4 connects to the other end of the phase winding that starts with terminal 7. Terminal 5 connects to the other end of winding 8, and terminal 6 connects to the other end of winding 9. In actual motor construction, the opposite ends of windings 7, 8, and 9 are connected together inside the motor and are not

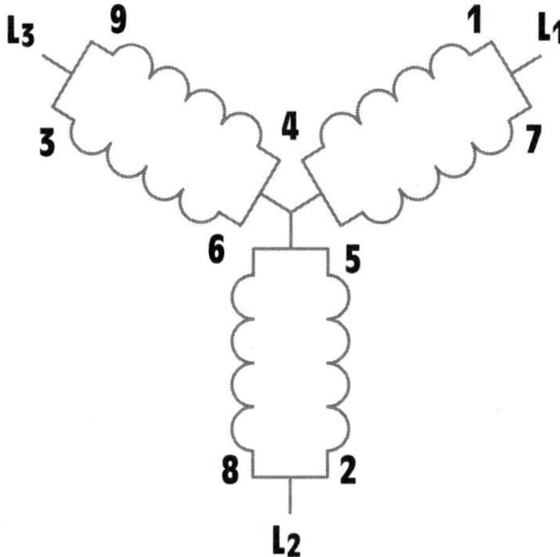

Figure 16-7 Stator windings connected in parallel.

brought outside the motor case. The problem is solved, however, by forming a second wye connection by connecting terminals 4, 5, and 6 as shown in *Figure 16-8*.

The phase windings of a delta-connected motor must also be connected in parallel for use on low voltage. A schematic for this connection is shown in *Figure 16-9*. A connection diagram and terminal connection chart for this hook-up is shown in *Figure 16-10*.

Some dual voltage motors will contain twelve T leads instead of nine. In this instance, the opposite ends of terminals 7, 8, and 9 are brought out

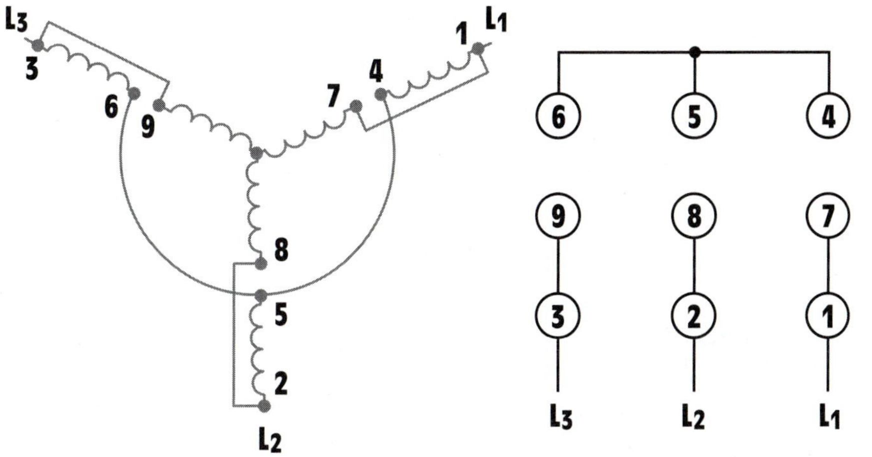

Figure 16-8 Low-voltage wye connection.

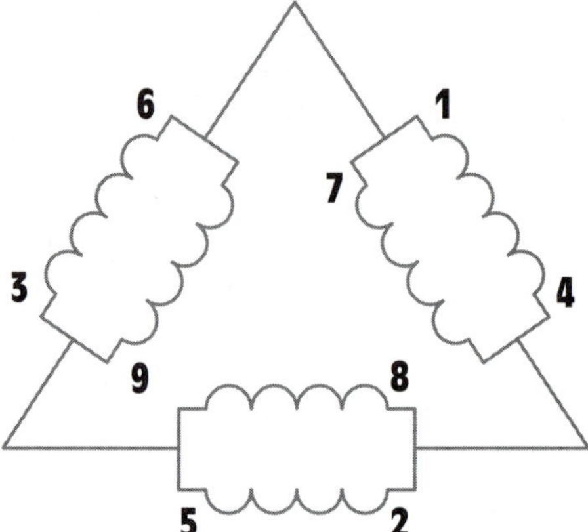

Figure 16-9 Parallel delta connection.

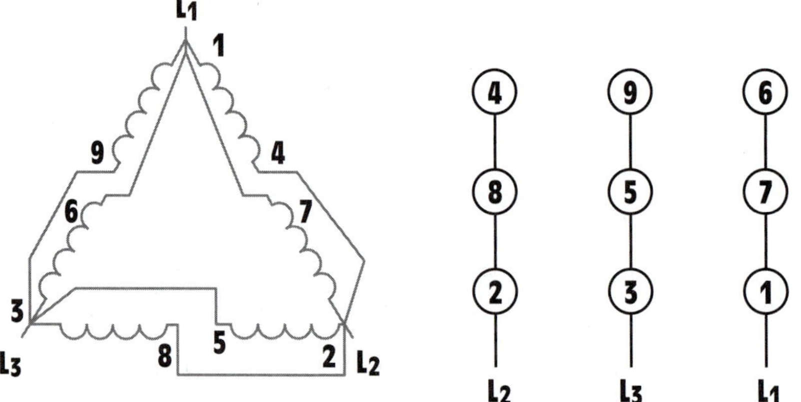

Figure 16-10 Low-voltage delta connection.

for connection. *Figure 16-11* shows the standard numbering for both delta- and wye-connected motors. Twelve leads are brought out if the motor is intended to be used for wye-delta starting. When this is the case, the motor must be designed for normal operation with its windings connected in delta. If the windings are connected in wye during starting, the starting current of the motor is reduced to one-third of what it will be if the motor is started as a delta.

VOLTAGE AND CURRENT RELATIONSHIPS FOR DUAL VOLTAGE MOTORS

When a motor is connected to the higher voltage, the current flow will be half as much as when it is connected for low-voltage operation. The reason is that when the windings are connected in series for high voltage op-

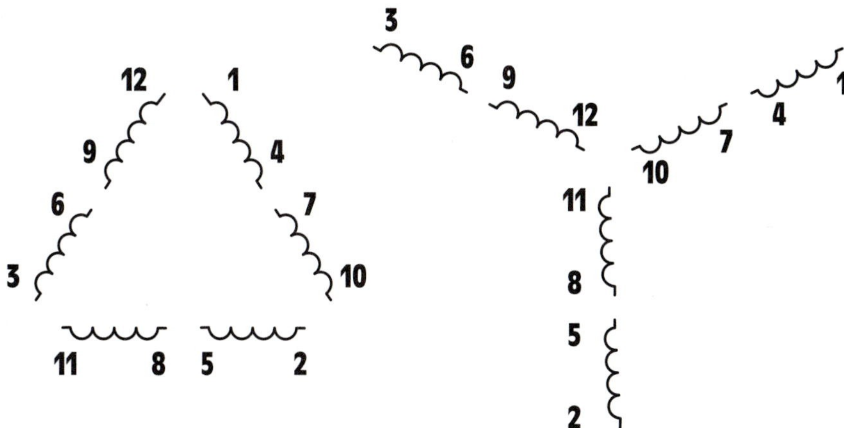

Figure 16-11 A twelve-lead motor.

eration, the impedance will be four times greater than when the windings are connected for low voltage operation. For example, assume a dual voltage motor is intended to operate on 480 or 240 V. Also assume that during full load, the motor's windings exhibit an impedance of 10 Ω each. When the winding is connected in series, *Figure 16-12,* the impedance per phase will be 20 Ω (10 + 10 = 20).

If a voltage of 480 V is connected to the motor, the phase voltage will be:

$$E_{PHASE} = \frac{E_{LINE}}{1.732}$$

$$E_{PHASE} = \frac{480}{1.732}$$

$$E_{PHASE} = 277\text{ V}$$

Figure 16-12 Impedance adds in series.

The amount of current flow through the phase can be computed using Ohm's Law.

$$I = \frac{E}{Z}$$

$$I = \frac{277}{20}$$

$$I = 13.85\ \text{A}$$

If the stator windings are connected in parallel, the total impedance will be found by adding the reciprocals of the impedances of the windings *(Figure 16-13)*.

$$Z_T = \frac{1}{\frac{1}{Z_1} + \frac{1}{Z_2}}$$

$$Z_T = 5\ \Omega$$

If a voltage of 240 V is connected to the motor, the voltage applied across each phase will be 138.6 V (240/1.732 = 138.6). The amount of phase current can now be computed using Ohm's Law.

$$I = \frac{E}{Z}$$

$$I = \frac{138.6}{5}$$

$$I = 27.12\ \text{A}$$

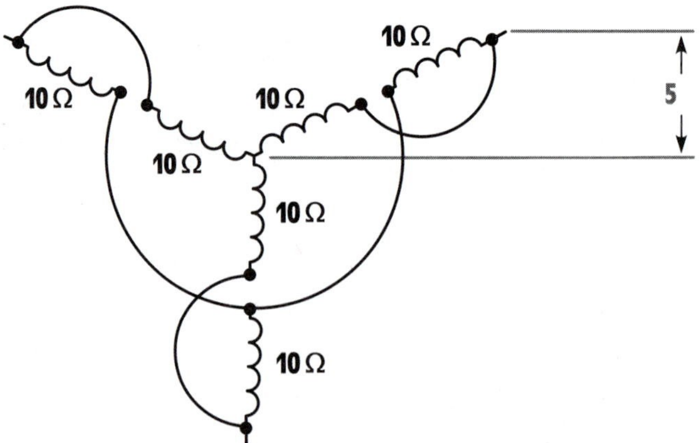

Figure 16-13 Impedance is less when in parallel.

SQUIRREL-CAGE INDUCTION MOTORS

squirrel-cage motor

The **squirrel-cage motor** receives its name from the type of rotor used in the motor. A squirrel-cage rotor is made by connecting bars to two end rings. If the metal laminations were to be removed from the rotor, the result would look very similar to a squirrel-cage, *Figure 16-14*. A squirrel-cage is a cylindrical device constructed of heavy wire. A shaft is placed through the center of the cage. This permits the cage to spin around the shaft. A squirrel-cage is placed inside the cage of small pets, such as squirrels and hamsters, to permit them to exercise by running inside of the squirrel-cage. A cutaway view of a squirrel-cage motor is shown in *Figure 16-15*. A cutaway view of a squirrel-cage rotor is shown in *Figure 16-16*. The major parts of a squirrel-cage motor are shown in *Figure 16-17*.

PRINCIPLES OF OPERATION

The squirrel-cage motor is an induction motor. This means that the current flow in the rotor is produced by induced voltage from the rotating magnetic field of the stator. In *Figure 16-18*, a squirrel-cage rotor is shown inside the stator of a three-phase motor. It will be assumed that the motor shown in Figure 16-18 contains four poles per phase, which produces a rotating magnetic field with a synchronous speed of 1,800 RPM when the stator is connected to a 60-Hz line. When power is first connected to the stator, the rotor is not turning. The magnetic field of the stator cuts the rotor bars at a rate of 1,800 RPM. This cutting action induces a voltage into the rotor bars.

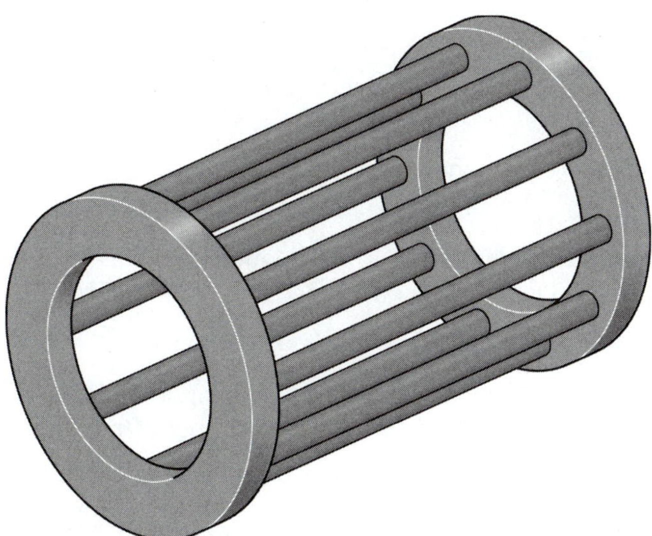

Figure 16-14 Basic squirrel-cage rotor without laminations.

Figure 16-15 Cutaway view of a squirrel-cage motor.

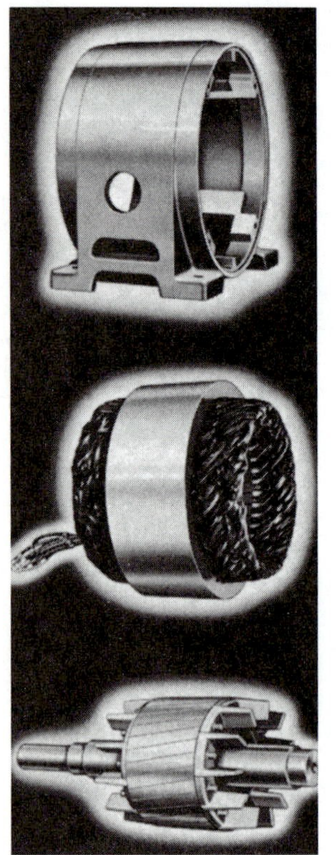

Figure 16-17 Squirrel-cage induction motor frame, stator winding, and rotor.

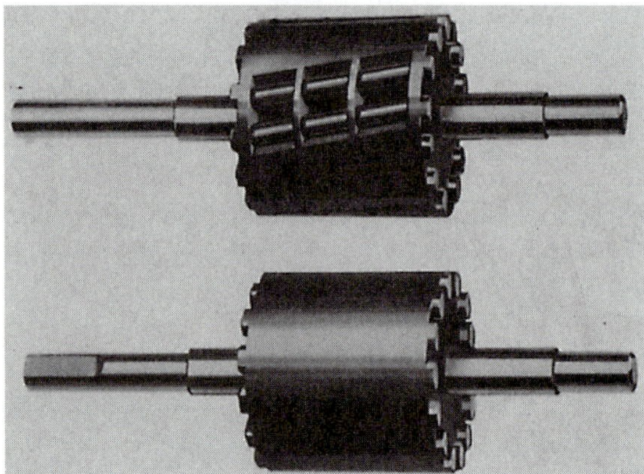

Figure 16-16 Cutaway view of a squirrel-cage rotor.

This induced voltage will be the same frequency as the voltage applied to the stator. The amount of induced voltage is determined by three factors:

1. the strength of the magnetic field of the stator.
2. the number of turns of wire cut by the magnetic field. (In the case of a squirrel-cage rotor this will be the number of bars in the rotor.)
3. the speed of the cutting action.

Since the rotor is stationary at this time, maximum voltage is induced into the rotor. The induced voltage causes current to flow through the ro-

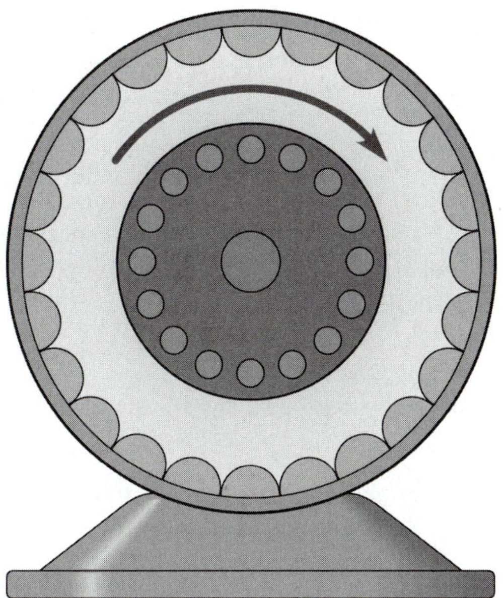

Figure 16-18 Voltage is induced in the rotor by the rotating magnetic field.

tor bars. As current flows through the rotor, a magnetic field is produced around each bar, *Figure 16-19.*

The magnetic field of the rotor is attracted to the magnetic field of the stator, and the rotor begins to turn in the same direction as the rotating magnetic field.

As the speed of the rotor increases, the rotating magnetic field cuts the rotor bars at a slower rate. For example, assume the rotor has accelerated to a speed of 600 RPM. The synchronous speed of the rotating magnetic field is 1,800 RPM. Therefore, the rotor bars are now being cut at a rate of

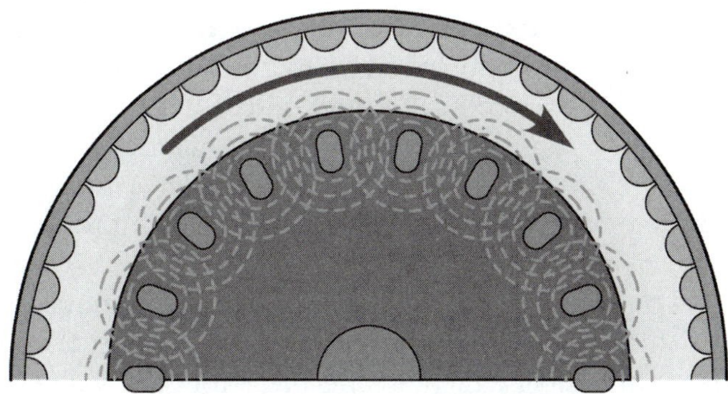

Figure 16-19 A magnetic field is produced around each rotor bar.

1,200 RPM instead of 1,800 RPM (1800 RPM − 600 RPM = 1200 RPM). Since the rotor bars are being cut at a slower rate, less voltage is induced in the rotor, reducing rotor current. When the rotor current decreases, the stator current decreases also.

As the rotor continues to accelerate, the rotating magnetic field cuts the rotor bars at a decreasing rate. This reduces the amount of induced voltage, and therefore, the amount of rotor current. If the motor is operating without a load, the rotor will continue to accelerate until it reaches a speed close to that of the rotating magnetic field.

torque

TORQUE

The amount of **torque** produced by an AC induction motor is determined by three factors:

1. the strength of the magnetic field of the stator.
2. the strength of the magnetic field of the rotor.
3. the phase angle difference between rotor and stator fields.

$$T = K_T \times \varphi_S \times I_R \times \text{COS}\ \theta_R$$

where

T = torque in lb-ft

K_T = torque constant

φ_S = stator flux (constant at all speeds)

I_R = rotor current

$\text{COS}\ \theta_R$ = rotor power factor

An induction motor can never reach synchronous speed.

Notice that one of the factors that determines the amount of torque produced by an induction motor is the strength of the magnetic field of the rotor. **An induction motor can never reach synchronous speed.** If the rotor were to turn at the same speed as the rotating magnetic field, there would be no induced voltage in the rotor, and consequently no rotor current. Without rotor current there could be no magnetic field developed by the rotor and, therefore, no torque or turning force. A motor operating at no load will accelerate until the torque developed is proportional to the windage and bearing friction losses.

If a load is connected to the motor, it must furnish more torque to operate the load. This causes the motor to slow down. When the motor speed decreases, the rotating magnetic field cuts the rotor bars at a faster rate. This causes more voltage to be induced in the rotor and, therefore, more current. The increased current flow produces a stronger magnetic field in the rotor, which causes more torque to be produced. The increased current flow in the rotor causes increased current flow in the stator. This is why motor current will increase as load is added.

Another factor that determines the amount of torque developed by an induction motor is the phase angle difference between stator and rotor field flux. **Maximum torque is developed when the stator and rotor flux are in phase with each other.** Note in the previous formula that one of the factors that determines the torque developed by an induction motor is the cosine of the rotor power factor. The cosine function reaches its maximum value of 1 when the phase angle is 0 (COS 0° = 1).

Maximum torque is developed when the stator and rotor flux are in phase with each other.

STARTING CHARACTERISTICS

When a squirrel-cage motor is first started, it will have a current draw several times greater than its normal running current. The actual amount of starting current is determined by the type of rotor bars, the horsepower rating of the motor, and the applied voltage. The type of rotor bars is indicated by the code letter found on the nameplate of a squirrel-cage motor. Table 430-7b of the *National Electrical Code*® can be used to compute the locked rotor current (starting current) of a squirrel-cage motor when the applied voltage, horsepower, and code letter are known. The table shown in *Figure 16-20* lists values for locked currents.

Example 3

An 800-hp. three-phase squirrel-cage motor is connected to 2,300 V. The motor has a code letter of J. What is the starting current of this motor?

Solution:

The table in Figure 16-20 lists a value of 7.1 to 7.99 kilo-volt-amperes per horsepower as the locked rotor current of a motor with a code letter J. An average value of 7.5 will be used for this calculation. The apparent power can be computed by multiplying the 7.5 times the horsepower rating of the motor.

$$\text{kVA} = 7.5 \times 800$$

$$\text{kVA} = 6000$$

The line current supplying the motor can now be computed using the formula:

$$I_{(LINE)} = \frac{VA}{E_{(LINE)} \times 1.732}$$

$$I_{(LINE)} = \frac{6{,}000{,}000}{2300 \times 1.732}$$

$$I_{(LINE)} = 1506.175 \text{ A}$$

Code Letter	Kilo-volt-Amperes per Horsepower with Locked Rotor
A	0 - 3.14
B	3.15 - 3.54
C	3.55 - 3.99
D	4.0 - 4.49
E	4.5 - 4.99
F	5.0 - 5.59
G	5.6 - 6.29
H	6.3 - 7.09
J	7.1 - 7.99
K	8.0 - 8.99
L	9.0 - 9.99
M	10.0 - 11.9
N	11.2 - 12.49
P	12.5 - 13.49
R	14.0 - 15.99
S	16.0 - 17.99
T	18.0 - 19.99
U	20.0 - 22.39
V	22.4 - and up

Figure 16-20 Table for determining starting current for squirrel-cage motors.

This large starting current is caused by the fact that the rotor is not turning when power is first applied to the stator. Since the rotor is not turning, the squirrel-cage bars are cut by the rotating magnetic field at a fast rate. Remember that one of the factors that determines the amount of induced voltage is the speed of the cutting action. This high induced voltage causes a large amount of current to flow in the rotor. The large current flow in the rotor causes a large amount of current flow in the stator. Since a large amount of current flows in both the stator and rotor, a strong magnetic field is established in both.

The starting torque of a squirrel-cage motor is high since the magnetic field of both the stator and rotor are strong at this point. The starting torque of a typical squirrel-cage motor can range from 200% to 300% of its full load-rated torque. Although the starting torque is high, the squirrel-cage motor does not develop as much starting torque per ampere of starting current as other types of three-phase motors. Recall that the third factor for determining the torque developed by an induction motor is the difference in phase angle between stator flux and rotor flux. Since the rotor is being cut at a high rate of speed by the rotating stator field, the bars in the squirrel-cage rotor appear to be very inductive at this point because of the high frequency of the induced voltage. This causes the phase angle difference between the induced voltage in the rotor and rotor current to be almost 90° out of phase with each other, producing a lagging power factor for the rotor, *Figure 16-21.* This causes the rotor flux to lag the stator flux by a large amount, and consequently a relatively weak starting torque per ampere of starting current, compared to other types of three-phase motors, is developed, *Figure 16-22.* A typical torque curve for a squirrel-cage motor is shown in *Figure 16-23.*

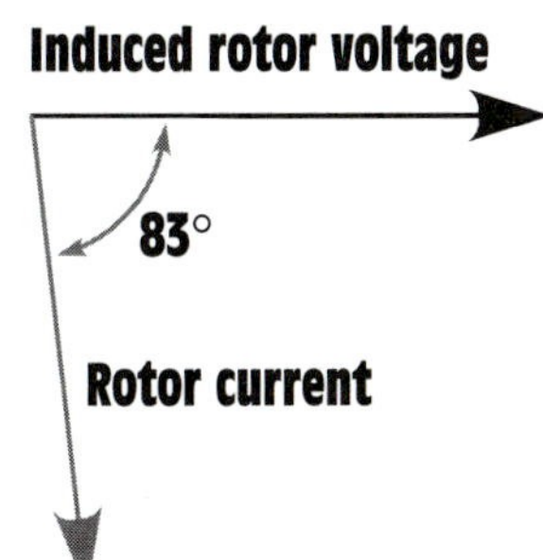

Figure 16-21 Rotor current is almost 90° out of phase with the induced voltage at the moment of starting.

PERCENT SLIP

The speed performance of an induction motor is measured in **percent slip.** The percent slip can be determined by subtracting the synchronous speed from the speed of the rotor. For example, assume an induction motor has a synchronous speed of 1800 RPM and at full load the rotor turns at

percent slip

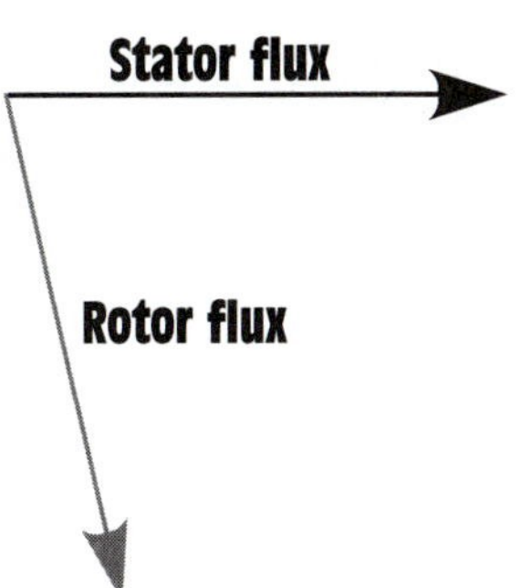

Figure 16-22 Rotor flux lags the stator flux by a large amount during starting.

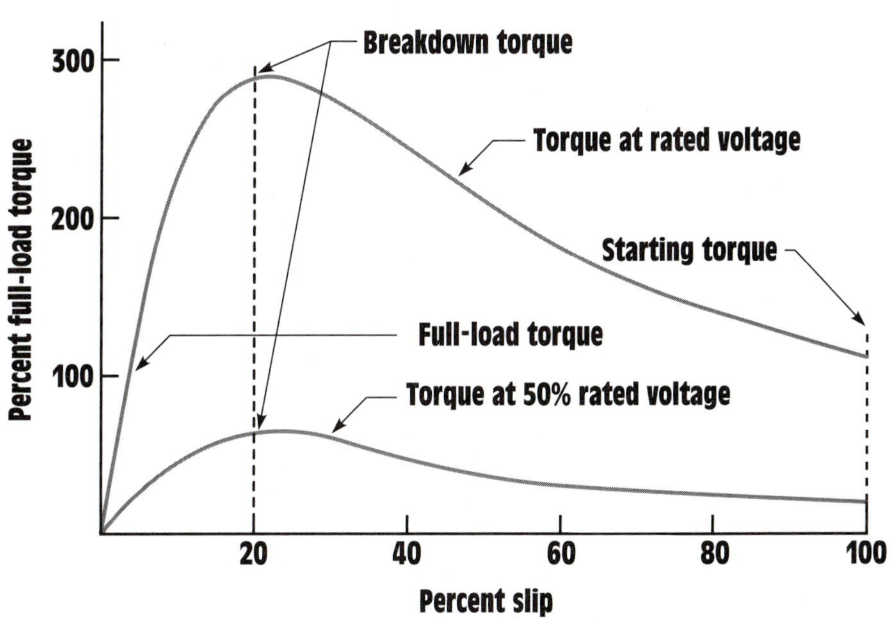

Figure 16-23 Typical torque curves for a squirrel-cage motor.

a speed of 1725 RPM. The difference between the two speeds is 75 RPM (1800 − 1725 = 75). The percent slip can be determined using the formula:

$$\text{percent slip} = \frac{\text{synchronous speed} - \text{rotor speed}}{\text{synchronous speed}} \times 100$$

$$\text{percent slip} = \frac{75}{1800} \times 100$$

$$\text{percent slip} = 4.16\%$$

A rotor slip of 2% to 5% is common for most squirrel-cage induction motors. The amount of slip for a particular motor is greatly affected by the type of rotor bars used in the construction of the rotor. Squirrel-cage motors are considered to be constant speed motors because there is a small difference between no-load speed and full-load speed.

ROTOR FREQUENCY

rotor frequency

In the previous example, the rotor slips behind the rotating magnetic field by 75 RPM. This means that at full load, the bars of the rotor are being cut by magnetic lines of flux at a rate of 75 RPM. Therefore, the voltage being induced in the rotor at this point in time is at a much lower frequency than when the motor was started. The **rotor frequency** can be determined using the formula:

$$F = \frac{P \times S_R}{120}$$

where

F = frequency in Hz

P = number of stator poles

S_R = rotor slip in RPM

$$F = \frac{4 \times 75}{120}$$

$$F = 2.5\ \text{Hz}$$

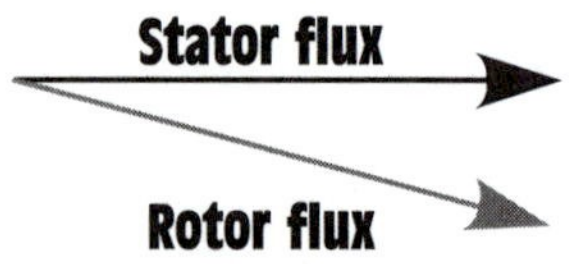

Stator flux

Rotor flux

Figure 16-24 Rotor and stator flux become more in phase with each other as motor speed increases.

Because the frequency of the current in the rotor decreases as the rotor approaches synchronous speed, the rotor bars become less inductive. The current flow through the rotor becomes limited more by the resistance of the bars and less by inductive reactance. The current flow in the rotor becomes more in phase with the induced voltage, which causes less phase angle shift between stator and rotor flux, *Figure 16-24*. This is the reason why squirrel-cage motors generally have a relatively poor starting torque per ampere of starting current but a good running torque, as compared to other types of three-phase motors.

REDUCED VOLTAGE STARTING

Due to the large amount of starting current required for many squirrel-cage motors, it is sometimes necessary to reduce the voltage during the starting period. When the voltage is reduced, the starting torque is reduced also. If the applied voltage is reduced to 50% of its normal value, the magnetic fields of both the stator and rotor are reduced to 50% of normal. This causes the starting torque to be reduced to 25% of normal. A chart of typical torque curves for squirrel-cage motors is shown in Figure 16-23.

The torque formula given earlier can be used to show why this large reduction of torque occurs. Both the stator flux, φ_S, and the rotor current, I_R, are reduced to half their normal value. The product of these two values, torque, is reduced to one-fourth. The torque varies as the square of the applied voltage for any given value of slip.

CODE LETTERS

Squirrel-cage rotors are not all the same. Rotors are made with different types of bars. The type of rotor bars used in the construction of the rotor determines the operating characteristics of the motor. AC squirrel-cage motors are given a code letter on their nameplate. The code letter indicates the type of bars used in the rotor. *Figure 16-25* shows a rotor with type A bars. A type A rotor has the highest resistance of any squirrel-cage rotor. This means that the starting torque per ampere of starting current will be high

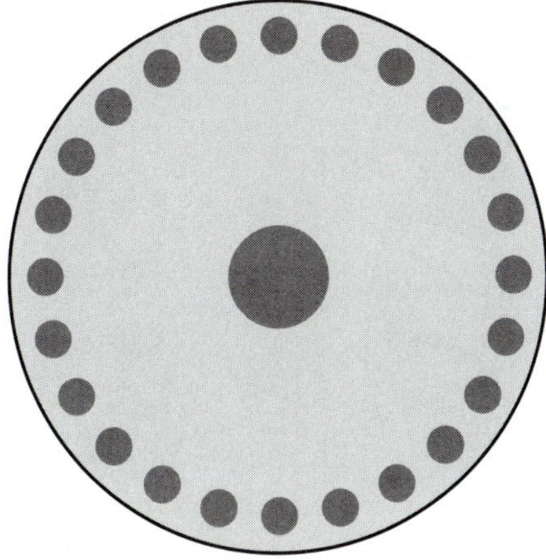

Figure 16-25 Type A rotor.

since the rotor current is closer to being in phase with the induced voltage than any other type of rotor. Also, the high resistance of the rotor bars limits the amount of current flow in the rotor when starting. This produces a low starting current for the motor. A rotor with type A bars has very poor running characteristics, however. Since the bars are resistive, a large amount of voltage will have to be induced into the rotor to produce an increase in rotor current and, therefore, an increase in the rotor magnetic field. This means that when load is added to the motor, the rotor must slow down a great amount to produce enough current in the rotor to increase the torque. Motors with type A rotors have the highest percent slip of any squirrel-cage motor. Motors with type A rotors are generally used in applications where starting is a problem, such as a motor that must accelerate a large flywheel from 0 RPM to its full speed. Flywheels can have a very large amount of inertia, which may require several minutes to accelerate them to their running speed when they are started.

Figure 16-26 shows a rotor with bars similar to those found in rotors with code letters B through E. These rotor bars have lower resistance than the type A rotor. Rotors of this type have fair starting torque, low starting current, and fair speed regulation.

Figure 16-27 shows a rotor with bars similar to those found in rotors with code letters F through V. This rotor has low starting torque, high starting current, and good running torque. Motors containing rotors of this type generally have very good speed regulation and low percent slip.

THE DOUBLE SQUIRREL-CAGE ROTOR

Some motors use a rotor that contains two sets of squirrel-cage windings, *Figure 16-28*. The outer winding consists of bars with a relatively high resistance located close to the top of the iron core. Since these bars are lo-

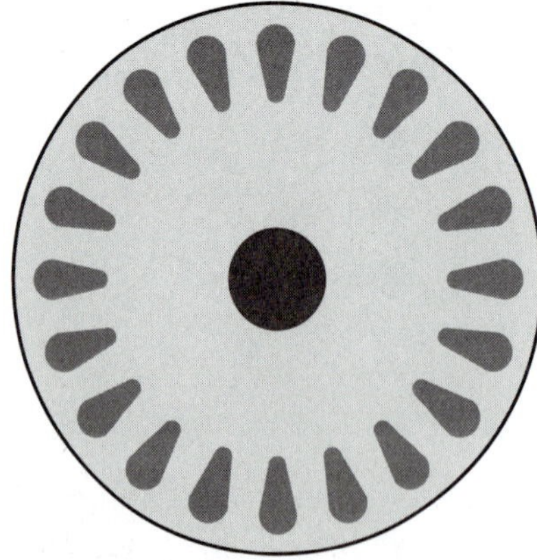

Figure 16-26 Type B–E rotor.

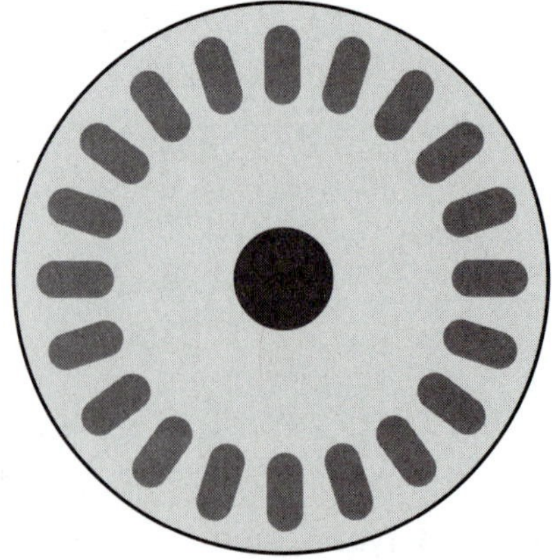

Figure 16-27 Type F–V rotor.

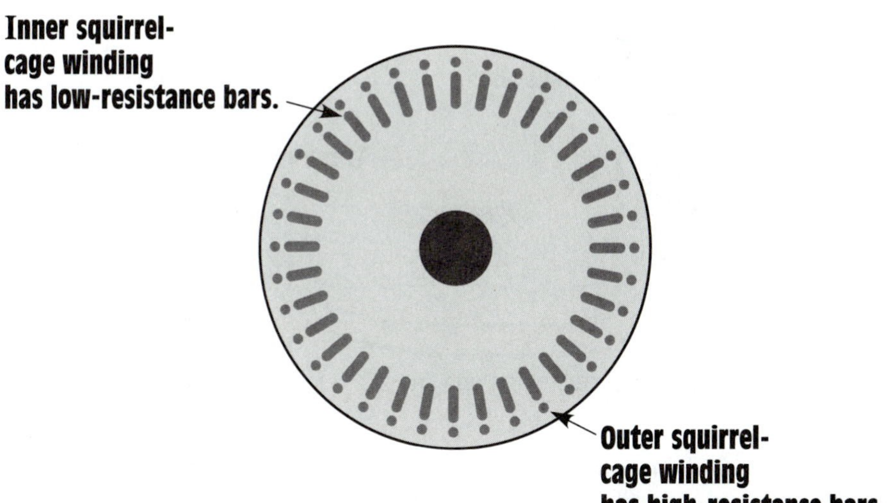

Figure 16-28 Double squirrel-cage rotor.

cated close to the surface, they have a relatively low reactance. The inner winding consists of bars with a large cross-sectional area, which gives them a low resistance. The inner winding is placed deeper in the core material, which causes it to have a much higher reactance.

When the double squirrel-cage motor is started, the rotor frequency is high. Since the inner winding is inductive, its impedance will be high as compared to the resistance of the outer winding. During this period, most of the rotor current flows through the outer winding. The resistance of the outer winding limits the current flow through the rotor, which limits the starting current to a relatively low value. Since the current is close to being in phase with the induced voltage, the rotor flux and stator flux are close to being in phase with each other and a strong starting torque is developed. The starting torque of a double squirrel-cage motor can be as high as 250% of rated full-load torque.

When the rotor reaches its full load speed, rotor frequency decreases to 2 or 3 Hz. The inductive reactance of the inner winding has now decreased to a low value. Most of the rotor current now flows through the low-resistance inner winding. This type of motor has good running torque and excellent speed regulation.

SINGLE PHASING

Three lines supply power to a three-phase motor. If one of these lines should open, the motor will be connected to single-phase power, *Figure 16-29*. This condition is known as **single phasing.**

single phasing

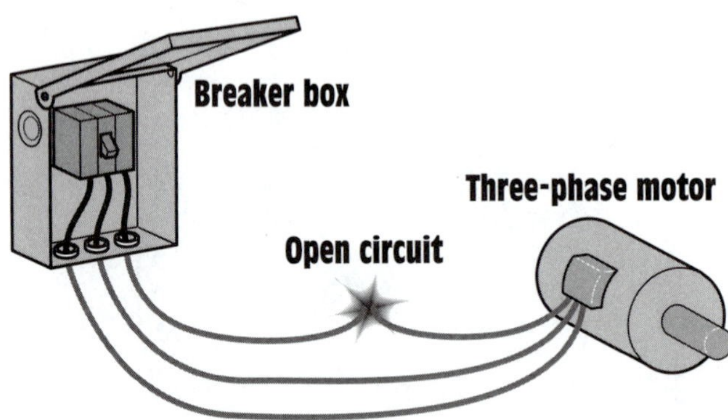

Figure 16-29 Single phasing occurs when one line of a three-phase system is open.

If the motor is not running and single-phase power is applied to the motor, the induced voltage in the rotor sets up a magnetic field in the rotor. This magnetic field opposes the magnetic field of the stator (Lenz's Law). As a result, practically no torque is developed in either the clockwise or counterclockwise direction, and the motor will not start. The current supplying the motor will be excessive, however, and damage to the stator windings can occur.

If the motor is operating under load at the time the single-phasing condition occurs, the rotor will continue to turn at a reduced speed. The moving bars of the rotor cut the stator field flux, which continues to induce voltage and current in the bars. Due to reduced speed, the rotor has high reactive and low resistive components, causing the rotor current to lag the induced voltage by almost 90°. This lagging current creates rotor fields midway between the stator poles, resulting in greatly reduced torque. The reduction in rotor speed causes high current flow and will most likely damage the stator winding if the motor is not disconnected from the power line.

THE NAMEPLATE

Electric motors have nameplates that give a great deal of information about the motor. *Figure 16-30* illustrates the nameplate of a three-phase squirrel-cage induction motor. The nameplate shows that the motor is 10 hp, it is a three-phase motor, and operates on 240 or 480 V. The full-load running current of the motor is 28 A when operated on 240 V and 14 A when operated on 480 V. The motor is designed to be operated on a 60-Hz AC voltage, and has a full-load speed of 1,745 RPM. This speed indicates that the motor has four poles per phase. Since the full-load speed is 1745 RPM, the synchronous speed would be 1,800 RPM. The motor contains a type J squirrel-cage rotor, and has a service factor of 1.25. The service fac-

Manufacturer	
HP 10	Phase 3
Volts 240/480	Amps 28/14
Hz 60	Fl speed 1745 RPM
Code J	SF 1.25
Frame XXXX	Model No. XXXX

Figure 16-30 Motor nameplate.

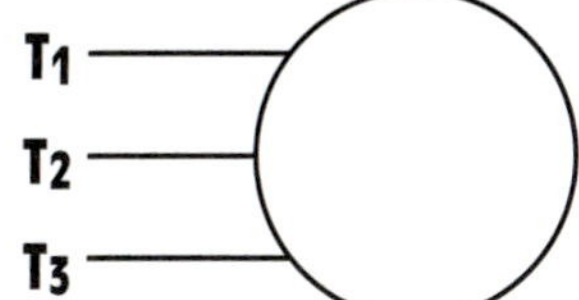

Figure 16-31 Schematic symbol of a three-phase squirrel-cage induction motor.

tor is used to determine the amperage rating of the overload protection for the motor. Some motors indicate a marked temperature rise in celsius degrees instead of a service factor. The frame number indicates the type of mounting the motor has. *Figure 16-31* shows the schematic symbol used to represent a three-phase squirrel-cage motor.

CONSEQUENT POLE SQUIRREL-CAGE MOTORS

consequent pole

Consequent pole squirrel-cage motors permit the synchronous speed to be changed by changing the number of stator poles. If the number of poles is doubled, the synchronous speed will be reduced by one-half. A two-pole motor has a synchronous speed of 3,600 RPM when operated at 60 Hz. If the number of poles is doubled to four, the synchronous speed becomes 1,800 RPM. The number of stator poles can be changed by changing the direction of current flow through alternate pairs of poles.

Figure 16-32 illustrates this concept. In Figure 16-32A, two coils are connected in such a manner that current flows through them in the same direction. Both poles will produce the same magnetic polarity and are essentially one pole. In Figure 16-32B, the coils have been reconnected in such a manner that current flows through them in opposite directions. Each coil now produces the opposite magnetic polarity, and are essentially two different poles.

Consequent pole motors with one stator winding bring out six leads labeled T_1 through T_6. Depending on the application, the windings will be connected as a series delta or a parallel wye. If it is intended that the motor maintain the same horsepower rating for both high and low speed, the

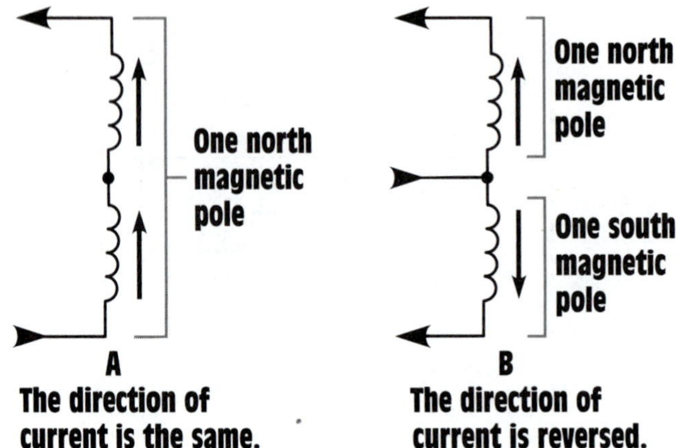

Figure 16-32 The number of poles can be changed by reversing the current flow through alternate poles.

high-speed connection will be a series delta, *Figure 16-33*. The low-speed connection will be a parallel wye, *Figure 16-34*.

If it is intended that the motor maintain constant torque for both low and high speeds, the series delta connection will provide low speed and the parallel wye will provide high speed.

Since the speed range of a consequent pole motor is limited to a 1:2 ratio, motors intended to operate at more than two speeds contain more than

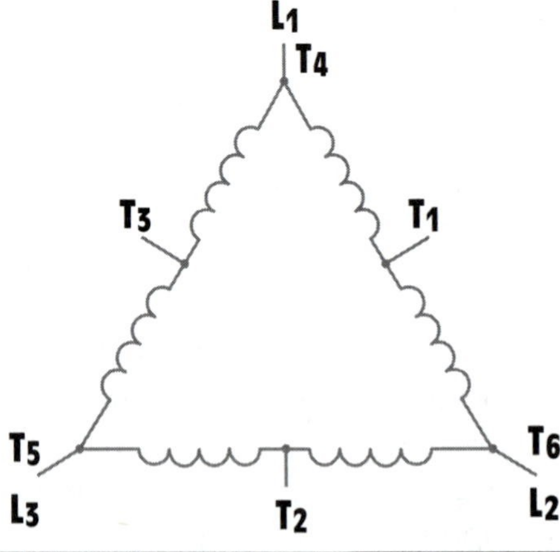

Speed	L1	L2	L3	Open	Together
Low	T1	T2	T3	————	T4 T5 T6
High	T6	T4	T5	All others	————

Figure 16-33 High-speed series delta connection.

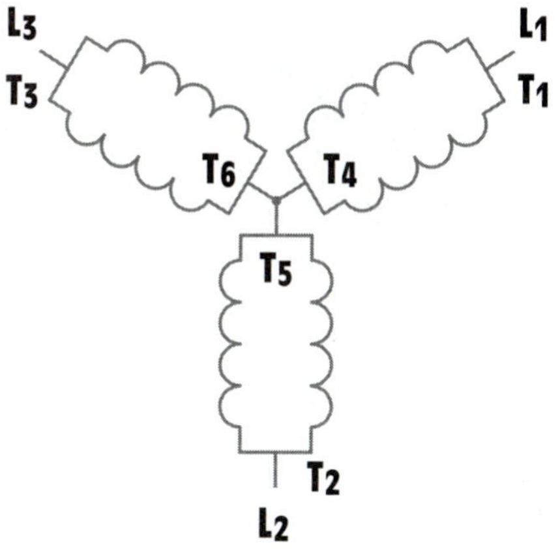

Speed	L1	L2	L3	Open	Together
Low	T1	T2	T3	————	T4 T5 T6
High	T6	T4	T5	All others	————

Figure 16-34 Low-speed parallel wye connection.

one stator winding. A consequent pole motor with three speeds, for example, will have one stator winding for one speed only and a second winding with taps. The tapped winding may provide synchronous speeds of 1,800 and 900 RPM, and the separate winding may provide a speed of 1,200 RPM. Consequent pole motors with four speeds contain two separate stator windings with taps. If the second stator winding of the motor in this example were to be tapped, the motor would provide synchronous speeds of 1,800, 1,200, 900, and 600 RPM.

THE WOUND-ROTOR INDUCTION MOTOR

wound-rotor motor

The **wound-rotor** induction **motor** is very popular in industry because of its high starting torque and low starting current. The stator winding of the wound-rotor motor is the same as the squirrel-cage motor. The difference between the two motors lies in the construction of the rotor. Recall that the squirrel-cage rotor is constructed of bars connected together at each end by a shorting ring as shown in Figure 16-14.

The rotor of a wound-rotor motor is constructed by winding three separate coils on the rotor 120° apart. The rotor will contain as many poles per phase as the stator winding. These coils are then connected to three sliprings located on the rotor shaft as shown in *Figure 16-35*. Brushes, connected to the sliprings, provide external connection to the rotor. This permits the rotor circuit to be connected to a set of resistors, as shown in *Figure 16-36*.

Figure 16-35 Rotor of a wound-rotor induction motor.

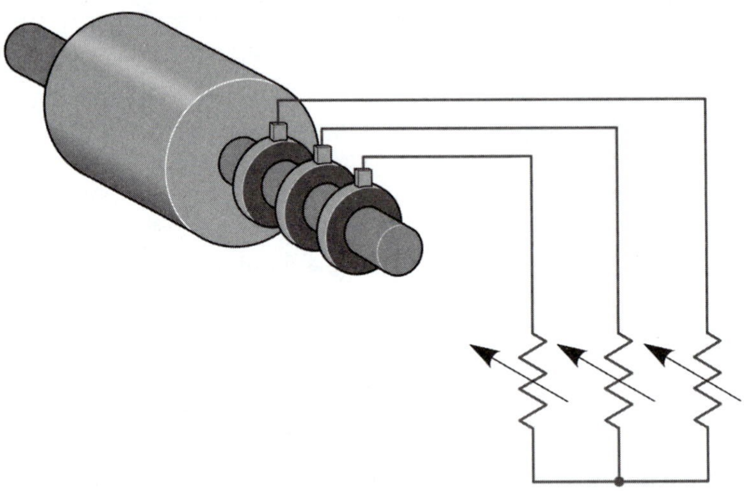

Figure 16-36 The rotor of a wound-rotor motor is connected to external resistors.

The stator terminal connections are generally labeled T_1, T_2, and T_3. The rotor connections are commonly labeled M_1, M_2, and M_3. The M_2 lead is generally connected to the middle slipring and the M_3 lead is connected close to the rotor windings. The direction of rotation for the wound-rotor motor is reversed by changing any two stator leads. Changing the M leads will have no effect on the direction of rotation. The schematic symbol for a wound-rotor motor is shown in *Figure 16-37*.

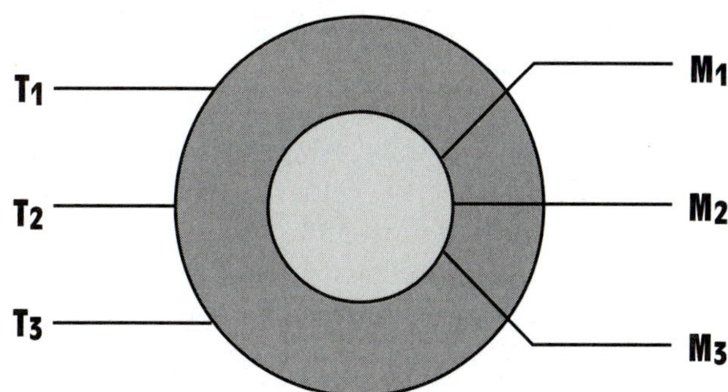

Figure 16-37 Schematic symbol for a wound-rotor induction motor.

WOUND-ROTOR MOTOR OPERATION

When power is applied to the stator winding, a rotating magnetic field is created in the motor. This magnetic field cuts through the windings of the rotor and induces a voltage into them. The amount of current flow in the rotor is determined by the amount of induced voltage and the total impedance of the rotor circuit ($I = E/Z$). The rotor impedance is a combination of inductive reactance created in the rotor windings and the external resistance. The impedance could be calculated using the formula for resistance and inductive reactance connected in series.

$$Z = \sqrt{R^2 + X_L^2}$$

As the rotor speed increases, the frequency of the induced voltage will decrease just as it does in the squirrel-cage motor. The reduction in frequency causes the rotor circuit to become more resistive and less inductive, decreasing the phase angle between induced voltage and rotor current.

When current flows through the rotor, a magnetic field is produced. This magnetic field is attracted to the rotating magnetic field of the stator. As the rotor speed increases, the induced voltage decreases because of less cutting action between the rotor windings and rotating magnetic field. This produces less current flow in the rotor and, therefore, less torque. If rotor circuit resistance is reduced, more current can flow, which will increase motor torque, and the rotor will increase in speed. This action continues until all external resistance has been removed from the rotor circuit by shorting the M leads together and the motor is operating at maximum speed. At this point, the wound-rotor motor is operating in the same manner as a squirrel-cage motor.

STARTING CHARACTERISTICS OF A WOUND-ROTOR MOTOR

The wound-rotor motor will have a higher starting torque and lower starting current per horsepower than a squirrel-cage motor. The starting current is less because resistance is connected in the rotor circuit during starting. This resistance limits the amount of current that can flow in the rotor circuit. Since the stator current is proportional to rotor current because of transformer action, the stator current is less also. The starting torque is high because of the resistance in the rotor circuit. Recall that one of the factors that determines motor torque is the phase angle difference between stator flux and rotor flux. Since resistance is connected in the rotor circuit, stator and rotor flux are close to being in phase with each other, producing a high starting torque for the wound-rotor induction motor. If an attempt is made to start the motor with no circuit connected to the rotor, the motor cannot start. If no resistance is connected to the rotor circuit, there can be no current flow, and consequently, no magnetic field developed in the rotor.

SPEED CONTROL

The speed of a wound-rotor motor can be controlled by permitting resistance to remain in the rotor circuit during operation. When this is done, the rotor and stator current is limited, which reduces the strength of both magnetic fields. The reduced magnetic field strength permits the rotor to slip behind the rotating magnetic field of the stator. The resistors of speed controllers must have higher power ratings than the resistors of starters because they operate for extended periods of time.

The operating characteristics of a wound-rotor motor with the sliprings shorted are almost identical to those of a squirrel-cage motor. The percent slip, power factor, and efficiency are very similar for motors of equal horsepower rating.

SYNCHRONOUS MOTORS

synchronous

The three-phase **synchronous** motor has several characteristics that separate it from other types of three-phase motors. Some of these characteristics are:

1. The synchronous motor is not an induction motor. It does not depend on induced current in the rotor to produce a torque.
2. It will operate at a constant speed from full load to no load.
3. The synchronous motor must have DC excitation to operate.

4. It will operate at the speed of the rotating magnetic field (synchronous speed).
5. It has the ability to correct its own power factor and the power factor of other devices connected to the same line.

ROTOR CONSTRUCTION

The synchronous motor has the same type of stator windings as the other two three-phase motors. The rotor of a synchronous motor, however, contains both wound-pole pieces and a squirrel-cage winding, *Figure 16-38*. The wound pole pieces become electromagnets when direct current is applied to them. The excitation current can be applied to the rotor through two sliprings located on the rotor shaft or by a brushless exciter.

STARTING A SYNCHRONOUS MOTOR

The rotor of a synchronous motor contains a set of squirrel-cage bars similar to those found in a type A rotor. This set of squirrel-cage bars is used to start the motor and is known as the **amortiseur winding,** Figure 16-38. When power is first connected to the stator, the rotating magnetic field cuts through the squirrel-cage bars. The cutting action of the field induces a current into the squirrel-cage winding. The current flow through the amortiseur winding produces a rotor magnetic field that is attracted to the rotating magnetic field of the stator. This causes the rotor to begin turning in the direction of rotation of the stator field. When the rotor has accelerated to a speed that is close to the synchronous speed of the field, direct current is connected to the rotor through the sliprings on the rotor shaft or by a brushless exciter, *Figure 16-39*.

amortiseur winding

Figure 16-38 Rotor of a synchronous motor.

Figure 16-39 DC excitation current supplied through sliprings.

When DC current is applied to the rotor, the windings of the rotor become electromagnets. The electromagnetic field of the rotor locks in step with the rotating magnetic field of the stator. The rotor will now turn at the same speed as the rotating magnetic field. When the rotor turns at the synchronous speed of the field, there is no more cutting action between the stator field and the amortiseur winding. This causes the current flow in the amortiseur winding to cease.

Notice that the synchronous motor starts as a squirrel-cage induction motor. Since the rotor uses bars that are similar to those used in a type A rotor, they have a relatively high resistance, which gives the motor good starting torque and low starting current. **A synchronous motor must never be started with DC current connected to the rotor.** If DC current is applied to the rotor, the field poles of the rotor become electromagnets. When the stator is energized, the rotating magnetic field begins turning at synchronous speed. The electromagnets are alternately attracted and repelled by the stator field. As a result, the rotor does not turn. The rotor and power supply can be damaged by high induced voltages, however.

A synchronous motor must never be started with DC current connected to the rotor.

THE FIELD DISCHARGE RESISTOR

When the stator winding is first energized, the rotating magnetic field cuts through the rotor winding at a fast rate of speed. This causes a large amount of voltage to be induced into the winding of the rotor. To prevent this from becoming excessive, a resistor is connected across the winding. This resistor is known as the field discharge resistor, *Figure 16-40*. It also helps to reduce the voltage induced into the rotor by the collapsing magnetic field when the DC current is disconnected from the rotor. The field discharge resistor is connected in parallel with the rotor winding during starting. If the motor is manually started, a field discharge switch is used to

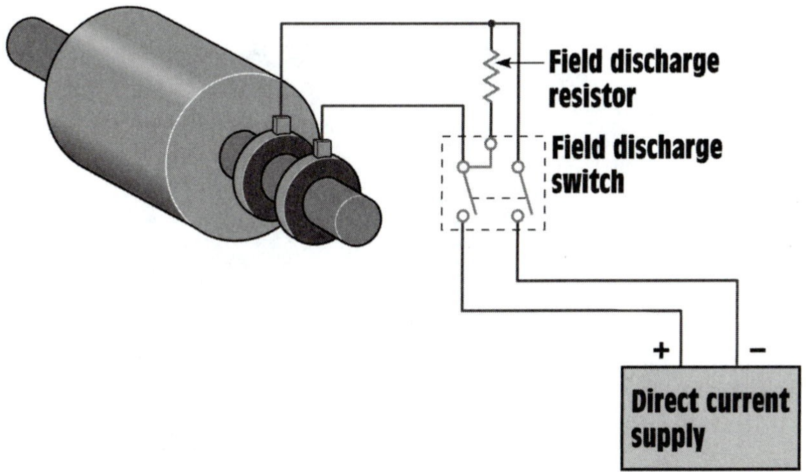

Figure 16-40 The field discharge resistor is connected in parallel with the rotor winding during starting.

connect the excitation current to the rotor. If the motor is automatically started, a special type of relay is used to connect excitation current to the rotor and disconnect the field discharge resistor.

CONSTANT SPEED OPERATION

Although the synchronous motor starts as an induction motor, it does not operate as one. After the amortiseur winding has been used to accelerate the rotor to about 95% of the speed of the rotating magnetic field, direct current is connected to the rotor, and the electromagnets lock in step with the rotating field. Notice that the synchronous motor does not depend on induced voltage from the stator field to produce a magnetic field in the rotor. The magnetic field of the rotor is produced by external DC current applied to the rotor. This is the reason why the synchronous motor has the ability to operate at the speed of the rotating magnetic field.

As load is added to the motor, the magnetic field of the rotor remains locked with the rotating magnetic field of the stator and the rotor continues to turn at the same speed. The added load, however, causes the magnetic fields of the rotor and stator to become stressed, *Figure 16-41*. The action is similar to connecting the north and south ends of two magnets and then trying to pull them apart. If the force being used to pull the magnets apart becomes greater than the strength of the magnetic attraction, the magnetic coupling will be broken and the magnets can be separated. The same is true for the synchronous motor. If the load on the motor becomes too great, the rotor will be pulled out of sync with the rotating magnetic field. The amount

Figure 16-41 The magnetic field becomes stressed as load is added.

pull-out torque

of torque necessary to cause this condition is called the **pull-out torque.** The pull-out torque for most synchronous motors will range from 150% to 200% of rated full-load torque. If this should happen, the motor must be stopped and restarted. The schematic symbol for a synchronous motor is shown in *Figure 16-42*.

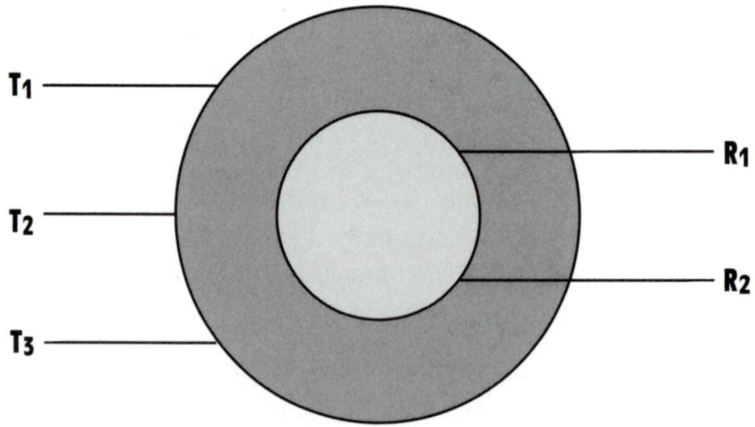

Figure 16-42 Schematic symbol for a synchronous motor.

THE POWER SUPPLY

The DC power supply of a synchronous motor can be provided by several methods. The most common are to use either a small DC generator mounted to the shaft of the motor or an electronic power supply that converts the AC line voltage into DC voltage.

SUMMARY

1. Three basic types of three-phase motors are:

   A. squirrel-cage induction motor.
   B. wound-rotor induction motor.
   C. synchronous motor.

2. All three-phase motors operate on the principle of a rotating magnetic field.

3. Three factors that cause a magnetic field to rotate are:

   A. the voltages of a three-phase system are 120° out of phase with each other.
   B. the voltages change polarity at regular intervals.
   C. the arrangement of the stator windings.

4. The speed of the rotating magnetic field is called the synchronous speed.

5. Two factors that determine the synchronous speed are:

   A. number of stator poles per phase.
   B. frequency of the applied voltage.

6. The direction of rotation of any three-phase motor can be changed by reversing the connection of any two stator leads.

7. The direction of rotation of a three-phase motor can be determined with a phase rotation meter before power is applied to the motor.

8. Dual-voltage motors will have nine or twelve leads brought out at the terminal connection box.

9. Dual-voltage motors intended for high-voltage connection have their phase windings connected in series.

10. Dual-voltage motors intended for low-voltage connection have their phase windings connected in parallel.

11. Motors that bring out twelve leads are generally intended for wye-delta starting.

12. Three factors that determine the torque produced by an induction motor are:

    A. the strength of the magnetic field of the stator.
    B. the strength of the magnetic field of the rotor.
    C. the phase angle difference between rotor and stator flux.

13. Maximum torque is developed when stator and rotor flux are in phase with each other.
14. The code letter on the nameplate of a squirrel-cage motor indicates the type of rotor bars used in the construction of the rotor.
15. The type A rotor has the lowest starting current, highest starting torque, and poorest speed regulation of any type of squirrel-cage rotor.
16. The double squirrel-cage rotor contains two sets of squirrel-cage windings in the same rotor.
17. Consequent pole squirrel-cage motors change speed by changing the number of stator poles.
18. Wound-rotor induction motors have wound rotors that contain three-phase windings.
19. Wound-rotor motors have three sliprings on the rotor shaft to provide external connection to the rotor.
20. Wound-rotor motors have higher starting torque and lower starting current than squirrel-cage motors of the same horsepower.
21. The speed of a wound-rotor motor can be controlled by permitting resistance to remain in the rotor circuit during operation.
22. Synchronous motors operate at synchronous speed.
23. Synchronous motors operate at a constant speed from no load to full load.
24. When load is connected to a synchronous motor, stress develops between the magnetic fields of the rotor and stator.
25. Synchronous motors must have DC excitation for an external source.
26. DC excitation is provided to some synchronous motors through two sliprings located on the rotor shaft, and other motors use a brushless exciter.
27. Synchronous motors have the ability to produce a leading power factor by overexcitation of the DC current supplied to the rotor.

28. Synchronous motors have a set of type A squirrel-cage bars used for starting. This squirrel-cage winding is called the amortiseur winding.
29. A field discharge resistor is connected across the rotor winding during starting to prevent high voltage in the rotor due to induction.
30. Changing the DC excitation current does not affect the speed of the motor.

REVIEW QUESTIONS

1. What are the three basic types of three-phase motors?
2. What is the principle of operation of all three-phase motors?
3. What is synchronous speed?
4. What two factors determine synchronous speed?
5. Name three factors that cause the magnetic field to rotate.
6. Name three factors that determine the torque produced by an induction motor.
7. Is the synchronous motor an induction motor?
8. What is the amortiseur winding?
9. Why must a synchronous motor never be started when DC excitation current is applied to the rotor?
10. Name three characteristics that make the synchronous motor different from an induction motor.
11. What is the function of the field discharge resistor?
12. Why can an induction motor never operate at synchronous speed?
13. What is the difference between a squirrel-cage motor and a wound-rotor motor?
14. What is the advantage of the wound-rotor motor over the squirrel-cage motor?
15. Name three factors that determine the amount of voltage induced in the rotor of a wound-rotor motor.
16. Why will the rotor of a wound-rotor motor not turn if the rotor circuit is left open with no resistance connected to it?

17. Why is the starting torque of a wound-rotor motor higher per ampere of starting current than that of a squirrel-cage motor?
18. What determines when a synchronous motor is at normal excitation?
19. How can a synchronous motor be made to have a leading power factor?
20. Is the excitation current of a synchronous motor AC or DC?
21. How is the speed of a consequent pole squirrel-cage motor changed?
22. A three-phase squirrel-cage motor is connected to a 60-Hz line. The full load speed is 870 RPM. How many poles per phase does the stator have?

Unit 17

Single-Phase Motors

objectives

*fter studying this unit, you should be able to:*

- List the different types of split-phase motors.
- Discuss the operation of split-phase motors.
- Reverse the direction of rotation of a split-phase motor.
- Discuss the operation of multispeed split-phase motors.
- Discuss the operation of shaded pole type motors.
- Discuss the operation of universal motors.

Although most of the large motors used in industry are three-phase, there are times when single-phase motors must be used. Single-phase motors are used almost exclusively to operate home appliances such as air conditioners, refrigerators, well pumps, and fans. They are designed to operate on 120 V or 240 V. They range in size from fractional horsepower to several horsepower depending on the application.

SINGLE-PHASE MOTORS

In Unit 16, it was stated that there are three basic types of three-phase motors and that all operate on the principle of a rotating magnetic field. While this is true for three-phase motors, it is not true for single phase motors. There are not only many different types of single-phase motors, but they have different operating principles.

SPLIT-PHASE MOTORS

Split-phase motors fall into three general classifications:

1. the resistance-start induction-run motor,
2. the capacitor-start induction-run motor,
3. the capacitor-start capacitor-run motor. (This motor is also known as a permanent-split capacitor motor in the air conditioning and refrigeration industry.)

split-phase motors

Although all of these motors have different operating characteristics, they are similar in construction and use the same operating principle. **Split-phase** motors receive their name from the manner in which they operate. Like three-phase motors, split-phase motors operate on the principle of a rotating magnetic field. A rotating magnetic field, however, cannot be produced with only one phase. Split-phase motors, therefore, split the current flow through two separate windings to simulate a two-phase power system. A rotating magnetic field can be produced with a two-phase system.

THE TWO-PHASE SYSTEM

two-phase

In some parts of the world two-phase power is produced. A **two-phase** system is produced by having an alternator with two sets of coils wound 90° apart, *Figure 17-1*. The voltages of a two-phase system are, therefore, 90° out of phase with each other. These two out-of-phase voltages can be used to produce a rotating magnetic field in a manner similar to that of producing a rotating magnetic field with the voltages of a three-phase system. Since there have to be two voltages or currents out of phase with each other to produce a rotating magnetic field, split-phase motors use two separate windings to create a phase difference between the currents in each of these windings. These motors literally split one phase and produce a second phase, hence the name split-phase motor.

start winding

run winding

STATOR WINDINGS

The stator of a split-phase motor contains two separate windings, the **start winding** and the **run winding**. The start winding is made of small

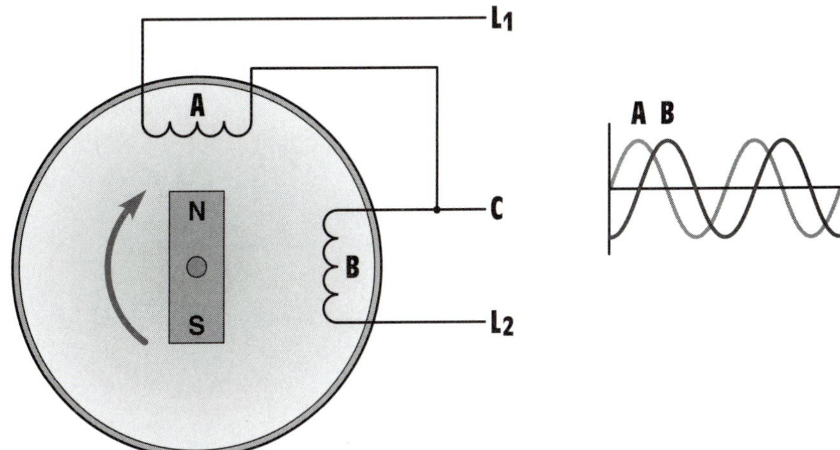

Figure 17-1 A two-phase alternator produces voltages that are 90° out of phase with each other.

wire and is placed near the top of the stator core. The run winding is made of relatively large wire and is placed in the bottom of the stator core. *Figure 17-2* shows a photograph of two split-phase stators. The stator on the left is used for a resistance-start induction-run motor, or a capacitor-start induction-run motor. The stator on the right is used for a capacitor-start capacitor-run motor. Both stators contain four poles, and the start winding is placed at a 90° angle from the run winding.

Notice the difference in size and position of the two windings of the stator shown on the left. The start winding is made from small wire and is placed near the top of the stator core. This causes it to have a higher resistance than the run winding. The start winding is located between the poles

Figure 17-2 Stator windings used in single-phase motors. *(Courtesy of Bodine Electric Co.)*

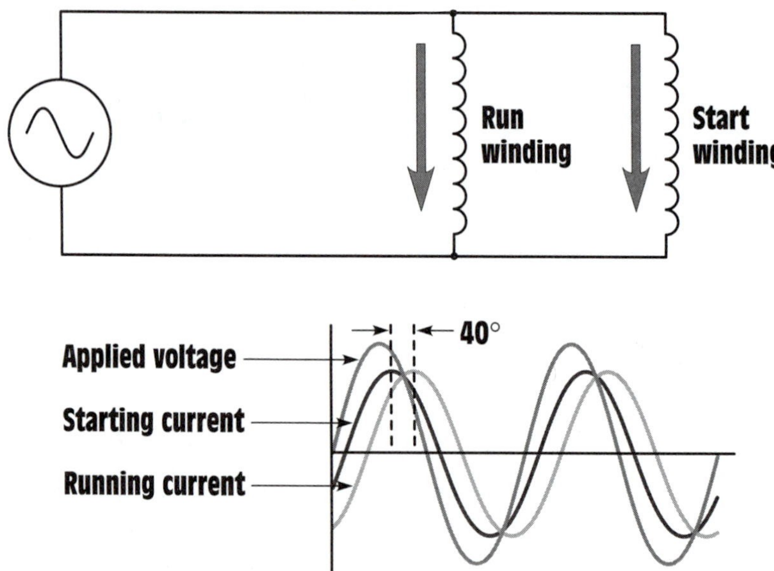

Figure 17-3 The start and run windings are connected in parallel with each other.

of the run winding. The run winding is made with larger wire and placed near the bottom of the core. This gives it higher inductive reactance and less resistance than the start winding. These two windings are connected in parallel with each other, *Figure 17-3.*

When power is applied to the stator, current will flow through both windings. Since the start winding is more resistive, the current flow will be more in phase with the applied voltage. The current flow through the run winding will lag the applied voltage due to inductive reactance. These two out-of-phase currents are used to create a rotating magnetic field in the stator. The speed of this rotating magnetic field is called **synchronous speed** and is determined by the same two factors that determined the synchronous speed for a three-phase motor:

synchronous speed

1. number of stator poles per phase.
2. frequency of the applied voltage.

THE RESISTANCE-START INDUCTION-RUN MOTOR

The resistance-start induction-run motor receives its name from the fact that the out-of-phase condition between start and run winding current is caused by the start winding being more resistive than the run winding. The amount of starting torque produced by a split-phase motor is determined by three factors:

1. the strength of the magnetic field of the stator.
2. the strength of the magnetic field of the rotor.
3. the phase angle difference between current in the start winding and current in the run winding. (Maximum torque is produced when these two currents are 90° out of phase with each other.)

Although these two currents are out of phase with each other, they are not 90° out of phase. The run winding is more inductive than the start winding, but it does have some resistance, which prevents the current from being 90° out of phase with the voltage. The start winding is more resistive than the run winding, but it does have some inductive reactance, preventing the current from being in phase with the applied voltage. Therefore, a phase angle difference of 35° to 40° is produced between these two currents, resulting in a rather poor starting torque, *Figure 17-4*.

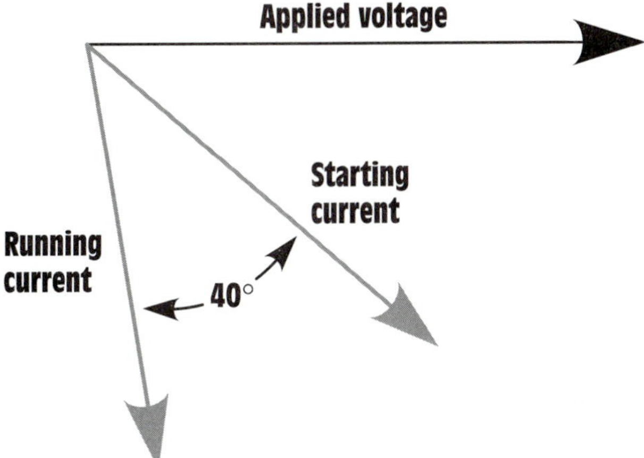

Figure 17-4 The running current and starting current are 35° to 40° out of phase with each other.

DISCONNECTING THE START WINDING

A stator rotating magnetic field is necessary only to start the rotor turning in both the resistance-start induction-run and capacitor-start induction-run motors. Once the rotor has accelerated to approximately 75% of rated speed, the start winding can be disconnected from the circuit, and the motor will continue to operate with only the run winding energized. Motors that are not hermetically sealed (most refrigeration and air conditioning compressors are hermetically sealed), use a **centrifugal switch** to disconnect the start windings from the circuit. The contacts of the centrifugal switch are connected in series with the start winding, *Figure 17-5*. The

centrifugal switch

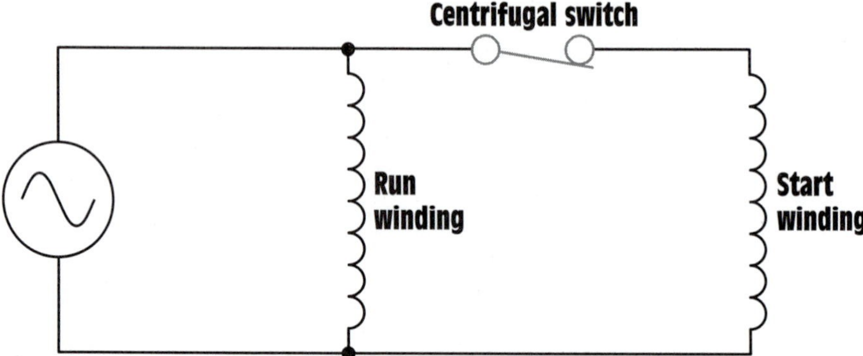

Figure 17-5 A centrifugal switch is used to disconnect the start winding from the circuit.

centrifugal switch contains a set of spring-loaded weights that control the operation of a fiber brushing and a switch assembly, *Figure 17-6*. When the shaft is not turning, the springs hold a fiber washer in contact with the movable contact of the switch, *Figure 17-7*. The fiber washer causes the movable contact to complete a circuit with a stationary contact.

When the rotor accelerates to about 75% of rated speed, centrifugal force causes the weights to overcome the force of the springs. The fiber washer retracts and permits the contacts to open and disconnect the start winding from the circuit. The start winding of this type motor is intended to be energized only during the period of time that the motor is actually starting. If the start winding is not disconnected, it will be damaged by excessive current flow.

STARTING HERMETICALLY SEALED MOTORS

Resistance-start induction-run and capacitor-start induction-run motors are sometimes hermetically sealed as is the case with air conditioning and refrigeration compressors. When these motors are hermetically sealed, a centrifugal switch cannot be used to disconnect the start winding. A device that can be mounted externally must be used to disconnect the start windings from the circuit. Starting relays are used to perform this function. There are two basic types of starting relays used with the resistance start and capacitor start motors:

1. the hot wire relay
2. the current relay

Although the *hot wire relay* is seldom used, it is still found on some older units that are still in service. The hot wire relay functions as both a

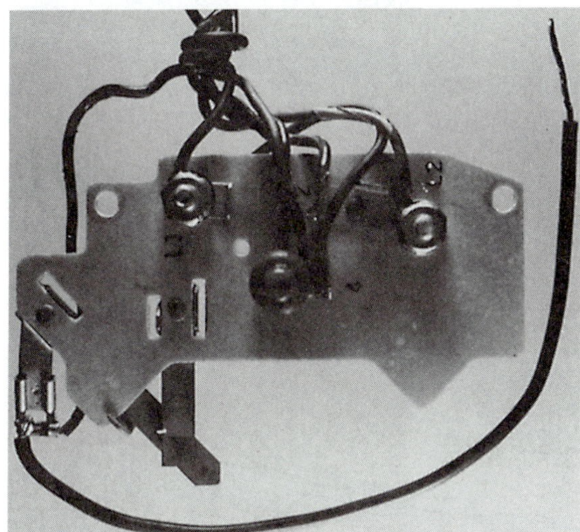

Figure 17-6 Centrifugal switch.

starting relay and an overload relay. In the circuit shown in *Figure 17-8*, it is assumed that a thermostat controls the operation of the motor. When the thermostat closes, current flows through a resistive wire and two normally closed contacts connected to the start and run windings of the motor. The starting current of the motor is high and rapidly heats the resistive wire, causing it to expand. The expansion of the wire causes the spring-loaded start winding contact to open and disconnect the start winding from the circuit, reducing motor current. If the motor is not overloaded, the resistive wire never becomes hot enough to cause the overload contact to open and the motor continues to run. If the motor should become overloaded, however, the resistive wire will expand enough to open the overload contact

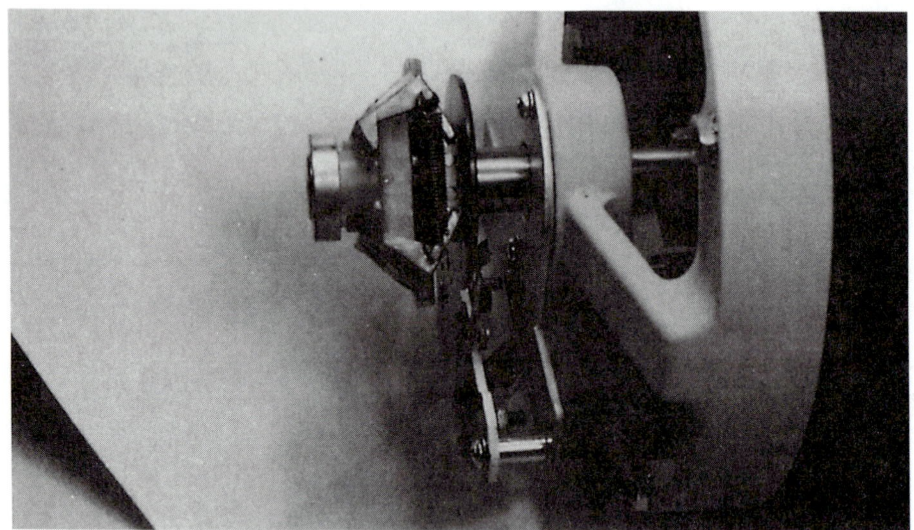

Figure 17-7 The centrifugal switch is closed when the rotor is not turning.

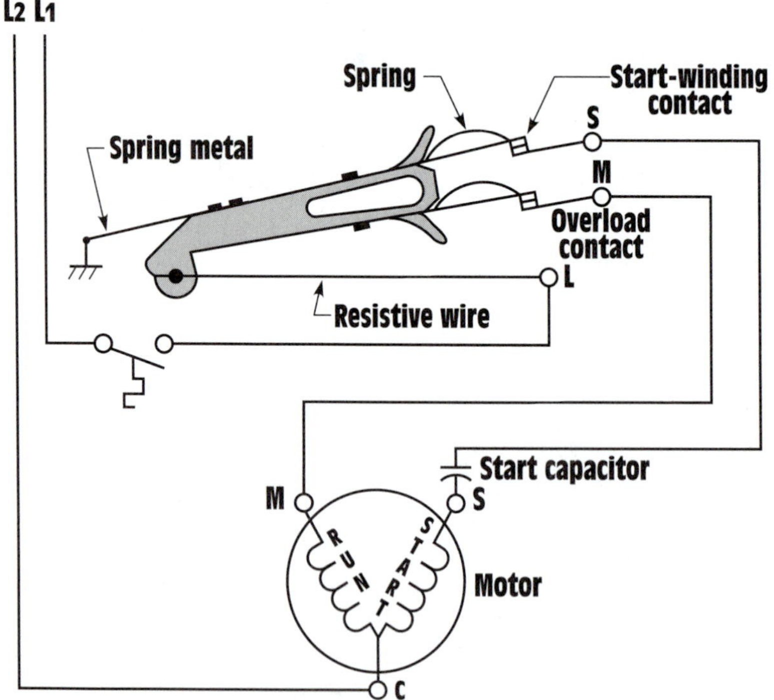

Figure 17-8 Hot wire relay connection.

and disconnect the motor from the line. A photograph of a hot wire starting relay is shown in *Figure 17-9.*

current relay

The **current relay** also operates by sensing the amount of current flow in the circuit. This type of relay operates on the principle of a magnetic field instead of expanding metal. The current relay contains a coil with a few turns of large wire and a set of normally open contacts, *Figure 17-10.* The coil of the relay is connected in series with the run winding of the motor, and the contacts are connected in series with the start winding as shown in *Figure 17-11.* When the thermostat contact closes, power is applied to the run winding of the motor. Since the start winding is open, the motor cannot start. This causes a high current to flow in the run winding circuit. This

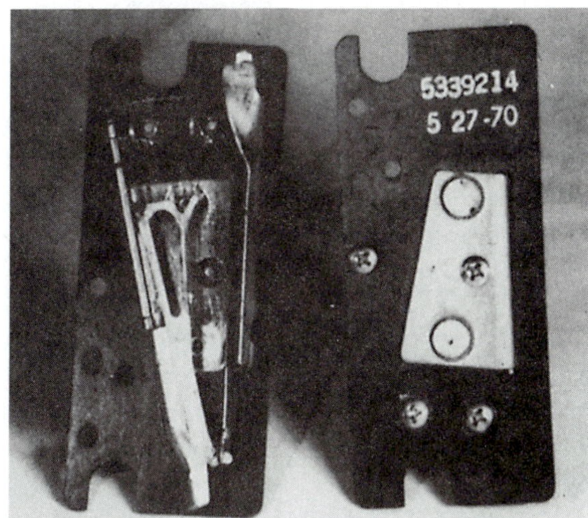

Figure 17-9 Hot wire type of starting relay.

Figure 17-10 Current type of starting relay.

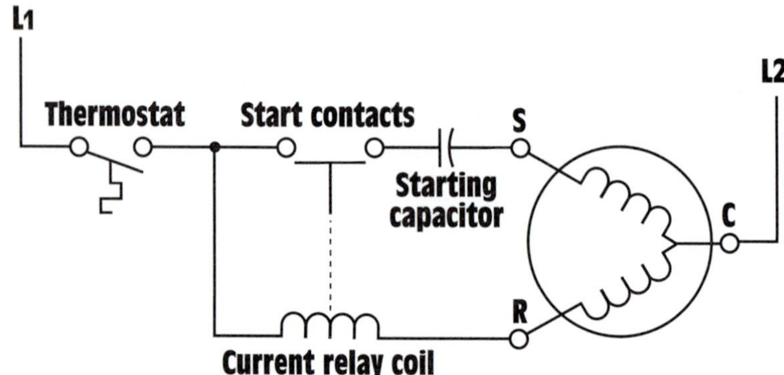

Figure 17-11 Current relay connection.

high current flow produces a strong magnetic field in the coil of the relay, causing the normally open contacts to close and connect the start winding to the circuit. When the motor starts, the run winding current is greatly reduced, permitting the start contacts to reopen and disconnect the start winding from the circuit.

RELATIONSHIP OF STATOR AND ROTOR FIELDS

The split-phase motor contains a squirrel-cage rotor very similar to those used with three-phase squirrel-cage motors. When power is connected to the stator windings, the rotating magnetic field induces a voltage into the bars of the squirrel-cage rotor. The induced voltage causes current to flow in the rotor, and a magnetic field is produced around the rotor bars. The magnetic field of the rotor is attracted to the stator field and the rotor begins to turn in the direction of the rotating magnetic field. After the centrifugal switch opens, only the run winding induces voltage into the rotor. This induced voltage is in phase with the stator current. The inductive reactance of the rotor is high, causing the rotor current to be almost 90° out of phase with the induced voltage. This causes the pulsating magnetic field of the rotor to lag the pulsating magnetic field of the stator by 90°. Magnetic poles are created in the rotor and are located midway between the stator poles, *Figure 17-12*. These two pulsating magnetic fields produce a rotating magnetic field of their own, and the rotor continues to rotate.

DIRECTION OF ROTATION

The direction of rotation for the motor is determined by the direction of rotation of the rotating magnetic field created by the run and start windings when the motor is first started. The direction of motor rotation can be changed by reversing the connection of either the start winding or the run

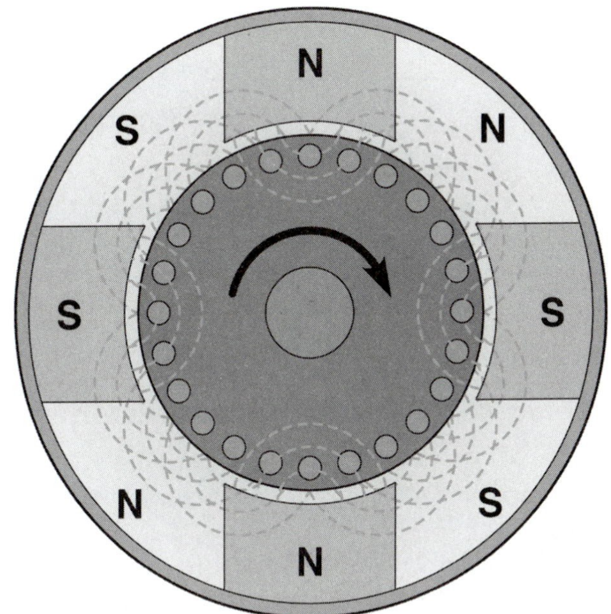

Figure 17-12 A rotating magnetic field is produced by the stator and rotor flux.

winding, but not both. If the start winding is disconnected, the motor can be operated in either direction by manually turning the rotor shaft in the desired direction of rotation.

THE CAPACITOR-START INDUCTION-RUN MOTOR

The capacitor-start induction-run motor is very similar in construction and operation to the resistance-start induction-run motor. The capacitor-start induction-run motor, however, has an AC electrolytic capacitor connected in series with the centrifugal switch and start winding, *Figure 17-13*. Although the running characteristics of both the capacitor-start induction-run motor and the resistance-start induction-run motor are identical, the starting characteristics are not. The capacitor-start induction-run motor produces a starting torque that is substantially higher than the resistance-start induction-run motor. Recall that one of the factors that determines the starting torque for a split-phase motor is the phase angle difference between start-winding current and run-winding current. The starting torque of a resistance-start induction-run motor is low because the phase angle difference between these two currents is only about 40°, Figure 17-4.

When a capacitor of the proper size is connected in series with the start winding, it causes the start winding current to lead the applied voltage. This leading current produces a 90° phase shift between run-winding current

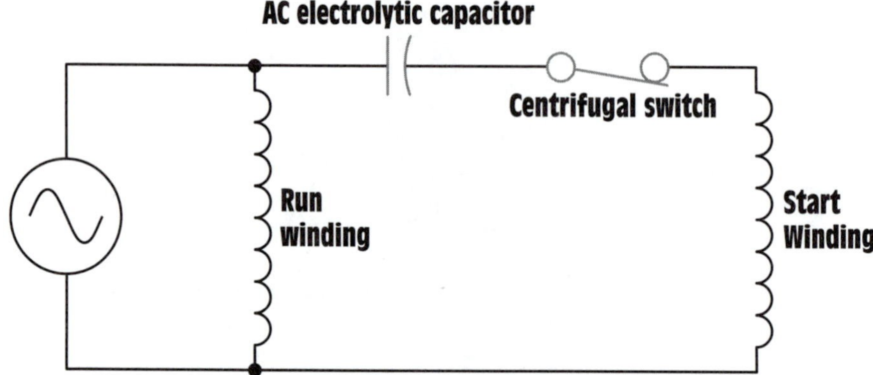

Figure 17-13 An AC electrolytic capacitor is connected in series with the start winding.

Although the capacitor-start induction-run motor has a high starting torque, the motor should not be started more than about eight times per hour.

and start-winding current, *Figure 17-14.* Maximum starting torque is developed at this point.

Although the capacitor-start induction-run motor has a high starting torque, the motor should not be started more than about eight times per hour. Frequent starting can damage the start capacitor due to overheating. If the capacitor must be replaced, care should be taken to use a capacitor of the correct microfarad rating. If a capacitor with too little capacitance is used, the starting current will be less than 90° out of phase with the running current, and the starting torque will be reduced. If the capacitance value is too great, the starting current will be more than 90° out of phase with the running current, and the starting

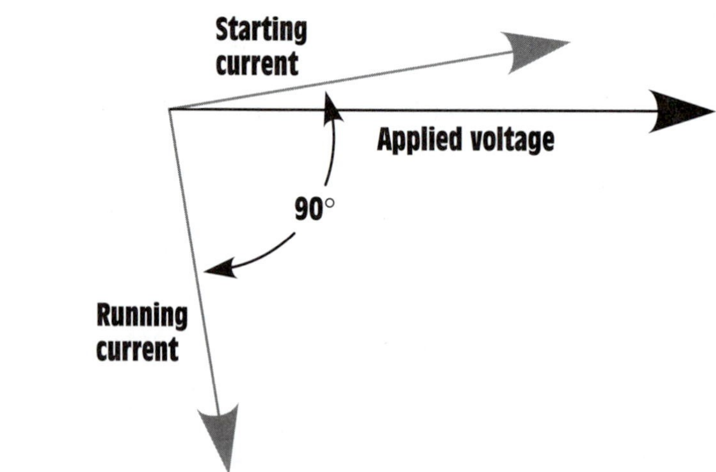

Figure 17-14 Run-winding current and start-winding current are 90° out of phase with each other.

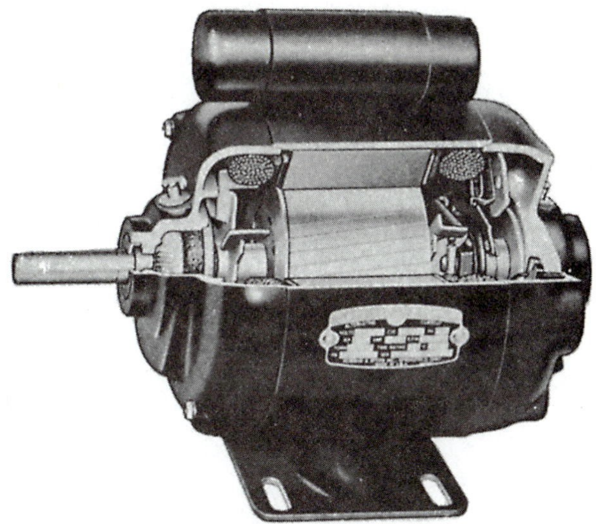

Figure 17-15 Capacitor-start induction-run motor.

torque will again be reduced. A capacitor-start induction-run motor is shown in *Figure 17-15*.

DUAL-VOLTAGE SPLIT-PHASE MOTORS

Many split-phase motors are designed for operation on 120 or 240 V. *Figure 17-16* shows the schematic diagram of a split-phase motor designed for dual voltage operation. This particular motor contains two run windings and two start windings. The lead numbers for single-phase motors are numbered in a standard manner also. One of the run windings has lead numbers of T_1 and T_2. The other run winding has its leads numbered T_3 and T_4. This particular motor uses two different sets of start winding leads. One set is labeled T_5 and T_6, and the other set is labeled T_7 and T_8.

If the motor is to be connected for high-voltage operation, the run windings and start windings will be connected in series as shown in *Figure 17-17*. The start windings are then connected in parallel with the run windings. It should be noted that if the opposite direction of rotation is desired, T_5 and T_8 will be changed.

For low-voltage operation, the windings must be connected in parallel as shown in *Figure 17-18*. This connection is made by first connecting the run windings in parallel by hooking T_1 and T_3 together, and T_2 and T_4 together. The start windings are paralleled by connecting T_5 and T_7 together, and T_6 and T_8 together. The start windings are then connected in parallel with the run windings. If the opposite direction of rotation is desired, T_5 and T_6 should be reversed.

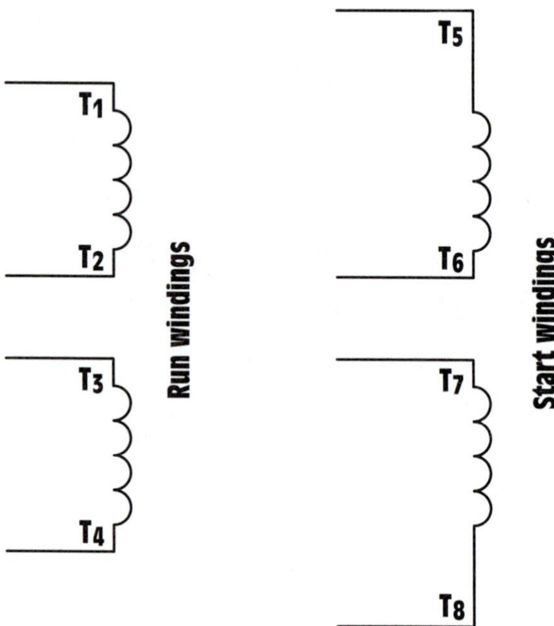

Figure 17-16 Dual-voltage windings for a split-phase motor.

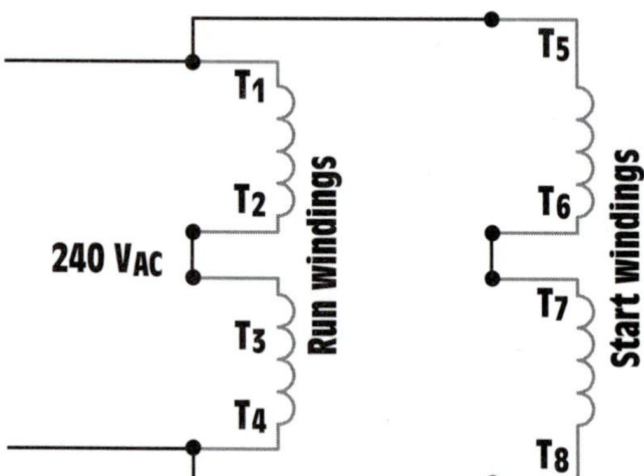

Figure 17-17 High-voltage connection for a split-phase motor with two run and two start windings.

Not all dual-voltage single-phase motors contain two sets of start windings. *Figure 17-19* shows the schematic diagram of a motor that contains two sets of run windings and only one start winding. In this illustration, the start winding is labeled T_5 and T_6. It should be noted, however, that some motors identify the start winding by labeling it T_5 and T_8 as shown in *Figure 17-20*.

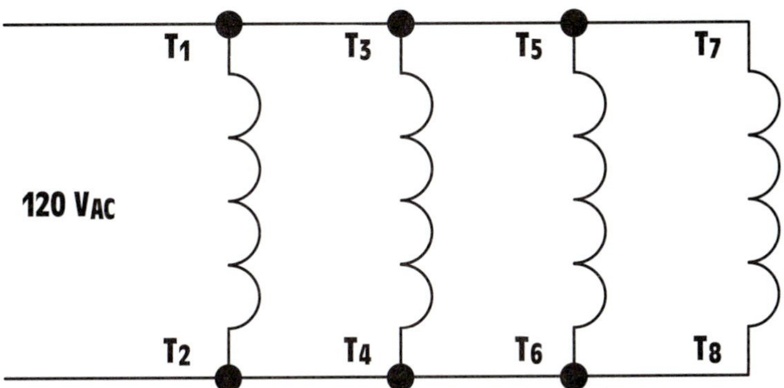

Figure 17-18 Low-voltage connection for a split-phase motor with two run and two start windings.

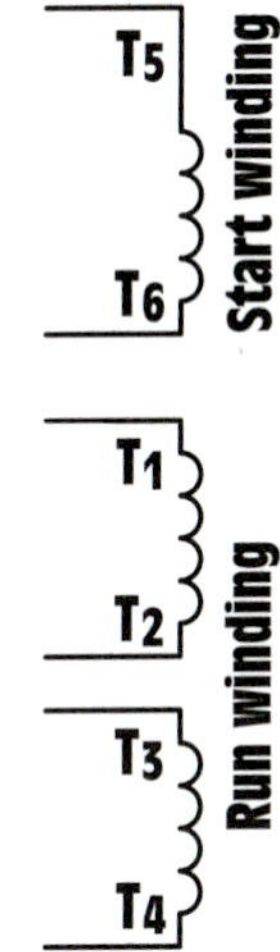

Figure 17-19 Dual-voltage motor with one start winding labeled T_5 and T_6.

Regardless of which method is used to label the terminal leads of the start winding, the connection will be the same. If the motor is to be connected for high-voltage operation, the run windings will be connected in series, and the start winding will be connected in parallel with one of the run windings, as shown in *Figure 17-21*. In this type of motor, each winding is rated at 120 V. If the run windings are connected in series across 240 V, each winding will have a voltage drop of 120 V. By connecting the start winding in parallel across only one run winding, it will receive only 120 V when power is applied to the motor. If the opposite direction of rotation is desired, T_5 and T_8 should be changed.

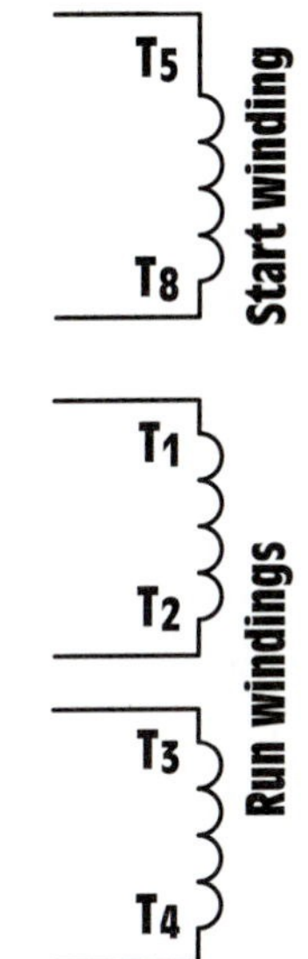

Figure 17-20 Dual-voltage motor with one start winding labeled T_5 and T_8.

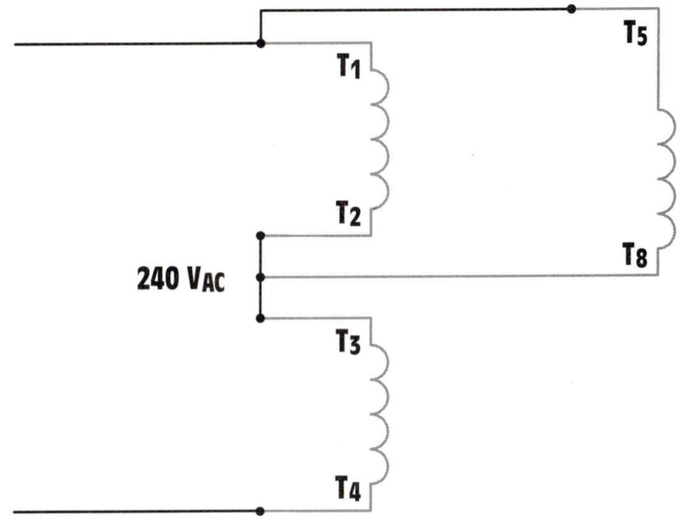

Figure 17-21 High-voltage connection with one start winding.

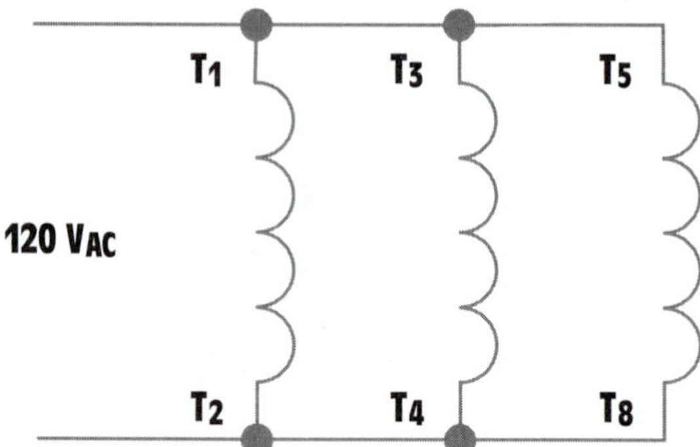

Figure 17-22 Low-voltage connection for a split-phase motor with one start winding.

If the motor is to be operated on low voltage, the windings are connected in parallel as shown in *Figure 17-22*. Since all windings are connected in parallel, each will receive 120 V when power is applied to the motor.

DETERMINING THE DIRECTION OF ROTATION FOR SPLIT-PHASE MOTORS

The direction of rotation of a single-phase motor can generally be determined when the motor is connected. The direction of rotation is determined by facing the back or rear of the motor. *Figure 17-23* shows a connection diagram for rotation. If clockwise rotation is desired, T_5 should be connected to T_1. If counterclockwise rotation is desired, T_8 (or T_6) should be connected to T_1. It should be noted that this connection diagram assumes the motor contains two sets of run and two sets of start windings. The type

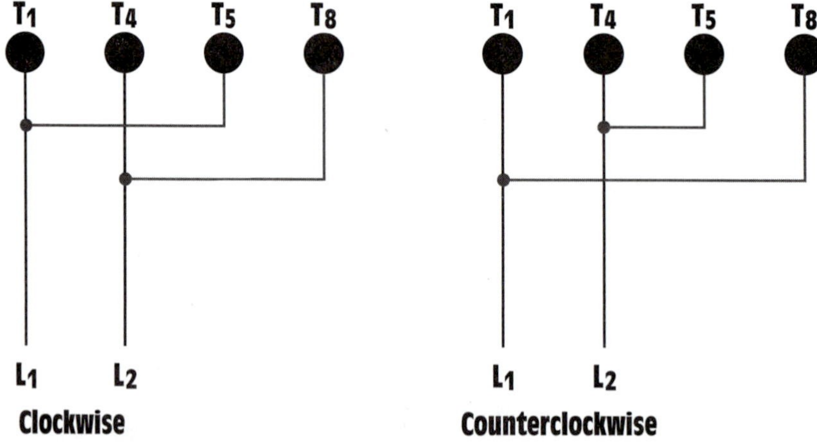

Figure 17-23 Determining the direction of rotation for a split-phase motor.

motor used will determine the actual connection. For example, Figure 17-21 shows the connection of a motor with two run windings and only one start winding. If this motor were to be connected for clockwise rotation, terminal T_5 would be connected to T_1 and terminal T_8 would be connected to T_2 and T_3. If counterclockwise rotation is desired, terminal T_8 would be connected to T_1 and terminal T_5 would be connected to T_2 and T_3.

CAPACITOR-START CAPACITOR-RUN MOTOR (PERMANENT SPLIT-CAPACITOR MOTOR)

Although the capacitor-start capacitor-run motor is a split-phase motor, it operates on a different principle than the resistance-start induction-run motor or the capacitor-start induction-run motor. The capacitor-start capacitor-run motor is designed so that its start winding remains energized at all times. A capacitor is connected in series with the winding to provide a continuous leading current in the start winding, *Figure 17-24.* Since the start winding remains energized at all times, no centrifugal switch is needed to disconnect the start winding as the motor approaches full speed. The capacitor used in this type motor will generally be of the oil-filled type since it is intended for continuous use. An exception to this general rule are small fractional horsepower motors used in reversible ceiling fans. These fans have a low current draw and use an AC electrolytic capacitor to help save space.

The capacitor-start capacitor-run motor actually operates on the principle of a rotating magnetic field in the stator. Since both run and start windings remain energized at all times, the stator magnetic field continues to rotate and the motor operates as a two-phase motor. This motor has excellent starting and running torque. It is quiet in operation and has a high efficiency. Since the capacitor remains connected in the circuit at all times, the motor power factor is close to unity.

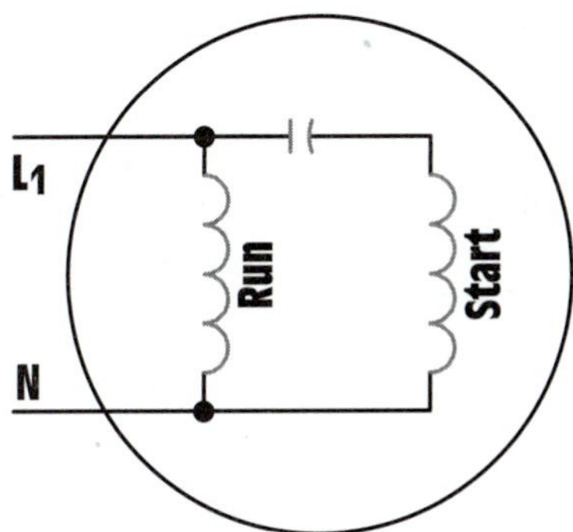

Figure 17-24 A capacitor-start capacitor-run motor.

Although the capacitor-start capacitor-run motor does not require a centrifugal switch to disconnect the capacitor from the start winding, there are some motors that use a second capacitor during the starting period to help improve starting torque, *Figure 17-25*. Capacitor-start capacitor-run motors that are not hermetically sealed contain a centrifugal switch to disconnect the start capacitor when the motor reaches about 75% of its full load speed.

Hermetically sealed motors, such as the compressor of a central air conditioning unit designed for operation on single-phase power, use an external starting relay to disconnect the starting capacitor. This type of motor generally employs a potential starting relay, *Figure 17-26*. The potential starting relay operates by sensing an increase in the voltage developed in the start winding when the motor is operating. A schematic diagram of a potential starting relay circuit is shown in *Figure 17-27*. In this circuit, the potential relay is used

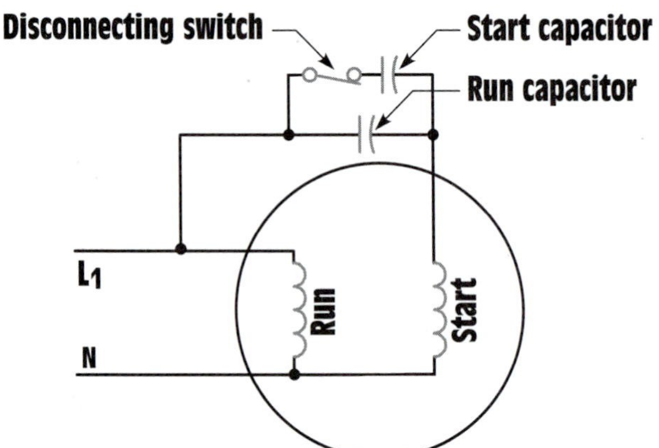

Figure 17-25 Capacitor-start capacitor-run motor with additional starting capacitor.

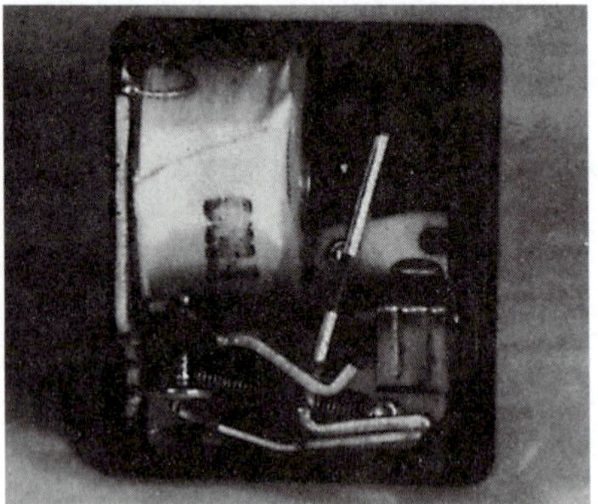

Figure 17-26 Potential starting relay.

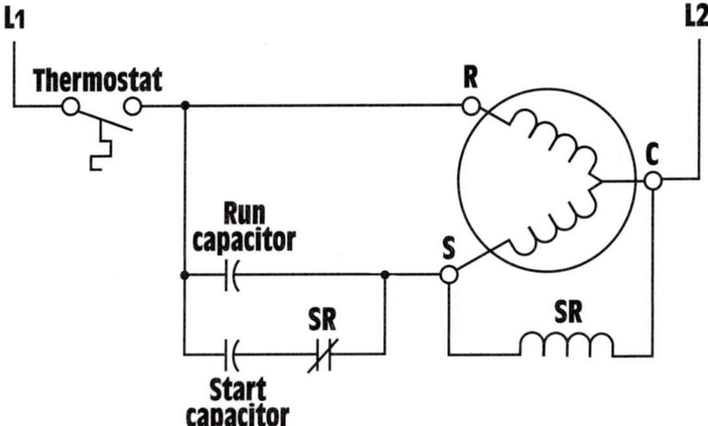

Figure 17-27 Potential relay connection.

to disconnect the starting capacitor from the circuit when the motor reaches about 75% of its full speed. The starting relay coil, SR, is connected in parallel with the start winding of the motor. A normally closed SR contact is connected in series with the starting capacitor. When the thermostat contact closes, power is applied to both the run and start windings. At this point in time, both the start and run capacitors are connected in the circuit.

As the rotor begins to turn, its magnetic field induces a voltage into the start winding, producing a higher voltage across the start winding than the applied voltage. When the motor has accelerated to about 75% of its full speed, the voltage across the start winding is high enough to energize the coil of the potential relay. This causes the normally closed SR contact to open and disconnect the start capacitor from the circuit.

SHADED-POLE INDUCTION MOTORS

The **shaded-pole induction motor** is popular because of its simplicity and long life. This motor contains no start windings or centrifugal switch. It contains a squirrel-cage rotor and operates on the principle of a rotating magnetic field. The rotating magnetic field is created by a **shading coil** wound on one side of each pole piece. Shaded-pole motors are generally fractional-horsepower motors and used for low-torque applications such as operating fans and blowers.

shaded-pole induction motor

shading coil

THE SHADING COIL

The shading coil is wound around one end of the pole piece, *Figure 17-28*. The shading coil is actually a large loop of copper wire or a copper band. Both ends are connected to form a complete circuit. The shading coil

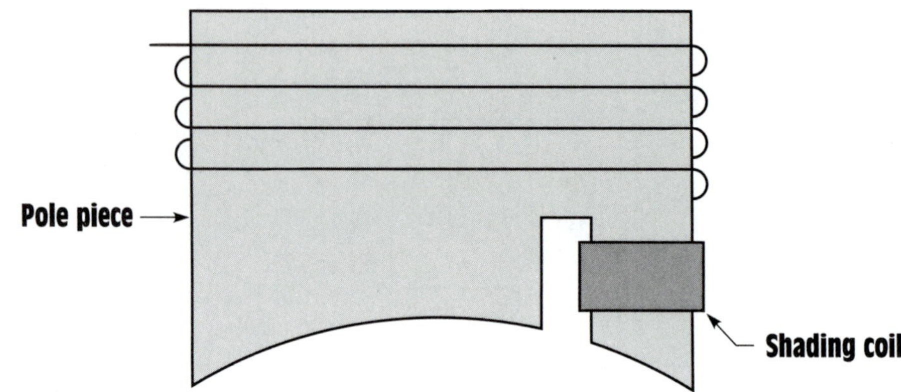

Figure 17-28 A shaded pole.

acts in the same manner as a transformer with a shorted secondary winding. When the current of the AC waveform increases from zero toward its positive peak, a magnetic field is created in the pole piece. As magnetic lines of flux cut through the shading coil, a voltage is induced in the coil. Since the coil is a low resistance short circuit, a large amount of current flows in the loop. This current causes an opposition to the change of magnetic flux, *Figure 17-29.* As long as voltage is induced into the shading coil, there will be an opposition to the change of magnetic flux.

When the AC current reaches its peak value, it is no longer changing, and no voltage is being induced into the shaded coil. Since there is no current flow in the shading coil, there is no opposition to the magnetic flux. The magnetic flux of the pole piece is now uniform across the pole face, *Figure 17-30.*

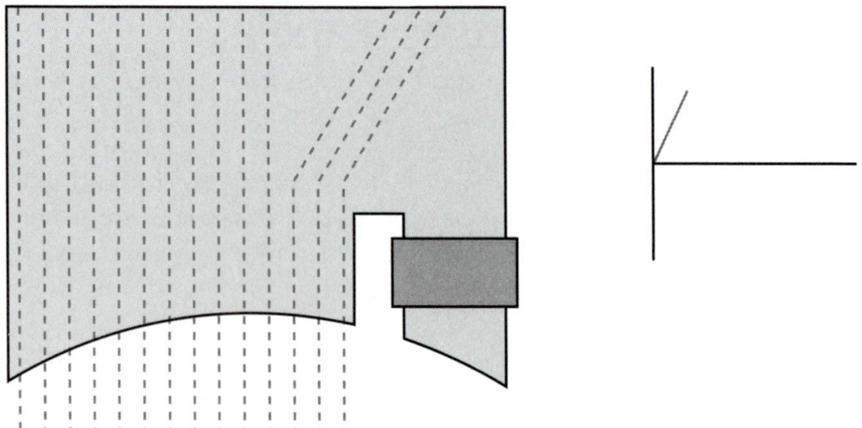

Figure 17-29 The shading coil opposes a change of flux as current increases.

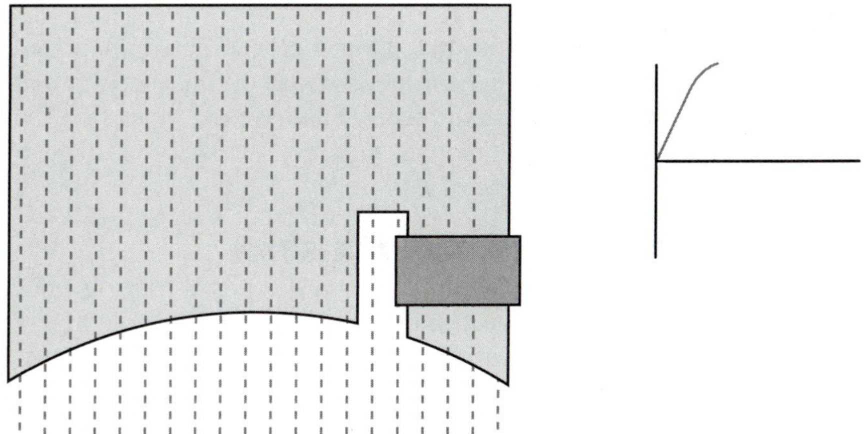

Figure 17-30 There is no opposition to magnetic flux when the current is not changing.

When the AC current begins to decrease from its peak value back toward zero, the magnetic field of the pole piece begins to collapse. A voltage is again induced into the shading coil. This induced voltage creates a current that opposes the change of magnetic flux, *Figure 17-31.* This causes the magnetic flux to be concentrated in the shaded section of the pole piece.

When the AC current passes through zero and begins to increase in the negative direction, the same set of events happen, except that the polarity of the magnetic field is reversed. If these events were to be viewed in rapid order, the magnetic field would be seen to rotate across the face of the pole piece.

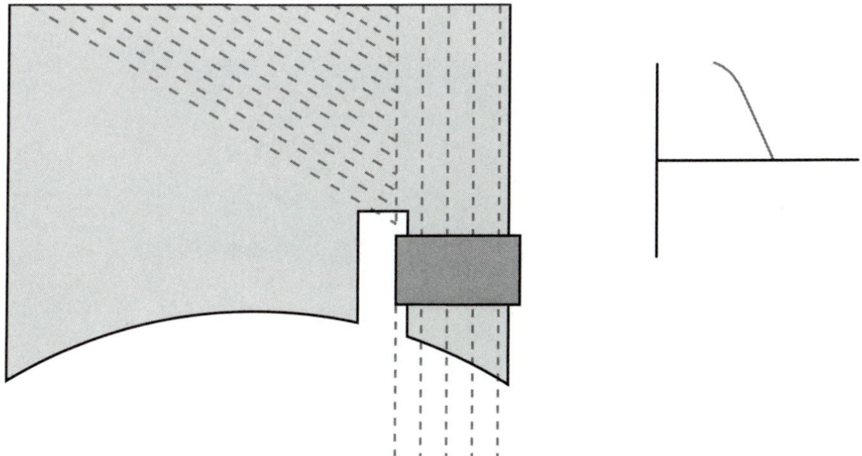

Figure 17-31 The shading coil opposes a change of flux when the current decreases.

SPEED

The speed of the shaded-pole induction motor is determined by the same factors that determine the synchronous speed of other induction motors: frequency and number of stator poles. Shaded-pole motors are commonly wound as four or six pole motors. *Figure 17-32* shows a drawing of a four-pole shaded-pole induction motor.

GENERAL OPERATING CHARACTERISTICS

The shaded-pole motor contains a standard squirrel-cage rotor. The amount of torque produced is determined by the strength of the magnetic field of the stator, the strength of the magnetic field of the rotor, and the phase angle difference between rotor and stator flux. The shaded-pole induction motor has low starting and running torque.

The direction of rotation is determined by the direction in which the rotating magnetic field moves across the pole face. The rotor will turn in the direction shown by the arrow in Figure 17-32. If the direction must be changed, it can be done by removing the stator winding and turning it around. This is not a common practice, however. As a general rule the shaded-pole induction motor is considered to be nonreversible. *Figure 17-33* shows a photograph of the stator winding and rotor of a shaded-pole induction motor.

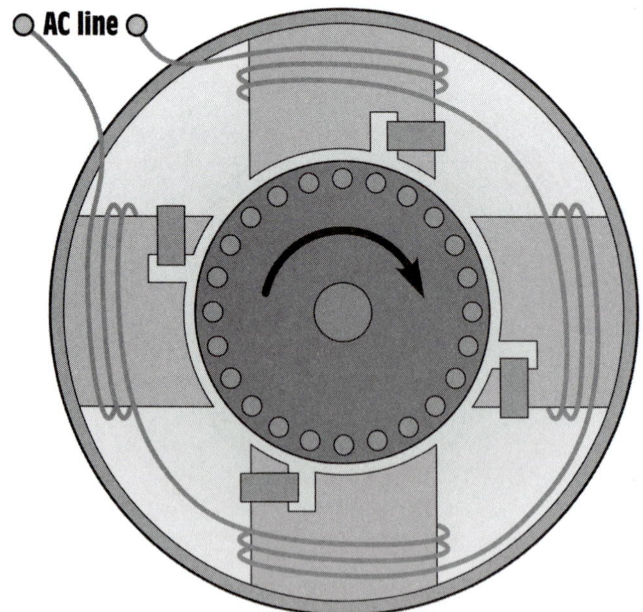

Figure 17-32 Four-pole shaded-pole induction motor.

Figure 17-33 Stator winding and rotor of a shaded-pole induction motor.

MULTISPEED MOTORS

multispeed motors

consequent pole motor

There are two basic types of **multispeed** single-phase **motors**. One is the **consequent pole motor** and the other is a specially wound capacitor-start capacitor-run motor or shaded-pole induction motor. The consequent pole single-phase motor operates in the same basic way as the three-phase consequent pole discussed in Unit 16. The speed is changed by reversing the current flow through alternate poles and increasing or decreasing the total number of stator poles. The consequent pole motor is used where high running torque must be maintained at different speeds. A good example of where this type of motor is used is in two-speed compressors for central air conditioning units.

MULTISPEED FAN MOTORS

Multispeed fan motors have been used for many years. These motors are generally wound for two to five steps of speed and operate fans and squirrel-cage blowers. A schematic drawing of a three-speed motor is shown in *Figure 17-34*. Notice that the run winding has been tapped to produce low, medium, and high speed. The start winding is connected in parallel with the run winding section. The other end of the start winding lead is connected to an external oil-filled capacitor. This motor obtains a change of speed by inserting inductance in series with the run winding. The actual run winding for this motor is between the terminals marked *high* and *common*. The winding shown between *high* and *medium* is connected in series with the main run winding. When the rotary switch is connected to the medium speed position, the inductive reactance of this coil limits the amount of current flow through the run winding. When the current of the run winding is reduced, the strength of the magnetic field of the run winding is reduced and the motor produces

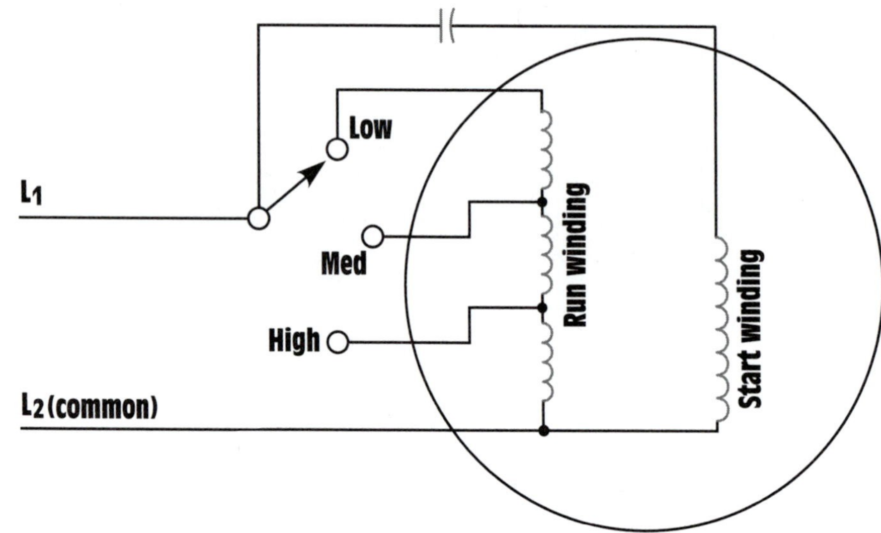

Figure 17-34 A three-speed motor.

less torque. This causes a greater amount of slip and the motor speed to decrease.

If the rotary switch is changed to the low position, more inductance is inserted in series with the run winding. This causes less current to flow through the run winding and another reduction in torque. When the torque is reduced the motor speed decreases again.

Common speeds for a four-pole motor of this type are 1,625, 1,500, and 1,350 RPM. Notice that this motor does not have wide ranges between speeds as would be the case with a consequent pole motor. Most induction motors would overheat and damage the motor winding if the speed were to be reduced to this extent. This type motor, however, has much higher impedance windings than most motors. The run windings of most split-phase motors have a wire resistance of 1–4 Ω. This motor will generally have a resistance of 10–15 Ω in its run winding. It is the high impedance of the windings that permits the motor to be operated in this manner without damage.

Since this motor is designed to slow down when load is added, it is not used to operate high-torque loads. This type of motor is generally used to operate only low-torque loads such as fans and blowers.

synchro-nous motors

SINGLE-PHASE SYNCHRONOUS MOTORS

Single-phase **synchronous motors** are small and develop only fractional horsepower. They operate on the principle of a rotating magnetic field developed by a shaded-pole stator. Although they will operate at synchronous speed, they do not require DC excitation current. They are used in applications where constant speed is required such as clock motors, timers, and recording instruments. They also are used as the driving force for small fans because they are small and inexpensive to manufacture. There are two basic types of synchronous motors: the Warren or General Electric motor, and the Holtz motor. These motors are also referred to as hysteresis motors.

WARREN MOTORS

Warren motor

The **Warren motor** is constructed with a laminated stator core and a single coil. The coil is generally wound for 120-V AC operation. The core contains two poles, which are divided into two sections each. One-half of each pole piece contains a shading coil to produce a rotating magnetic field, *Figure 17-35*. Since the stator is divided into two poles, the synchronous field speed will be 3,600 RPM when connected to 60 Hz.

The difference between the Warren and Holtz motors is the type of rotor used. The rotor of the Warren motor is constructed by stacking hardened steel laminations onto the rotor shaft. These disks have high hysteresis loss.

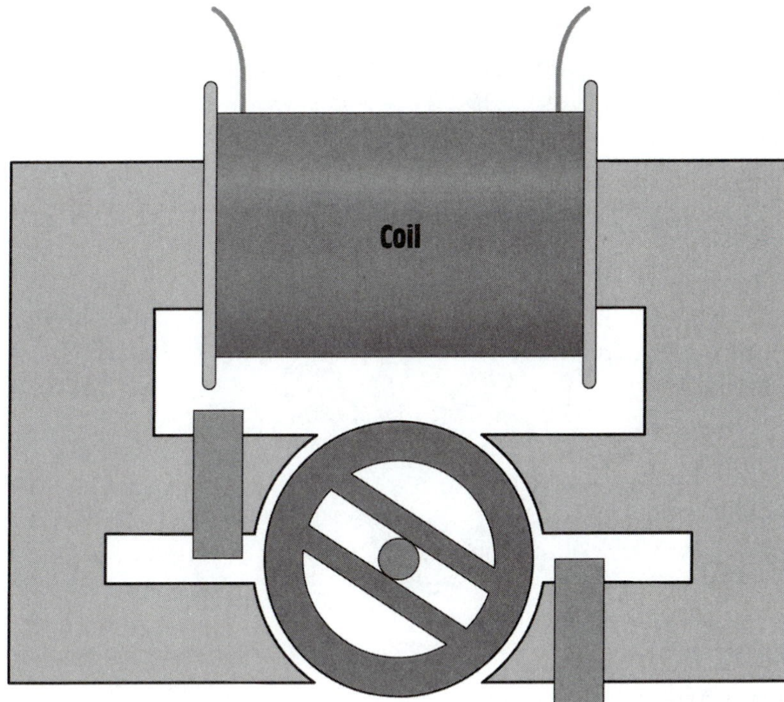

Figure 17-35 A Warren motor.

The laminations form two crossbars for the rotor. When power is connected to the motor, the rotating magnetic field induces a voltage into the rotor and a strong starting torque is developed, causing the rotor to accelerate to near synchronous speed. Once the motor has accelerated to near synchronous speed, the flux of the rotating magnetic field will follow the path of minimum reluctance (magnetic resistance) through the two crossbars. This causes the rotor to lock in step with the rotating magnetic field and the motor operates at 3,600 RPM. These motors are often used with small gear trains to reduce the speed to the desired level.

Holtz motor

HOLTZ MOTORS

The **Holtz motor** uses a different type of rotor, *Figure 17-36*. This rotor is cut in such a manner that six slots are formed. These slots form six salient (projecting or jutting) poles for the rotor. A squirrel-cage winding is constructed by inserting a metal bar at the bottom of each slot. When power is connected to the motor, the squirrel-cage winding provides the torque

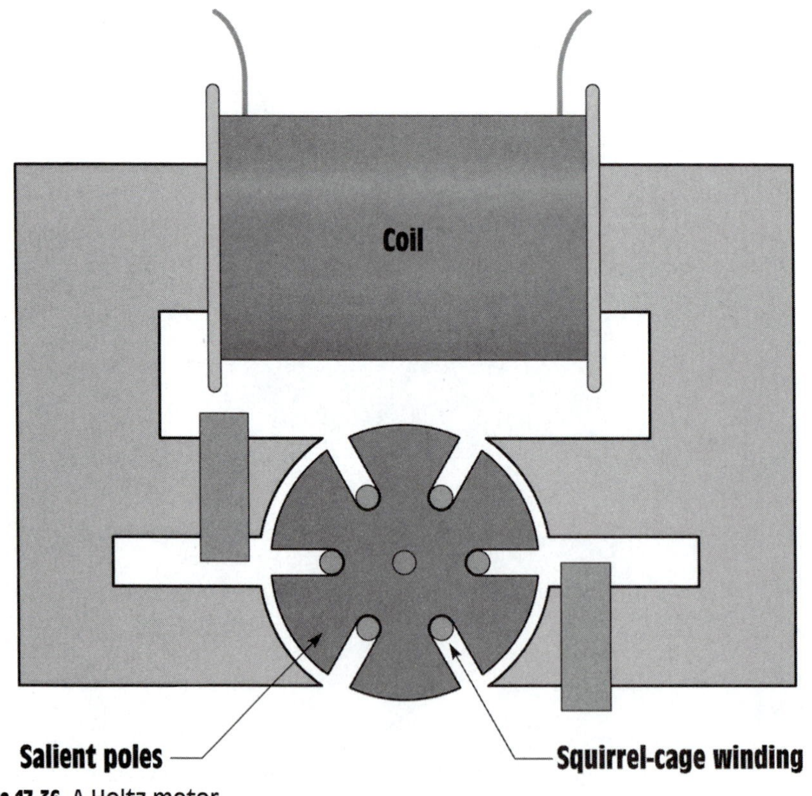

Figure 17-36 A Holtz motor.

necessary to start the rotor turning. When the rotor approaches synchronous speed, the salient poles will lock in step with the field poles each half cycle. This produces a rotor speed of 1,200 RPM (one-third of synchronous speed) for the motor.

UNIVERSAL MOTORS

universal motor

compensating winding

The **universal motor** is often referred to as an AC series motor. This motor is very similar to a DC series motor in its construction because it contains a wound armature and brushes, *Figure 17-37*. The universal motor, however, has the addition of a **compensating winding**. The universal motor is so named because it can be operated on AC or DC voltage.

SPEED REGULATION

The speed regulation of the universal motor is very poor. If the universal motor is connected to a light load or no load, its speed is almost unlimited. It is not unusual for this motor to be operated at several thousand RPM. Universal motors are used in a number of portable appliances where high horsepower and light weight is needed, such as drill motors, skill saws, and vacuum cleaners. The universal motor is able to produce a high horsepower for its size and weight. To change the direction of rotation, change the armature leads with respect to the field leads.

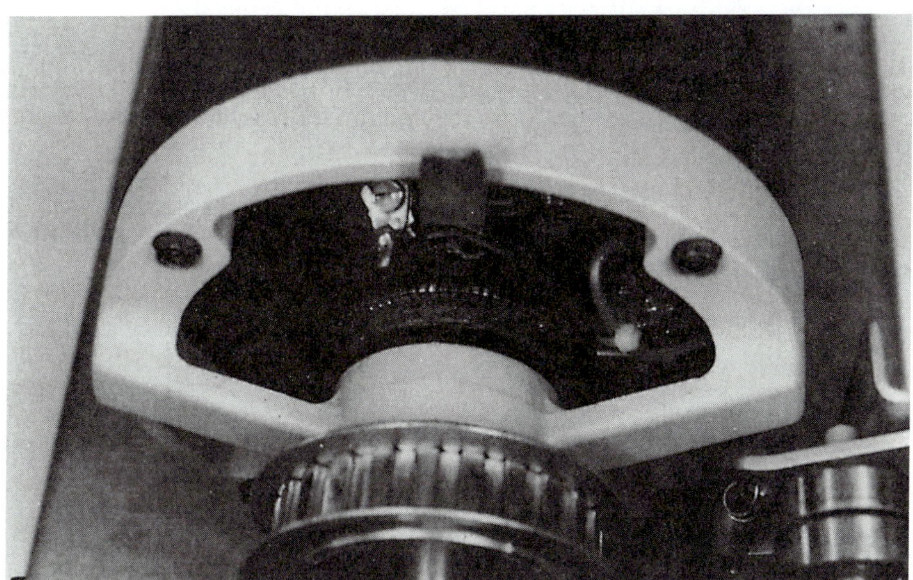

Figure 17-37 Armature and brushes of a universal motor.

SUMMARY

1. Not all single-phase motors operate on the principle of a rotating magnetic field.
2. Split-phase motors start as two-phase motors by producing an out of phase condition for the current in the run winding and the current in the start winding.
3. The resistance of the wire in the start winding of a resistance-start induction-run motor is used to produce a phase angle difference between the current in the start winding and the current in the run winding.
4. The capacitor-start induction-run motor uses an AC electrolytic capacitor to increase the phase angle difference between starting and running current. This causes an increase in starting torque.
5. Maximum starting torque for a split-phase motor is developed when the start-winding current and run-winding current are 90° out of phase with each other.
6. Most resistance-start induction-run motors and capacitor-start induction-run motors use a centrifugal switch to disconnect the start windings when the motor reaches approximately 75% of full load speed.
7. The capacitor-start capacitor-run motor operates like a two-phase motor because both the start and run windings remain energized during motor operation.
8. Most capacitor-start capacitor-run motors use an AC oil-filled capacitor connected in series with the start winding.
9. The capacitor of the capacitor-start capacitor-run motor does help to correct the power factor.
10. Shaded-pole induction motors operated on the principle of a rotating magnetic field.
11. The rotating magnetic field of a shaded-pole induction motor is produced by placing shading loops or coils on one side of the pole piece.
12. The synchronous field speed of a single-phase motor is determined by the number of stator poles and the frequency of the applied voltage.

13. Consequent pole motors are used when a change of motor speed is desired and high torque must be maintained.
14. Multispeed fan motors are constructed by connecting windings in series with the main run winding.
15. Multispeed fan motors have high-impedance stator windings to prevent them from overheating when their speed is reduced.
16. Hermetically sealed resistance-start induction-run and capacitor-start induction-run motors use a starting relay to disconnect the start windings from the circuit when the motor approaches about 75% of its full speed.
17. Two types of starting relays used for the resistance-start induction-run and capacitor-start induction-motors are the hot wire relay and the current relay.
18. A potential starting relay is generally used to disconnect the start capacitor from the line for a hermetically sealed capacitor-start capacitor-run motor.
19. The capacitor-start capacitor-run motor is also known as a permanent-split capacitor motor.
20. Shaded-pole motors are generally considered to be not reversible.
21. There are two type of single-phase synchronous motors: the Warren and the Holtz.
22. Single-phase synchronous motors are sometimes called hysteresis motors.
23. The Warren motor operates at a speed of 3,600 RPM.
24. The Holtz motor operates at a speed of 1,200 RPM.
25. Universal motors operate on direct or alternating current.
26. Universal motors contain a wound armature and brushes.
27. Universal motors are also called AC series motors.
28. The direction of rotation for a universal motor can be changed by reversing the armature leads with respect to the field leads.
29. Universal motors develop a large amount of horsepower for their size and weight because of their high operating speed.

REVIEW QUESTIONS

1. What are the three basic types of split-phase motors?
2. The voltages of a two-phase system are how many degrees out of phase with each other?
3. How are the start and run windings of a split-phase motor connected in relation to each other?
4. In order to produce maximum starting torque in a split-phase motor, how many degrees out of phase should the start and run winding currents be with each other?
5. What is the advantage of the capacitor-start induction-run motor over the resistance-start induction-run motor?
6. On the average, how many degrees out of phase with each other are the start and run winding current in a resistance-start induction-run motor?
7. What device is used to disconnect the start windings for the circuit in most non–hermetically sealed capacitor-start induction-run motors?
8. Why does a split-phase motor continue to operate after the start windings have been disconnected from the circuit?
9. How can the direction of rotation of a split-phase motor be reversed?
10. If a dual-voltage split-phase motor is to be operated on high voltage, how are the run windings connected in relation to each other?
11. When determining the direction of rotation for a split-phase motor, should that motor be faced from the front or from the rear?
12. What type of split-phase motor does not generally contain a centrifugal switch?
13. What type of single-phase motor develops the highest starting torque?
14. Name two types of starting relays used with hermetically sealed capacitor-start induction-run motors.
15. What is the principle of operation of a capacitor-start capacitor-run motor?
16. What causes the magnetic field to rotate in a shaded-pole induction motor?

17. How can the direction of rotation of a shaded-pole induction motor be changed?
18. How is the speed of a consequent pole motor changed?
19. Why can a multispeed fan motor be operated at lower speed than most induction motors without harm to the motor windings?
20. What is the speed of operation of the Warren motor?
21. What is the speed of operation of the Holtz motor?
22. Why is the AC series motor often referred to as a universal motor?
23. How is the direction of rotation of the universal motor reversed?

18 **Unit**

Schematics and Wiring Diagrams

objectives

fter studying this unit, you should be able to:

- Place wire numbers on a schematic diagram.
- Place the corresponding numbers on control components.
- Draw a wiring diagram from a schematic diagram.
- Define the difference between a schematic, or ladder, diagram and a wiring diagram.
- Discuss the operation of a start-stop pushbutton circuit.
- Connect a start-stop pushbutton control circuit.

A schematic diagram shows components in their electrical sequence, without regard for the physical location of any component, *Figure 18-1.* **A wiring diagram is a pictorial representation of components with connecting wires.**

The pictorial representation of the components is shown in *Figure 18-2*. Schematic diagrams are generally preferred over wiring diagrams for troubleshooting a circuit because they illustrate circuit logic better. However, some people prefer to use wiring diagrams when making actual circuit connections because they show the placement of the wires.

When viewing a schematic diagram, there are several rules that should be understood.

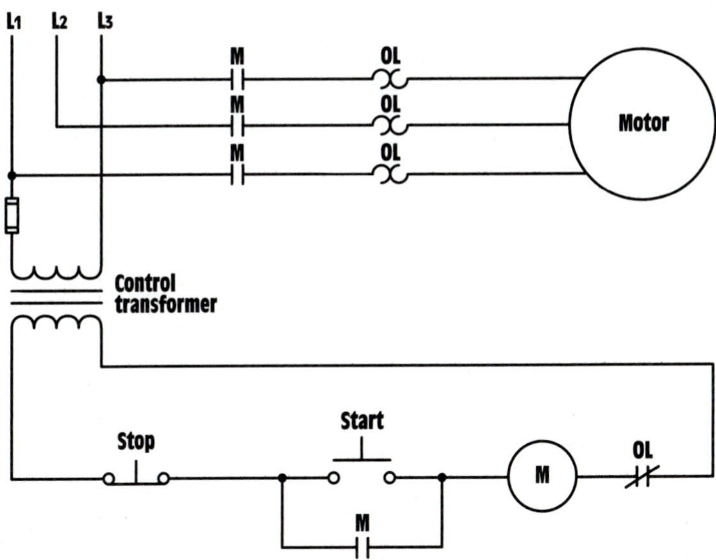

Figure 18-1 Basic diagram of a basic start-stop pushbutton control circuit.

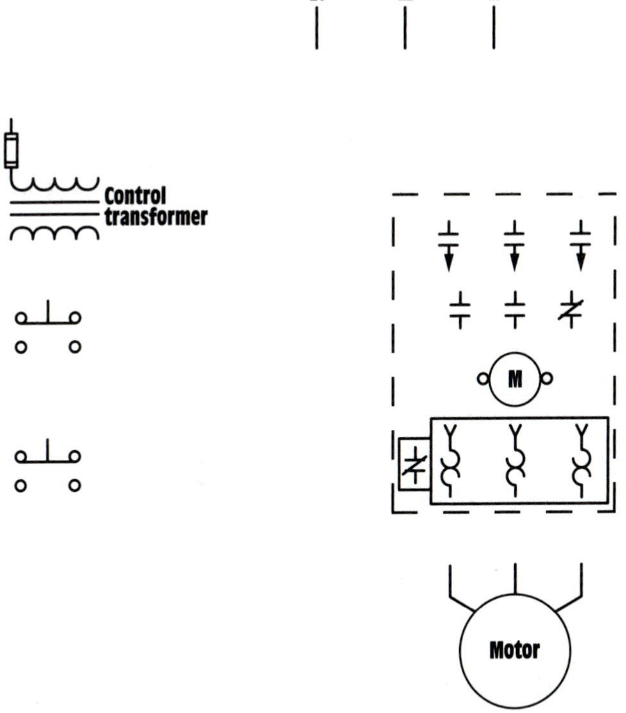

Figure 18-2 Components of the basic start-stop control circuit.

1. Schematics are always shown in the off, or de-energized, condition. The states of all contacts and components are shown as they would be if the machine were turned off and not operating.
2. Schematic diagrams are generally read like a book, from left to right and from top to bottom.
3. Schematic symbols are generally drawn to represent the components they symbolize. A list of symbols is shown in *Figure 18-3*. While there are no actual standards for symbols, the most frequently employed are those used by the *National Electrical Manufacturers Association* (NEMA).
4. Coils are generally labeled with letters and/or numbers. Contacts with the same labels are controlled by one coil.
5. When a coil is energized, all contacts controlled by that coil will change condition. Contacts shown normally open will close and contacts shown normally closed will open.

ANALYZING THE CIRCUIT

The circuit shown in Figure 18-1 is a basic start-stop pushbutton control circuit. This circuit is the basis on which many other control circuits are built. Understanding how this circuit works is imperative to understanding more complicated control circuits. The schematic in Figure 18-1 shows both the load and control circuits. The load circuit consists of the motor, overload heaters, load contacts controlled by the M starter, and the three phase power lines labeled L_1, L_2, and L_3. Motor starters are used to connect a motor to the power line. They contain several parts:

COIL

In the schematic, the coil is shown as a circle with an M in the center of it. A motor starter is basically an electrically operated switch or solenoid with multiple contacts connected to it. When the coil is energized (power applied), all contacts controlled by the coil change position. When the coil is de-energized, all the contacts return to their normal position.

LOAD CONTACTS

The load contacts are large and intended to control large amounts of current. The load contacts are used to connect the motor to the power line.

AUXILIARY CONTACTS

Most motor starters contain small contacts intended to be used as part of the control circuit. These contacts are generally called auxiliary contacts.

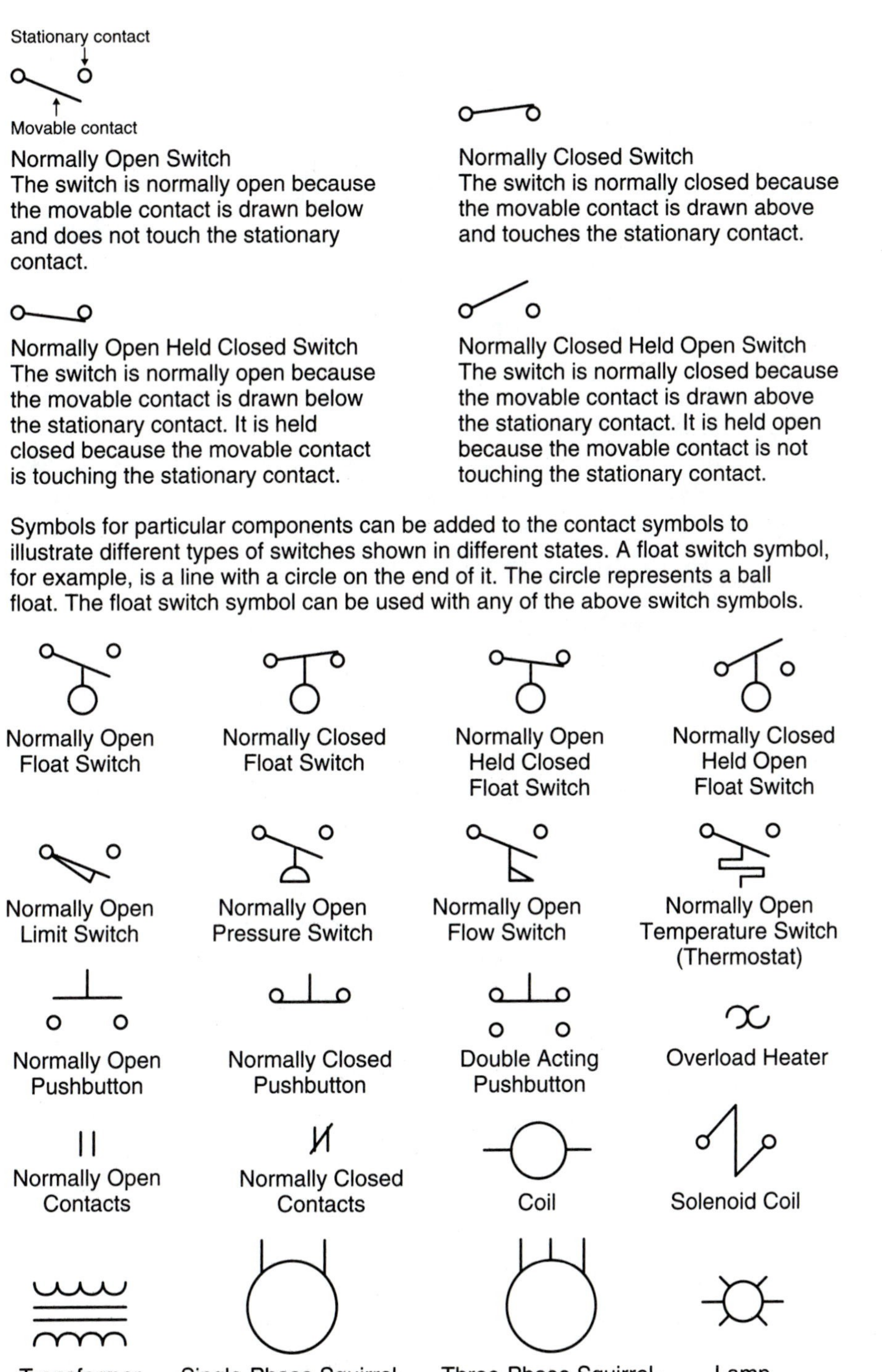

Figure 18-3 Basic schematic symbols.

They can be used to control small current loads such as those required for pilot lights and the coils of motor starters and relays.

OVERLOAD RELAY

The overload relay is used to sense motor current. It is divided into two separate sections: the heater section and the contact section. The heater section employs nichrome heaters that are inserted in series with the motor. Since motor current must pass through them, the amount of heat they produce is proportional to motor current. The contact section contains a normally closed contact that is connected in series with the starter coil. If the motor current becomes too great, the overload contact will open and de-energize the motor starter. The load contacts then open and disconnect the motor from the power line.

The schematic in Figure 18-1 also contains a transformer that is used to change the line voltage to control voltage. In this circuit it will be assumed that the three phase lines are providing 480 volts to operate the motor and that the control circuit operates on 120 volts. The control transformer changes the 480 volts to 120 volts.

The control circuit also contains a stop pushbutton and a start pushbutton. The stop button is normally closed. This means that when the button is not pressed, there will be a circuit through the button. When the button is pressed, the circuit will open. The start button is shown normally open. When the button is not pressed, there is no circuit across the stationary contacts. When the button is pressed, the movable contact bridges the gap between the two stationary contacts and completes a circuit to the M coil.

To understand circuit operation, assume that the start button is pressed, *Figure 18-4*. This completes a circuit to the M coil, causing all M contacts to close. The three M load contacts connect the motor to the line. The M auxiliary contact connected in parallel with the start button closed in order to maintain a circuit to the M coil when the start button was released, *Figure 18-5*. Since this contact holds the current path to the starter coil, it is often referred to as the holding, maintaining, or sealing contact. The circuit will continue to operate until the stop button is pressed, de-energizing the M coil.

If the motor should become overloaded while in operation, the overload contact connected in series with the starter coil will open and de-energize the starter. This will cause the load contacts to open and disconnect the motor from the power line.

CONNECTING THE CIRCUIT

As stated previously, schematic diagrams show components in their electrical sequence, without regard for physical location. They are superior

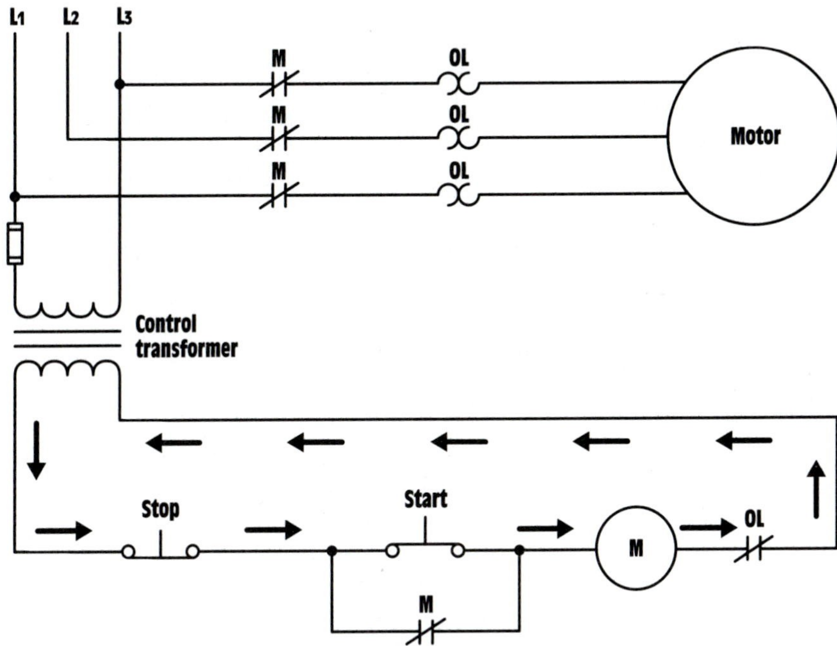

Figure 18-4 Pressing the start button completes a circuit to M coil.

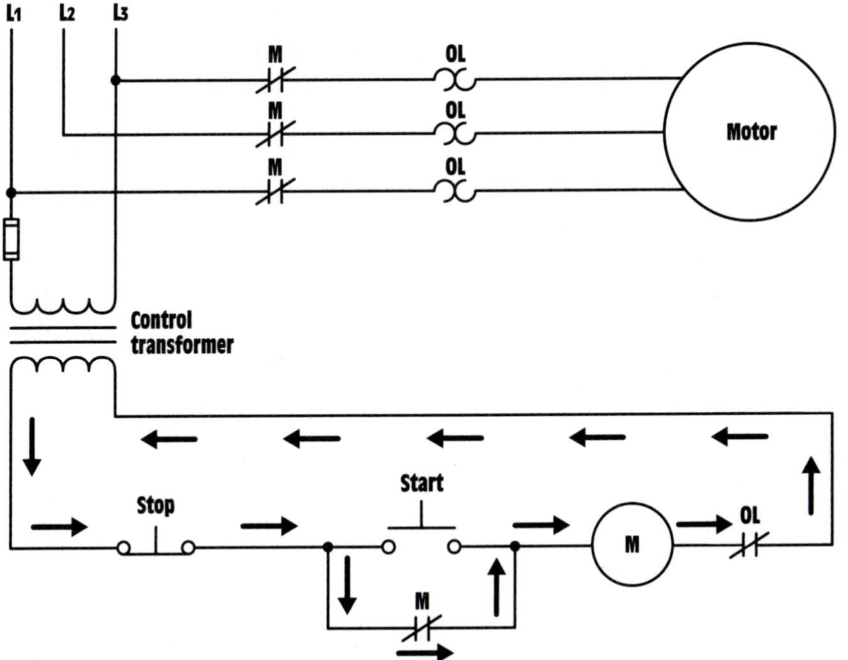

Figure 18-5 M auxiliary contacts maintain the circuit when the start button is released.

for interpreting the logic of a circuit but bear no resemblance to a connected circuit. Experienced electricicans generally prefer using schematic diagrams for connecting circuits, but students find this difficult because they are not familiar with looking at a component on a schematic and recognizing where that component is physically located. The best method for learning to connect control circuits is to learn how to change a schematic diagram into a wiring diagram.

To simplify the task of converting a schematic diagram into a wiring diagram, wire numbers will be added to the schematic diagram. These numbers will then be transferred to the control components shown in Figure 18-2. The rules for numbering a schematic diagram are as follows:

1. A set of numbers can be used only once.
2. Each time you go through a component, the number set must change.
3. All components that are connected together will have the same number.

To begin the numbering procedure, start at Line 1 (L_1) with the number 1 and place a number 1 beside each component connected to L_1, *Figure 18-6.* The number 2 is placed beside each component connected to L_2, *Figure 18-7,* and a 3 is placed beside each component connected to L_3, *Figure 18-8.* The number 4 is placed on the other side of the M load contact that already has a number 1 on one side and also on one side of the overload heater, *Figure 18-9.* The number 5 is placed on the other side of the M load contact that has one side already numbered with a 2 and also beside the second overload heater. The other side of the M load contact that has been numbered with a 3 is numbered with a 6, as is one side of the third overload heater. The numbers 7, 8, and 9 are placed between the other sides of the overload heaters and the motor T leads.

The number 10 begins at one side of the control transformer secondary and goes to one side of the normally closed stop pushbutton. The number 11 is placed on the other side of the stop button; it is also placed on one side of the normally open start pushbutton and the normally open M auxiliary contact. The number 12 is placed on the other side of the start button and the M auxiliary contact and on one side of the M coil. The number 13 is placed on the other side of the coil, to one side of the normally closed overload contact. The number 14 is placed on the other side of the normally closed overload contact and on the other side of the control transformer secondary winding, *Figure 18-10.*

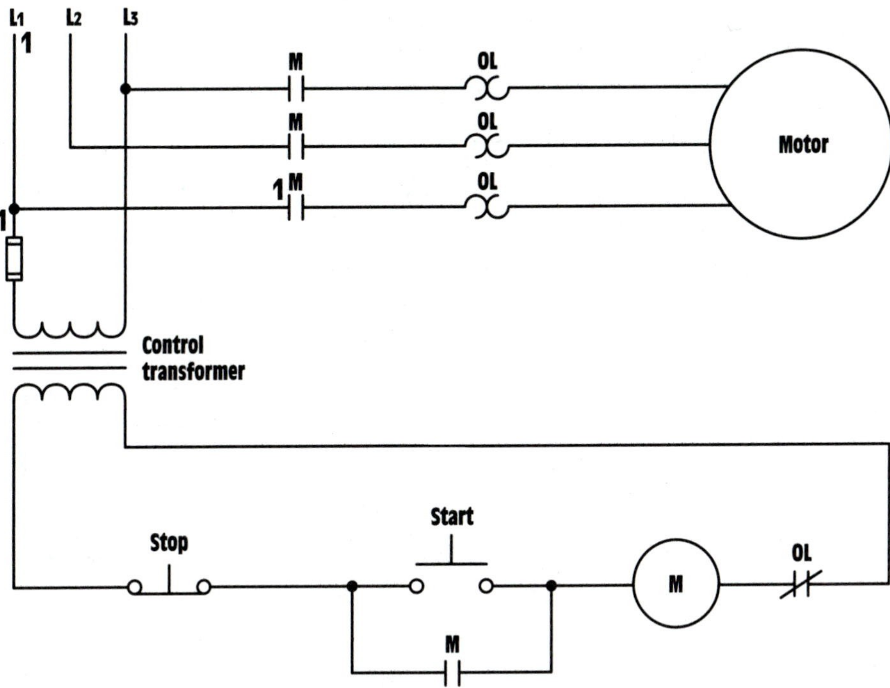

Figure 18-6 The number 1 is placed beside each component connected to L_1.

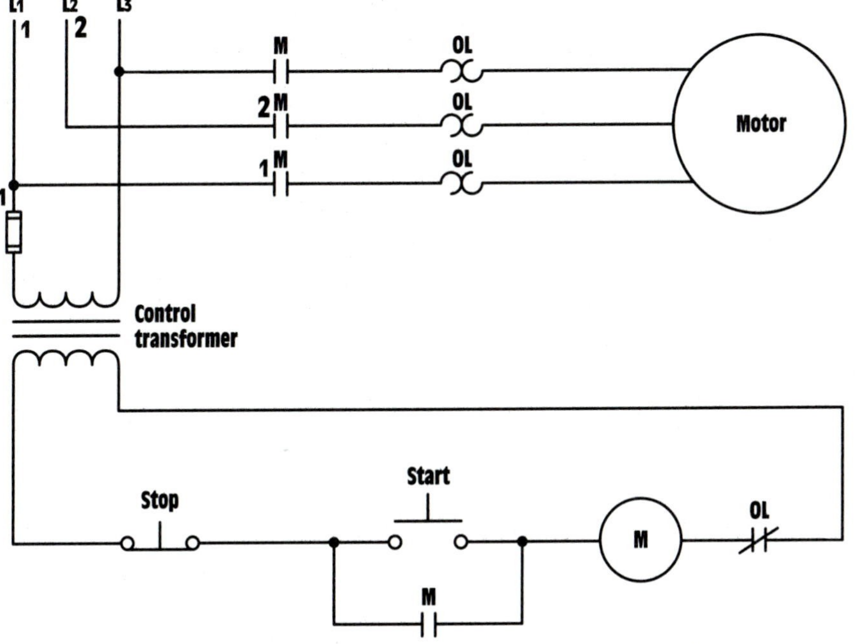

Figure 18-7 The number 2 is placed beside each component connected to L_2.

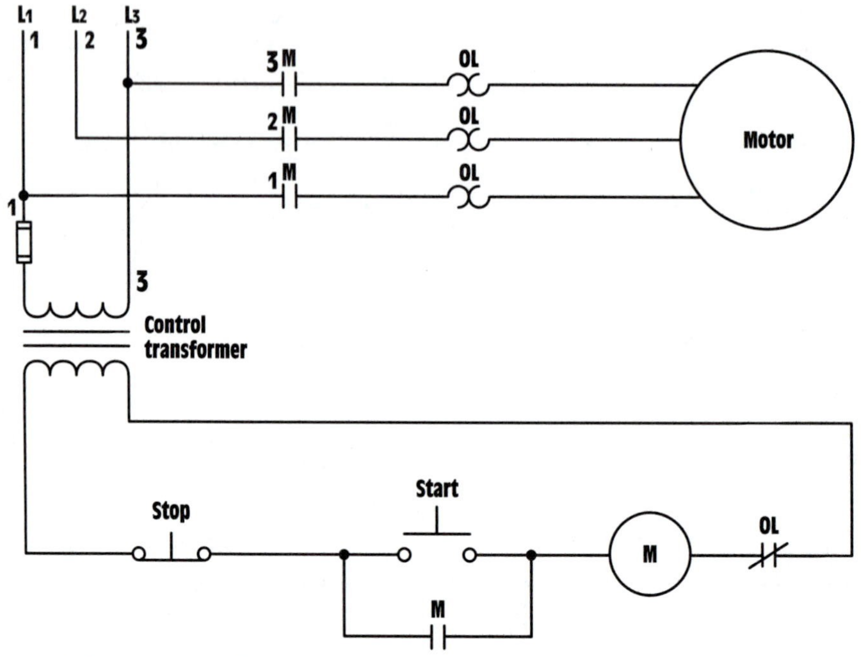

Figure 18-8 The number 3 is placed beside each component connected to L_3.

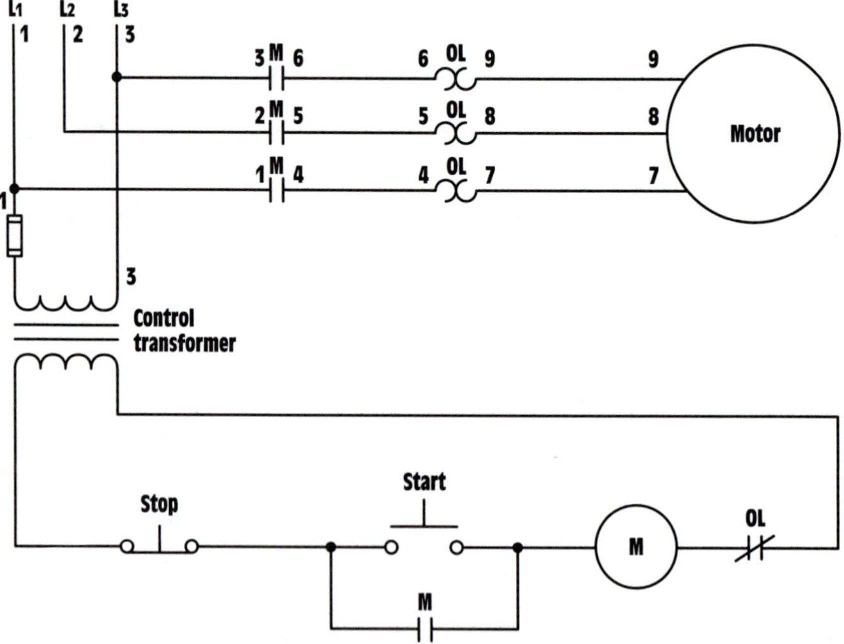

Figure 18-9 The number changes each time you proceed across a component.

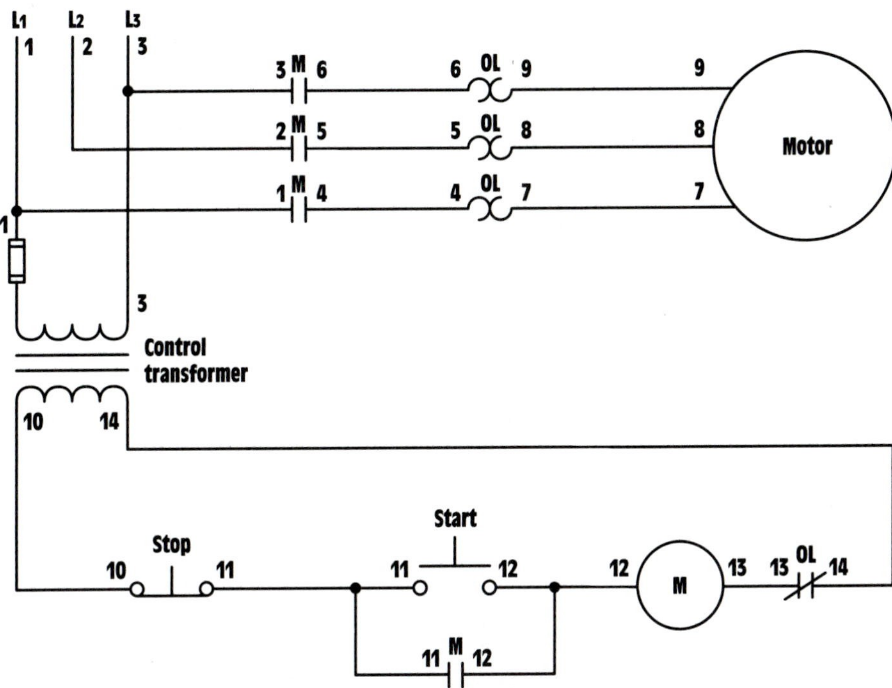

Figure 18-10 Numbers are placed beside all components.

NUMBERING THE COMPONENTS

Now that the components on the schematic have been numbered, the next step is to place the same numbers on the corresponding components of the wiring diagram. The schematic diagram in Figure 18-10 shows that the number 1 has been placed beside L_1, the fuse on the control transformer, and one side of a load contact on the M starter, *Figure 18-11*. The number 2 is placed beside L_2 and on the second load contact on the M starter, *Figure 18-12*. The number 3 is placed beside L_3, the third load contact on the M starter, and on the other side of the primary winding on the control transformer. The numbers 4, 5, 6, 7, 8, and 9 are placed beside the corresponding components on the schematic diagram, *Figure 18-13*. Note on connection points 4, 5, and 6, from the output of the load contacts to the overload heaters, that these connections are factory made on a motor starter and do not have to be made in the field. For the sake of simplicity, these connections are not shown in the diagram. If a separate contactor and overload relay are being used, however, these connections will have to be made. Recall that a motor starter is a contactor and an overload relay combined.

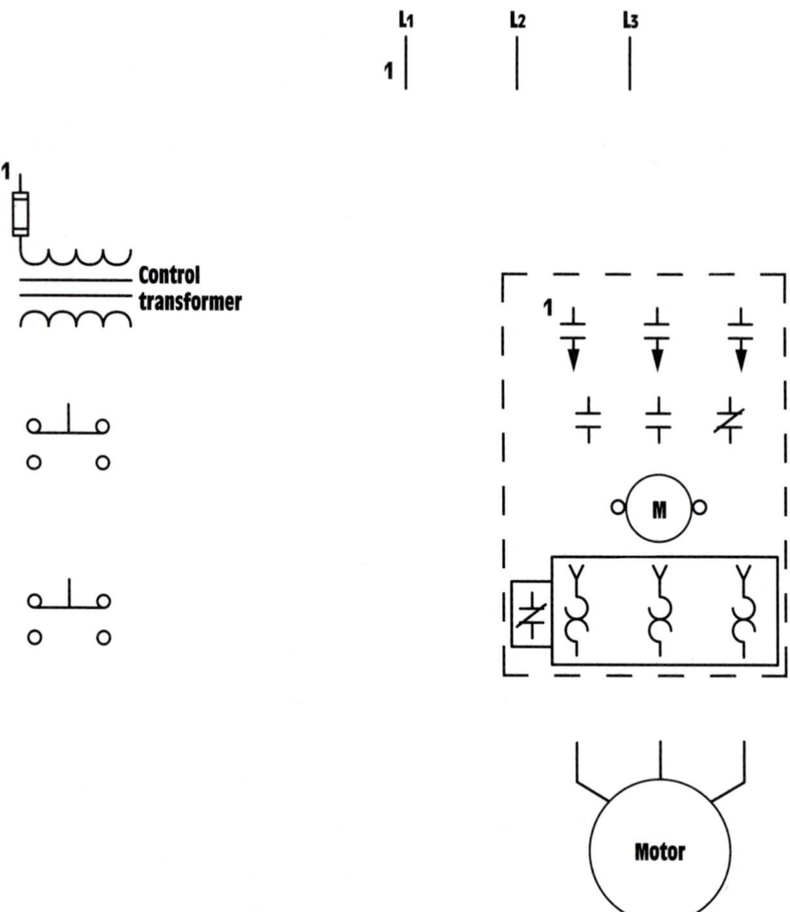

Figure 18-11 The number is placed beside L_1, the control transformer fuse, and M load contacts.

The number 10 starts at the secondary winding of the control transformer and goes to one side of the normally closed stop pushbutton. When making this connection, care must be taken to make certain that connection is made to the normally closed side of the pushbutton. Since this is a double-acting pushbutton, it contains both normally closed and normally open contacts, *Figure 18-14*.

The number 11 starts at the other side of the normally closed stop-button and goes to one side of the normally open start pushbutton and to

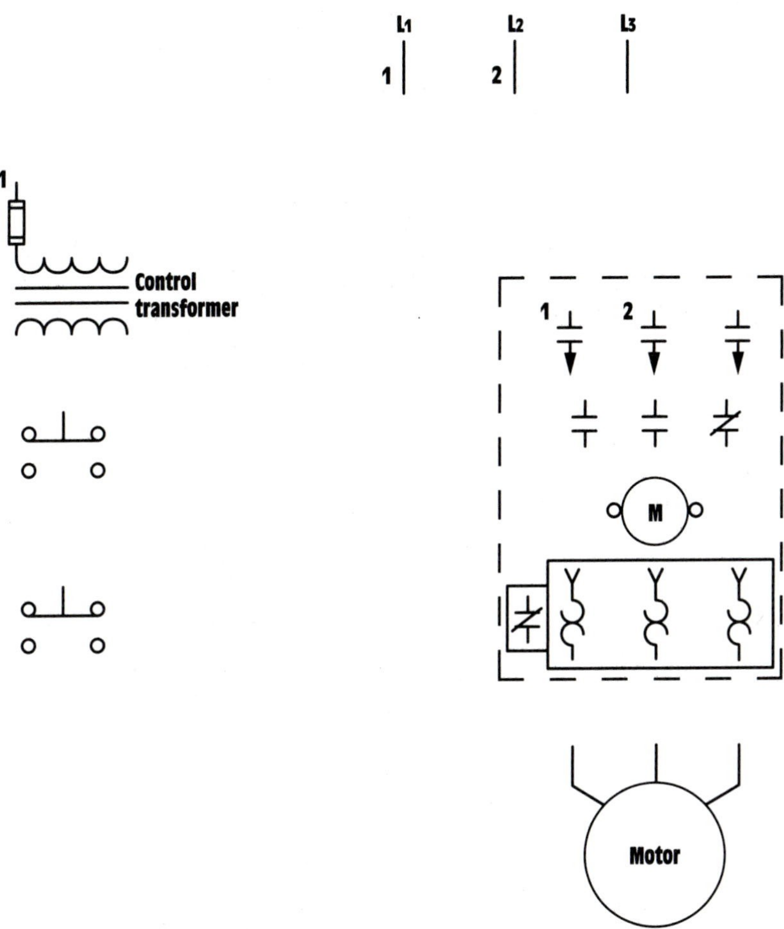

Figure 18-12 The number 2 is placed beside L_2 and the second load contact on M starter.

one side of a normally open M auxiliary contact, *Figure 18-15*. The starter in this example shows three auxiliary contacts: two normally open and one normally closed. It makes no difference which normally open contact is used.

This same procedure is followed until all circuit components have been numbered with the number that corresponds to the same component on the schematic diagram, *Figure 18-16*.

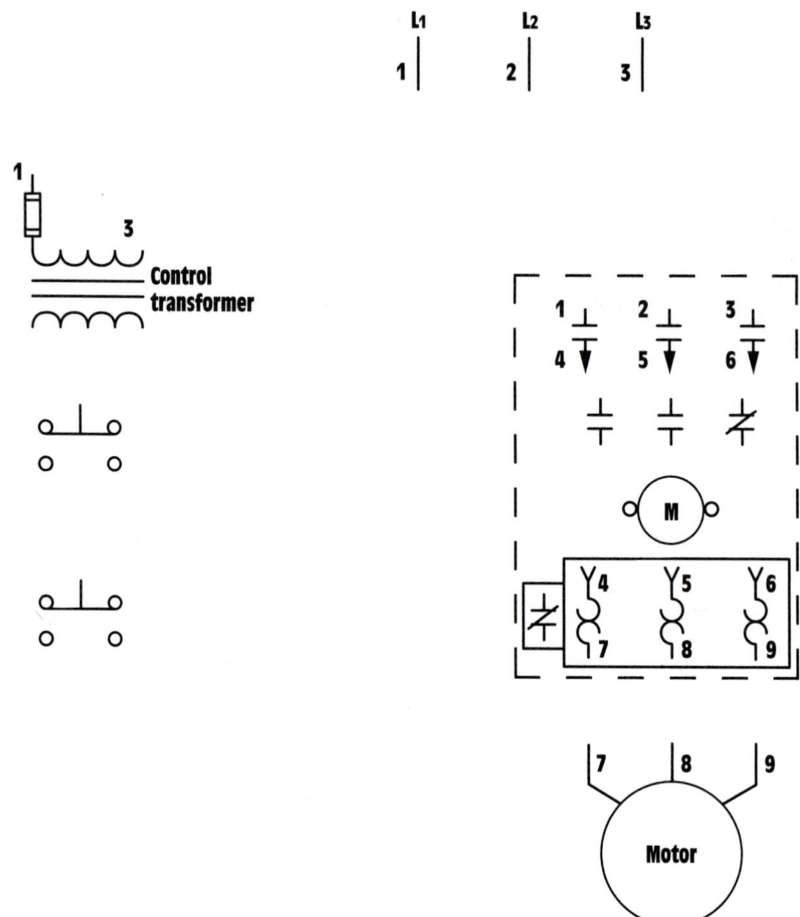

Figure 18-13 Placing numbers 3, 4, 5, 6, 7, 8, and 9 beside the proper components.

CONNECTING THE WIRES

Now that numbers have been placed beside the components, wiring the circuit becomes a matter of connecting numbers. First, connect all components labeled with a number 1 together, *Figure 18-17.* Then, all components numbered with a 2 are connected together, *Figure 18-18.* Next, all components numbered with a 3 are connected together, *Figure 18-19.* This procedure is followed until all the numbered components are connected together with the exception of 4, 5, and 6 (which are assumed to be factory connected), *Figure 18-20.*

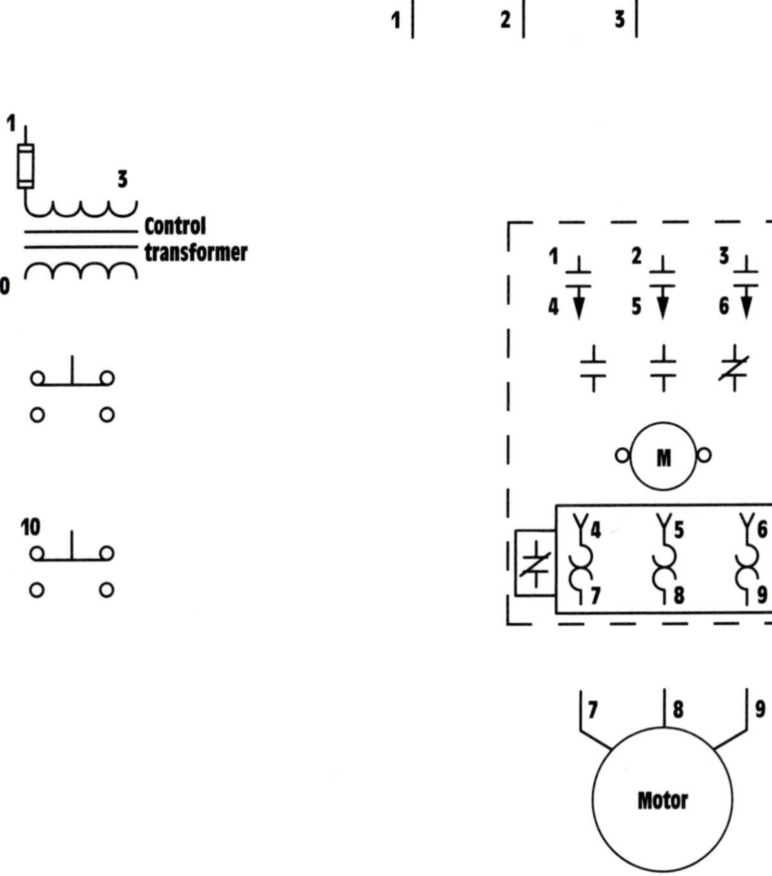

Figure 18-14 Wire number 10 connects from the transformer secondary to the stop button.

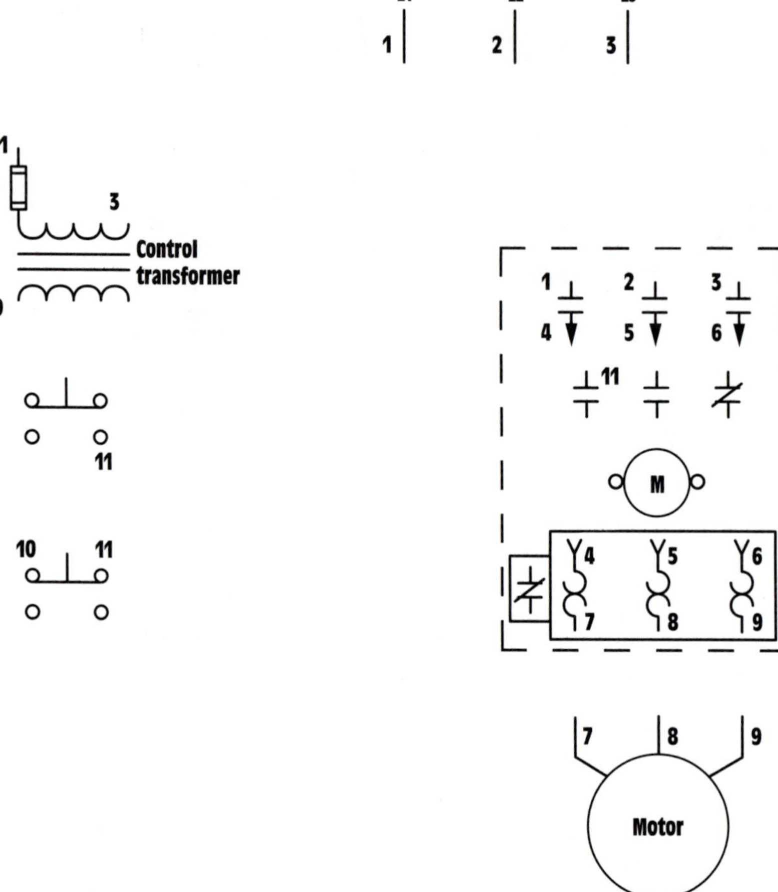

Figure 18-15 Number 11 connects to the stop button, start button, and holding contact.

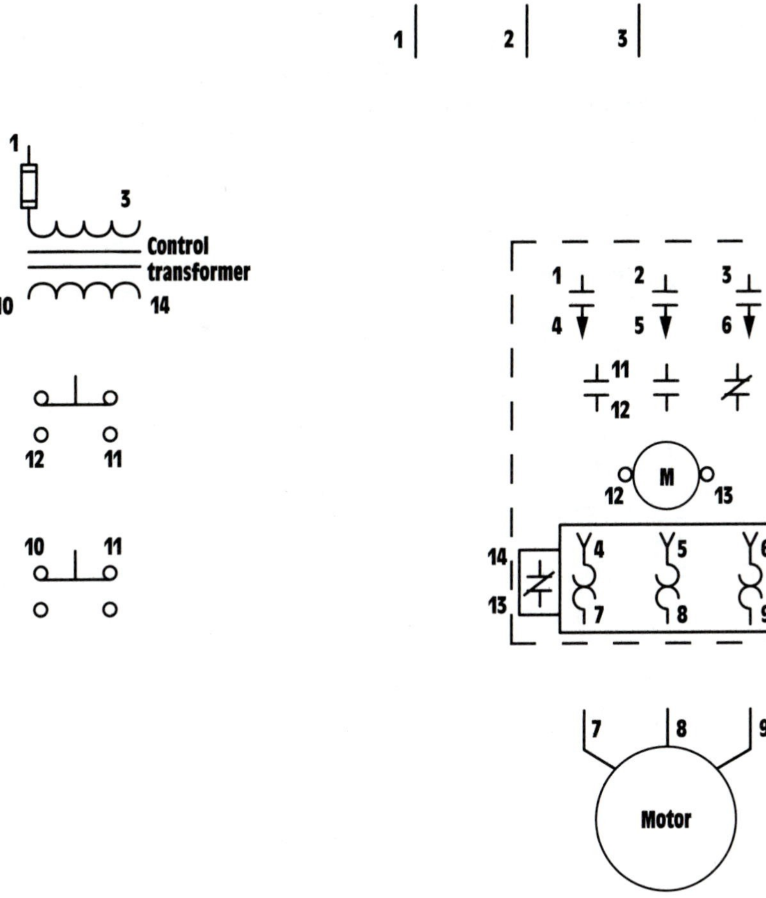

Figure 18-16 All components have been numbered.

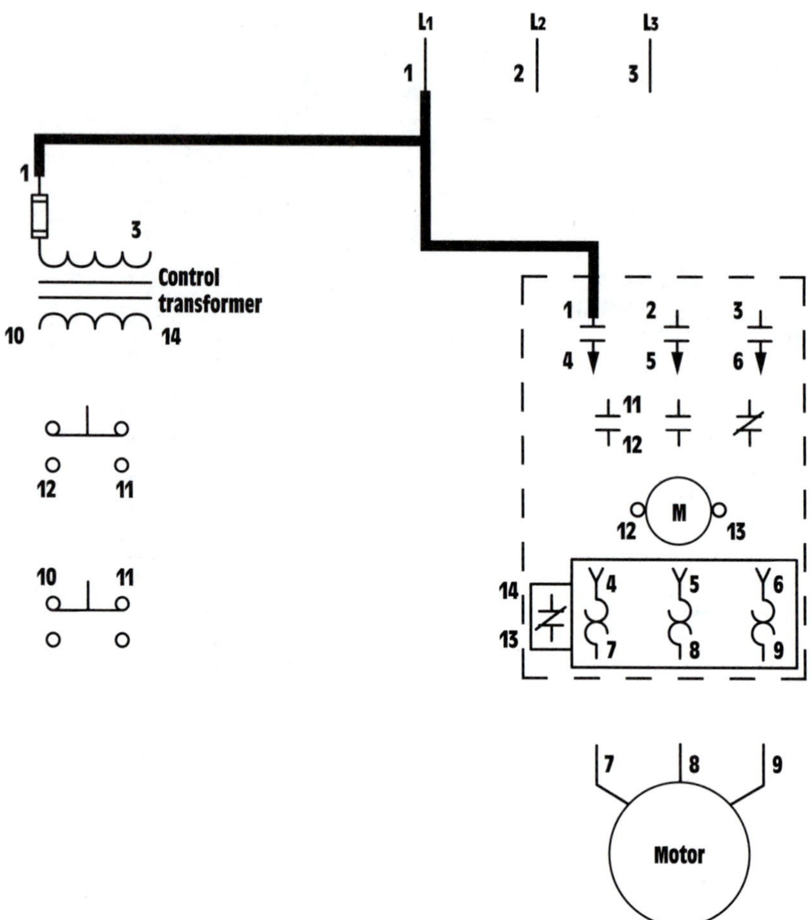

Figure 18-17 Connecting all components numbered with a 1.

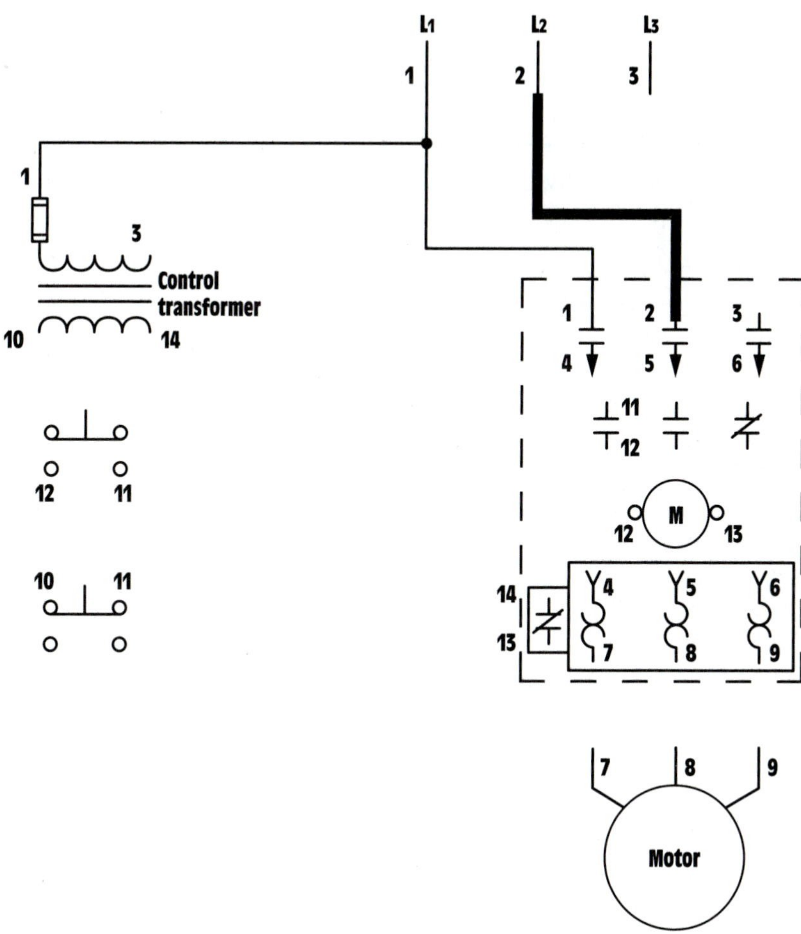

Figure 18-18 Connecting all components numbered with a 2.

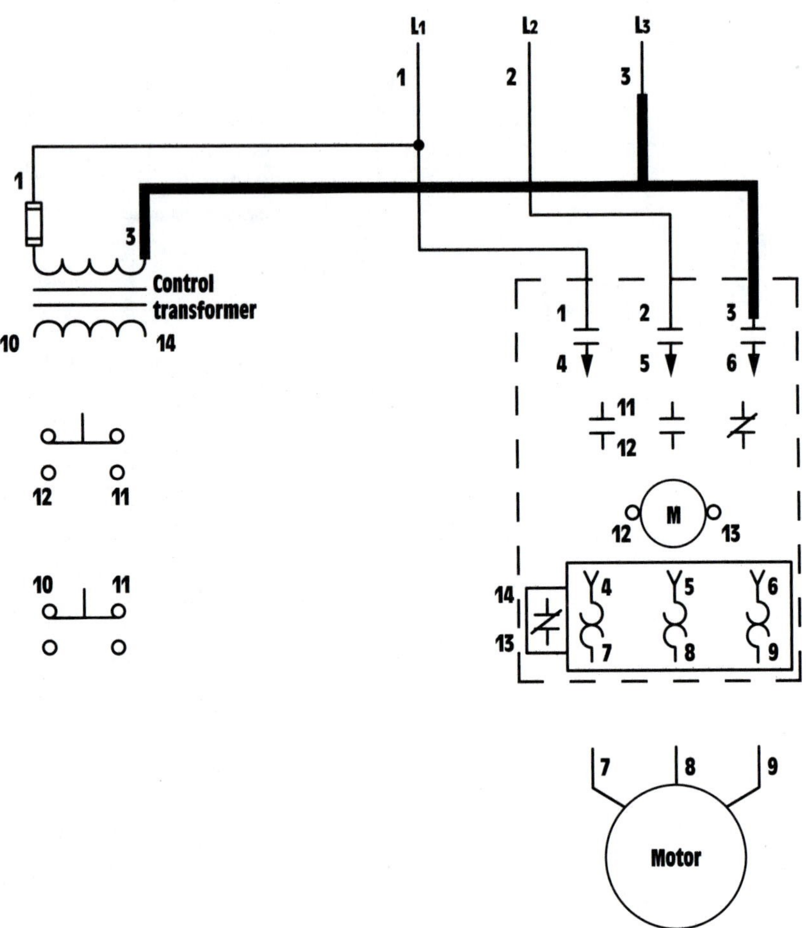

Figure 18-19 Connecting all components numbered with a 3.

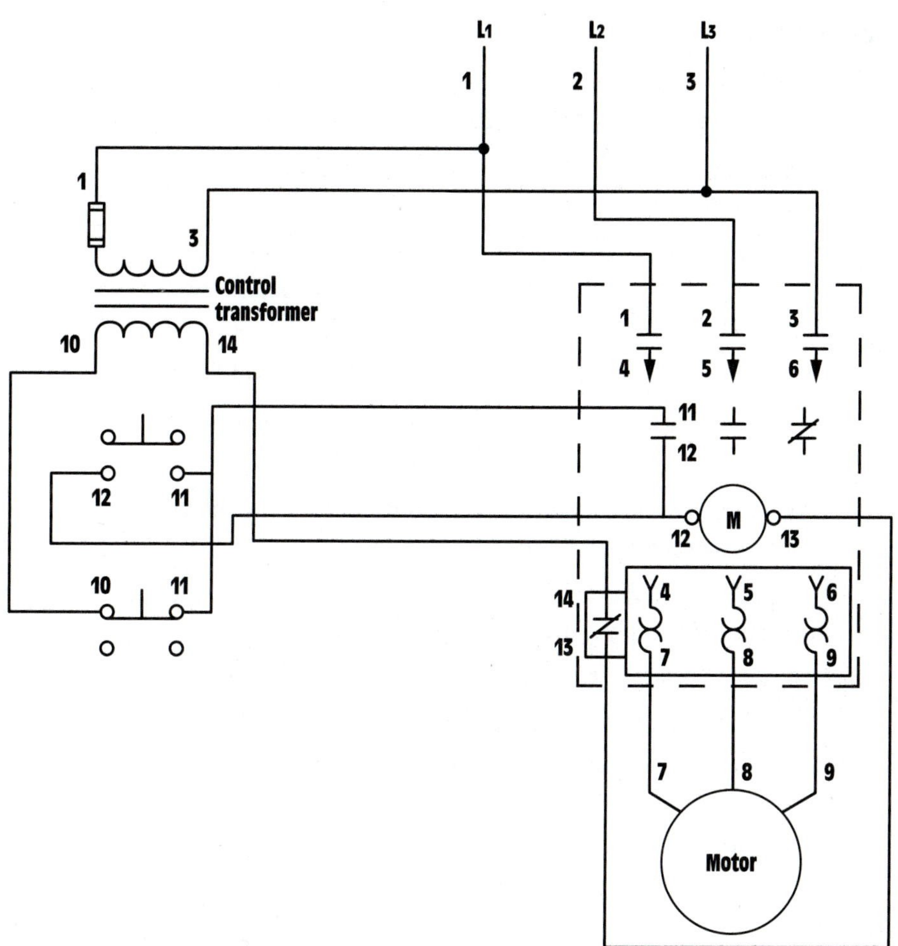

Figure 18-20 Completing the wiring diagram.

SUMMARY

1. Schematic diagrams show components in their electrical sequence, without regard for physical location.
2. Wiring diagrams show a pictorial representation of the components, with connecting wires.
3. Schematic, or ladder, diagrams are generally used for troubleshooting because they illustrate the circuit logic better than a wiring diagram.
4. The most often used control symbols are the NEMA symbols.
5. Control symbols are generally drawn to represent the device they depict.
6. Normally open switches are drawn with their movable contact below and not touching the stationary contact.
7. Normally closed switches are drawn with their movable contact above and touching the stationary contact.
8. Normally open, held closed switches are drawn with the movable contact below and touching the stationary contact.
9. Normally closed, held open switches are shown with the movable contact above and not touching the stationary contact.
10. Motor starters generally contain both load and auxiliary contacts.
11. Load contacts are large and are designed to connect the load to the line.
12. Auxiliary contacts are small and operate on a small current load.
13. Motor starters contain overload relays.
14. Most overload relays sense motor current by connecting an electric heating element in series with the motor.

REVIEW QUESTIONS

1. Refer to the circuit shown in Figure 18-10. If wire number 11 were disconnected at the normally open auxiliary M contact, how would the circuit operate?
2. Assume that when the start button is pressed, the M starter does not energize. List seven possible causes for this problem.

   A.____

   B.____

   C.____

   D. _______________________________________

   E.____

   F. _______________________________________

   G. _______________________________________

3. Explain the difference between a motor starter and a contactor.
4. Refer to the schematic in Figure 18-10. Assume that when the start button is pressed, the control transformer fuse blows. What is the most likely cause of this trouble?
5. Explain the difference between load and auxiliary contacts.

A

Appendix

American Wire Gauge Table

AMERICAN WIRE GAUGE TABLE							
			Ohms per 1000 Ft. (ohms per 100 m)			Pounds per 1,000 Ft. (kg per 100 m)	
B & S Gauge No.	Diam. in Mils	Area in Circular Mils	Copper* 68°F (20°C)	Copper* 167°F (75°C)	Aluminum 68°F (20°C)	Copper	Aluminum
0000	460	211 600	0.049 (0.016)	0.0596 (0.0195)	0.0804 (0.0263)	640 (95.2)	195 (29.0)
000	410	167 800	0.0618 (0.020)	0.0752 (0.0246)	0.101 (0.033)	508 (75.5)	154 (22.9)
00	365	133 100	0.078 (0.026)	0.0948 (0.031)	0.128 (0.042)	403 (59.9)	122 (18.1)
0	325	105 500	0.0983 (0.032)	0.1195 (0.0392)	0.161 (0.053)	320 (47.6)	97 (14.4)
1	289	83 690	0.1239 (0.0406)	0.151 (0.049)	0.203 (0.066)	253 (37.6)	76.9 (11.4)
2	258	66 370	0.1563 (0.0512)	0.190 (0.062)	0.526 (0.084)	201 (29.9)	61.0 (9.07)
3	229	52 640	0.1970 (0.0646)	0.240 (0.079)	0.323 (0.106)	159 (23.6)	48.4 (7.20)
4	204	41 740	0.2485 (0.0815)	0.302 (0.099)	0.408 (0.134)	126 (18.7)	38.4 (5.71)
5	182	33 100	0.3133 (0.1027)	0.381 (0.125)	0.514 (0.168)	100 (14.9)	30.4 (4.52)
6	162	26 250	0.395 (1.29)	0.481 (0.158)	0.648 (0.212)	79.5 (11.8)	24.1 (3.58)
7	144	20 820	0.498 (0.163)	0.606 (0.199)	0.817 (0.268)	63.0 (9.37)	19.1 (2.84)
8	128	16 510	0.628 (0.206)	0.764 (0.250)	1.03 (0.338)	50.0 (7.43)	15.2 (2.26)
9	114	13 090	0.792 (0.260)	0.963 (0.316)	1.30 (0.426)	39.6 (5.89)	12.0 (1.78)
10	102	10 380	0.999 (0.327)	1.215 (0.398)	1.64 (0.538)	31.4 (4.67)	9.55 (1.42)
11	91	8 234	1.260 (0.413)	1.532 (0.502)	2.07 (0.678)	24.9 (3.70)	7.57 (1.13)
12	81	6 530	1.588 (0.520)	1.931 (0.633)	2.61 (0.856)	19.8 (2.94)	6.00 (0.89)
13	72	5 178	2.003 (0.657)	2.44 (0.80)	3.29 (1.08)	15.7 (2.33)	4.8 (0.71)
14	64	4 107	2.525 (0.828)	3.07 (1.01)	4.14 (1.36)	12.4 (1.84)	3.8 (0.56)
15	57	3 257	3.184 (0.043)	3.98 (1.27)	5.22 (1.71)	9.86 (1.47)	3.0 (0.45)
16	51	2 583	4.016 (0.316)	4.88 (1.60)	6.59 (2.16)	7.82 (1.16)	2.4 (0.36)
17	45.3	2 048	5.06 (1.66)	6.16 (2.02)	8.31 (2.72)	6.20 (0.922)	1.9 (0.28)
18	40.3	1 624	6.39 (2.09)	7.77 (2.55)	10.5 (3.44)	4.92 (0.731)	1.5 (0.22)
19	35.9	1 288	8.05 (2.64)	9.79 (3.21)	13.2 (4.33)	3.90 (0.580)	1.2 (0.18)
20	32.0	1 022	10.15 (3.33)	12.35 (4.05)	16.7 (5.47)	3.09 (0.459)	0.94 (0.14)
21	28.5	810	12.8 (4.2)	15.6 (5.11)	21.0 (6.88)	2.45 (0.364)	0.745 (0.110)
22	25.4	642	16.1 (5.3)	19.6 (6.42)	26.5 (8.69)	1.95 (0.290)	0.591 (0.09)
23	22.6	510	20.4 (6.7)	24.8 (8.13)	33.4 (10.9)	1.54 (0.229)	0.468 (0.07)
24	20.1	404	25.7 (8.4)	31.2 (10.2)	42.1 (13.8)	1.22 (0.181)	0.371 (0.05)
25	17.9	320	32.4 (10.6)	39.4 (12.9)	53.1 (17.4)	0.97 (0.14)	0.295 (0.04)
26	15.9	254	40.8 (13.4)	49.6 (16.3)	67.0 (22.0)	0.77 (0.11)	0.234 (0.03)
27	14.2	202	51.5 (16.9)	62.6 (20.5)	84.4 (27.7)	0.61 (0.09)	0.185 (0.03)
28	12.6	160	64.9 (21.3)	78.9 (25.9)	106 (34.7)	0.48 (0.07)	0.147 (0.02)
29	11.3	126.7	81.8 (26.8)	99.5 (32.6)	134 (43.9)	0.384 (0.06)	0.117 (0.02)
30	10.0	100.5	103.2 (33.8)	125.5 (41.1)	169 (55.4)	0.304 (0.04)	0.092 (0.01)
31	8.93	79.7	130.1 (42.6)	158.2 (51.9)	213 (69.8)	0.241 (0.04)	0.073 (0.01)
32	7.95	63.2	164.1 (53.8)	199.5 (65.4)	269 (88.2)	0.191 (0.03)	0.058 (0.01)
33	7.08	50.1	207 (68)	252 (82.6)	339 (111)	0.152 (0.02)	0.046 (0.01)
34	6.31	39.8	261 (86)	317 (104)	428 (140)	0.120 (0.02)	0.037 (0.01)
35	5.62	31.5	329 (108)	400 (131)	540 (177)	0.095 (0.01)	0.029
36	5.00	25.0	415 (136)	505 (165)	681 (223)	0.076 (0.01)	0.023
37	4.45	19.8	523 (171)	636 (208)	858 (281)	0.0600 (0.01)	0.0182
38	3.96	15.7	660 (216)	802 (263)	1080 (354)	0.0476 (0.01)	0.0145
39	3.53	12.5	832 (273)	1012 (332)	1360 (446)	0.0377 (0.01)	0.0115
40	3.15	9.9	1049 (344)	1276 (418)	1720 (564)	0.0299 (0.01)	0.0091
41							
42	2.50	6.3					
43							
44	1.97	3.9					

*Resistance figures are given for standard annealed copper. For hard-drawn copper add 2%

Identification of Mica and Tubular Capacitors

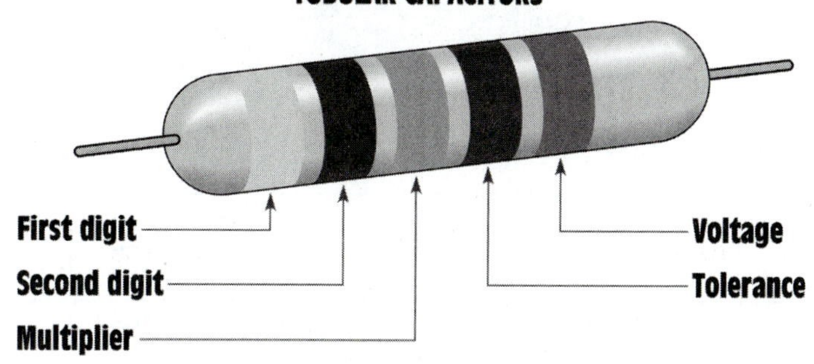

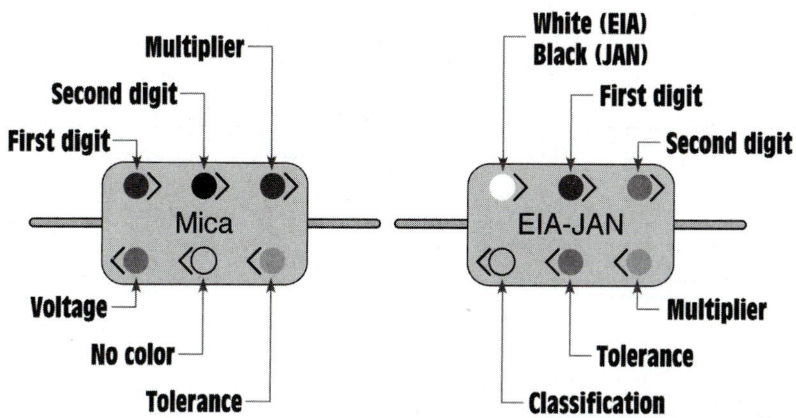

COLOR	NUMBER	MULTIPLIER	TOLERANCE (%)	VOLTAGE
No color			20	500
Black	0	1		
Brown	1	10	1	100
Red	2	100	2	200
Orange	3	1000	3	300
Yellow	4	10,000	4	400
Green	5	100,000	5 (EIA)	500
Blue	6	1,000,000	6	600
Violet	7	10,000,000	7	700
Gray	8	100,000,000	8	800
White	9	1,000,000,000	9	900
Gold		0.1	5 (JAN)	1000
Silver		0.01	10	2000

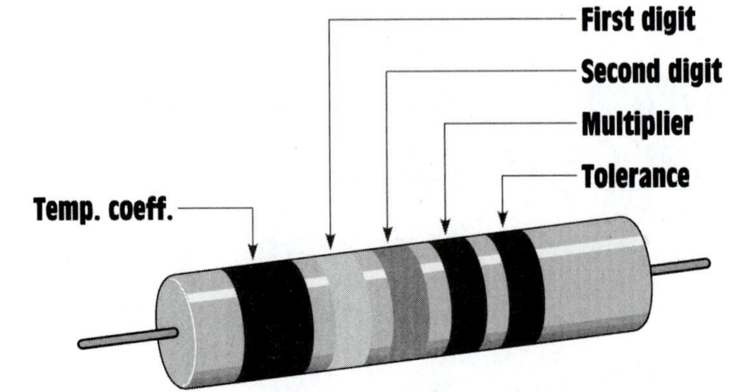

COLOR	NUMBER	MULTIPLIER	TOLERANCE OVER 10 pF	TOLERANCE 10 pF OR LESS	TEMP. COEFF.
Black	0	1	20%	2.0 pF	0
Brown	1	10	1%		N30
Red	2	100	2%		N80
Orange	3	1000			N150
Yellow	4				N220
Green	5				N330
Blue	6		5%	0.5 pF	N470
Violet	7				N750
Gray	8	0.01		0.25 pF	P30
White	9	0.1	10%	1.0 pF	P500

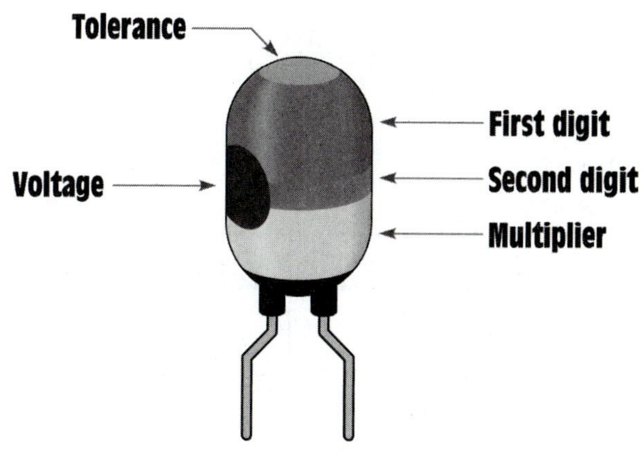

COLOR	NUMBER	MULTIPLIER	TOLERANCE (%)	VOLTAGE
			No dot 20	
Black	0			4
Brown	1			6
Red	2			10
Orange	3			15
Yellow	4	10,000		20
Green	5	100,000		25
Blue	6	1,000,000		35
Violet	7	10,000,000		50
Gray	8			
White	9			3
Gold			5	
Silver			10	

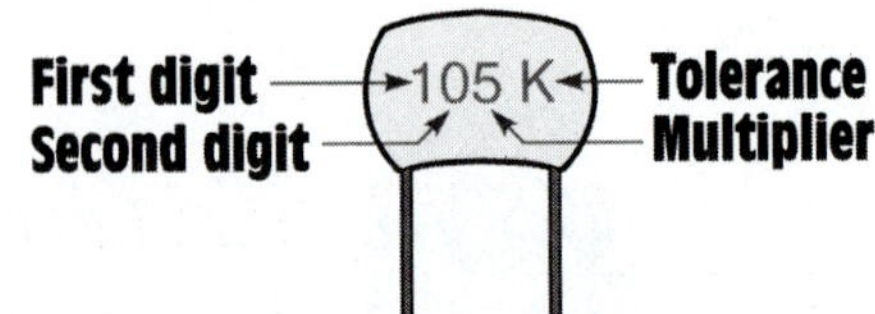

NUMBER	MULTIPLIER	TOLERANCE		
			10 pf or less	Over 10 pf
0	1	B	0.1 pf	
1	10	C	0.25 pf	
2	100	D	0.5 pf	
3	1000	F	1.0 pf	1%
4	10,000	G	2.0 pf	2%
5	100,000	H		3%
6		J		5%
7		K		10%
8	0.01	M		20%
9	0.1			

Alternating Current Formulas

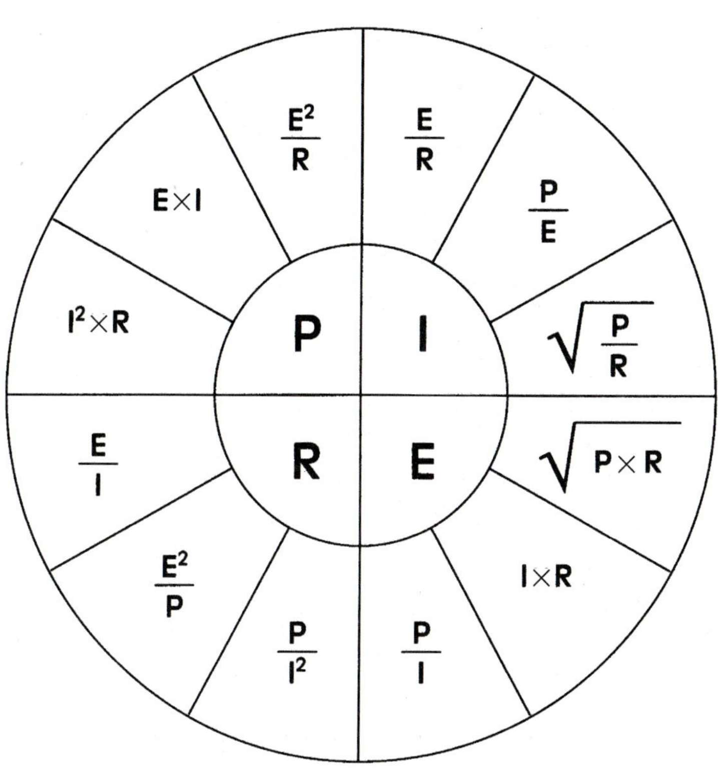

INSTANTANEOUS AND MAXIMUM VALUES

The instantaneous value of voltage and current for a sine wave is equal to the peak, or maximum, value of the waveform times the sine of the angle. For example, a waveform has a peak value of 300 V. What is the voltage at an angle of 22°?

$$E_{(INST)} = E_{(MAX)} \times \text{SIN} \angle$$

$$E_{(INST)} = 300 \times 0.3746 \text{ (SIN of } 22°)$$

$$E_{(INST)} = 112.328 \text{ V}$$

$$E_{MAX} = \frac{E_{INST}}{\text{SIN} \angle}$$

$$\text{SIN} \angle = \frac{E_{INST}}{E_{MAX}}$$

CHANGING PEAK, RMS, AND AVERAGE VALUES

To change	To	Multiply by
Peak	RMS	0.707
Peak	Average	0.637
Peak	Peak-to-Peak	2
RMS	Peak	1.414
Average	Peak	1.567
RMS	Average	0.9
Average	RMS	1.111

PURE RESISTIVE CIRCUIT

$E = I \times R$ $\quad I = \frac{E}{R}$ $\quad R = \frac{E}{I}$ $\quad P = E \times I$

$E = \frac{P}{I}$ $\quad I = \frac{P}{R}$ $\quad R = \frac{E^2}{P}$ $\quad P = I^2 \times R$

$E = \sqrt{P \times R}$ $\quad I = \sqrt{\frac{P}{R}}$ $\quad R = \frac{P}{I^2}$ $\quad P = \frac{E^2}{R}$

SERIES RESISTIVE CIRCUITS

$R_T = R_1 + R_2 + R_3$

$I_T = I_1 = I_2 = I_3$

$E_T = E_1 + E_2 + E_3$

$P_T = P_1 + P_2 + P_3$

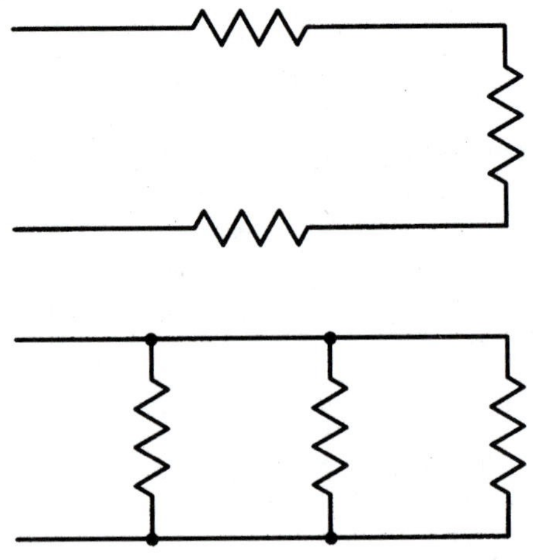

PARALLEL RESISTIVE CIRCUITS

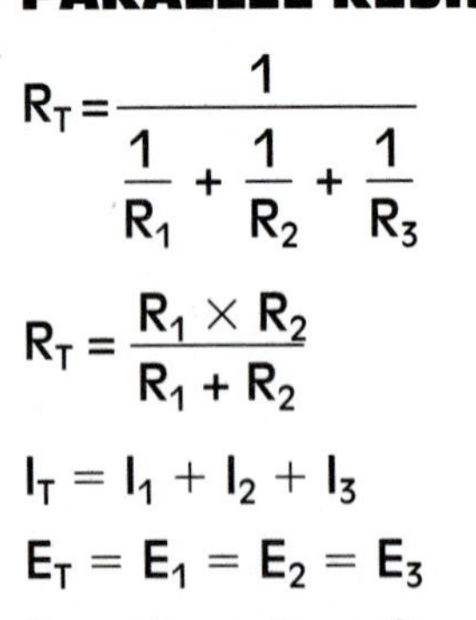

$$R_T = \frac{1}{\frac{1}{R_1} + \frac{1}{R_2} + \frac{1}{R_3}}$$

$$R_T = \frac{R_1 \times R_2}{R_1 + R_2}$$

$I_T = I_1 + I_2 + I_3$

$E_T = E_1 = E_2 = E_3$

$P_T = P_1 + P_2 + P_3$

PURE INDUCTIVE CIRCUITS

In a pure inductive circuit, the current lags the voltage by 90°. Therefore, there is no true power, or watts, and the power factor is 0. VARs is the inductive equivalent of watts.

$$L = \frac{X_L}{2\pi F} \qquad X_L = 2\pi FL$$

$$E_L = I_L \times X_L \qquad I_L = \frac{E_L}{X_L} \qquad X_L = \frac{E_L}{I_L} \qquad VARs_L = E_L \times I_L$$

$$E_L = \sqrt{VARs_L \times X_L} \qquad I_L = \frac{VARs_L}{E_L} \qquad X_L = \frac{E_L^2}{VARs_L} \qquad VARs_L = I_L^2 \times X_L$$

$$E_L = \frac{VARs_L}{I_L} \qquad I_L = \sqrt{\frac{VARs_L}{X_L}} \qquad X_L = \frac{VARs_L}{I_L^2} \qquad VARs_L = \frac{E_L^2}{X_L}$$

SERIES INDUCTIVE CIRCUITS

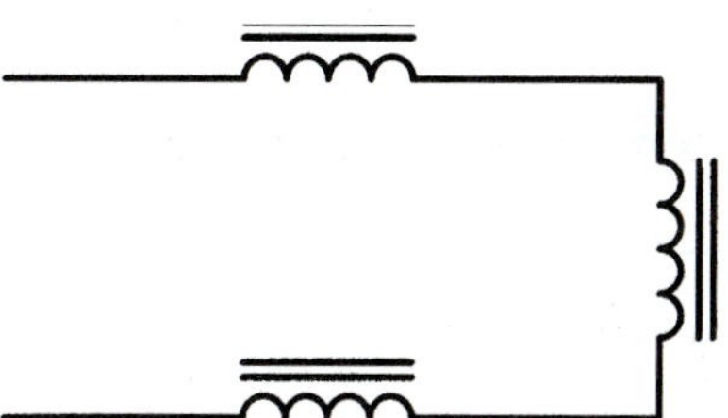

$$E_{LT} = E_{L1} + E_{L2} + E_{L3}$$

$$X_{LT} = X_{L1} + X_{L2} + X_{L3}$$

$$I_{LT} = I_{L1} = I_{L2} = I_{L3}$$

$$VARs_{LT} = VARs_{L1} + VARs_{L2} + VARs_{L3}$$

$$L_T = L_1 + L_2 + L_3$$

PARALLEL INDUCTIVE CIRCUITS

$$E_{LT} = E_{L1} = E_{L2} = E_{L3}$$

$$I_{LT} = I_{L1} + I_{L2} + I_{L3}$$

$$X_{LT} = \frac{X_{L1} \times X_{L2}}{X_{L1} + X_{L2}}$$

$$X_{LT} = \frac{1}{\frac{1}{X_{L1}} + \frac{1}{X_{L2}} + \frac{1}{X_{L3}}} \qquad L_T = \frac{1}{\frac{1}{L_1} + \frac{1}{L_2} + \frac{1}{L_3}} \qquad L_T = \frac{L_1 \times L_2}{L_1 + L_2}$$

$$VARs_{LT} = VARs_{L1} + VARs_{L2} + VARs_{L3}$$

PURE CAPACITIVE CIRCUITS

In a pure capacitive circuit, the current leads the voltage by 90°. For this reason there is no true power, or watts, and no power factor. VARs is the equivalent of watts in a pure capacitive circuit.

The value of C in the formula for finding capacitance is in farads and must be changed into the capacitive units being used.

$$C = \frac{1}{2\pi F X_C} \qquad X_C = \frac{1}{2\pi F C}$$

$$E_C = I_C \times X_C \qquad I_C = \frac{E_C}{X_C} \qquad X_C = \frac{E_C}{I_C} \qquad VARS_C = E_C \times I_C$$

$$E_C = \sqrt{VARS_C \times X_C} \qquad I_C = \frac{VARS_C}{E_C} \qquad X_C = \frac{E_C^2}{VARS_C} \qquad VARS_C = I_C^2 \times X_C$$

$$E_C = \frac{VARS_C}{I_C} \qquad I_C = \sqrt{\frac{VARS_C}{X_C}} \qquad X_C = \frac{VARS_C}{I_C^2} \qquad VARS_C = \frac{E_C^2}{X_C}$$

SERIES CAPACITIVE CIRCUITS

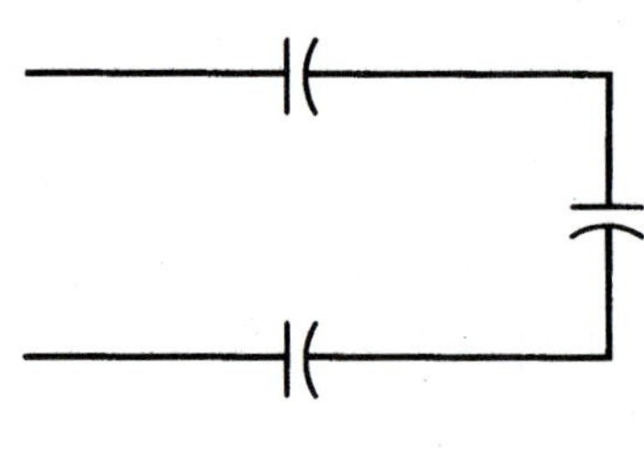

$$E_{CT} = E_{C1} + E_{C2} + E_{C3}$$

$$I_{CT} = I_{C1} = I_{C2} = I_{C3}$$

$$X_{CT} = X_{C1} + X_{C2} + X_{C3}$$

$$VARS_{CT} = VARS_{C1} + VARS_{C2} + VARS_{C3}$$

$$C_T = \frac{1}{\frac{1}{C_1} + \frac{1}{C_2} + \frac{1}{C_3}} \qquad C_T = \frac{C_1 \times C_2}{C_1 + C_2}$$

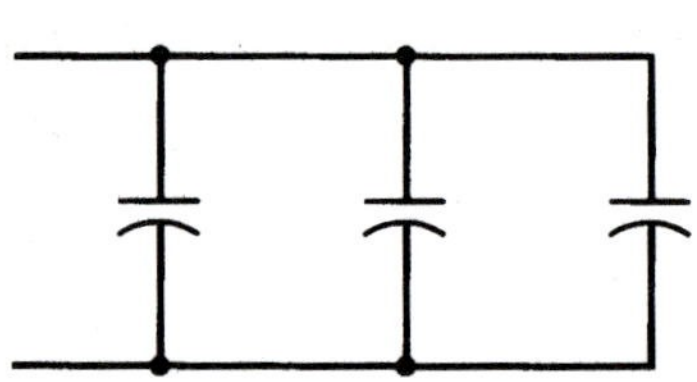

PARALLEL CAPACITIVE CIRCUITS

$$E_{CT} = E_{C1} = E_{C2} = E_{C3}$$

$$X_{CT} = \frac{1}{\frac{1}{X_{C1}} + \frac{1}{X_{C2}} + \frac{1}{X_{C3}}} \qquad X_{CT} = \frac{X_{C1} \times X_{C2}}{X_{C1} + X_{C2}} \qquad I_{CT} = I_{C1} + I_{C2} + I_{C3}$$

$$C_T = C_1 + C_2 + C_3 \qquad VARS_{CT} = VARS_{C1} + VARS_{C2} + VARS_{C3}$$

RESISTIVE-INDUCTIVE SERIES CIRCUITS

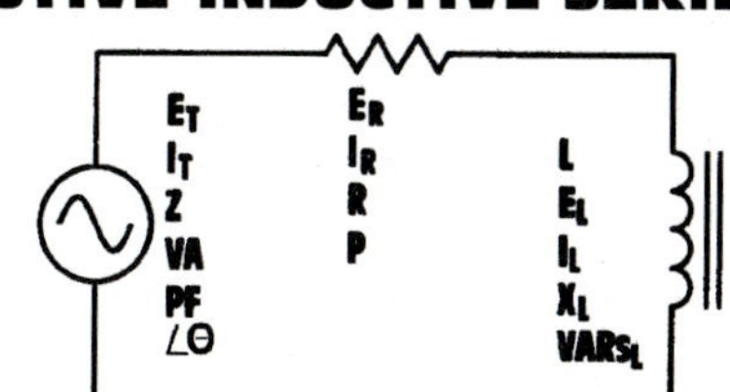

To find values for the resistor, use the formulas in the "Pure Resistive" section.

To find values for the inductor, use the formulas in the "Pure Inductive" section.

$$E_T = \sqrt{E_R^2 + E_L^2}$$

$$E_T = I_T \times Z$$

$$E_T = \frac{VA}{I_T}$$

$$E_T = \frac{E_R}{PF}$$

$$PF = \frac{R}{Z}$$

$$PF = \frac{P}{VA}$$

$$PF = \frac{E_R}{E_T}$$

$$PF = \cos \angle\theta$$

$$R = \sqrt{Z^2 - X_L^2}$$

$$R = \frac{E_R}{I_R}$$

$$R = \frac{E_R^2}{P}$$

$$R = \frac{P}{I_R^2}$$

$$R = Z \times PF$$

$$VARS_L = \sqrt{VA^2 - P^2}$$

$$Z = \sqrt{R^2 + X_L^2}$$

$$Z = \frac{E_T}{I_T}$$

$$Z = \frac{VA}{I_T^2}$$

$$Z = \frac{R}{PF}$$

$$P = E_R \times I_R$$

$$P = \sqrt{VA^2 - VARS_L^2}$$

$$P = \frac{E_R^2}{R}$$

$$P = I^2_R \times R$$

$$P = VA \times PF$$

$$E_L = I_L \times X_L$$

$$E_L = \sqrt{E_T^2 - E_R^2}$$

$$E_L = \sqrt{VARS_L \times X_L}$$

$$E_L = \frac{VARS_L}{I_L}$$

$$VARS_L = E_L \times I_L$$

$$VA = E_T \times I_T$$

$$VA = I_T^2 \times Z$$

$$VA = \frac{E_T^2}{Z}$$

$$VA = \sqrt{P^2 + VARS_L^2}$$

$$Z = \frac{E_T^2}{VA}$$

$$E_R = I_R \times R$$

$$E_R = \sqrt{P \times R}$$

$$E_R = \frac{P}{I_R}$$

$$E_R = \sqrt{E_T^2 - E_L^2}$$

$$E_R = E_T \times PF$$

$$I_L = I_R = I_T$$

$$I_L = \frac{E_L}{X_L}$$

$$I_L = \frac{VARS_L}{E_L}$$

$$I_L = \sqrt{\frac{VARS_L}{X_L}}$$

$$VARS_L = \frac{E_L^2}{X_L}$$

$$L = \frac{X_L}{2\pi F}$$

$$I_T = I_R = I_L$$

$$I_T = \frac{E_T}{Z}$$

$$I_T = \frac{VA}{E_T}$$

$$VA = \frac{P}{PF}$$

$$I_R = I_T \times I_L$$

$$I_R = \frac{E_R}{R}$$

$$I_R = \frac{P}{E_R}$$

$$I_R = \sqrt{\frac{P}{R}}$$

$$X_L = \sqrt{Z^2 - R^2}$$

$$X_L = \frac{E_L}{I_L}$$

$$X_L = \frac{E_L^2}{VARS_L}$$

$$X_L = \frac{VARS_L}{I_L^2}$$

$$X_L = 2\pi FL$$

$$VARS_L = I_L^2 \times X_L$$

RESISTIVE-INDUCTIVE PARALLEL CIRCUITS

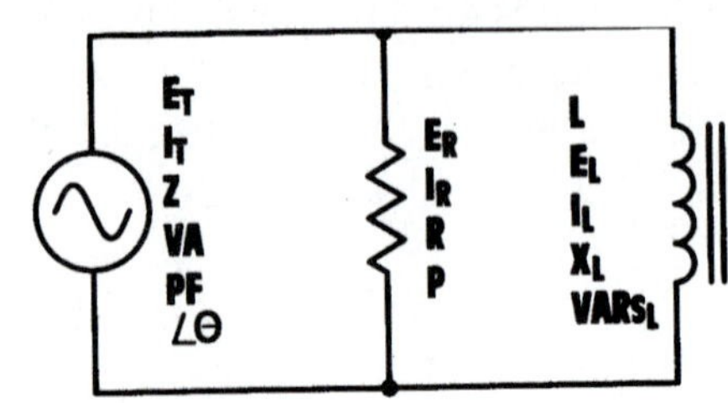

$E_T = E_R = E_L$

$E_T = I_T \times Z$

$L = \frac{X_L}{2\pi F}$

$E_T = \frac{VA}{I_T}$

$E_T = \sqrt{VA \times Z}$

$Z = \frac{1}{\sqrt{\left(\frac{1}{R}\right)^2 + \left(\frac{1}{X_L}\right)^2}}$

$Z = \frac{VA}{I_T^2}$

$I_T = \sqrt{I_R^2 + I_L^2}$

$I_T = \frac{VA}{E_T}$ $\quad I_T = \sqrt{\frac{VA}{Z}}$

$Z = \frac{E_T}{I_T}$ $\quad Z = \frac{E_T^2}{VA}$

$Z = R \times PF$

$I_T = \frac{E_T}{Z}$

$I_T = \frac{I_R}{PF}$

$VA = E_T \times I_T$

$PF = \frac{Z}{R}$

$E_L = I_L \times X_L$

$I_L = \sqrt{I_T^2 - I_R^2}$

$VA = I_T^2 \times Z$

$PF = \frac{P}{VA}$

$E_L = E_T = E_R$

$I_L = \frac{E_L}{X_L}$

$VA = \frac{E_T^2}{Z}$

$PF = \frac{I_R}{I_T}$

$E_L = \sqrt{VARs_L \times X_L}$

$I_L = \frac{VARs_L}{E_L}$

$VA = \sqrt{P^2 + VARs_L^2}$

$PF = COS \angle\theta$

$E_L = \frac{VARs_L}{I_L}$

$I_L = \sqrt{\frac{VARs_L}{X_L}}$

$VA = \frac{P}{PF}$ $\quad VA = \frac{E_T^2}{Z}$

$VARs_L = \sqrt{VA^2 - P^2}$

$VARs_L = E_L \times I_L$

$VARs_L = \frac{E_L^2}{X_L}$

$VARs_L = I_L^2 \times X_L$

$E_R = I_R \times R$

$I_R = \sqrt{I_T^2 - I_L^2}$

$X_L = \frac{E_L}{I_L}$

$X_L = \frac{1}{\sqrt{\left(\frac{1}{Z}\right)^2 - \left(\frac{1}{R}\right)^2}}$

$E_R = \sqrt{P \times R}$

$I_R = \frac{E_R}{R}$

$X_L = \frac{E_L^2}{VARs_L}$

$X_L = 2\pi FL$

$E_R = \frac{P}{I_R}$

$I_R = \frac{P}{E_R}$

$X_L = \frac{VARs_L}{I_L^2}$

$R = \frac{E_R}{I_R}$

$E_R = E_T = E_L$

$R = \frac{P}{I_R^2}$

$I_R = \sqrt{\frac{P}{R}}$

$I_R = I_T \times PF$

RESISTIVE-INDUCTIVE PARALLEL CIRCUITS (CONTINUED)

$R = \frac{1}{\sqrt{\left(\frac{1}{Z}\right)^2 - \left(\frac{1}{X_L}\right)^2}}$ $R = \frac{E_R^2}{P}$ $R = \frac{Z}{PF}$ $P = \sqrt{VA^2 - VARs_L^2}$

$P = E_R \times I_R$ $P = \frac{E_R^2}{R}$ $P = I_R^2 \times R$ $P = VA \times PF$

RESISTIVE-CAPACITIVE SERIES CIRCUITS

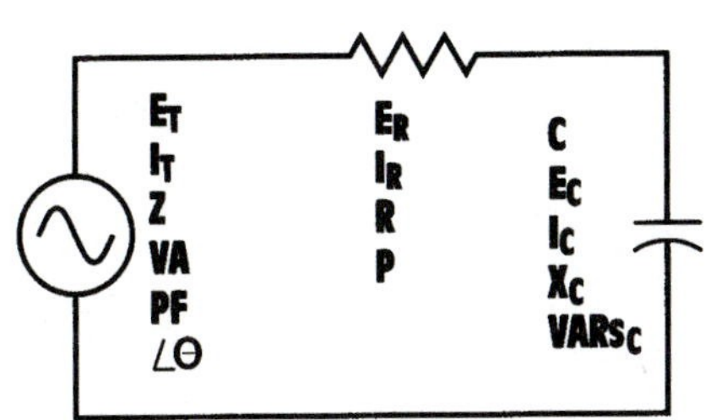

$E_T = \sqrt{E_R^2 + E_C^2}$

$E_T = I_T \times Z$ $I_T = I_R = I_C$

$E_T = \frac{VA}{I_T}$ $I_T = \frac{E_T}{Z}$

$E_T = \frac{E_R}{PF}$ $I_T = \frac{VA}{E_T}$

$Z = \sqrt{R^2 + X_C^2}$ $Z = \frac{VA}{I_T^2}$ $Z = \frac{E_T^2}{VA}$

$Z = \frac{E_T}{I_T}$ $Z = \frac{R}{PF}$ $C = \frac{1}{2\pi F X_C}$

$VA = E_T \times I_T$ $PF = \frac{R}{Z}$ $P = E_R \times I_R$ $E_R = I_R \times R$

$VA = I_T^2 \times Z$ $PF = \frac{P}{VA}$ $P = \sqrt{VA^2 - VARs_C^2}$ $E_R = \sqrt{P \times R}$

$VA = \frac{E_T^2}{Z}$ $PF = \frac{E_R}{E_T}$ $P = \frac{E_R^2}{R}$ $E_R = \frac{P}{I_R}$

$VA = \sqrt{P^2 + VARs_C^2}$ $PF = COS \angle\theta$ $P = I_R^2 \times R$ $E_R = \sqrt{E_T^2 - E_C^2}$

$VA = \frac{P}{PF}$ $R = \sqrt{Z^2 - X_C^2}$ $P = VA \times PF$ $E_R = E_T \times PF$

$I_R = I_T = I_C$ $R = \frac{E_R}{I_R}$ $E_C = I_C \times X_C$ $I_C = I_R = I_T$

$I_R = \frac{E_R}{R}$ $R = \frac{E_R^2}{P}$ $E_C = \sqrt{E_T^2 - E_R^2}$ $I_C = \frac{E_C}{X_C}$

RESISTIVE-CAPACITIVE SERIES CIRCUITS (CONTINUED)

$$I_R = \frac{P}{E_R}$$

$$I_R = \sqrt{\frac{P}{R}}$$

$$X_C = \sqrt{Z^2 - R^2}$$

$$X_C = \frac{E_C}{I_C}$$

$$VARs_C = \sqrt{VA^2 - P^2}$$

$$R = \frac{P}{I_R^2}$$

$$R = Z \times PF$$

$$X_C = \frac{E_C^2}{VARs_C}$$

$$X_C = \frac{VARs_C}{I_C^2}$$

$$E_C = \sqrt{VARs_C \times X_C}$$

$$E_C = \frac{VARs_C}{I_C}$$

$$X_C = \frac{1}{2\pi FC}$$

$$VARs_C = E_C \times I_C$$

$$I_C = \frac{VARs_C}{E_C}$$

$$I_C = \sqrt{\frac{VARs_C}{X_C}}$$

$$VARs_C = I_C^2 \times X_C$$

$$VARs_C = \frac{E_C^2}{X_C}$$

RESISTIVE-CAPACITIVE PARALLEL CIRCUITS

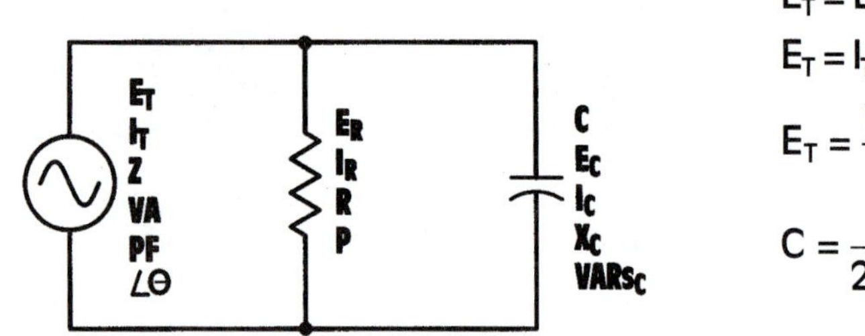

$$E_T = E_R = E_C$$

$$E_T = I_T \times Z$$

$$E_T = \frac{VA}{I_T}$$

$$C = \frac{1}{2\pi FX_C}$$

$$E_T = \sqrt{VA \times Z}$$

$$Z = \frac{1}{\sqrt{\left(\frac{1}{R}\right)^2 + \left(\frac{1}{X_C}\right)^2}}$$

$$I_T = \sqrt{I_R^2 + I_C^2}$$

$$I_T = \frac{E_T}{Z}$$

$$I_T = \frac{VA}{E_T}$$

$$I_T = \frac{I_R}{PF}$$

$$Z = \frac{VA}{I_T^2}$$

$$VA = E_T \times I_T$$

$$VA = I_T^2 \times Z$$

$$VA = \frac{E_T^2}{Z}$$

$$VA = \sqrt{P^2 + VARs_C^2}$$

$$Z = \frac{E_T}{I_T} \qquad Z = \frac{E_T^2}{VA}$$

$$PF = \frac{Z}{R}$$

$$PF = \frac{P}{VA}$$

$$PF = \frac{E_I}{E_R}$$

$$PF = \cos \angle\theta$$

$$Z = R \times PF$$

$$P = E_R \times I_R$$

$$P = \sqrt{VA^2 - VARs_C^2}$$

$$P = \frac{E_R^2}{R}$$

$$P = I_R^2 \times R$$

RESISTIVE-CAPACITIVE PARALLEL CIRCUITS (CONTINUED)

$$I_T = \sqrt{\frac{VA}{Z}}$$

$$VA = \frac{P}{PF}$$

$$P = VA \times PF$$

$$E_R = I_R \times R$$

$$I_R = \sqrt{I_T^2 - I_C^2}$$

$$R = \frac{1}{\sqrt{\left(\frac{1}{Z}\right)^2 - \left(\frac{1}{X_C}\right)^2}}$$

$$E_C = I_C \times X_C$$

$$E_R = \sqrt{P \times R}$$

$$I_R = \frac{E_R}{R}$$

$$R = \frac{E_R}{I_R}$$

$$E_C = E_T = E_R$$

$$E_R = \frac{P}{I_R}$$

$$I_R = \frac{P}{E_R}$$

$$R = \frac{E_R^2}{P}$$

$$E_C = \sqrt{VARs_C \times X_C}$$

$$E_R = E_T = E_C$$

$$I_R = \sqrt{\frac{P}{R}}$$

$$R = \frac{P}{I_R^2}$$

$$E_C = \frac{VARs_C}{I_C}$$

$$E_R = E_T \times PF$$

$$R = \frac{Z}{PF}$$

$$VARs_C = I_C^2 \times X_C$$

$$I_C = \sqrt{I_T^2 - I_R^2}$$

$$X_C = \frac{E_C^2}{VARs_C}$$

$$VARs_C = \frac{E_C^2}{X_C}$$

$$I_C = \frac{E_C}{X_C}$$

$$X_C = \frac{1}{\sqrt{\left(\frac{1}{Z}\right)^2 - \left(\frac{1}{R}\right)^2}}$$

$$X_C = \frac{VARs_C}{I_C^2}$$

$$VARs_C = E_C \times I_C$$

$$I_C = \frac{VARs_C}{E_C}$$

$$X_C = \frac{E_C}{I_C}$$

$$X_C = \frac{1}{2\pi FC}$$

$$VARs_C = \sqrt{VA^2 - P^2}$$

$$I_C = \sqrt{\frac{VARs_C}{X_C}}$$

RESISTIVE-INDUCTIVE-CAPACITIVE SERIES CIRCUITS

$$E_T = \sqrt{E_R^2 + (E_L - E_C)^2}$$

$$E_T = \frac{VA}{I_T}$$

$$Z = \sqrt{R^2 + (X_L - X_C)^2}$$

$$Z = \frac{VA}{I_T^2}$$

$$E_T = I_T \times Z$$

$$E_T = \frac{E_R}{PF}$$

$$Z = \frac{E_R}{I_T}$$

$$Z = \frac{R}{PF}$$

RESISTIVE-INDUCTIVE-CAPACITIVE SERIES CIRCUITS (CONTINUED)

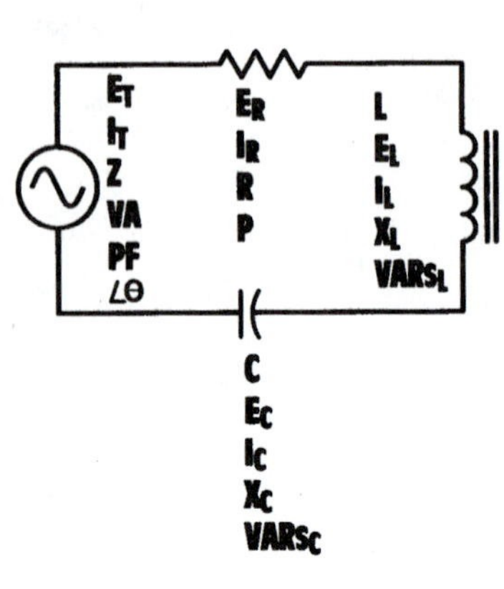

$I_T = I_R = I_L = I_C$

$I_T = \frac{VA}{E_T}$

$I_T = \frac{E_T}{Z}$

$I_T = \sqrt{\frac{VA}{Z}}$

$VA = E_T \times I_T$

$VA = \frac{P}{PF}$

$VA = I_T^2 \times Z$

$VA = \frac{E_T^2}{Z}$

$VA = \sqrt{P^2 + (VARs_L - VARs_C)^2}$

$PF = \frac{R}{Z}$

$P = E_R \times I_R$

$P = VA \times PF$

$E_R = I_R \times R$

$PF = \frac{P}{VA}$

$P = \sqrt{VA^2 - (VARs_L - VARs_C)^2}$

$E_R = \frac{P}{I_R}$

$PF = \frac{E_R}{E_T}$

$P = \frac{E_R^2}{R}$

$E_R = \sqrt{P \times R}$

$E_R = \sqrt{E_T^2 - (E_L - E_C)^2}$

$PF = COS \angle\theta$

$P = I_R^2 \times R$

$E_R = E_T \times PF$

$I_R = I_T = I_C = I_L$

$R = \sqrt{Z^2 - (X_L - X_C)^2}$

$E_C = I_C \times X_C$

$I_R = \frac{E_R}{R}$

$R = \frac{E_R}{I_R}$

$R = Z \times PF$

$E_C = \sqrt{VARs_C \times X_C}$

$I_R = \frac{P}{E_R}$

$R = \frac{E_R^2}{P}$

$R = \frac{P}{I_R^2}$

$E_C = \frac{VARs_C}{I_C}$

$I_R = \sqrt{\frac{P}{R}}$

$X_C = \frac{1}{2\pi FC}$

$C = \frac{1}{2\pi FX_C}$

$I_C = I_R = I_T = I_L$

$X_C = \frac{E_C^2}{VARs_C}$

$X_C = \frac{VARs_C}{I_C^2}$

$VARs_C = E_C \times I_C$

$I_C = \frac{E_C}{X_C}$

$X_C = \frac{E_C}{I_C}$

$L = \frac{X_L}{2\pi F}$

$X_L = 2\pi FL$

$I_C = \frac{VARs_C}{E_C}$

$VARs_C = I_C^2 \times X_C$

$X_L = \frac{E_L}{I_L}$

$X_L = \frac{VARs_L}{I_L^2}$

$I_C = \sqrt{\frac{VARs_C}{X_C}}$

$VARs_C = \frac{E_C^2}{X_C}$

$X_L = \frac{E_L^2}{VARs_L}$

RESISTIVE-INDUCTIVE-CAPACITIVE SERIES CIRCUITS (CONTINUED)

$E_L = I_L \times X_L$

$E_L = \sqrt{VARs_L \times X_L}$

$E_L = \dfrac{VARs_L}{I_L}$

$I_L = I_R = I_T = I_C$

$I_L = \dfrac{E_L}{X_L}$

$I_L = \dfrac{VARs_L}{E_L}$

$I_L = \sqrt{\dfrac{VARs_L}{X_L}}$

$VARs_L = E_L \times I_L$

$VARs_L = \dfrac{E_L^2}{X_L}$

$VARs_L = I_L^2 \times X_L$

RESISTIVE-INDUCTIVE-CAPACITIVE PARALLEL CIRCUITS

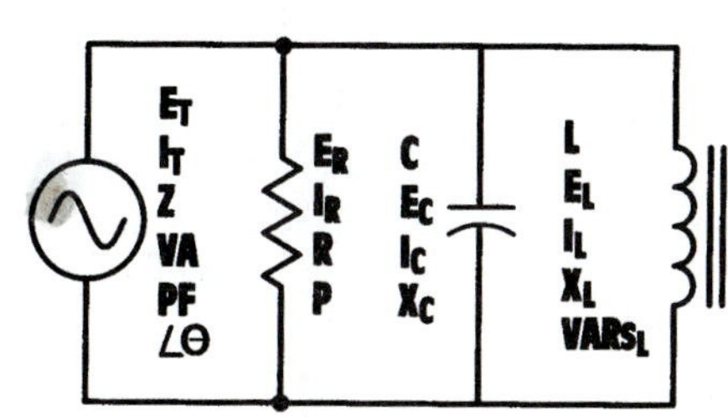

$E_T = E_R = E_L = E_C$

$E_T = I_T \times Z$

$E_T = \dfrac{VA}{I_T}$

$E_T = \sqrt{VA \times Z}$

$Z = \dfrac{1}{\sqrt{\left(\dfrac{1}{R}\right)^2 + \left(\dfrac{1}{X_L} - \dfrac{1}{X_C}\right)^2}}$

$Z = \dfrac{VA}{I_T^2}$

$I_T = \sqrt{I_R^2 + (I_L - I_C)^2}$

$I_T = \dfrac{VA}{E_T}$

$I_T = \sqrt{\dfrac{VA}{Z}}$

$Z = \dfrac{E_T}{I_T}$ $\quad Z = \dfrac{E_T^2}{VA}$

$Z = R \times PF$

$I_T = \dfrac{E_T}{Z}$

$I_T = \dfrac{I_R}{PF}$

$VA = E_T \times I_T$

$PF = \dfrac{Z}{R}$

$PF = COS \angle\theta$

$E_L = E_T = E_R = E_C$

$VA = I_T^2 \times Z$

$PF = \dfrac{P}{VA}$

$E_L = I_L \times X_L$

$E_L = \sqrt{VARs_L \times X_L}$

$VA = \dfrac{E_T^2}{Z}$

$PF = \dfrac{I_R}{I_T}$

$E_L = \dfrac{VARs_L}{I_L}$

$L = \dfrac{X_L}{2\pi F}$

$VA = \sqrt{P^2 + (VARs_L - VARs_C)^2}$

$X_L = \dfrac{E_L}{I_L}$

$X_L = 2\pi FL$

$VARs_L = \dfrac{E_L^2}{X_L}$

$VA = \dfrac{P}{PF}$

$I_L = \dfrac{VARs_L}{E_L}$

$X_L = \dfrac{E_L^2}{VARs_L}$

$VARs_L = I_L^2 \times X_L$

$I_L = \dfrac{E_L}{X_L}$

$I_L = \sqrt{\dfrac{VARs_L}{X_L}}$

$X_L = \dfrac{VARs_L}{I_L^2}$

$VARs_L = E_L \times I_L$

RESISTIVE-INDUCTIVE-CAPACITIVE PARALLEL CIRCUITS (CONTINUED)

$E_R = I_R \times R$

$I_R = \sqrt{I_T^2 - (I_L - I_C)^2}$

$I_R = I_T \times PF$

$E_R = \sqrt{P \times R}$

$I_R = \frac{E_R}{R}$

$R = \frac{E_R}{I_R}$

$R = \frac{1}{\sqrt{\left(\frac{1}{Z}\right)^2 - \left(\frac{1}{X_L} - \frac{1}{X_C}\right)^2}}$

$E_R = \frac{P}{I_R}$

$I_R = \frac{P}{E_R}$

$E_R = E_T = E_L = E_C$

$I_R = \sqrt{\frac{P}{R}}$

$R = \frac{P}{I_R^2}$

$R = \frac{Z}{PF}$

$R = \frac{E_R^2}{P}$

$P = \sqrt{VA^2 - (VARs_L - VARs_C)^2}$

$P = E_R \times I_R$

$P = I_R^2 \times R$

$E_C = \frac{VARs_C}{I_C}$

$P = VA \times PF$

$P = \frac{E_R^2}{R}$

$X_C = \frac{1}{2\pi FC}$

$E_C = I_C \times X_C$

$I_C = \frac{E_C}{X_C}$

$X_C = \frac{E_C}{I_C}$

$VARs_C = E_C \times I_C$

$E_C = E_T = E_R = E_L$

$I_C = \frac{VARs_C}{E_C}$

$X_C = \frac{E_C^2}{VARs_C}$

$VARs_C = I_C^2 \times X_C$

$E_C = \sqrt{VARs_C \times X_C}$

$I_C = \sqrt{\frac{VARs_C}{X_C}}$

$X_C = \frac{VARs_C}{I_C^2}$

$VARs_C = \frac{E_C^2}{X_C}$

TRANSFORMERS

$\frac{E_P}{E_S} = \frac{N_P}{N_S}$

$\frac{E_P}{E_S} = \frac{I_S}{I_P}$

$\frac{N_P}{N_S} = \frac{I_S}{I_P}$

$Z_P = Z_S \left(\frac{N_P}{N_S}\right)^2$

$Z_S = Z_P \left(\frac{N_P}{N_S}\right)^2$

E_P—Voltage of the primary
E_S—Voltage of the secondary
I_P—Current of the primary
I_S—Current of the secondary
N_P—Number of turns of the primary
N_S—Number of turns of the secondary
Z_P—Impedance of the primary
Z_S—Impedance of the secondary

THREE-PHASE CONNECTIONS

WYE CONNECTION

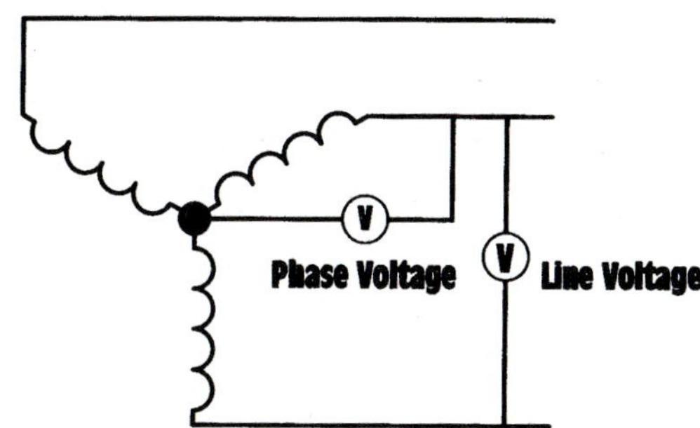

In a *wye* connection, the line current and phase current are the same.

$$I_{(LINE)} = I_{(PHASE)}$$

In a *wye* connection, the line voltage is higher than the phase voltage by a factor of the square root of 3.

$$E_{(LINE)} = E_{(PHASE)} \times \sqrt{3}$$

$$E_{(PHASE)} = \frac{E_{(LINE)}}{\sqrt{3}}$$

DELTA CONNECTION

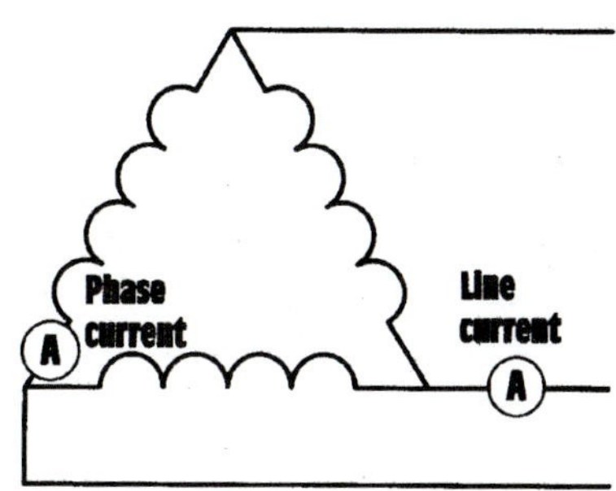

In a *delta*-connected system, the line voltage and phase voltage are the same.

$$E_{(LINE)} = E_{(PHASE)}$$

In a *delta*-connected system the line current is higher than the phase current by a factor of the square root of 3.

$$I_{(LINE)} = I_{(PHASE)} \times \sqrt{3}$$

$$I_{(PHASE)} = \frac{I_{(LINE)}}{\sqrt{3}}$$

OPEN DELTA CONNECTION

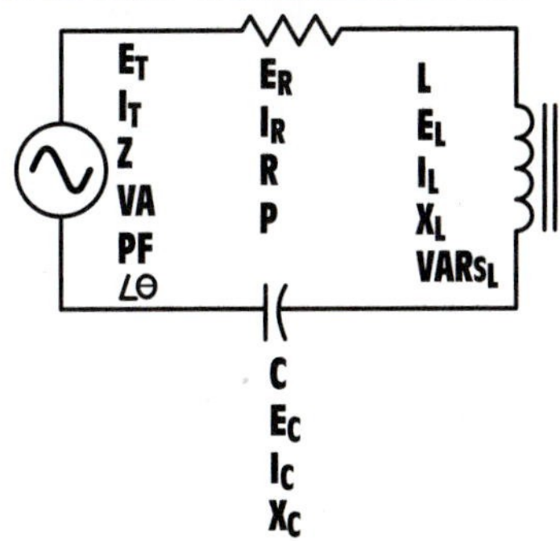

Open delta connections provide 87% of the sum of the power rating of the two transformers. Example: The two transformers shown are rated at 60 kVA each.

The total power rating for this connection would be

$$60 + 60 = 120 \text{ kVA}$$

$$120 \text{ kVA} \times 0.87 = 104.4 \text{ kVA}$$

All values of voltage and current are computed in the same manner as for a closed delta connection.

APPARENT AND TRUE POWER

$$VA = \sqrt{3} \times E_{(LINE)} \times I_{(LINE)}$$

$$VA = 3 \times E_{(PHASE)} \times I_{(PHASE)}$$

$$P = \sqrt{3} \times E_{(LINE)} \times I_{(LINE)} \times PF$$

$$P = 3 \times E_{(PHASE)} \times I_{(PHASE)} \times PF$$

D

Appendix

Greek Alphabet

Name of letter	Upper case	Lower case	Designates
Alpha	Α	α	Angles
Beta	Β	β	Angles; flux density
Gamma	Γ	γ	Conductivity
Delta	Δ	δ	Change of a quantity
Epsilon	Ε	ε	Base of natural logarithms
Zeta	Ζ	ζ	Impedance; coefficients; coordinates
Eta	Η	η	Hysteresis coefficient; efficiency
Theta	Θ	θ	Phase angle
Iota	Ι	ι	
Kappa	Κ	κ	Dielectric constant; coefficient of coupling; susceptibility
Lambda	Λ	λ	Wavelength
Mu	Μ	μ	Permeability; the prefix micro-, amplification factor
Nu	Ν	ν	Change of a quantity
Xi	Ξ	ξ	
Omicron	Ο	ο	
Pi	Π	π	3.1416
Rho	Ρ	ρ	Resistivity
Sigma	Σ	σ	
Tau	Τ	τ	Time constant; time phase displacement
Upsilon	Υ	υ	
Phi	Φ	φ	Angles; magnetic flow
Chi	Χ	χ	
Psi	Ψ	ψ	Dielectric flux; phase difference
Omega	Ω	ω	Resistivity

Appendix E

Answers to Practice Problems

UNIT 1 OHM'S LAW

1. 6.25 amperes $I = \sqrt{\frac{P}{R}}$ $\mathrm{I} = \sqrt{\frac{1500}{38.4}}$
2. 240 volts $\mathrm{E} = 6.25 \times 38.4$
3. 0.8 amperes (100/120)
4. 960 Ω $R = \frac{E^2}{P}$
5. 4.332 amperes (300 × 4 = 1200) (1200/277)
6. 12 volts (1500 × 0.008)
7. 45 mA
8. 450 μA
9. 61,594.203 amperes (850,000,000/13,800)
10. 2.542 kV

UNIT 2 RESISTOR

1st Band	2nd Band	3rd Band	4th Band	Value	% Tol.
Red	Yellow	Brown	Silver	240 Ω	10
Blue	Gray	Red	Gold	6800 Ω	5
Orange	Orange	Orange	Gold	33k Ω	5
Brown	Red	Black	Red	12 Ω	2
Brown	Green	Silver	Silver	0.15 Ω	10
Brown	Gray	Green	Silver	1.8 M Ω	10
Brown	Black	Yellow	None	100k Ω	20
Brown	Black	Orange	Gold	10k Ω	5
Violet	Green	Black	Red	75 Ω	2
Yellow	Violet	Red	None	4.7k Ω	20
Gray	Red	Green	Red	8.2M Ω	2
Green	Blue	Gold	Red	5.6 Ω	2

UNIT 5 SERIES CIRCUITS

PRACTICE PROBLEM 1

E_T 116.624	E_1 15.882	E_2 13.299	E_3 27.698	E_4 **38**	E_5 21.714
I_T 0.0391	R_1 406.189	R_2 340.128	R_3 708.394	R_4 971.083	R_5 555.334
R_T 2,982.712	P_1 **0.621**	P_2 **0.52**	P_3 **1.083**	P_4 1.487	P_5 **.849**
P_T **4.56**					

PRACTICE PROBLEM 2

E_T 50	E_1 **6**	E_2 16.5	E_3 11	E_4 9	E_5 7.5
I_T .005	R_1 1.2k	R_2 **3.3k**	R_3 **2.2k**	R_4 **1.8k**	R_5 **1.5k**
R_T **10k**	P_1 0.03	P_2 0.0825	P_3 0.055	P_4 0.045	P_5 0.0375
P_T 0.25					

PRACTICE PROBLEM 3

E_T **340**	E_1 **44**	E_2 **94**	E_3 60	E_4 **40**	E_5 **102**
I_T 0.002	R_1 22k	R_2 47k	R_3 30k	R_4 20k	R_5 51k
R_T 170k	P_1 0.088	P_2 0.188	P_3 **0.12**	P_4 0.08	P_5 0.204
P_T 0.68					

PRACTICE PROBLEM 4

E_T **36**	E_1 7.398	E_2 8.13	E_3 4.553	E_4 6.992	E_5 8.943
I_T 0.0813	R_1 **91**	R_2 **100**	R_3 **56**	R_4 **86**	R_5 **110**
R_T 443	P_1 0.601	P_2 0.661	P_3 0.37	P_4 0.568	P_5 0.727
P_T 2.927					

UNIT 6 PARALLEL CIRCUITS

PRACTICE PROBLEM 1

E_T 119.524	I_1 **0.176**	I_2 0.147	I_3 **0.255**	I_4 **0.364**
R_T 126.883	R_1 679.114	R_2 813.088	R_3 468.722	R_4 328.363
P_T 110.44	P_1 21.036	P_2 **17.57**	P_3 30.479	P_4 43.507
I_T **0.942**	$I_{,2,3,4}$ 0.766	$I_{,3,4}$ 0.619		

PRACTICE PROBLEM 2

E_T 277.154	I_1 2.31 mA	I_2 1.39 mA	I_3 1.54 mA	I_4 1.15 mA
R_T 43,373.494	R_1 **120k**	R_2 **200k**	R_3 **180k**	R_4 **240k**
P_T **1.771**	P_1 0.64	P_2 0.385	P_3 0.427	P_4 0.319
I_T 6.39 mA	$I_{,2,3,4}$ 4.08 mA	$I_{,3,4}$ 2.69 mA		

PRACTICE PROBLEM 3

E_T 47.994	I_1 3	I_2 4.799	I_3 **3.2**	I_4 2.34
R_T **3.582**	R_1 **16**	R_2 **10**	R_3 14.998	R_4 **20**
P_T 642.88	P_1 143.982	P_2 230.323	P_3 153.581	P_4 112.306
I_T 13.395	$I_{,2,3,4}$ 10.339	$I_{,3,4}$ 5.54		

PRACTICE PROBLEM 4

E_T 240.98	I_1 2.92 mA	I_2 3.24 mA	I_3 4.29 mA	I_4 3.86 mA
R_T 16,851.748	R_1 82,527.397	R_2 74,376.543	R_3 56,172.494	R_4 62,430.052
P_T 3.446	P_1 **0.704**	P_2 **0.78**	P_3 **1.031**	P_4 **0.931**
I_T **0.0143**	$I_{,2,3,4}$ 11.39 mA	$I_{,3,4}$ 8.15 mA		

UNIT 7 COMBINATION CIRCUITS

PRACTICE PROBLEM 1

A–B	55.636	A–C	93.497	A–D	73.262
B–C	81.497	B–D	93.497	C–D	66.225

PRACTICE PROBLEM 2

A–B	767.742	A–C	1535.484	A–D	838.71
B–C	1187.097	B–D	1535.484	C–D	1470.968

PRACTICE PROBLEM 3

A–B	390,539.568	A–C	648,381.295	A–D	640,287.77
B–C	447,194.245	B–D	648,381.295	C–D	547,661.871

PRACTICE PROBLEM 4

A–B	31,109.434	A–C	55,449.057	A–D	58,090.566
B–C	36,022.642	B–D	55,449.057	C–D	62,264.151

PRACTICE PROBLEM 5

E_T **36**	E_1 16.68	E_2 19.315	E_3 19.315
I_T 0.695	I_1 0.695	I_2 0.284	I_3 0.411
R_T 51.791	R_1 **24**	R_2 **68**	R_3 **47**
P_T 25.02	P_1 11.593	P_2 5.485	P_3 7.938

PRACTICE PROBLEM 6

E_T 47.895	E_1 24.72	E_2 23.173	E_3 23.173
I_T 2.06 mA	I_1 2.06 mA	I_2 0.772 mA	I_3 1.29 mA
R_T **23,250**	R_1 **12,000**	R_2 30,000	R_3 **18,000**
P_T 0.0987	P_1 0.0509	P_2 **0.0179**	P_3 0.03

PRACTICE PROBLEM 7

E_T 24.006	E_1 9.441	E_2 14.58	E_3 14.736	E_4 9.271
I_T 0.0629	I_1 0.0286	I_2 0.0286	I_3 0.0343	I_4 0.0343
R_T 381.653	R_1 330.105	R_2 509.79	R_3 **430**	R_4 270.288
P_T **1.510**	P_1 **0.270**	P_2 **0.417**	P_3 0.505	P_4 **0.318**

PRACTICE PROBLEM 8

E_T 120.062	E_1 68.64	E_2 51.58	E_3 80	E_4 40
I_T **0.00829**	I_1 0.00429	I_2 0.00429	I_3 0.004	I_4 0.004
R_T 14,482.759	R_1 **16k**	R_2 **12k**	R_3 **20k**	R_4 **10k**
P_T 0.995	P_1 0.294	P_2 0.221	P_3 0.343	P_4 0.172

PRACTICE PROBLEM 9

E_T 207.869	E_1 155.366	E_2 155.366	E_3 52.503	E_4 75.682	E_5 132.187
I_T **0.351**	I_1 0.0973	I_2 0.0779	I_3 0.175	I_4 0.176	I_5 0.0695
R_T 592.219	R_1 1596.773	R_2 1994.429	R_3 300.017	R_4 430.011	R_5 1901.971
P_T 72.962	P_1 **15.123**	P_2 **12.099**	P_3 **9.188**	P_4 **13.32**	P_5 **8.712**
E_6 132.187	I_6 0.11	R_6 1201.7	P_6 **14.52**		

PRACTICE PROBLEM 10

E_T **208**	E_1 48	E_2 64	E_3 38.4	E_4 57.6	E_5 96
I_T 0.64	I_1 0.64	I_2 0.64	I_3 0.32	I_4 0.32	I_5 0.32
R_T 325	R_1 **75**	R_2 **100**	R_3 **120**	R_4 **180**	R_5 **300**
P_T 133.12	P_1 30.72	P_2 40.96	P_3 12.288	P_4 18.432	P_5 30.72

UNIT 10 ALTERNATING CURRENT LOADS

Inductance (H)	Frequency (Hz)	Induct. Rct. (Ω)
1.2	60	452.4
0.085	400	213.628
0.75	1000	4712.389
0.65	600	2450.442
3.6	30	678.584
2.62	25	411.459
0.5	60	188.5
0.85	1200	6408.849
1.6	20	201.062
0.45	400	1130.973
4.8	80	2412.743
0.0065	1000	40.841

UNIT 11 CAPACITIVE CIRCUITS

Capacitance	Cap. Rect. (Ω)	Frequency
38 μF	69.803 Ω	60 Hz
5.049 μF	78.8 Ω	400 Hz
250 pF	4.5 k Ω	141.47 Hz
234 μF	0.068 Ω	10 k Hz
13.263 μF	240 Ω	50 Hz
10 μF	36.8 Ω	432.486 Hz
560 nF	0.142 Ω	2 MHz
176.84	15 k Ω	60 Hz
75 μF	560 Ω	3.789 Hz
470 pF	1.693 k Ω	200 kHz
58.513 nF	6.8 k Ω	400 Hz
34 μF	450 Ω	10.402 Hz

UNIT 12 THREE-PHASE CIRCUITS

PRACTICE PROBLEM 1

$E_{P(Alt.)}$	138.57	$E_{P(Line)}$	240
$I_{P(Alt.)}$	34.64	$I_{P(Line)}$	20
$E_{L(Alt.)}$	**240**	$E_{L(Line)}$	240
$I_{L(Alt.)}$	34.64	$I_{L(Line)}$	34.64
P	14,399.6 W	$Z_{(Phase)}$	**12 Ω**

PRACTICE PROBLEM 2

$E_{P(Alt.)}$	4160	$E_{P(Line)}$	2401.85
$I_{P(Alt.)}$	23.11	$I_{P(Line)}$	40.03
$E_{L(Alt.)}$	**4160**	$E_{L(Line)}$	4160
$I_{L(Alt.)}$	40.03	$I_{L(Line)}$	40.03
P	288,420.95	$Z_{(Phase)}$	**60 Ω**

PRACTICE PROBLEM 3

$E_{P(Alt.)}$	323.33	$E_{P(Line\ 1)}$	323.33	$E_{P(Line\ 2)}$	560
$I_{P(Alt.)}$	185.91	$I_{P(Line\ 1)}$	64.67	$I_{P(Line\ 2)}$	70
$E_{L(Alt.)}$	**560**	$E_{L(Line\ 1)}$	560	$E_{L(Line\ 2)}$	560
$I_{L(Alt.)}$	185.91	$I_{L(Line\ 1)}$	64.67	$I_{L(Line\ 2)}$	121.24
P	180,317.83	$Z_{(Phase\ 1)}$	**5 Ω**	$Z_{(Phase\ 2)}$	**8 Ω**

UNIT 13 TRANSFORMERS

1.

E_P 120	E_S 24
I_P 1.6	I_S 8
N_P 300	N_S 60
Ration 5:1	Z = 3 Ω

2.

E_P 240	E_S 320
I_P 0.853	I_S 0.643
N_P 210	N_S 280
Ration 1:1.333	Z = 500 Ω

3.

E_P 64	E_S 160
I_P 33.333	I_S 13.333
N_P 32	N_S 80
Ration 1:2.5	Z = 12 Ω

4.

E_P 48	E_S 240
I_P 3.333	I_S 0.667
N_P 220	N_S 1100
Ration 1:5	Z = 360 Ω

5.

E_P 35.848	E_S 182
I_P 16.5	I_S 3.25
N_P 87	N_S 450
Ration 1:5.077	Z = 56 Ω

6.

E_P 480	E_S 916.346
I_P 1.458	I_S 0.764
N_P 275	N_S 525
Ration 1:1.909	Z = 1.2 Ω

7.

E_P 208	E_S 1 320	E_{S2} 120	E_{S3} 24
I_P 11.93	I_{S1} 0.0267	I_{S2} 20	I_{S3} 3
N_P 800	N_{S1} 1231	N_{S2} 462	N_{S3} 92
	Ratio 1 1:1.54	Ratio 2 1.73:1	Ratio 3 1:8.67
	R_1 = 12 kΩ	R_2 6 Ω	R_3 8 Ω

8.

E_P 277	E_{S1} 480	E_{S2} 208	E_{S3} 120
I_P 8.93	I_{S1} 2.4	I_{S2} 3.47	I_{S3} 5
N_P 350	N_{S1} 606	N_{S2} 263	N_{S3} 152
	Ratio 1 1:1.73	Ratio 2 1.33:1	Ration 3 2.31:1
	R_1 = 200 Ω	R_2 60 Ω	R_3 24 Ω

Glossary

AC (alternating current) current that reverses its direction of flow periodically. Reversals generally occur at regular intervals.

across-the-line method of motor starting that connects the motor directly to the supply line on starting or running. (Also known as full-voltage starting.)

air gap space between two magnetically related components.

alternating current (AC) current that reverses its direction of flow periodically. Reversals generally occur at regular intervals.

alternator machine used to generate alternating current by rotating conductors through a magnetic field.

ambient temperature temperature surrounding a device.

American Wire Gauge (AWG) measurement of the diameter of wire. The gauge scale was formerly known as the Brown and Sharp scale. The scale has a fixed constant of 1.123 between gauge sizes.

ammeter instrument used to measure the flow of current.

amortiseur winding squirrel-cage winding on the rotor of a synchronous motor used for starting purposes only.

ampacity maximum current rating of a wire or device.

amp (ampere) unit of measure for the rate of current flow.

ampere-turns basic unit for the measurement of magnetism (the product of turns of wire and current flow).

amp-hour unit of measure for describing the capacity of a battery.

amplifier device used to increase a signal.

amplitude highest value reached by a signal, voltage, or current.

analog voltmeter voltmeter that uses a meter movement to indicate the voltage value. Analog meters use a pointer and scale.

anode positive terminal of an electrical device.

apparent power value found by multiplying the applied voltage and total current of an AC circuit. Apparent power should not be confused with true power, or watts.

applied voltage amount of voltage connected to a circuit or device.

armature rotating member of a motor or generator. The armature generally contains windings and a commutator.

armature reaction twisting or bending of the main magnetic field of a motor or generator. Armature reaction is proportional to armature current.

ASA American Standards Association.

Askarel special type of dielectric oil used to cool electrical equipment such as transformers. Some types of Askarel contain polychlorinated biphenyl (PCB).

atom smallest part of an element that contains all the properties of that element.

attenuator device that decreases the amount of signal voltage or current.

automatic self-acting; operation of a device by its own mechanical or electrical mechanism.

autotransformer transformer that uses only one winding for both primary and secondary.

back voltage induced voltage in the coil of an inductor or generator that opposes the applied voltage.

base semiconductor region between the collector and emitter of a transistor. The base controls the current flow through the collector-emitter circuit.

battery device used to convert chemical energy into electrical energy; a group of voltaic cells connected together in a series or parallel connection.

bias DC voltage applied to the base of a transistor to preset its operating point.

bimetallic strip strip made by bonding together two unlike metals that expand at different temperatures when heated. This causes a bending or warping action.

branch circuit portion of a wiring system that extends beyond the circuit-protective device, such as a fuse or circuit breaker.

breakdown torque maximum amount of torque that can be developed by a motor at a rated voltage and frequency before an abrupt change in speed occurs.

bridge circuit circuit that consists of four sections connected in series to form a closed loop.

bridge rectifier device constructed with four diodes that converts both positive and negative cycles of AC voltage into DC voltage. The bridge rectifier is one type of full-wave rectifier.

brush sliding contact, generally made of carbon, used to provide connection to rotating parts of machines.

bus way enclosed system used for power transmission that is voltage and current rated.

capacitance electrical size of a capacitor.

capacitive reactance (X_C) current-limiting property of a capacitor in an alternating-current circuit.

capacitor device made with two conductive plates separated by an insulator or dielectric.

capacitor-start motor single-phase induction motor that uses a capacitor connected in series with the start winding to increase starting torque.

cathode negative terminal of an electrical device.

center-tapped transformer transformer that has a wire connected to the electrical midpoint of its winding. Generally the secondary winding is tapped.

charging current current flowing from an electrical source to a capacitor.

choke inductor designed to present an impedance to AC current or to be used as the current filter of a DC-power supply.

circuit electrical path between two points.

circuit breaker device designed to open under an abnormal amount of current flow. The device is not damaged and may be used repeatedly. It is rated by voltage, current, and horsepower.

clock timer time-delay device that uses an electric clock to measure delay time.

collapse (of a magnetic field) occurs when a magnetic field suddenly changes from its maximum value to a zero value.

collector semiconductor region of a transistor that must be connected to the same polarity as the base.

commutating field field used in direct-current machines to help overcome the problems of armature reaction. The commutating field connects in series with the armature and is also known as the interpole winding.

commutator strips or bars of metal insulated from each other and arranged around an armature. They provide connection between the armature windings and the brushes. The commutator is used to ensure proper direction of current flow through the armature windings.

comparator device or circuit that compares two like quantities such as voltage levels.

compensating winding winding embedded in the main field poles of a DC machine. The compensating winding is used to help overcome armature reaction.

compound DC machine generator or motor that uses both series and shunt-field windings. DC machines may be connected long-shunt compound, short-shunt compound, cumulative compound, or differential compound.

conduction level point at which an amount of voltage or current will cause a device to conduct.

conductor device or material that permits current to flow through it easily.

contact conducting part of a relay that acts as a switch to connect or disconnect a circuit or component.

continuity complete path for current flow.

conventional current flow theory considers current to flow from the most positive source to the most negative source.

copper losses power loss due to current flowing through wire. Copper loss is proportional to the resistance of the wire and the square of the current.

core magnetic material used to form the center of a coil or transformer. The core may be made of a nonmagnetic conductor (air core), iron, or some other magnetic material.

core losses power loss in the core material due to eddy-current induction and hysteresis loss.

cosine from trigonometry; it is the ratio of the adjacent side of the angle and the hypotenuse.

coulomb quantity of electrons equal to 6.25×10^{18}.

counter EMF (CEMF) voltage induced in the armature of a DC motor, which opposes the applied voltage and limits armature current.

counter torque magnetic force developed in the armature of a generator, which makes the shaft difficult to turn. Counter torque is proportional to armature current and is a measure of the electrical energy produced by the generator.

current rate of flow of electrons.

current rating amount of current flow a device is designed to withstand.

current relay relay that is operated by a predetermined amount of current flow. Current relays are often used as one type of starting relay for air conditioning and refrigeration equipment.

cycle one complete AC waveform.

d'Arsonival meter meter movement using a permanent magnet and a coil of wire; this is the basic meter movement used in many analog-type voltmeters, ammeters, and ohmmeters.

DC (direct current) current that does not reverse its direction of flow.

delta connection circuit formed by connecting three electrical devices in series to form a closed loop. It is used most often in three-phase connections.

diac bidirectional diode.

diamagnetic material that will not conduct magnetic lines of flux. Diamagnetic materials have a permeability rating less than that of air (1).

dielectric electrical insulator.

dielectric breakdown point at which the insulating material separating two electrical charges permits current to flow between the two charges. Dielectric breakdown is often caused by excessive voltage, excessive heat, or both.

digital device device that has only two states of operation, on or off.

digital logic circuit elements connected in such a manner as to solve problems using components that have only two states of operation.

digital voltmeter voltmeter that uses a direct reading numerical display as opposed to a meter movement.

diode two-element device that permits current to flow through it in only one direction.

direct current (DC) current that flows in only one direction.

disconnecting means (disconnect) device or group of devices used to disconnect a circuit or device from its source of supply.

domain group of atoms aligning themselves north and south to create a magnetic material.

dot notation dots placed beside transformer windings on a schematic to indicate relative polarity between different windings.

DVM abbreviation for digital voltmeter.

dynamic braking (1) Using a DC motor as a generator to produce counter torque and thereby produce a braking action. (2) Applying direct current to the stator winding of an AC induction motor to cause a magnetic braking action.

eddy current circular induced current contrary to the main currents. Eddy currents are a source of heat and power loss in magnetically operated devices.

effective valve *See* RMS valve.

electrical interlock when the contacts of one device or circuit prevent the operation of some other device or circuit.

electric controller device or group of devices used to govern in some predetermined manner the operation of a circuit or piece of electrical apparatus.

electrodynamometer machine used to measure the torque developed by a motor or engine for the purpose of determining output horsepower.

electrolysis decomposition of a chemical compound or metals caused by an electric current.

electrolyte chemical compound capable of conducting electric current by being broken down into ions. Electrolytes can be acids or alkalines.

electron one of the three major parts of an atom. The electron carries a negative charge.

electronic control control circuit that uses solid-state devices as control components.

electrostatic field of force that surrounds a charged object. The term is often used to describe the force of a charged capacitor.

element (1) One of the basic building blocks of nature. An atom is the smallest part of an element. (2) One part of a group of devices.

electromotive force (EMF) voltage; electrical pressure.

emitter semiconductor region of a transistor that must be connected to a polarity different than the base.

enclosure mechanical, electrical, or environmental protection for components used in a system.

eutectic alloy metal with a low and sharp melting point used in a thermal-overload relay.

excitation current direct current used to produce electromagnetism in the fields of a DC motor or generator or in the rotor of an alternator or synchronous motor.

farad basic unit of capacitance.

feeder circuit conductor between the service equipment, or the generator switchboard of an isolated plant and the branch circuit overcurrent-protective device.

ferromagnetic material that will conduct magnetic lines of force easily, such as iron (ferris). Ferromagnetic materials have a permeability much greater than that of air (1).

field loss relay (FLR) current relay connected in series with the shunt field of a direct-current motor. The relay causes power to be disconnected from the armature in the event that field current drops below a certain level.

filter device used to remove the ripple produced by a rectifier.

flashing the field method used to produce residual magnetism in the pole pieces of a DC machine, done by applying full voltage to the field winding for a period of not less than 30 seconds.

flat compounding setting the strength of the series field in a DC generator so that the output voltage will be the same at full load as it is at no load.

flux (ϕ) magnetic lines of force.

flux density number of magnetic lines contained in a certain area. The area measurement depends on the system of measurement.

frequency number of complete cycles of AC voltage that occur in 1 second.

full-load torque amount of torque necessary to produce the full horsepower of a motor at rated speed.

fuse device used to protect a circuit or electrical device from excessive current. Fuses operate by melting a metal link when the current becomes excessive.

gain increase in signal power produced by an amplifier.

galvanometer meter movement requiring microamperes to cause a full-scale deflection. Many galvanometers utilize a zero center, which permits them to measure both positive and negative values.

gate (1) A device that has multiple inputs and a single output. There are five basic types of gates, the and, or, nand, nor, and inverter. (2) One terminal of some electronic devices such as SCRs, triacs, and field-effect transistors.

gauss unit of measure in the centimeter-gram-second (CGS) system. One gauss equals one maxwell per square centimeter.

generator device used to convert mechanical energy into electrical energy.

giga metric prefix meaning one billion ($\times 10^9$).

gilbert basic unit of magnetism in the CGS system.

heat sink metallic device designed to increase the surface area of an electronic component for the purpose of removing heat at a faster rate.

henry (H) basic unit of inductance.

hermetic completely enclosed; air tight.

hertz (Hz) international unit of frequency.

holding contacts contacts used for the purpose of maintaining current flow to the coil of a relay.

holding current amount of current needed to keep an SCR or triac turned on.

horsepower measure of power for electrical and mechanical devices.

hydrometer device used to measure the specific gravity of a fluid, such as the electrolyte used in a battery.

hypotenuse longest side of a right triangle.

hysteresis loop graphic curve that shows the value of magnetizing force for a particular type of material.

hysteresis loss power loss in a conductive material due to molecular friction. Hysteresis loss is proportional to frequency.

impedance total opposition to current flow in an electrical circuit.

incandescent ability to produce light as a result of heating an object.

induced current current produced in a conductor by the cutting action of a magnetic field.

inductive reactance (X_L) current-limiting property of an inductor in an alternating-current circuit.

inductor coil.

input voltage amount of voltage connected to a device or circuit.

insulator material used to electrically isolate two conductive surfaces.

interlock device used to prevent some action from taking place in a piece of equipment or circuit until some other action has occurred.

interpole small pole piece placed between the main field poles of a DC machine to reduce armature reaction.

ion charged atom.

isolation transformer transformer whose secondary winding is electrically isolated from its primary winding.

joule basic unit of electrical energy. A joule is the amount of power used when 1 amp flows through 1 ohm for 1 second. A joule is equal to 1 watt/second.

jumper short piece of conductor used to make connection between components or a break in a circuit.

junction diode made by joining together two pieces of semiconductor material.

kick-back diode used to eliminate the voltage spike induced in a coil by the collapse of a magnetic field.

kilo metric measure prefix meaning thousand ($\times 10^3$).

kinetic energy the energy of a moving object such as the energy of a flywheel in motion.

lamination one thickness of the sheet material used to construct the core material for transformers, inductors, and alternating-current motors.

LED (light-emitting diode) diode that produces light when current flows through it.

Leyden jar glass jar used to store electrical charges in the very early days of electrical experimentation. The Leyden jar was constructed by lining the inside and outside of the jar with metal foil. The Leyden jar was a basic capacitor.

limit switch mechanically operated switch that detects the position or movement of an object.

linear when used in comparing electrical devices or quantities, one unit is equal to another.

load center generally the service entrance. A point from which branch circuits originate.

locked-rotor current amount of current produced when voltage is applied to a motor and the rotor is not turning.

locked-rotor torque amount of torque produced by a motor at the time of starting.

lockout mechanical device used to prevent the operation of some component.

long-shunt compound connection of field windings in a DC machine where the shunt field is connected in parallel with both the armature and series field.

low-voltage protection magnetic relay circuit so connected that a drop in voltage causes the motor starter to disconnect the motor from the line.

magnetic contactor contactor operated electromechanically.

magnetic field space in which a magnetic force exists.

magnetimotive force magnetic force produced by current flowing through a conductor or coil.

maintaining contact also known as a holding or sealing contact. It is used to maintain the coil circuit in a relay control circuit. The contact is connected in parallel with the start pushbutton.

manual controller controller operated by hand at the location of the controller.

maxwell measure of magnetic flux in the centimeter-gram-second (CGS) system.

mica mineral used as an electrical insulator.

micro metric prefix meaning one millionth. ($\times 10^{-6}$).

microfarad measurement of capacitance.

microprocessor small computer. The central processing unit is generally made from a single integrated circuit.

mill unit for measuring the diameter of a wire equal to one thousandth of an inch.

mill-foot standard for measuring the resistivity of wire. A mill-foot is the resistance of a piece of wire 1 mill in diameter and 1 foot in length.

milli metric prefix for one thousandth ($\times 10^{-3}$).

mode state or condition.

motor device used to convert electrical energy into rotating motion.

motor controller device used to control the operation of a motor.

multispeed motor motor that can be operated at more than one speed.

nano metric prefix meaning one billionth. ($\times 10^{-9}$).

negative one polarity of a voltage, current, or charge.

NEMA National Electrical Manufacturers Association.

NEMA ratings electrical control device ratings of voltage, current, horsepower, and interrupting capability given by NEMA.

neutron one of the principal parts of an atom. The neutron has no charge and is part of the nucleus.

noninductive load electrical load that does not have induced voltages caused by a coil. Noninductive loads are generally considered to be resistive, but can be capacitive.

nonreversing device that can be operated in only one direction.

normally closed The contact of a relay that is closed when the coil is deenergized.

normally open contact of a relay that is open when the coil is de-energized.

off-delay timer timer that delays changing its contacts back to their normal position when the coil is de-energized.

ohm unit of measure for electrical resistance.

ohmmeter device used to measure resistance.

on-delay timer timer that delays changing the position of its contacts when the coil is energized.

operational amplifier (OP AMP) integrated circuit used as an amplifier.

opto-isolator device used to connect different sections of a circuit by means of a light beam.

oscillator device used to change DC voltage into AC voltage.

oscilloscope voltmeter that displays a waveform of voltage in proportion to its amplitude with respect to time.

out of phase condition in which two components do not reach their positive or negative peaks at the same time.

overcompounded condition of a DC generator when the series field is too strong. It is characterized by the output voltage being greater at full load than it is at no load.

overexcited condition that occurs when the DC current supplying excitation current to the rotor of a synchronous motor is greater than necessary.

overload relay relay used to protect a motor from damage due to overloads. The overload relay senses motor current and disconnects the motor from the line if the current is excessive for a certain length of time.

panelboard metallic or nonmetallic panel used to mount electrical controls, equipment, or devices.

parallel circuit circuit that contains more than one path for current flow.

paramagnetic material that has a permeability slightly greater than that of air (1).

peak-inverse/peak-reverse voltage rating of a semiconductor device that indicates the maximum amount of voltage in the reverse direction that can be applied to the device.

peak-to-peak voltage amplitude of AC voltage measured from its positive peak to its negative peak.

peak voltage amplitude of voltage measured from zero to its highest value.

permalloy alloy used in the construction of electromagnets (approximately 78% nickel and 21% iron).

permeability measurement of a material's ability to conduct magnetic lines of flux. The standard is air, which has a permeability of 1.

phase shift change in the phase relationship between two quantities of voltage or current.

photoconductive material that changes its resistance due to the amount of light.

photovoltaic material or device that produces a voltage in the presence of light.

pico unit of metric measure for one trillionth ($\times 10^{-12}$).

piezoelectric production of electricity by applying pressure to a crystal.

pilot device control component designed to control small amounts of current. Pilot devices are used to control larger control components.

pneumatic timer device that used the displacement of air in a bellows or diaphragm to produce a time delay.

polarity characteristic of a device that exhibits opposite quantities within itself: positive and negative.

potentiometer variable resistor with a sliding contact that is used as a voltage divider.

power factor comparison of the true power (watts) to the apparent power (volt amps) in an AC circuit.

power rating rating of a device that indicates the amount of current flow and voltage drop that can be permitted.

pressure switch device that senses the presence or absence of pressure and causes a set of contacts to open or close.

primary cell voltaic cell that cannot be recharged.

primary winding winding of a transformer to which power is applied.

prime mover device supplying the turning force necessary for turning the shaft of a generator or alternator (steam turbine, diesel engine, water wheel, etc.)

printed circuit board on which a predetermined pattern of printed connections has been made.

proton one of the three major parts of an atom. The proton has a positive charge.

pushbutton pilot control device operated manually by being pushed or pressed.

reactance (X) opposition to current flow in an AC circuit offered by pure inductance or pure capacitance.

rectifier device or circuit used to change AC voltage into DC voltage.

regulator device that maintains a quantity at a predetermined level.

relay magnetically operated switch that may have one or more sets of contacts.

reluctance resistance to magnetism.

remote control controls the functions of some electrical device from a distant location.

residual magnetism amount of magnetism left in an object after the magnetizing force has been removed.

resistance opposition to current flow in an AC or DC circuit.

resistance-start induction-run motor one type of split-phase motor that uses the resistance of the start winding to produce a phase shift between the current in the start winding and the current in the run winding.

resistor device used to introduce some amount of resistance into an electrical circuit.

retentivity a material's ability to retain magnetism after the magnetizing force has been removed.

rheostat variable resistor.

RMS value value of AC voltage that will produce as much power when connected across a resistor as a like amount of DC voltage.

rotor rotating member of an alternating-current machine.

saturation maximum amount of magnetic flux a material can hold.

schematic electrical drawing showing components in their electrical sequence without regard for physical location.

SCR (silicone-controlled rectifier) semiconductor device that can be used to change AC voltage into DC voltage. The gate of the SCR must be triggered before the device will conduct current.

sealing contacts Contacts connected in parallel with the start button and used to provide a continued path for current flow to the coil of the contactor when the start button is released. *See also* holding contacts and maintaining contacts.

secondary cell voltaic cell that can be recharged.

secondary winding winding of a transformer to which the load is connected.

semiconductor material that contains four valence electrons and is used in the production of solid-state devices.

sensing device pilot device that detects some quantity and converts it into an electrical signal.

series circuit circuit that contains only one path for current flow.

series field winding of large wire and few turns designed to be connected in series with the armature of a DC machine.

series machine direct-current motor or generator that contains only a series field winding connected in series with the armature.

service conductors and equipment necessary to deliver energy from the electrical supply system to the premises served.

service factor allowable overload for a motor indicated by a multiplier, which, when applied to a normal horsepower rating, indicates the permissible loading.

shaded-pole motor AC induction motor that develops a rotating magnetic field by shading part of the stator windings with a shading loop.

shading loop large copper wire or band connected around part of a magnetic pole piece to oppose a change of magnetic flux.

short circuit electrical circuit that contains no resistance to limit the flow of current.

shunt field coil wound with small wire having many turns designed to be connected in parallel with the armature of a DC machine.

shunt machine DC motor or generator that contains only a shunt field connected in parallel with the armature.

sine-wave voltage voltage waveform whose value at any point is proportional to the trigonometric sine of the angle of the generator producing it.

slip difference in speed between the rotating magnetic field and the speed of the rotor in an induction motor.

sliprings circular bands of metal placed on the rotating part of a machine. Carbon brushes riding in contact with the sliprings provide connection to the external circuit.

snap-action quick opening and closing action of a spring-loaded contact.

solenoid magnetic device used to convert electrical energy into linear motion.

solenoid valve valve operated by an electric solenoid.

solid-state device electronic component constructed from semiconductor material.

specific gravity ratio of the volume and weight of a substance as compared to an equal volume and weight of water. Water has a specific gravity of 1.

split-phase motor type of single-phase motor that uses resistance or capacitance to cause a shift in the phase of the current in the run winding and the current in the start winding. The three primary types of split-phase motors are: resistance-start induction-run, capacitor-start induction-run, and permanent split-capacitor motor.

squirrel-cage rotor rotor of an AC induction motor constructed by connecting metal bars together at each end.

star connection *see* wye connection.

starter relay used to connect a motor to the power line.

stator stationary winding of an AC motor.

step-down transformer transformer that produces a lower voltage at its secondary than is applied to its primary.

step-up transformer transformer that produces a higher voltage at its secondary than is applied to its primary.

surge transient variation in the current or voltage at a point in the circuit. Surges are generally unwanted and temporary.

switch mechanical device used to connect or disconnect a component or circuit.

synchronous speed speed of the rotating magnetic field of an AC induction motor.

synchroscope instrument used to determine the phase-angle difference between the voltage of two alternators.

temperature relay relay that functions at a predetermined temperature. Generally used to protect some other component from excessive temperature.

terminal fitting attached to a device for the purpose of connecting wires to it.

tesla unit of magnetic measure in the meter-kilogram-second (MKS) system. (1 tesla = 1 weber per square meter).

thermistor resistor that changes its resistance with a change of temperature.

thyristor electronic component that has only two states of operation, on or off.

time constant amount of time required for the current flow through an inductor or for the voltage applied to a capacitor to reach 63.2% of its total value.

torque turning force developed by a motor.

torroid doughnut-shaped electromagnet, with a hole through which conductors are wound.

transducer device that converts one type of energy into another type. For example, a solar cell converts light into electricity.

transformer electrical device that changes one value of AC voltage into another value of AC voltage.

transistor solid-state device made by combining together three layers of semiconductor material. A small amount of current flow through the base-emitter can control a larger amount of current flow through the collector-emitter.

triac bidirectional thyristor used to control AC voltage.

troubleshoot to locate and eliminate problems in a circuit.

turns ratio (transformer) The ratio of the number of primary turns of wire as compared to the number of secondary turns.

undercompounded condition of a direct-current generator when the series field is too weak. The condition is characterized by the fact that the output voltage at full load will be less than the output voltage at no load.

unity power factor power factor of 1 (100%). Unity power factor is accomplished when the applied voltage and circuit current are in phase with each other.

valence electrons electrons located in the outer orbit of an atom.

variable resistor resistor whose resistance value can be varied between its minimum and maximum values.

varistor resistor that changes its resistance value with a change of voltage.

vector line having a specific length and direction.

voltage electrical measurement of potential difference, electrical pressure, or electromotive force (EMF).

voltage drop amount of voltage required to cause an amount of current to flow through a certain resistance.

voltage rating rating that indicates the amount of voltage that can be safely connected to a device.

voltage regulator device or circuit that maintains a constant value of voltage.

voltaic cell device that converts chemical energy into electrical energy.

voltmeter instrument used to measure a level of voltage.

volt-ohm-milliammeter (VOM) a test instrument designed to measure voltage, resistance, or milliamperes.

watt measure of true power.

waveform shape of a wave as obtained by plotting a graph with respect to voltage and time.

weber (Wb) measure of magnetic lines of flux in the meter-kilogram-second (MKS) system (1 weber = 100 million [$\times 10^8$] lines of flux).

windage loss loss encountered by the armature or rotor of a rotating machine caused by the friction of the surrounding air.

wiring diagram electrical diagram used to show components in their approximated physical location with connecting wires.

wound-rotor motor three-phase motor containing a rotor with windings and sliprings. This rotor permits control of rotor current by connecting external resistance in series with the rotor winding.

wye connection connection of three components made in such a manner that one end of each component is connected. This connection is generally used to connect devices to a three-phase power system.

zener diode diode that has a constant voltage drop when operated in the reverse direction. Zener diodes are commonly used as voltage regulators in electronic circuits.

Index